PLURIPOTENT STEM CELLS - FROM THE BENCH TO THE CLINIC

Edited by **Minoru Tomizawa**

Pluripotent Stem Cells - From the Bench to the Clinic
http://dx.doi.org/10.5772/61549
Edited by Minoru Tomizawa

Contributors

Naoki Nakayama, John Lee, Nadine Matthias, Katsutsugu Umeda, Qing Yan, Johnny Huard, Wei Seong Toh, Karim Benabdel Lah Elkhlanji, Roxana Nat, Jari Koistinaho, Takaaki Noguchi, Vittorio Sebastiano, Tapash Jay Sarkar, Daman Kumari, Keiichi Fukuda, Tomohisa Seki, Srdjan Antic, Pallavi Limaye, Michele McGovern, Mandakini Singh, Katerina Oikonomou, Glenn Belinsky, Erika Pedrosa, Herbert Lachman, Vilas Wagh, Marisa Ojala, Katriina Aalto-Setälä, Liesbeth Vossaert, Wilairat Leeanansaksiri, Teresa De Souza Fernandez, Cecilia De Souza Fernandez, Mohammad Haque, Praneet Sandhu, Jianxun Song, Yvonne Reinwald, Jessica Bratt, Alicia El Haj, Kamthorn Pruksananonda, Ruttachuk Rungsiwiwut, Una Chen, Domenico Maiorano, Elena Lo Furno, Siem Van Der Laan, Vanessa Hall, Qiao Li, Jihong Chen, Minoru Tomizawa

Published by InTech

Janeza Trdine 9, 51000 Rijeka, Croatia

Publishing Process Manager Andrea Koric

Technical Editor SPi Global

Cover InTech Design team

First published August, 2016

Printed in Croatia

Additional hard copies can be obtained from orders@intechopen.com

Pluripotent Stem Cells - From the Bench to the Clinic, Edited by Minoru Tomizawa

p. cm.

Print ISBN 978-953-51-2471-9
Online ISBN 978-953-51-2472-6

For EU product safety concerns:
IN TECH d.o.o., Prolaz Marije Krucifikse Kozulić 3, 51000 Rijeka, Croatia,
info@intechopen.com or visit our website at intechopen.com.

Meet the editor

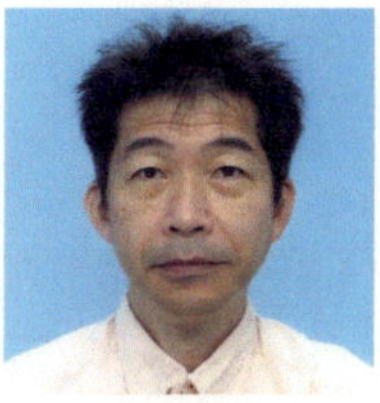

Minoru Tomizawa, MD, PhD, is the head of the Department of Gastroenterology at National Hospital Organization Shimoshizu Hospital in Yotsukaido City, Chiba, Japan. He completed his PhD in Second Department of Pathology at Chiba University Graduate School of Medicine in Chiba City, Japan. He was appointed as a post-doctoral fellowship at National Human Genome Research Institute, NIH, Bethesda, USA, and subsequently at National Institute of Diabetes Digestive and Kidney Diseases, NIH, Bethesda, USA. He held a faculty position at the Department of Medicine and Clinical Oncology in Chiba University Graduate School of Medicine in Chiba City, Japan. He has a broad interest of research: hepatocyte differentiation, signal transduction of cancer, and sonoporation.

Contents

Preface

Pluripotent stem cells, including embryonic stem cells and induced pluripotent stem cells, have the capacity to proliferate and differentiate into various types of somatic cells. These cells share a unique characteristic: pluripotency. Pluripotency is not only interesting from a scientific research point of view but also useful for regenerative medicine. After the discovery of embryonic stem cells, research on pluripotency and the mechanisms of cell differentiation has substantially increased and is still continuing. When embryonic stem cells are applied clinically, however, two major problems arise: ethics and graft-versus-host disease. These difficulties with embryonic stem cells have been overcome by Professor Shinya Yamanaka through the establishment of induced pluripotent stem cells. This innovation paves a way for the clinical application of pluripotent stem cells. Since its introduction, the field of pluripotent stem cells research has been expanding with an increasing amount of literature reported each year. One type of pluripotent stem cells is mesenchymal stem cells, which has the potential to differentiate into certain types of somatic cells. Mesenchymal stem cells are discussed in this book with respect to their pluripotent potential.

This book overviews the state-of-the-art knowledge and technology of pluripotent stem cells. The contributing authors are experts in their fields and proudly introduce this book to the readers.

Dr. Minoru Tomizawa
Department of Gastroenterology
National Hospital Organization Shimoshuzu Hospital
Yotsukaido City, Chiba, Japan

Section 1

Introduction

Chapter 1

Introductory chapter: Pluripotent stem cells

Minoru Tomizawa

Additional information is available at the end of the chapter

http://dx.doi.org/10.5772/64462

Stem cells are defined as those that have potential of self-renewal and differentiation into specific cells. In 1961, Till et al. transplanted bone marrow cells to recipient mice whose bone marrow is damaged with irradiation [1]. After bone marrow cells are transplanted to the irradiated mice, they colonize and recapitulate the bone marrow of the recipient mice. These results suggest that stem cells perform self-renewal and differentiation. This type of stem cells, hematopoietic stem cells, produces only hematopoietic cells. This concept has been developed and now is clinically applied to hematopoietic stem cell transplantation [2].

Pluripotent stem cells have two properties: self-renewal and pluripotency. Self-renewal has the potency to produce daughter cells with the same character through infinite cell division. Pluripotency is the characteristic to produce specialized cells of three layers: endoderm, mesoderm, and ectoderm.

Pluripotent stem cells are totally different from other cells, such as cancer cell lines and primary cells. Cancer cells divide infinitely, but their differentiation status does not change. Primary cells divide only into limited numbers and change their characteristics as they divide. Pluripotency, therefore, has been the main focus of stem cell research.

Regenerative medicine is to replace or transplant cells or tissues to restore normal function. For the generation of cells or tissues, cell source has been a major problem. Pluripotent stem cells produce cells of three germ layers. That is all the somatic cells could be derived from pluripotent stem cells virtually. Embryonic stem (ES) cells have been expected to be a cell source for the transplantation of patients with organ failure.

Embryonic carcinoma (EC) cells, extracted from teratocarcinoma, have such characters. EC cells not only divide infinitely, but also produce cells of three germ layers [3]. These characters fulfill the criteria of pluripotency. EC cells are the first cells that have been reported to exhibit pluripotency. EC cells are successful in the formation of chimera in germ line and in the production of chimera mice. Researchers were excited about the results expecting that genetically modified mice would be produced. Disappointingly, cancer develops in the chimera mice. Researchers speculate that one of the major reasons of cancer is the formation of EC cells.

With the experience of EC cells, researchers attempted to derive cells from a healthy embryo. Mouse embryonic stem cells were first extracted in 1981, and human ES cells were first extracted in 1997 [4, 5]. ES cells are derived from the inner cell mass of preimplantation embryos. ES cells have been used to produce knockout mice.

To produce ES cells, embryos need to be broken. This raises ethical issues. The transplantation of cells or tissues produced from ES cells causes graft-versus-host disease (GVHD). Ethical issues and GVHD are innate obstacles to develop regenerative medicine.

To overcome these problems, methods to produce human-induced pluripotent stem (iPS) cells have been developed by Professor Yamanaka et al. with the introduction of reprogramming factors, such as Oct3/4, Sox2, Klf4, and c-Myc [6]. One of the most important features of iPS cells is that they can be derived from cells of the patients who are potential recipients. Ethical issues and GVHD, thus, do not arise from iPS cells. Currently, a clinical trial is underway for retinitis pigmentosa with iPS cells derived from the patients [7].

iPS cells are useful not only in the application of regenerative medicine but also in stem cell research. Researches on the networks of expressed genes and stem cell niche are underway [8].

For the clinical application of iPS cells, new problems arise, that is, culture methods. To transplant somatic cells differentiated from iPS cells, the iPS cells should be manipulated under xeno-free condition. Another problem with manipulation is that a good-medical-practice (GMP) level should be realized.

Practically, the extraction of iPS cells from a patient requires plenty of time and labor. If iPS cells are stored and provided upon request, time and labor would be saved. Patients would be subjected to regenerative medicine promptly and smoothly. iPS cell banks have been under development with dental pulp [9].

From a pharmacological point of view, somatic cells and tissues are necessary for the evaluation of toxicity of certain types of drugs. Human cells or tissues are very hard to obtain for the investigation of toxicity. iPS cells are expected to be useful for "toxicology" because they would produce somatic cells in the toxicological experiments.

To innovate a drug for the treatment of a disease, the mechanism of the disease should be clarified. As mentioned earlier, it is difficult to obtain somatic cells or tissues from the patients. iPS cells are again expected to be useful for the investigation of the mechanism of the disease because they would differentiate into somatic cells.

With the above discussions, the differentiation of iPS cells has been the main focus of the investigation. There have been plenty of literatures on differentiation protocols of iPS cells. Unfortunately, some somatic cells are difficult to obtain from iPS cells.

During the investigation of differentiation protocols, a new concept arises, that is, self-organization [10]. iPS cells differentiate into various types of cells forming an organ. The

http://dx.doi.org/10.5772/64462

resultant cells finally form the organ. This procedure is particularly useful for not only the production of specific types of cells but also for the production of organs.

Mesenchymal stem cells (MSCs) are usually obtained from bone marrow. MSCs have the potential to differentiate into adipose cells, chondrocytes, and hematopoietic cells. MSCs can be obtained from a potential recipient and differentiated into target cells. MSCs, therefore, are considered interesting and useful for the scientific and clinical applications.

This book demonstrates the current status of stem cell research, new concepts, and solutions to the problems of pluripotent stem cells. Finally, it is shown how the issues are addressed, and horizons of accumulated and broadened knowledge.

Author details

Minoru Tomizawa

Address all correspondence to: nihminor-cib@umin.ac.jp

Department of Gastroenterology, National Hospital Organization Shimoshizu Hospital, Yotsukaido City, Chiba, Japan

References

[1] Till JE, McCulloch EA: A direct measurement of the radiation sensitivity of normal mouse bone marrow cells. Radiat Res 1961: 14, 213–22.

[2] Imamura M, Shigematsu A: Allogeneic hematopoietic stem cell transplantation in adult acute lymphoblastic leukemia: potential benefit of medium-dose etoposide conditioning. Exp Hematol Oncol 2015: 4, 20.

[3] Papaioannou VE, Gardner RL, McBurney MW, Babinet C, Evans MJ: Participation of cultured teratocarcinoma cells in mouse embryogenesis. J Embryol Exp Morphol 1978: 44, 93–104.

[4] Evans MJ, Kaufman MH: Establishment in culture of pluripotential cells from mouse embryos. Nature 1981: 292, 154–6.

[5] Thomson JA, Itskovitz-Eldor J, Shapiro SS, Waknitz MA, Swiergiel JJ, Marshall VS, Jones JM: Embryonic stem cell lines derived from human blastocysts. Science 1998: 282, 1145–7.

[6] Takahashi K, Tanabe K, Ohnuki M, Narita M, Ichisaka T, Tomoda K, Yamanaka S: Induction of pluripotent stem cells from adult human fibroblasts by defined factors. Cell 2007: 131, 861–72.

[7] Sun J, Mandai M, Kamao H, Hashiguchi T, Shikamura M, Kawamata S, Sugita S, Takahashi M: Protective effects of human iPS-derived retinal pigmented epithelial cells in comparison with human mesenchymal stromal cells and human neural stem cells on the degenerating retina in rd1 mice. Stem Cells 2015: 33, 1543–53.

[8] He S, Nakada D, Morrison SJ: Mechanisms of stem cell self-renewal. Annu Rev Cell Dev Biol 2009: 25, 377–406.

[9] Tamaoki N, Takahashi K, Tanaka T, Ichisaka T, Aoki H, Takeda-Kawaguchi T, Iida K, Kunisada T, Shibata T, Yamanaka S, Tezuka K: Dental pulp cells for induced pluripotent stem cell banking. J Dent Res 2010: 89, 773–8.

[10] Eiraku M, Takata N, Ishibashi H, Kawada M, Sakakura E, Okuda S, Sekiguchi K, Adachi T, Sasai Y: Self-organizing optic-cup morphogenesis in three-dimensional culture. Nature 2011: 472, 51–6.

Section 2

Pluripotency

Chapter 2

Embryonic Stem Cells: Keeping Track of the Pluripotent Status

Liesbeth Vossaert, Ellen Scheerlinck and
Dieter Deforce

Additional information is available at the end of the chapter

http://dx.doi.org/10.5772/62651

Abstract

Embryonic stem cells are defined by their pluripotent status, which allows them to differentiate toward all cell types of an adult organism. This pluripotency can be characterized through many parameters, ranging from morphological traits, over certain enzymatic activities, to the expression of specific pluripotency factors, taken into account that these parameters may vary depending on the pluripotent stem cell type. As such, considerable differences are seen between human and mouse embryonic stem cell (ESC), or more generally stated, between primed and naïve pluripotent stem cells. This chapter offers an overview of the markers involved and the molecular biology techniques to monitor them during both ESC culture maintenance or differentiation experiments.

Keywords: Embryonic stem cells, Pluripotency, Pluripotency assessment, Noninvasive monitoring, RT-qPCR

1. Introduction

The hallmarks of embryonic stem cells (ESC) are their potential for self-renewal and their pluripotent status. The latter ensures that they can differentiate toward the three germ layers (endoderm, mesoderm, and ectoderm) and primordial germ cells, and eventually to all different cell types of an adult organism.

The first mouse ESC (mESC) were derived in 1981, independently by two different research groups [1, 2]. The first human ESC (hESC) in culture followed almost 20 years later [3]. Up to date, also for the derivatization of ESC from other species considerable research efforts have been made [4]. As ESC are being derived from the inner cell mass of a blastocyst, they provide a good model to study fundamental processes in early development and cellular differentiation.

Equally important is that they also served as a template for the generation of induced pluripotent stem cells (iPSC) and contribute to a better (clinical) application of these cells in the future. As such, also in the field of induced pluripotency, tremendous progress has been made over the last decade. In 2006 and 2007, respectively, the research group led by Shinya Yamanaka developed a reprogramming cocktail for the establishment of mouse and human iPSC [5, 6].

Nevertheless, as the comparison of the different ESC and iPSC types teaches us, pluripotency comes in different intensities, instead of a single definition as was firstly assumed [7]. mESC are considered as 'true' pluripotent ESC and find themselves in a ground pluripotent state, as does the early preimplantation epiblast. hESC on the other hand, are primed pluripotent cells and resemble more to mouse epiblast stem cells (mEpiSC), which are derived from a postimplantation epiblast, in terms of culture requirements and molecular profile among other factors.

This chapter will give an overview on those different types of pluripotency and their according characteristics, and will then further elaborate on how these features can be monitored in order to keep track of the pluripotent status of ESC in culture.

2. Primed versus naïve pluripotency

hESC are considered to be primed, in that they are already predestined for one lineage or another, despite their remaining broad differentiation potential. In contrast, mESC are termed as 'naïve' pluripotent cells, while mEpiSC form the primed counterpart. It is clear that these different types possess different characteristics, which can be monitored via a range of different features and techniques, as is outlined below in more detail.

2.1. Pluripotent ESC morphology

A first distinctive feature is the difference in morphology. Naïve mESC form dome-shaped colonies, as do miPSC. However, primed hESC show a more flat morphology, with round individual cells having a high nucleus-to-cytoplasm ratio, as such resembling mEpiSC. These distinct appearances thus illustrate the different developmental states of the different cell types. The conversion of hiPSC to their naïve form changes the morphology from flat to domed [8]. It is remarkable how well these iPSCs resemble ESC in terms of morphology, even on an ultrastructural level [9, 10]. The latter reference describes the comparison of mESC, mouse embryonic fibroblasts (MEF), and iPSC derived thereof: after reprogramming, the accompanying morphological changes that the iPSC undergo make them virtually indistinguishable from ESC. Nevertheless, within the population of morphologically similar miPSC colonies, there still appears to be considerable variation in terms of molecular pluripotency, in contrast

http://dx.doi.org/10.5772/62651

to hiPSC, which show a much higher homogeneity among the colonies that are selected on analogous morphology [11]. Moreover, true hiPSC colonies have a typical hESC morphology that is very distinct from non-hiPSC colonies, hence morphology can very well serve as a criterion for identifying the real hiPSC colonies.

It is important to monitor the colonies' morphology in culture, as this will quickly visualize potential spontaneous differentiation. E.g. for hESC, if two or more colonies come into contact with one another, differentiation sets in at the contact area, and cells will start to pile up and acquire a more lengthened shape. Also in the center of a colony, cells might start to accumulate in multiple layers. As hESCs typically form flat colonies, timely passaging (at least on weekly basis, dependent on the culture system) is crucial to prevent overgrowing colonies and consequential spontaneous differentiation [12].

2.2. Pluripotency on a molecular level

2.2.1. Signalling pathways

The pluripotent state is regulated and maintained via several signaling pathways. As such, four major pathways involved in pluripotency can be distinguished for hESC: (1) the transforming growth factor β (TGFβ)-Activin-Nodal pathway, (2) the phosphatidylinositol 3-kinase (PI3K) pathway, (3) the Ras-Raf-mitogen activated protein kinase (MEK)-extracellular signal-regulated kinase (ERK) (or MAPK/ERK) pathway, and (4) Wnt signaling [13–16].

TGFβ activates the TGFβ-Activin-Nodal pathway through the signal transducer SMAD2/3. The latter forms a complex with Smad4 and then translocates to the nucleus to trigger the expression of *NODAL* among other factors, which in turn stimulates self-renewal and inhibits differentiation. Also, the addition of activin A to hESC culture enables the activation of this pathway [14].

Addition of fibroblast growth factor (FGF; basic FGF; bFGF) on the other hand activates the PI3K pathway and the Ras-Raf-MEK-ERK pathway. In short, for the PI3K pathway, phosphatidylinositol 4,5-bisphosphate becomes phosphorylated by means of PI3K during activation. The resulting phosphatidylinositol 3,4,5-triphosphate subsequently binds with Akt (also known as protein kinase B). Activation of this pathway results in increased concentrations of Oct3/4, Nanog and Sox2 and thus in the maintenance of pluripotency [16, 17]. Additionally, activation of the Ras-Raf-MEK-ERK pathway leads to the activation of Ras (a GTPase), which in turn binds with Raf. This kinase phosphorylates another kinase, MEK, which then phosphorylates ERK (or MAPK). The latter translocates to the nucleus, leading to the phosphorylation of c-Myc, c-Jun, and c-Fos, factors involved in stem cell renewal [18, 19].

The Wnt signaling pathway can be activated by several Wnt ligands and is stimulated in culture e.g. by by 6-brominedirubin-3′-oxime. This agent inhibits glycogensynthase kinase-3 β which normally promotes the degradation of β-catenin in a complex with axin and adenomatous polyposis coli protein, by making it a target for the proteasome [19, 20]. As a consequence of this inhibition, β-catenin accumulates in the cytoplasm and a portion of this pool translocates to the cell nucleus and interacts with genes important for keeping hESC undifferentiated [15].

Importantly, reasonable differences exist between mESC and hESC in terms of signaling pathways. In hESC, endogenous BMP signals (interacting with the TGFβ-Activin-Nodal pathway) need to be suppressed [21], while on the contrary this pathway in mESC helps in maintaining the pluripotent state and BMP-4 can be added to their culture medium. Analogously, leukemia inhibitory factor (LIF) is added to mESC media, which promotes self-renewal by influencing Jak/Stat3 signaling, one of the downstream FGF/ERK pathways [13, 21]. hESC on the other hand, are not dependent on LIF supplementation.

2.2.2. *Transcription factors*

As was already pointed out in the previous paragraph, the transcription factors Oct3/4, Nanog, and Sox2 are part of the core pluripotency network. Oct3/4 (Octamer4 or Oct4, encoded by *POU5F1*) has been termed the gatekeeper at the start of mammalian development. This member of the Pit-Oct-Unc (POU) transcription factor family can activate its target genes' expression through binding on an octameric sequence motif (consensus sequence AGT-CAAAT) [22]. Although not being totally exclusive for ESC, it is nevertheless considered as one of the most important features to define a pluripotent cell state. In early embryos, loss of *POU5F1* expression causes the cells that were predisposed to form the ICM, to differentiate toward trophectoderm cells. Also, in ESC, lowering Oct3/4 levels to ≤50% or an increased expression above 150% leads to differentiation to trophectoderm or primitive endoderm cells, respectively [13]. Oct3/4 also appears to support the maintenance of mammalian germ cells, as apoptosis is induced when its expression is abrogated. Sox2 (*SOX2*) is a member of the SRY-related high-mobility group box-containing family and cooperatively functions with Oct3/4. However, its role goes further than just being a synergistic factor. Sox2-null embryos fail to give rise to ESC, but differentiate primarily to trophectoderm instead. Deletion of *SOX2* in hESC and mESC leads to differentiation, confirming Sox2 to be essential in itself to maintain the pluripotent state [23]. Nanog (*NANOG*) is another crucial pluripotency- and self-renewal-maintaining factor in both mESC and hESC. It was for instance shown that deletion of *NANOG* resulted in loss of pluripotency and induced differentiation, both in mouse ICM and ESC [24]. Its ability to maintain pluripotency is found to be independent of LIF supplementation to the cell culture. One of the possible mechanisms for self-renewal maintenance may be that Nanog represses the transcription of differentiation-promoting genes such as *GATA4* and *GATA6* [13, 24, 25].

Oct3/4, Nanog and Sox2 are involved in an intensive autoregulatory loop, as was for instance shown by the fact that knockdown of Sox2 led to a reduced expression of Oct3/4 and Nanog [23]. Additionally, they share a whole number of target genes, both in an activating or inhibitory manner. They stimulate genes involved in chromatin remodeling, histone post-translational modifications, TGFβ signaling, etc., and other ESC transcription factors, among which also themselves. They are able to inhibit the expression of many genes promoting ectoderm, mesoderm, or endoderm differentiation [26]. It was found recently that among other factors, the hypoxia inducible factor 2-α (HIF 2-α) is an upstream regulator of these key transcription factors, as it binds directly to predicted hypoxic response elements in the proximal promotor of *POU5F1*, *NANOG*, and *SOX2* [27].

This list of core pluripotency factors can be completed with Klf4, a member of the Krupple-like Factor family. Together with other members of its family, it helps in maintaining ESC self-renewal and regulating the expression of several other genes such as *NANOG*. It is also one of the components included for iPSC generation starting from human or mouse cells, in combination with *POU5F1, SOX2,* and *c-MYC*. Alternatively, a combination of *POU5F1, NANOG, LIN28,* and *SOX2* is applied for reprogramming.

Besides the aforementioned factors, there is a whole series of other transcription factors that are involved in pluripotency regulation and maintenance. For instance, for hESC, the expression of *REX1, FOXD3, GDF3, GABRB3, EBAF, POXDL, NODAL, ZFP42, LIN28, TCF3, EOMES,* and *SFPR2* is highly correlated to the expression of *NANOG* or are potential Nanog targets, as was elaborately described in [28]. Of note, there was also a number of genes found to be highly negatively correlated with *NANOG* expression, including *CDX2* and *CGB* (associated with trophectoderm differentiation), *GATA6* and *AFP* (extraembryonic endoderm), and *PAX6* and *NEUROD1* (neural lineage), once more confirming the role of Nanog in pluripotency regulation [28]. In mESC, the sustaining of the pluripotent status relies both on similar (e.g. *NANOG, POU5F1, SOX2, REX1,* and *FOXD3*) and different (e.g. *PECAM1* and *STELLA*) factors.

Yeo et al. [29] proposed the monitoring of DNA methylation patterns as a possible means to study the extent of hESC differentiation. DNA methylation is mostly associated with transcriptional repression, whether established via chromatin remodeling complexes or by direct blocking of transcription factor binding, and is as such involved in cellular programming events [30, 31]. During the hESC differentiation process toward embryoid bodies, the promotor regions of *POU5F1* and *NANOG* undergo substantial methylation, in contrast to those of *SOX2, REX1,* and *FOXD3* [29].

2.2.3. Pluripotency on RNA level

Besides the well-described transcription factors and cell surface markers, also long noncoding RNAs (lncRNAs) take up their role in pluripotency maintenance [32]. lncRNA molecules are transcripts from RNA polymerase II, that are over 200 nucleotides in size and that do not serve as a template for protein production. Instead, they prove to be involved in processes such as mRNA stability and translation modulation, and related epigenetic regulatory processes [33]. Both for mESC and hESC a characteristic set of lncRNAs has yet been identified, of which some are under the direct control of the core pluripotency transcription factor network [32]. The authors of the latter publication provide an elaborate overview of the different lncRNAs involved in the maintenance of ESC self-renewal and the preventing of differentiation. The interaction of lncRNAs with histone modifiers and other chromatin-associated proteins has been described in ESC, and they might serve as a scaffold to connect different chromatin modifying complexes.

2.2.4. Lamins

Also the expression of certain lamin proteins may serve as a marker for ESC differentiation [34]. Nuclear lamins are intermediate filament proteins within the nuclear lamina, that not only fulfill their role in the structural organization and support to the nuclear envelope but also participate in processes such as DNA replication and transcription. This family of proteins can be divided into 2 subgroups: lamins of the B-type (B1, B2, and B3) and of the A-type (A, AΔ10, C, and C2). Lamins B1 and B2 are expressed in both embryonic and adult cells, while A-type lamins are primarily found in differentiated cells, as was described for mouse and human cells. Nevertheless, when neuronal differentiation was induced in hESC, the expression of lamins A/C increased, as was also the case when differentiating toward cardiomyocytes. The authors hypothesize that the mechanism of action can be two-fold. A/C lamins might keep the differentiated state by directly influencing the nuclear structure and making it more rigid and thus less prone to chromatin remodeling, a process occurring during cellular differentiation. Alternatively or in complement, these lamins might also indirectly lock a specific gene expression pattern, by affecting the expression of other genes, as their interactions with several transcription factors have yet been described. In those hESC differentiation experiments, the authors could not point out a direct link with the expression of Oct3/4, as the expression of both markers overlapped in most cells, both in a high or low expression level or as one marker being more abundant than the other. The fact that A-type lamins are already present before total Oct3/4 decrease, makes these lamins a good marker for the indication of early differentiation. Of note, there was no overlapping expression found with TRA-1-60, TRA-1-81 or SSEA-4. For mESC, it has been described that they do show very low but yet detectable levels of Lamin A/C even when pluripotent, albeit in a much lower pattern than for differentiated cells [35].

2.2.5. Cell surface markers

Surface antigens are a valuable tool not only for the monitoring of the (non-)differentiation of stem cells but also for the isolation of a subset of cells. The function of most of these markers is not yet fully elucidated, a search that is further hampered because of their different expression patterns in different pluripotent cell types. Some antigens are associated with carbohydrate epitopes, linked with glycolipids (e.g. SSEA-3) or with glycoproteins (e.g. TRA-1-60), and it has been hypothesized that the core structures of these antigens are essential for the cellular function. For instance, the Lewis-X carbohydrate structure recognized as SSEA-1 may be important for compaction at the morula stage during mouse embryonic development [36].

One of the most well-known types of ESC surface markers is the class of stage-specific embryonic antigens (SSEA), which are of the glycosphingolipid-type. Undifferentiated hESC express SSEA-3 and -4 (globoseries structure), while SSEA-1 (lactoseries structure) is only expressed in low levels [36, 37]. Upon hESC differentiation, the expression of these markers is quickly downregulated, with SSEA-3 disappearing faster than SSEA-4. On top of that, there is a significant increase in SSEA-1 [38]. Contrastingly, the reversed pattern is seen in pluripotent mESC [36, 39]. Also another SSEA-molecule, namely, SSEA-5, has been identified on the surface of hESC, which undergoes an even larger reduction upon differentiation than SSEA-3 or -4 [40]. Their expression depends on the combined actions of the different enzymes that are involved in their synthesis. Also the modified expression profile upon differentiation is primarily directed by an altered expression of the key glycosyltransferases (GT), with an upregulation of ganglio-related GT and simultaneous downregulation of globo- and lacto-series-related GT [41]. Of note, this list of glycosphingolipids is far from complete; hESC additionally express several other, less well-known markers of this type, of which the expression rapidly diminishes upon differentiation [38]. Although the SSEA have been challenged not to be essential for pluripotency maintenance [42], but to fulfill a role in cellular differentiation instead, their presence is nevertheless still considered as one of the criteria that must be fulfilled to categorize a cell as pluripotent.

TRA-1-60 and TRA-1-81 are surface carbohydrate antigens present on hESC [36, 37], but not in mESC [36, 39]. Retinoic acid-induced hESC differentiation significantly downregulates their expression [37]. They have been shown to interact with keratan sulphated proteoglycans, although the exact structural determinants of their epitopes remain unknown [43]. Their expression is related to that of podocalyxin, a transmembrane glycoprotein, which has yet been found to be highly expressed in both ESC and iPSC, and might serve as a carrier for the TRA-antigens [44]. Cell surface glycans on ESC might play a role in the modulation of multiple signaling pathways [43].

Besides the more frequently indicated SSEA- and TRA-markers, also several other categories of surface molecules have been described. Cluster of differentiation (CD) antigens can be subdivided into several classes such as integrins, adhesion molecules, glycoproteins, and receptors [36]. Some of them such as CD9, CD24, and CD133 are associated with mESC and hESC [36]. Of note, CD133 is also expressed in other cell lines like embryonic carcinoma cells and hematopoietic stem cells [36]. Integrins are important for keeping ESC undifferentiated. These cell surface receptors can bind several extracellular matrix proteins such as fibronectin, vitronectin, collagen, and laminin and provide in this way the cell matrix interaction [36, 45].

Next to the pluripotency cell surface markers, a range of markers for differentiation has been described. Depending on the differentiation protocol applied, specific markers can be investigated. On top of the markers already mentioned under paragraph 2.2.2, SSEA-1 and Gata4 for example indicate hESC endoderm differentiation, and the possibilities are still expanding. For instance, Holtzinger *et al.* recently described new markers for hepatocyte differentiation from hESC and hiPSC [46].

2.2.6. Alkaline phosphatase

Tissue-nonspecific alkaline phosphatase (TNAP or AP) is a membrane-bound glycosylated enzyme which is highly expressed in pluripotent ESC [3, 47], and is rapidly upregulated upon the reprogramming of somatic cells into iPSC. Of note, the expression of AP is absent in mEpiSC [48]. AP expression significantly decreases during differentiation, making it a suitable marker for pluripotency assessment. One way to monitor AP in hESC is detection with TRA-2-29 and TRA-2-54 antibodies [36, 49]. Its specific function is not totally elucidated, but its importance for ESC is ascribed to its role in the metabolism of vitamin B6 and thus also of the neurotransmitter γ-aminobutyric acid (GABA), a process that is considered to be imperative for ESC proliferation and self-renewal regulation. Additionally, it is very likely that because of the high proliferative rate of ESC, their need for substrate dephosphorylation is considerably higher than in somatic cells [48]. The activity of AP is highly correlated with its expression, which is mainly driven by the actual microenvironment of the cell, instead of via specific signaling pathways. MAPK p38 might play a role in AP's expression regulation, as deletion of p38 in ESC led to a decreased AP expression and activity, but the precise mechanism remains unknown.

2.3. Other pluripotency features

A high level of telomerase activity is also one of the typical ESC characteristics [3]. Telomerase is involved in maintaining telomere length and is able to add telomere repeats to chromosome ends. As such it plays an important role in sustaining the replicative time-span and self-renewal of ESC. Somatic cells do not display any telomerase activity, leading to shortened telomeres over time and entering senescence after a certain number of cell divisions. Even in adult stem cells only low levels of telomerase activity can be found, in contrast to ESC [50].

Pluripotent hESC also display an abbreviated cell cycle of about 16 h, in comparison to differentiated cells, due to a shortened G1 phase [51, 52]. Similarly, also mESC proliferate quickly and have a lengthening cell cycle upon differentiation. Because of their high proliferative pace, both mESC and hESC cultures need to be passaged very frequently, although the passaging technique is different between those two culture types: naïve cells are way more tolerant for single cell passaging than primed cells and thus allow better for bulk culture.

Besides confirming the presence or absence of certain characteristic traits, the most stringent way to show pluripotent potential is chimera formation. However, as hereby ESC are injected into a developing embryo, this is for obvious ethical reasons not possible for hESC. Additionally, pluripotency can be investigated through spontaneous ESC differentiation *in vivo* (teratoma formation) and/or *in vitro* (embryoid body formation). Alternatively, also the directed differentiation under stimulation by specific growth factors, small molecules, gene manipulation, etc. toward specific cell types can be assessed.

3. Pluripotency monitoring

3.1. Overview

The pluripotent state and cellular differentiation can be followed by means of multiple techniques, depending on the cell lines used, the experimental set-up, etc. As a start, microscopy is a very important asset in this monitoring process. Cell cultures' performance, colony, and individual cell morphology (whether or not during differentiation) can be checked by means of light microscopy. With the implementation of additional stainings with fluorescently labelled antibodies and nuclear stains (e.g. DAPI), a whole range of other parameters like nuclear morphology, the presence of core transcription factors, cell surface molecules, and intracellular markers can be assessed with fluorescence microscopy. Often specific kits are available for a defined marker panel for pluripotency assessment of both ESC and iPSC lines, such as the Fluorescent hESC/hiPSC Characterization Kit provided by Millipore [53]. A noninvasive method for daily check-up of hESC cultures with microscopy is further described in Section 3.2. As an alternative, flow cytometry analysis after immunostaining can be applied for monitoring those markers. This technique is also applied for the analysis of cell proliferation rates and the distribution of a cell population across the different cell cycle phases.

A relatively quick way to obtain gene expression data of the described pluripotency factors, differentiation markers, lncRNAs, lamin proteins, etc., is the application of reverse transcription-quantitative PCR (RT-qPCR); one such protocol for relative quantification is described under paragraph 3.3. Larger transcriptomics experiments can be set up by means of microarray analysis and next-generation sequencing. To detect these markers' expression on a protein level, also multiple techniques are available such mass spectrometry, or 1D or 2D gel electrophoresis combined with Western blotting.

Enzymatic activities of e.g. alkaline phosphatase and telomerase can be determined with specific kits available for cells of both human and murine origin. Telomerase activity is most often checked by means of a TRAP assay, or telomeric repeat amplification protocol, in which the telomerase activity is (semi-)quantified with qPCR after an initial enzymatic incubation step. Also ELISA-based methods have been described, whether or not in combination with TRAP (elaborately reviewed in [54]). AP activity detection is often based on colorimetric assays, e.g. performed on the supernatant of cell cultures. Even live stains are available, that allow noninvasive monitoring with microscopy. Hereby a non-fluorescent substrate is added to the cell culture, that becomes fluorescent after dephosphorylation by AP. Such a cell-permeable and non-toxic fluorescent substrate does not accumulate in the cells but diffuses out of the cells after 2 h [49], and thus does not yield any problems for following analyses.

3.2. Noninvasive monitoring

3.2.1. Background

As described above, immunostaining and RT-qPCR are most often used methods for pluripotency and differentiation screening. Nevertheless, for those methods, ESC need to be harvested, which thus makes it impossible to monitor the same colony over and over again on a daily basis, as new samples have to be collected for each analysis. We present a method enabling pluripotency monitoring of the same specific colonies during a time-lapse experiment by using a reporter ESC line, as was published by Scheerlinck *et al.* [55].

3.2.2. Experimental aspects and workflow

A commercially available Oct3/4-eGFP Knock-In hESC line was used, in which the transcription of enhanced green fluorescent protein (eGFP) is coupled to the transcription of *POU5F1* [56]. After eGFP translation, a fluorescent signal (ex. 489 nm, em. 511 nm) is detected, which can be measured using a flow cytometer (FC) or fluorescence microscope (FM) [57]. When using the latter, not only the pluripotency can be measured but also the morphology can be examined. FC can then be used to validate the FM results obtained and was thus used as a reference. However, FC was only used at the end of the experiments, because it requires cell harvesting and fresh samples, making the monitoring of the same specific colonies on a daily basis impossible.

This method can be applied both for feeder and feeder-free hESC culture conditions. Hence, a comparison was made between hESC cultured on MEF with regular DMEM/F-12 medium (with 4 ng/mL bFGF) on one hand and hESC on vitronectin-coated plates in combination with Essential 8 medium on the other. Both culture conditions were kept undifferentiated as for a regular ESC culture. Simultaneously, a differentiation experiment was set up, where differentiation was induced in case of MEF culture by omitting bFGF from the culture medium (spontaneous differentiation), whether or not combined with the addition of 2 μM retinoic acid (forced ectoderm differentiation). For feeder-free culture, only the latter condition with retinoic acid was applied. As such, there were in total three MEF conditions and two feeder-free conditions analyzed. The resolving power of the FM to compare differentiation status differences was determined by the comparison of those five conditions, of which only the ones containing bFGF are assumed to keep the hESC pluripotent. The resolving power of FC was determined by a complete 15-day differentiation of the Oct3/4 reporter cell line: the fluorescent signal quickly decreased during the first week and completely disappeared after 15 days in culture, falling back to the same level as UGENT2-cells, a non-reporter hESC line.

Importantly, possible auto-fluorescence of both the medium and the cells themselves needs to be investigated. Several amino acids (tryptophan, tyrosine, and phenylalanine) and vitamins (riboflavin) among other factors, all present in DMEM/F-12, are known to cause auto-fluorescence [58–60]. Subsequently, also non-reporter hESC (in our case the UGENT2 cell line) and MEF in case of feeder cultures need to be analyzed. When using FM, the signal-to-noise (S/N) ratio for each hESC colony can be determined by dividing the densitometric mean of the colony (signal) by that one of its background (noise). For FC analysis of feeder cultures, it was not possible to distinguish the MEF population from the hESC in terms of FS/SS. Ideally, these MEF should be isolated by means of fluorescence-activated cell sorting after immunostaining for a specific marker such as vimentin or CD90 [61]. However, as FC is only used at the end of the experiment only a small contribution of the MEF (<10%) to the fluorescence histogram is expected, as the relative portion of the inactivated MEF gradually reduces over time compared to the growing hESC colonies.

3.2.3. Outcome

As stated above, the UGENT2 cell line and MEF were included in the analysis to determine the background auto-fluorescence. For FM, the MEF did not yield a detectable auto-fluorescence signal and did thus not impact the S/N ratio in comparison to feeder-free cultured hESC, although FC analysis revealed a low but present MEF auto-fluorescence intensity (**Figure 1A**).

	Experiment 1		Experiment 2		Experiment 3	
Rank	*Ref.*	*M value*	*Ref.*	*M value*	*Ref.*	*M value*
1	B2M	0,277	B2M	0,203	B2M	0,378
2	AluSq	0,284	AluSq	0,213	RPL13A	0,378
3	RPL13A	0,290	RPL13A	0,221	AluSq	0,386
4	HPRT1	0,472	GAPDH	0,267	PPIA	0,480
5	YWHAZ	0,593	SDHA	0,360	GAPDH	0,609
6	TBP	0,696	TBP	0,409	HPRT1	0,672
7	HMBS	0,886	YWHAZ	0,451	TBP	0,813
8	ACTB	1,151	ACTB	0,480	YWHAZ	0,903
9	GAPDH	1,350	HPRT1	0,506	HMBS	1,088
10	PPIA	1,476	PPIA	0,524	SDHA	1,216
11	SDHA	1,598	HMBS	0,550	ACTB	1,347
12	UBC	1,697	UBC	0,579	UBC	1,484

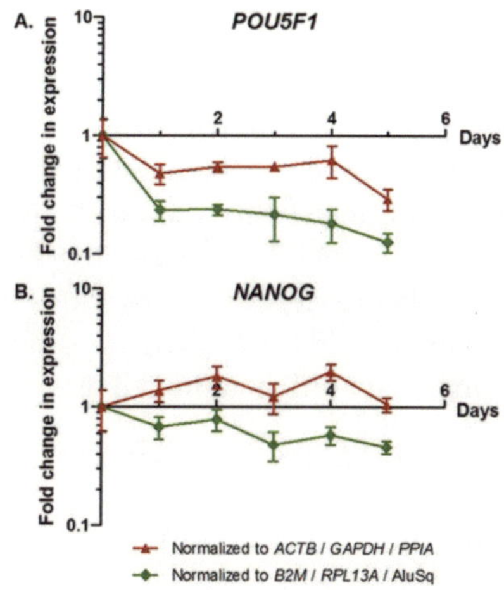

Figure 1. Noninvasive monitoring of hESC colonies. (A) Auto-fluorescence assessment by FC. (B) Results obtained by FM, expressed as S/N ratios. The differentiating conditions clearly show a decreased S/N ratio in comparison to the cells kept pluripotent. (C) FC results obtained at the end of the differentiation experiment, confirming the FM results. (D) Microscopic fluorescent images of a hESC colony in MEF culture conditions, under spontaneous (left) and retinoic acid-induced (right) differentiation.

For FM assessment, colony fluorescence was measured daily for 5 or 6 colonies per condition, during 6 days. The experiment was performed in triplicate and the obtained results are expressed as S/N ratios. As expected, addition of retinoic acid (directed differentiation) caused a significant decrease in S/N ratio over time both in feeder and feeder-free culture conditions, indicating a lowering expression of Oct3/4 in the course of the differentiation process. Also only omitting bFGF from the culture medium definitely leads to differentiation, as was confirmed by the decreasing S/N ratios, albeit in a slower rate than when also retinoic acid is added. No significant S/N ratio differences were found between feeder and feeder-free hESC cultures for both the non-differentiating and the differentiating conditions, confirming the low impact of MEF auto-fluorescence on FM measurements. Of note, for both undifferentiated culture conditions, an increase in fluorescence was seen toward the end of the experiment, which might be explained by an increase in eGFP/cell or more plausibly by the formation of multilayers (3D growth) resulting in an accumulation of fluorescent signal (**Figure 1B**).

The FM data were compared to the results of a FC hESC analysis (**Figure 1C**). As mentioned above, because of the destructive nature of this technique, this measurement is only performed at the end of the experiment. Both feeder-free and MEF grown hESC in the presence of bFGF retained their undifferentiated status. Of note, a small portion of cells in the latter population had a lower eGFP expression, most probably MEF as mentioned earlier (auto-fluorescence between 10^0 and 10^1, asterisk in **Figure 1C**). The finding that the eGFP signal/cell remained constant in the undifferentiated conditions indeed confirms that the daily increase in fluorescence as observed by FM is rather due to a multilayer effect resulting in an accumulated fluorescent signal. In the MEF condition without bFGF, most of the cells were still undifferentiated after 6-day culture (signal >10^1) but in comparison, in the condition with bFGF a significantly higher number of cells with a 10^0 and 10^1 eGFP expression were observed. These results are in line with the FM measurements, in which it was shown that there is indeed a mix of differentiated (low fluorescence; S/N ratio = 2.26) and undifferentiated hESC (high fluorescence, S/N ratio = 13.75) on day 5 of spontaneous differentiation (**Figure 1D** left). The conditions with retinoic acid showed a clear drop in fluorescence toward the end of the experiment, as also confirmed by the FM results. Remarkably, FM images of retinoic acid differentiated hESC colonies on MEF revealed the existence of demarcated zones with highly accumulated fluorescence (S/N ratio = ca. 24) (**Figure 1D** right). This small population of high fluorescent 'islands' could not be discriminated using FC as these individual highly fluorescent cells were somewhat hidden in the tail of the fluorescence histogram.

In conclusion, it is important to bear in mind that for FM assessment, an increasing fluorescence intensity of a hESC colony does not correlate with an increased eGFP signal per cell, but with hESC growing in multiple layers. As such, only a decrease in signal can be directly interpreted as ongoing differentiation. A flat signal can be considered as an hESC culture with both pluripotent and differentiating cells [55].

Despite its usefulness, this method still has some limitations. One issue is the auto-fluorescence of the medium, necessitating the use of a different medium than that used for regular culture. In this experiment, hESC colonies were analyzed by using no medium at all, although this evokes stress to the cells. ThermoFisher has recently developed an auto-fluorescence-free medium called FluoroBrite DMEM, which can be used as basal medium for analysis during fluorescence [58]. It should nevertheless be further investigated whether this medium can also be used for hESC. Secondly, ideally the analysis should be done in the same conditions as during culture, meaning that a cell imaging system with regulated O_2 and CO_2 supply would be more appropriate. Different companies such as Zeiss could offer a solution in this regard. Nevertheless, even if a microscope as described in the experiments above is used, the fact that the cells are monitored in culture enables an immediate follow-up of their behavior.

3.2.4. Applications

The application of this method is definitely not restricted to an Oct3/4-eGFP ESC line but can be used for any reporter cell line, as long as potential auto-fluorescence is overcome, e.g. by the implementation of an auto-fluorescence-free cell culture medium. For instance, also Nanog-eGFP reporter hESC lines are available [62], as is an hRex-GFP hESC line [63]. Additionally, reporters for specific differentiation markers have been generated, with an increasing GFP expression during the course of the differentiation process, e.g. [64, 65]. The production of a Nanog-eGFP reporter to monitor fibroblast reprogramming has been published, too [66]. A reporter cell line including multiple markers clearly strengthens the use of the method described above. For example, Maass *et al.* created a reporter mESC line harboring Cntn2-eGFP and MHCα-mCherry, which will be upregulated during directed cardiac differentiation, as such allowing to monitor the development of cardiac progenitors [67]. However, to our knowledge, the development of a dual reporter ESC line combining two pluripotency markers has not been published. Furthermore, most reporter ESC lines described express (e)GFP, thus ruling out a possible combination with GFP-based alkaline phosphatase live stain [68].

However, ActivMotiv provides a CDy1 Dye (ex. 544 nm/em. 577 nm) to detect pluripotent stem cells live in culture, that can be combined with GFP expression [69, 70].

3.3. Quantitative PCR monitoring

3.3.1. Background

As mentioned above, one other efficient method to monitor the expression profile of multiple target genes (e.g. the core pluripotency transcription factors) is RT-qPCR. This technique allows to analyze samples in high throughput, at a relatively low cost. In this way, it can easily be applied for e.g. pluripotency monitoring of (long term) ESC cultures, as was recently elaborately described for hESC by Galán and Simón [71]. Besides culture monitoring, it can also be applied for the comparison of pluripotent and differentiating ESC. Importantly, as goes for all experimental work, also here an adequate set-up is of utmost importance. During RT-qPCR data analysis, a suitable normalization factor should be taken into account, to correct for potential technical variabilities that were included along the experimental process. To this end, multiple strategies have been used. Normalization to the number of cells might not be accurate and cell enumeration is particularly less easy when dealing with adherent cell cultures such as ESC [72]. Moreover, this normalization strategy does not take into account the variability that might have been included during sample preparation such as potentially insufficient enzymatic reaction efficiencies. Alternatively, correcting toward RNA mass quantity has been described, but also here the same issue applies, as potential technical variation from the complementary DNA (cDNA) preparation from messenger RNA (mRNA) is not considered. Ribosomal RNA (rRNA) makes up the major part of the total RNA pool and may be prone to regulation, which will cause a variable ratio between rRNA and mRNA. Hence, normalization for the total RNA content may not be representative for the amount of mRNA [72, 73]. The by far mostly favored method for data normalization is correcting to one or preferentially multiple reference or so-called housekeeping genes. This allows the researcher to correct for any variation between samples that might have been implemented along the protocol. These reference genes are expected to be expressed in a stable manner throughout all samples of a given experimental set-up.

3.3.2. Experimental aspects and workflow

To determine which references need to be included, different algorithms are available such as BestKeeper or Normfinder, but the most widely used is the geNorm algorithm, included in the qBase software (Biogazelle) [74]. We applied the latter to find suitable reference genes to be used for the comparison between pluripotent and differentiating hESC [75]. After putting in all RT-qPCR data, the software calculates a so-called stability value (designated 'M value') for each candidate, with a low M value indicating a higher gene expression stability throughout all samples. Afterwards, all reference candidates are ranked according to that value and the most adequate reference genes can be selected.

A differentiation experiment was set up, in which hESC samples from two different cell lines (UGENT1 and UGENT2) were collected daily during 8 days after retinoic acid-induction, plus an extra sample on day 12 of differentiation (Experiment 1). This was repeated in two more experiments, albeit in a different time window: in Experiment 2 samples were collected daily during 6 days and in Experiment 3, cells were harvested every 4 h during day 3, 4, and 5 after onset of differentiation. After RNA isolation, quality control and cDNA preparation, RT-qPCR reactions were run in duplo, on an ABI Prism 7000 Sequence Detection System (Applied Biosystems) and a LightCycler 480 (Roche). Twelve reference loci candidates were included, mainly with SYBR Green detection (allowing melting curve analysis), except for *GAPDH* and *PPIA*, for which specific probes were used. All data were imported in geNorm and assessed for expression stability.

3.3.3. *Outcome*

According to the geNorm analysis performed for all three differentiation experiments, the most stable reference loci among the 12 included candidates were *B2M*, *RPL13A*, and Alu repeats. In contrast, the more classic reference genes such as *GAPDH* or *ACTB* did not perform so well (all M values listed in **Figure 2**) [75]. These findings were corroborated by the results obtained from the use of other algorithms for reference gene stability determination.

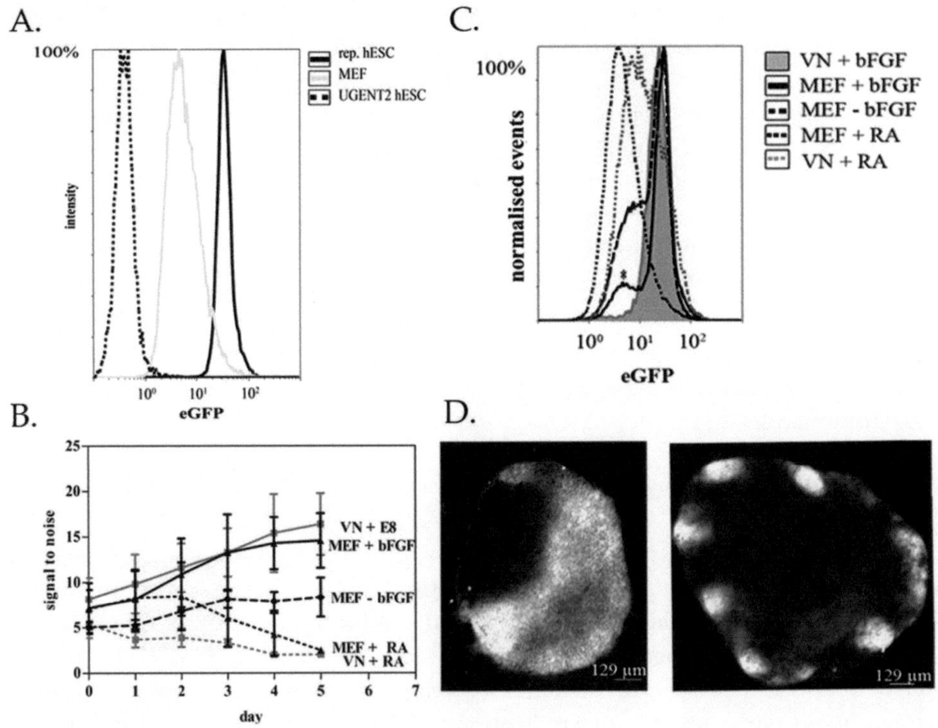

Figure 2. Reference genes selection and their application. A geNorm stability analysis was performed for 12 candidate reference genes; the table shows the resulting ranking. *B2M*, *RPL13A*, and Alu repeats appeared to be the most suitable references, performing significantly better than the more classic reference genes *GAPDH*, *ACTB*, and *PPIA* as is illustrated in the graph (fold change in expression over time during differentiation).

It is important to bear in mind that different experimental settings may require different references; e.g. one differentiation-inducing agent will not have the same influence on gene expression in general as another. Previously defined reference loci might thus not be blindly extrapolated to new experimental conditions. The difference in reference gene performance is illustrated by the comparison between to reference sets: *B2M, RPL13A,* and Alu repeats on one hand and *GAPDH, PPIA,* and *ACTB* on the other. These sets were used to normalize the expression of the pluripotency factors *POU5F1* and *NANOG,* which are supposed to decrease considerably after differentiation is induced. However, that decrease was significantly less pronounced (p-value = 1.30e-05) when applying the more classic reference genes, once more pointing out the importance of an adequate reference gene selection (graph **Figure 2**).

The fact that *B2M, RPL13A,* and Alu repeats are found to be the most stable genes is nevertheless not too surprising. *B2M* is expressed in every nucleated cell as it is a component of the major histocompatibility complex I. It is thus a very good candidate to apply as a normalization scalar for RT-qPCR analysis (e.g. [76]). Also *RPL13A,* involved in the protein translation process, has been widely used as a reference gene (e.g. [77]). The applicability of both genes was confirmed by our own data analysis. The Alu repeats are of particular interest, as their application provides a rather novel approach for data normalization [78]. These short interspersed elements are distributed genome-wide, which implies that a variation in the expression of a single gene will not substantially influence the total expression profile of these elements.

3.3.4. Applications

The setting described here specifies the monitoring of the pluripotency factors Oct3/4 and Nanog, but obviously any other marker of which the expression should be monitored and relatively quantified can be included. It is recommended to revaluate the reference genes' stability if an experimental set-up would be modified, but once the most suitable references have been established, any target gene can be implemented in this assay. As such, on top of the pluripotency markers, also the expression of specific differentiation markers can be monitored to follow development toward a certain cell lineage. This technique can as well be used for the assessment of noncoding RNA genes.

4. Conclusion

This chapter gives an overview on a number of typical characteristics which can be monitored to keep track of the pluripotent status of ESC cultures. Nevertheless, this list is far from complete and it is most likely that in time new biomarkers will be found while known features will be revaluated, as new culturing techniques are developed and the research field of induced pluripotency and naïve pluripotency further expands. Also from a technological point of view, novel methods will enhance these markers' detection and facilitate the discovery of new molecules.

Author details

Liesbeth Vossaert*, Ellen Scheerlinck and Dieter Deforce

*Address all correspondence to: liesbeth.vossaert@ugent.be

Laboratory of Pharmaceutical Biotechnology, Ghent University, Ottergemsesteenweg, Gent, Belgium

References

[1] Evans MJ, Kaufman MH. Establishment in culture of pluripotential cells from mouse embryos. Nature 1981;292:154–6.

[2] Martin GR. Isolation of a pluripotent cell line from early mouse embryos cultured in medium conditioned by teratocarcinoma stem cells. Proc Natl Acad Sci U S A 1981;78:7634–8.

[3] Thomson JA, Itskovitz-Eldor J, Shapiro SS, Waknitz MA, Swiergiel JJ, Marshall VS, et al. Embryonic stem cell lines derived from human blastocysts. Science (80-) 1998;282:1145–7.

[4] Ezashi T, Yuan Y, Roberts RM. Pluripotent stem cells from domesticated mammals. Annu Rev Anim Biosci 2016;4: 223–53. doi:10.1146/annurev-animal-021815-111202.

[5] Takahashi K, Yamanaka S. Induction of pluripotent stem cells from mouse embryonic and adult fibroblast cultures by defined factors. Cell 2006;126:663–76. doi:10.1016/j.cell.2006.07.024.

[6] Takahashi K, Tanabe K, Ohnuki M, Narita M, Ichisaka T, Tomoda K, et al. Induction of pluripotent stem cells from adult human fibroblasts by defined factors. Cell 2007;131:861–72. doi:10.1016/j.cell.2007.11.019.

[7] Nichols J, Smith A. Naive and primed pluripotent states. Cell Stem Cell 2009;4:487–92. doi:10.1016/j.stem.2009.05.015.

[8] Robinton DA, Daley GQ. The promise of induced pluripotent stem cells in research and therapy. Nature 2012;481:295–305. doi:10.1038/nature10761.

[9] Lai D, Bu S. Comparison of the ultrastructures of primed and naïve mouse embryonic stem cells. Cell Reprogram 2016;18(1): 48–53. doi:10.1089/cell.2015.0063.

[10] Zeuschner D, Mildner K, Zaehres H, Schöler HR. Induced pluripotent stem cells at nanoscale. Stem Cells Dev 2010;19:615–20. doi:10.1089/scd.2009.0159.

[11] Manian K V, Aalam SMM, Bharathan SP, Srivastava A, Velayudhan SR. Understanding the molecular basis of heterogeneity in induced pluripotent stem cells. Cell Reprogram 2015;17:427–40. doi:10.1089/cell.2015.0013.

[12] Hoffman LM, Carpenter MK. Characterization and culture of human embryonic stem cells. Nat Biotechnol 2005;23:699–708. doi:10.1038/nbt1102.

[13] Medvedev SP, Shevchenko a I, Mazurok N a, Zakiian SM. OCT4 and NANOG are the key genes in the system of pluripotency maintenance in mammalian cells. Genetika 2008;44:1589–608. doi:10.1134/S1022795408120016.

[14] James D, Levine AJ, Besser D, Hemmati-Brivanlou A. TGFbeta/activin/nodal signaling is necessary for the maintenance of pluripotency in human embryonic stem cells. Development 2005;132:1273–82. doi:10.1242/dev.01706.

[15] Sato N, Meijer L, Skaltsounis L, Greengard P, Brivanlou AH. Maintenance of pluripotency in human and mouse embryonic stem cells through activation of Wnt signaling by a pharmacological GSK-3-specific inhibitor. Nat Med 2004;10:55–63. doi:10.1038/nm979.

[16] McLean AB, D'Amour K a, Jones KL, Krishnamoorthy M, Kulik MJ, Reynolds DM, et al. Activin A efficiently specifies definitive endoderm from human embryonic stem cells only when phosphatidylinositol 3-kinase signaling is suppressed. Stem Cells 2007;25:29–38. doi:10.1634/stemcells.2006-0219.S.

[17] Armstrong L, Hughes O, Yung S, Hyslop L, Stewart R, Wappler I, et al. The role of PI3K/AKT, MAPK/ERK and NFκβ signalling in the maintenance of human embryonic stem cell pluripotency and viability highlighted by transcriptional profiling and functional analysis. Hum Mol Genet 2006;15:1894–913. doi:10.1093/hmg/ddl112.

[18] McCain J. The MAPK (ERK) Pathway: Investigational combinations for the treatment of BRAF-mutated metastatic melanoma. P T 2013;38:96–108.

[19] Ding VMY, Ling L, Natarajan S, Yap MGS, Cool SM, Choo ABH. FGF-2 modulates Wnt signaling in undifferentiated hESC and iPS cells through activated PI3-K/GSK3β signaling. J Cell Physiol 2010;225:417–28. doi:10.1002/jcp.22214.

[20] Walsh J, Andrews PW. Expression of Wnt and Notch pathway genes in a pluripotent human embryonal carcinoma cell line and embryonic stem cell. APMIS 2003;111:197–210; discussion 210–1. doi:10.1034/j.1600-0463.2003.1110124.x.

[21] Xu R-H, Peck RM, Li DS, Feng X, Ludwig T, Thomson J a. Basic FGF and suppression of BMP signaling sustain undifferentiated proliferation of human ES cells. Nat Methods 2005;2:185–90. doi:10.1038/nmeth744.

[22] Pan GJ, Chang ZY, Scholer HR, Pei D. Stem cell pluripotency and transcription factor Oct4. Cell Res 2002;12:321–9. doi:10.1038/sj.cr.7290134.

[23] Fong H, Hohenstein KA, Donovan PJ. Regulation of self-renewal and pluripotency by Sox2 in human embryonic stem cells. Stem Cells 2008;26:1931–8. doi:10.1634/stemcells.2007-1002.

[24] Mitsui K, Tokuzawa Y, Itoh H, Segawa K, Murakami M, Takahashi K, et al. The homeoprotein Nanog is required for maintenance of pluripotency in mouse epiblast and ES cells. Cell 2003;113:631–42.

[25] Singh AM, Hamazaki T, Hankowski KE, Terada N. A heterogeneous expression pattern for Nanog in embryonic stem cells. Stem Cells 2007;25:2534–42. doi:10.1634/stemcells.2007-0126.

[26] Boyer LA, Lee TI, Cole MF, Johnstone SE, Levine SS, Zucker JP, et al. Core transcriptional regulatory circuitry in human embryonic stem cells. Cell 2005;122:947–56. doi:10.1016/j.cell.2005.08.020.

[27] Petruzzelli R, Christensen DR, Parry KL, Sanchez-Elsner T, Houghton FD. HIF-2α regulates NANOG expression in human embryonic stem cells following hypoxia and reoxygenation through the interaction with an Oct-Sox Cis regulatory element. PLoS One 2014;9:e108309. doi:10.1371/journal.pone.0108309.

[28] Adewumi O, Aflatoonian B, Ahrlund-Richter L, Amit M, Andrews PW, Beighton G, et al. Characterization of human embryonic stem cell lines by the International Stem Cell Initiative. Nat Biotechnol 2007;25:803–16. doi:10.1038/nbt1318.

[29] Yeo S, Jeong S, Kim J, Han JS, Han YM, Kang YK. Characterization of DNA methylation change in stem cell marker genes during differentiation of human embryonic stem cells. Biochem Biophys Res Commun 2007;359:536–42. doi:10.1016/j.bbrc.2007.05.120.

[30] Tollervey J, Lunyak V V. Epigenetics: judge, jury and executioner of stem cell fate. Epigenetics 2012;7:823–40. doi:10.4161/epi.21141.

[31] Cheng X, Hashimoto H, Horton J, Zhang X. Mechanisms of DNA methylation, Methyl-CpG recognition and demethylation in mammals. In: Tollefsbol T, editor. Handbook of Epigenetics: The New Molecular and Medical Genetics, Academic Press. An imprint of Elsevier 525 B Street, Suite 1800, San Diego, CA 92101-4495, USA 2011, p. 9–24.

[32] Rosa A, Ballarino M. Long noncoding RNA regulation of pluripotency. Stem Cells Int 2016;2016:1797692. doi:10.1155/2016/1797692.

[33] Collins L, Schönfeld B, Chen XS. The epigenetics of non-coding RNA. In: Tollefsbol T, editor. Handbook of Epigenetics: The New Molecular and Medical Genetics, Academic Press. An imprint of Elsevier 525 B Street, Suite 1800, San Diego, CA 92101-4495, USA 2011, p. 49–61.

[34] Constantinescu D, Gray HL, Sammak PJ, Schatten GP, Csoka AB. Lamin A/C expression is a marker of mouse and human embryonic stem cell differentiation. Stem Cells 2006;24:177–85. doi:10.1634/stemcells.2004-0159.

[35] Eckersley-Maslin MA, Bergmann JH, Lazar Z, Spector DL. Lamin A/C is expressed in pluripotent mouse embryonic stem cells. Nucleus 4:53–60. doi:10.4161/nucl.23384.

[36] Zhao W, Ji X, Zhang F, Li L, Ma L. Embryonic stem cell markers. Molecules 2012;17:6196–236. doi:10.3390/molecules17066237.

[37] Draper JS, Pigott C, Thomson JA, Andrews PW. Surface antigens of human embryonic stem cells: changes upon differentiation in culture. J Anat 2002;200:249–58.

[38] Liang Y-J, Kuo H-H, Lin C-H, Chen Y-Y, Yang B-C, Cheng Y-Y, et al. Switching of the core structures of glycosphingolipids from globo- and lacto- to ganglio-series upon human embryonic stem cell differentiation. Proc Natl Acad Sci U S A 2010;107:22564–9. doi:10.1073/pnas.1007290108.

[39] Ginis I, Luo Y, Miura T, Thies S, Brandenberger R, Gerecht-Nir S, et al. Differences between human and mouse embryonic stem cells. Dev Biol 2004;269:360–80. doi:10.1016/j.ydbio.2003.12.034.

[40] Tang C, Lee AS, Volkmer J-P, Sahoo D, Nag D, Mosley AR, et al. An antibody against SSEA-5 glycan on human pluripotent stem cells enables removal of teratoma-forming cells. Nat Biotechnol 2011;29:829–34. doi:10.1038/nbt.1947.

[41] Hakomori S. Structure and function of glycosphingolipids and sphingolipids: recollections and future trends. Biochim Biophys Acta 2008;1780:325–46. doi:10.1016/j.bbagen.2007.08.015.

[42] Brimble SN, Sherrer ES, Uhl EW, Wang E, Kelly S, Merrill AH, et al. The cell surface glycosphingolipids SSEA-3 and SSEA-4 are not essential for human ESC pluripotency. Stem Cells 2007;25:54–62. doi:10.1634/stemcells.2006-0232.

[43] Lanctot PM, Gage FH, Varki AP. The glycans of stem cells. Curr Opin Chem Biol 2007;11:373–80. doi:10.1016/j.cbpa.2007.05.032.

[44] Schopperle WM, DeWolf WC. The TRA-1-60 and TRA-1-81 human pluripotent stem cell markers are expressed on podocalyxin in embryonal carcinoma. Stem Cells 2007;25:723–30. doi:10.1634/stemcells.2005-0597.

[45] Miyazaki T, Futaki S, Hasegawa K, Kawasaki M, Sanzen N, Hayashi M, et al. Recombinant human laminin isoforms can support the undifferentiated growth of human embryonic stem cells. Biochem Biophys Res Commun 2008;375:27–32. doi:10.1016/j.bbrc.2008.07.111.

[46] Holtzinger A, Streeter PR, Sarangi F, Hillborn S, Niapour M, Ogawa S, et al. New markers for tracking endoderm induction and hepatocyte differentiation from human pluripotent stem cells. Development 2015;142:4253–65. doi:10.1242/dev.121020.

[47] Palmqvist L, Glover CH, Hsu L, Lu M, Bossen B, Piret JM, et al. Correlation of murine embryonic stem cell gene expression profiles with functional measures of pluripotency. Stem Cells 2005;23:663–80. doi:10.1634/stemcells.2004-0157.

[48] Štefková K, Procházková J, Pacherník J. Alkaline phosphatase in stem cells. Stem Cells Int 2015;2015:628368. doi:10.1155/2015/628368.

[49] Singh U, Quintanilla RH, Grecian S, Gee KR, Rao MS, Lakshmipathy U. Novel live alkaline phosphatase substrate for identification of pluripotent stem cells. Stem Cell Rev 2012;8:1021–9. doi:10.1007/s12015-012-9359-6.

[50] Hiyama E, Hiyama K. Telomere and telomerase in stem cells. Br J Cancer 2007;96:1020–4. doi:10.1038/sj.bjc.6603671.

[51] Becker KA, Ghule PN, Therrien JA, Lian JB, Stein JL, van Wijnen AJ, et al. Self-renewal of human embryonic stem cells is supported by a shortened G1 cell cycle phase. J Cell Physiol 2006;209:883–93. doi:10.1002/jcp.20776.

[52] Kapinas K, Grandy R, Ghule P, Medina R, Becker K, Pardee A, et al. The abbreviated pluripotent cell cycle. J Cell Physiol 2013;228:9–20. doi:10.1002/jcp.24104.

[53] Fluorescent Human ES/iPSC characterization kit - Millipore n.d. http://www.merck-millipore.com/BE/en/product/Fluorescent-Human-ESiPS-Cell-Characterization-Kit-,MM_NF-SCR078.

[54] Skvortsov DA, Zvereva ME, Shpanchenko O V, Dontsova OA. Assays for detection of telomerase activity. Acta Naturae 2011;3:48–68.

[55] Scheerlinck E, Van Steendam K, Vandewoestyne M, Lepez T, Gobin V, Meert P, et al. Detailed method description for noninvasive monitoring of differentiation status of human embryonic stem cells. Anal Biochem 2014;461:60–6. doi:10.1016/j.ab.2014.05.026.

[56] Zwaka TP, Thomson JA. Homologous recombination in human embryonic stem cells. Nat Biotechnol 2003;21:319–21. doi:10.1038/nbt788.

[57] Heim R, Tsien RY. Engineering green fluorescent protein for improved brightness, longer wavelengths and fluorescence resonance energy transfer. Curr Biol 1996;6:178–82. doi:10.1016/S0960-9822(02)00450-5.

[58] Spencer VA, Kumar S, Paszkiet B, Fein J, Zmuda JF. Cell culture media for fluorescence imaging. Genet Eng Biotechnol News 2014;34:16, 18. doi:10.1089/gen.34.10.09.

[59] DMEM/F-12 - Thermo Scientific n.d. http://www.thermofisher.com/be/en/home/technical-resources/media-formulation.55.html.

[60] Aubin JE. Autofluorescence of viable cultured mammalian cells. J Histochem Cytochem 1979;27:36–43. doi:10.1177/27.1.220325.

[61] Yusuf B, Gopurappilly R, Dadheech N, Gupta S, Bhonde R, Pal R. Embryonic fibroblasts represent a connecting link between mesenchymal and embryonic stem cells. Dev Growth Differ 2013;55:330–40. doi:10.1111/dgd.12043.

[62] Fischer Y, Ganic E, Ameri J, Xian X, Johannesson M, Semb H. NANOG reporter cell lines generated by gene targeting in human embryonic stem cells. PLoS One 2010;5:e12533. doi:10.1371/journal.pone.0012533.

[63] Zhong JF, Weiner L, Jin Y, Lu W, Taylor CR. A real-time pluripotency reporter for human stem cells. Stem Cells Dev 2010;19:47–52. doi:10.1089/scd.2008.0363.

[64] Noisa P, Urrutikoetxea-Uriguen A, Li M, Cui W. Generation of human embryonic stem cell reporter lines expressing GFP specifically in neural progenitors. Stem Cell Rev 2010;6:438–49. doi:10.1007/s12015-010-9159-9.

[65] James D, Nam H, Seandel M, Nolan D, Janovitz T, Tomishima M, et al. Expansion and maintenance of human embryonic stem cell-derived endothelial cells by TGFbeta inhibition is Id1 dependent. Nat Biotechnol 2010;28:161–6. doi:10.1038/nbt.1605.

[66] Lei L, Li L, Du F, Chen C-H, Wang H, Keefer CL. Monitoring bovine fetal fibroblast reprogramming utilizing a bovine NANOG promoter-driven EGFP reporter system. Mol Reprod Dev 2013;80:193–203. doi:10.1002/mrd.22147.

[67] Maass K, Shekhar A, Lu J, Kang G, See F, Kim EE, et al. Isolation and characterization of embryonic stem cell-derived cardiac Purkinje cells. Stem Cells 2015;33:1102–12. doi: 10.1002/stem.1921.

[68] Alkaline Phosphatase Live Stain - ThermoFisher Scientific n.d. https://www.thermofisher.com/order/catalog/product/A14353.

[69] Stem Cell CDy1 Dye - ActiveMotif n.d. https://www.activemotif.com/catalog/895/stem-cell-cdy1-dye.

[70] Kang N-Y, Yun S-W, Ha H-H, Park S-J, Chang Y-T. Embryonic and induced pluripotent stem cell staining and sorting with the live-cell fluorescence imaging probe CDy1. Nat Protoc 2011;6:1044–52. doi:10.1038/nprot.2011.350.

[71] Galán A, Simón C. Monitoring stemness in long-term hESC cultures by real-time PCR. Methods Mol Biol 2016;1307:89–104. doi:10.1007/7651_2014_131.

[72] Huggett J, Dheda K, Bustin S, Zumla A. Real-time RT-PCR normalisation; strategies and considerations. Genes Immun 2005;6:279–84. doi:10.1038/sj.gene.6364190.

[73] Solanas M, Moral R, Escrich E. Unsuitability of using ribosomal RNA as loading control for Northern blot analyses related to the imbalance between messenger and ribosomal RNA content in rat mammary tumors. Anal Biochem 2001;288:99–102. doi:10.1006/abio.2000.4889.

[74] Vandesompele J, De Preter K, Pattyn F, Poppe B, Van Roy N, De Paepe A, et al. Accurate normalization of real-time quantitative RT-PCR data by geometric averaging of multiple internal control genes. Genome Biol 2002;3:Research0034.1–0034.11.

[75] Vossaert L, O Leary T, Van Neste C, Heindryckx B, Vandesompele J, De Sutter P, et al. Reference loci for RT-qPCR analysis of differentiating human embryonic stem cells. BMC Mol Biol 2013;14:21. doi:10.1186/1471-2199-14-21.

[76] Piehler AP, Grimholt RM, Ovstebø R, Berg JP. Gene expression results in lipopolysaccharide-stimulated monocytes depend significantly on the choice of reference genes. BMC Immunol 2010;11:21. doi:10.1186/1471-2172-11-21.

[77] Curtis KM, Gomez L a, Rios C, Garbayo E, Raval AP, Perez-Pinzon M a, et al. EF1alpha and RPL13a represent normalization genes suitable for RT-qPCR analysis of bone marrow derived mesenchymal stem cells. BMC Mol Biol 2010;11:61. doi: 10.1186/1471-2199-11-61.

[78] Marullo M, Zuccato C, Mariotti C, Lahiri N, Tabrizi SJ, Di Donato S, et al. Expressed Alu repeats as a novel, reliable tool for normalization of real-time quantitative RT-qPCR data. Genome Biol 2010;11:R9. doi:10.1186/gb-2010-11-1-r9.

Chapter 3

States of Pluripotency: Naïve and Primed Pluripotent Stem Cells

Daman Kumari

Additional information is available at the end of the chapter

http://dx.doi.org/10.5772/63202

Abstract

Pluripotent stem cells are classified into naïve and primed based on their growth characteristics **in vitro** and their potential to give rise to all somatic lineages and the germ line in chimeras. In this chapter, I describe the similarities and differences between the naïve and primed pluripotent states as exemplified by mouse embryonic stem cells (mESCs), mouse epiblast stem cells (mEpiSCs), human embryonic stem cells (hESCs), and human induced pluripotent stem cells (hiPSCs). I also review the efforts for derivation of naïve human pluripotent stem cells by manipulating culture conditions during reprogramming of somatic cells and attempts to revert primed hESCs to the naïve state. Understanding the requirements for induction and maintenance of the naïve pluripotent state will facilitate studies on early human embryonic development and understanding the mechanisms involved in X inactivation **in vitro**. In addition, the development of naïve hiPSCs will improve the efficiency of gene targeting for the purpose of modeling human diseases as well as for generating gene-corrected autologous pluripotent stem cells for regenerative medicine.

Keywords: naïve, primed, pluripotent stem cells, epigenetic, fragile X syndrome

1. Introduction

Two distinct pluripotent states are observed in embryonic stem cells from mice: the ground or naïve state, exemplified by the mouse embryonic stem cells (mESCs) [1], and the primed pluripotent state represented by mouse epiblast stem cells (mEpiSCs) [2,3]. The clearest difference between the two states is colony morphology, growth factor requirement for maintenance of the pluripotent state, and X inactivation status in female cells. Human embryonic stem cells (hESCs) and human induced pluripotent stem cells (hiPSCs) identify more closely with mEpiSCs than mESCs [4]. Understanding the requirements for induction and maintenance of the naïve pluripotent state for human stem cells will facilitate their application for studying early embryonic development, disease modeling, drug screening, and cell-based therapies.

In this chapter, I mention the similarities and differences between naïve and primed pluripotent stem cells (PSCs) and describe various strategies that have been utilized for the derivation of naïve human PSCs and/or reversion of primed state stem cells to the naïve-like ground state. I also discuss how the availability of naïve hESCs and hiPSCs will further the mechanistic studies of various biological phenomena and facilitate genetic manipulations of hPSCs for disease modeling as well as regenerative medicine.

2. Characteristic features of the naïve and primed pluripotent stem cell states

Embryonic stem cells in mice exist in two functionally distinct pluripotent stem cell states. Specifically, mESCs that are derived from the inner cell mass (ICM) of preimplantation embryos represent naïve pluripotent stem cells (PSCs). These cells have an unlimited self-renewal capacity when grown under appropriate conditions and are able to differentiate into tissues of all three germ layers **in vitro**. In addition, when injected back into the early embryos, naïve stem cells contribute to all somatic lineages including the germline. This ability to generate chimeras is indicative of their pluripotency **in vivo**. On the other hand, mEpiSCs that are derived from the epiblast of the post-implantation embryo typify the primed state. Also included in this category are hESCs and hiPSCs that resemble closely the mEpiSCs, even though they are isolated from preimplantation embryos. Like the naïve PSCs, the primed PSCs also have unlimited potential to self-renew and differentiate into three germ layers **in vitro**, but are limited in their pluripotency **in vivo**, as they cannot give rise to germline chimeras [5]. The naïve mESCs typically grow as small, compact, domed colonies, whereas the primed mEpiSC colonies are large and grow as monolayer similar to hESCs. In addition, naïve cells survive better than their primed counterparts when passaged as single cells and have a shorter doubling time [3]. The naïve and primed PSCs also use different modes of respiration for generating energy. While metabolism in naïve stem cells utilizes both oxidative phosphorylation (mitochondrial respiration) and glycolysis, primed cells preferentially generate energy through the glycolytic pathway [6].

Another key feature of the naïve state in mESCs is that both X chromosomes are active in female cells and undergo random X chromosome inactivation (XCI) upon differentiation **in vitro**. In contrast, XCI has already been established in primed mEpiSCs and this feature can thus be used as a reliable marker to distinguish between the two pluripotent states in female stem cells. The XCI status can also be used to identify the optimal culture conditions for maintenance of the naïve stem cell state and for monitoring the epigenetic stability of human PSCs [4]. Other epigenetic differences between the two murine pluripotent states include global DNA hypomethylation [7, 8], reduced prevalence of the repressive histone mark H3K27me3 at promoters, and fewer bivalent domains in naïve ESCs [9]. The key features of naïve and primed PSCs are summarized in **Table 1**.

Property	Naïve state	Primed state
Origin	ICM of early blastocyst	Post-implantation epiblast (Egg cylinder) or embryonic disc
Representative examples	mESCs, miPSCs	mEpiSCs, hESCs, hiPSCs
Expressed genes[1]	High expression of *Oct4, Nanog, Sox2, Klf2, Klf4, Klf5, Zpf42, Esrrb, Dppa3, Tfcp2l1, Fgf4, Tbx3, Cdh1*	*Oct4, Sox2, Dnmt3b, Fgf5, Pou3f1, Meis1, Otx2, Sox11, Gdf3*
Colony morphology	Compact dome shaped	Flattened
Differentiation bias	None	Variable
Tearatomas	Yes	Yes
Chimeric contribution in rodents[2]	Yes	No
Clonogenicity	High	Low
Single-cell mortality	Low	High
Growth factor dependence	LIF	ACTIVIN, FGF2
Respiration	Oxidative phosphorylation, glycolysis	Glycolysis
XCI status in female cells	XaXa	XaXi
Oct4 enhancer usage	Distal	Proximal
Global DNA methylation	Hypomethylated	Hypermethylated
Response to LIF/stat3	Self-renewal	None
Response to Fgf/Erk	Differentiation	Self-renewal
Response to 2i	Self-renewal	Differentiation
Level of HERVH[3] expression	High	Low
Efficiency of gene targeting	High	Low
H3K27Me3 over developmental regulators	Low	High
TFE3 localization	Nuclear	Cytoplasmic/absent

[1]Gene expression changes are based upon murine PSCs.

[2]Chimera assays are not always possible for hPSCs due to ethical concerns.

[3]HERVH expression is seen only in hPSCs.

Table 1. Characteristic features of naïve and primed pluripotent stem cell states.

Much of what we know about the growth factor dependence of naïve and primed PSCs is based on studies performed with mESCs and mEpiSCs. Mouse ESCs can be maintained long term in the naïve state when cultured in the presence of serum plus leukemia inhibitory factor

(LIF), which signals through a bipartite receptor to activate Janus-associated kinases (JAK) which leads to the activation of signal transducer and activator of transcription 3 (Stat3) [10]. However, in the absence of serum, LIF alone is unable to prevent differentiation of mESCs. This limitation is overcome by the addition of two small molecule kinase inhibitors termed "2i" with LIF. The 2i components include a specific inhibitor of extracellular signal-regulated kinase (ERK1/2)/mitogen-activated protein kinase (MAPK) signal transduction pathway (MEKi, PD0325901) and a specific inhibitor of glycogen synthase kinase 3 beta (GSK3βi, CHIR99021) that protect the ESCs from pro-differentiation stimuli and select against differentiating cells [11]. The most critical effects of GSK3i are mediated via β-catenin, which is the key mediator of Wnt/β-catenin signaling pathway. While fibroblast growth factor (FGF)-mediated activation of MEK signaling drives differentiation of mESCs, primed mEpiSCs require basic fibroblast growth factor (bFGF or FGF2) and are dependent on the activin/nodal pathway for the maintenance of pluripotency [3]. Similar to the mEpiSCs, human PSCs do not require LIF but are dependent on bFGF and activin/transforming growth factor β-1 (TGFβ1) for their long-term maintenance in culture [12, 13]. However, as discussed later, it is possible to revert primed hPSCs to the naïve state using LIF/2i.

Naïve mESCs and primed mEpiSCs also differ in the expression of pluripotency-associated transcription factors. While the pluripotency factors Oct4/Pou5f1 and Sox2 are expressed in both naïve and primed murine PSCs, factors like Nanog, Klf2, Klf4, Prdm14, Sall4, Tfcp2l1, Esrrb, and Tbx3 are preferentially expressed or upregulated in the naïve mESCs. In addition, Tfe3, which is localized in the nucleus of naïve PSCs, becomes cytoplasmic upon their conversion to the primed state and results in reduced expression of Esrrb [14]. The forced expression of naïvity-associated transcription factors has been shown to convert the primed iPSCs into to the naïve state suggesting their importance for the propagation of the ground state [15]. Although conventional hPSCs functionally resemble the primed pluripotent state of mEpiSCs, they are not identical and show specific differences in transcription regulation and the expression of various markers like FGF5, E-CADHERIN, and NANOG (reviewed in [16]).

3. Strategies for the generation of naïve human pluripotent stem cells

The naïve pluripotent state of mESCs is also observed in PSCs derived from rat embryos [17, 18]. However, naïve PSCs were not observed in other species in this study and it was not clear if this was a feature only of rodent ESCs or if it would be possible to capture it in other species. In recent years, naïve-like stem cells have been isolated from porcine fibroblasts [19], rhesus monkey fibroblasts [20], rabbit embryos, liver and stomach [21], finch embryos [22], and bovine amnion-derived cells [23]. In addition, several reports of the successful derivation of human PSCs resembling naïve mESCs in morphology and gene expression signatures have also been published [24–32]. This confirms that the naïve-like pluripotent state is not species specific as originally assumed to be limited to rodents, but may reflect an early developmental stage conserved across mammalian and possibly vertebrate evolution. The ability to capture human

PSCs in the naïve state opens up a plethora of opportunities for their use in disease modeling and regenerative medicine.

Reference	Method used	Growth factors in medium	XCI status	Chimera assay
Hanna et al. 2010 [24]	Sustained or transient expression of OCT4, SOX2, KLF4 or KLF2	LIF/2i or LIF/2i/FK or LIF/2i	Pre-XCI	ND
Gafni et al. 2013 [26]	Small molecules	LIF/2i/bFGF/TGFβ/JNKi/ p38i/ROCKi	Pre-XCI	Positive
Chan et al. 2013 [25]	Small molecules	LIF/2i/BMPi/ bFGF/TGFβ/ ROCKi	ND	ND
Takashima et al. 2014 [27]	Sustained or transient expression of NANOG and KLF2	LIF/2i/Go¨ 6983	Pre-XCI	ND
Valamehr et al. 2014 [29]	Small molecules	LIF/2i/bFGF	Pre-XCI	ND
Theunissen et al. 2014 [28]	Selection based on OCT4 distal enhancer activity	LIF/5i/bFGF/ ActivinA	XaXi	negative
Wang et al. 2014 [35]	Selection of HERVH expressing cells	LIF/2i/bFGF	Pre-XCI	ND
Ware et al. 2014 [30]	Small molecules	2i/bFGF	ND	ND
Duggal et al. 2015 [31]	Small molecules	LIF/2i/bFGF/FK/AA	ND	ND
Yang et al. 2016 [34]	Small molecules	LIF/5i/bFGF/ ActivinA	ND	Positive
Carter et al. 2016 [32]	Recombinant protein	$NME7_{AB}$	Pre-XCI	ND

FK, forskolin; AA, ascorbic acid; Xa, active X; Xi, inactive X; XCI, X chromosome inactivation; ND, not done.

Table 2. Summary of various protocols used for generating naïve human pluripotent stem cells.

Naïve-like human PSCs have been obtained either by forced expression of pluripotency transcription factors or by using various combinations of small molecules and/or growth factors to improve the culture conditions for inducing and maintaining naïve stem cell characteristics (**Table 2**). Buecker and colleagues described the derivation of hiPSCs in the presence of LIF and five ectopic reprogramming factors, OCT4, SOX2, KLF4, cMYC, and NANOG (called hLR5 iPSCs), that displayed morphological, molecular, and functional properties of mESCs [33]. In another study, the Jaenisch group showed that the ectopic induction of OCT4, KLF4, and KLF2 in hESCs grown in 2i/LIF media resulted in naïve stem cells that resembled mESCs in their gene expression profile and reactivation of both X chromosomes [24]. One limitation of this approach is that continuous expression of the pluripotency factors was required for maintenance of the ground state in the resultant PSCs thereby restricting their potential for downstream applications. To overcome this limitation, several groups have used small molecules to achieve transgene-independent derivation of naïve hPSCs. Using a reporter cell line containing a doxycycline (dox)-inducible OCT4-Green Fluorescent Protein (OCT4-GFP) reporter, Gafni et al. [26] identified a combination of factors that facilitated the derivation of genetically unmodified naïve hESCs from blastocysts, hiPSCs from somatic cells, as well as the reverse toggling and maintenance of primed hESCs into the naïve state. This naïve human stem cell medium (NHSM) included essential components [2i/LIF, TGFβ1, c-Jun N-terminal kinase inhibitor (JNKi, SP600125), p38i (SB203580)] and optimizing components [Rho kinase inhibitor (ROCKi, Y-27632) and protein kinase C inhibitor (PKCi, Go6983)] for supporting the growth of naïve hPSCs [26]. These naïve-like hiPSCs grown in NHSM showed higher integration into chimeras compared to the primed PSCs. In another study, Chan et al. [25] screened 11 small molecules for their ability to increase NANOG expression and identified a combination of three inhibitors [2i plus a bone morphogenetic protein (BMP) inhibitor, dorsomorphin] in TeSR1 base medium and LIF that maintained hESCs in the ground state. The authors observed an upregulation of a number of genes that are expressed in human preimplantation embryos in the hESCs grown in 3i/LIF medium compared to the conventionally cultured hESCs. Valamehr et al. [29] also reported the derivation and maintenance of transgene-free hiPSCs with naïve-like characteristics using fate maintenance medium (FMM) that contains 2i/LIF in conventional hESC medium. These cells showed high survival rates in single-cell dissociation and displayed the characteristics of a pre-XCI state with decreased levels of XIST and H3K27me3. In contrast, Ware et al. [30] were able to reverse toggle primed hESCs by preculture in the histone deacetylase inhibitors, sodium butyrate, and suberoylanilide hydroxamic acid, followed by culture in 2i with bFGF in the absence of LIF. In addition, they reported the derivation of naïve hESCs only in the presence of 2i and bFGF.

To more systematically identify optimal culture conditions for the induction and maintenance of naïve human pluripotency, Theunissen et al. [28] screened a kinase inhibitor library using a specific reporter system based on the activity of the endogenous OCT4 distal enhancer that is exclusively active in the naïve pluripotent state. They described a serum-free N2B27 medium supplemented with five kinase inhibitors (MEKi, GSK3βi, BRAFi, LCK/SRCi, ROCKi), LIF, bFGF, and activin A (5i/L/FA) that was sufficient to maintain naïve hESCs in culture. Cells grown under these conditions showed a dramatic upregulation of transcription factors typically associated with naïve pluripotency and human preimplantation development. However, abnormal karyotype in several converted and newly derived hESC lines in 5i/L/FA was observed leading to the authors' suggestion that naïve hPSCs may be more prone to acquiring chromosomal abnormalities. In addition, in this study, hESCs grown either in the 5i/L/FA medium or as per the Gafni et al. protocol [26] did not form interspecies chimeras efficiently. The authors attributed this discrepancy between the two studies to slight variations in culture conditions or embryo handling. Furthermore, female hESCs grown under 5i/L/FA conditions showed an upregulation of XIST and XCI upon conversion from the primed state. These features illustrate the limitation of hPSCs derived in 5i/L/FA medium or, as suggested by the authors, this may highlight the differences in the naïve pluripotent state between human and mouse. Using this 5i/L/FA medium, naïve iPSCs could be generated from fibroblasts derived from β-thalassemia patients in whom the genetic mutation was corrected by CRISPR/Cas9 gene editing [34]. More recently, Duggal et al. [31] reported the formulation of yet another culture medium that facilitated the rapid, robust, and efficient induction of naïve pluripotency in primed hESCs. In addition to the 2i/LIF, the authors included ascorbic acid (known to induce DNA demethylation and enhance reprogramming), forskolin (an activator of cAMP, shown to promote naïve pluripotency), and different concentrations of bFGF in the culture medium to identify optimal conditions for converting primed hESC lines toward a naïve state of pluripotency.

Another method, based on the selection of cells with naïve-like characteristics from a heterogeneous population of naïve and primed PSCs, has been published. Since LTR7/human endogenous retrovirus, HERVH, expression was found to be required for maintaining pluripotency in hPSC, Wang et al. developed an LTR7-driven GFP reporter construct to tag, select, and maintain naïve-like hPSCs [35]. By using this reporter, the authors were able to sort and enrich for LTR7-GFP-expressing cells and thus maintain a homogeneous population of naïve-like hPSCs over the long term. The hPSCs selected based on high expression of HERVH shared many features with naïve mESCs that included 3D rounded colonies, reactivation of the X chromosome in female cells, and uniform expression of core pluripotency transcription factors with downregulation of genes expressed in differentiated cells. Moreover, these cells grew faster and had improved cloning capacity making them suitable for genetic manipulation. However, the level of HERVH expression was found to be an important determinant of the ability of the stem cells to differentiate with higher levels of HERVH interfering with the differentiation potential. The LTR7-GFP reporter could be useful for further optimization of culture conditions for the maintenance of naïve pluripotent state for human stem cells.

Although it is possible to generate human PSCs with features that are characteristic of naïve mESCs by culturing them in the presence of exogenous factors or a cocktail of small molecules, their tendency to develop karyotypic abnormalities in long-term culture [28] raises concerns that these may not be the true, naturally occurring human naïve stem cells. To capture this natural naïve state of hPSCs without additional growth factors or small molecules, Carter et al. [32] developed a novel approach using a naturally occurring human growth factor, NME7, in minimal media without bFGF, LIF, or any other small molecule. NME7 belongs to the NME family of nucleotide phosphate kinases but lacks the kinase activity. It is proposed to act by binding to the MUC1* growth factor receptor that is expressed on the surface of stem cells and results in its dimerization which promotes growth and pluripotency. The hPSCs grown in the presence of $NME7_{AB}$, recombinant human NME7 protein, grew faster, showed resistance to differentiation, demonstrated increased cloning efficiency, had two active X chromosomes in the female cells, and had a normal karyotype after 30 or more passages. However, $NME7_{AB}$-derived cells did not form colonies and grew preferentially as a monolayer. The various strategies used for the derivation of naïve-like hPSCs are summarized in **Table 2**.

While all of the above-mentioned approaches claim to have successfully derived the so-called ground state naïve-like hPSC, the resultant cells do not share the same gene expression patterns, although there is some overlap [32]. Therefore, a set of core genes/markers that clearly define naïve hPSCs is not currently available but is highly desirable.

4. Naïve hPSCs for gene targeting and regenerative medicine

The ability to generate homogeneous populations of stable naïve hPSCs that have the potential to differentiate equally toward all lineages will be critical for their use in disease modeling and regenerative medicine. The faster doubling time of naïve PSCs and their capacity for clonal expansion is important for generating large number of cells required for screening applications. Marked differences in the propensity of hESCs to differentiate into certain lineages or cell types have been reported [36]. Additionally, hiPSCs generated using conventional methods were shown to be limited in their proficiency of lineage-specific differentiation that can be influenced by the donor cell type [37]. However, new data show that the differentiation propensities of hiPSCs are significantly biased by donor-dependent variability and not by cell type of origin [38]. It has been suggested that the conversion of hiPSCs to a more naïve-like state could improve their **in vitro** differentiation potential toward specific cell types by resetting their epigenetic status [39, 40]. However, a systematic study to directly compare the differentiation efficiency of isogenic naïve and primed hPSCs into different cell types has not been done and will be necessary to establish that naïve hiPSCs have a better differentiation ability than their primed counterparts. It has been observed that the efficiency of homologous recombination is significantly higher in mESCs in comparison to hESCs and hiPSCs [41, 26], suggesting that the PSCs in the naïve state might be more amenable to gene editing. In addition, the very low clonal efficiency and slow growth rate of primed hPSCs limit the full potential of gene targeting mediated by site-specific nucleases, thereby supporting the generation of naïve PSCs for efficient gene editing. Hu et al. [42] converted primed state iPSCs from Parkinson's disease patients to a so-called "naivetropic" state by dox-induced expression of transgenes (OCT4, SOX2, KLF4, c-MYC, and NANOG) and the use of LIF/2i. Using these cells, they were able to target GFP to the PITX3 locus by transcription activator-like effector nuclease (TALEN) very efficiently. Similarly, Yang et al. [34] showed significant improvement in the mutation correction efficiency in naïve hiPSCs derived from β-thalassemia patients using the 5i/L/FA system [28].

5. Naïve hPSCs for studying the process of X inactivation during early human embryonic development

In addition to the therapeutic potential of pluripotent stem cells for drug screening and regenerative medicine, their ability to recapitulate early embryonic differentiation is useful for study of the pathways involved in cell commitment and embryonic differentiation. One such process that is developmentally regulated and tightly linked with cell differentiation is the random inactivation of one of the two X chromosomes in all of the cells in female mammals that balances the sex difference in dosage of X-linked genes. This is a highly coordinated and stepwise process involving many noncoding RNAs that affect epigenetic modifications and cause transcriptional silencing of the entire X chromosome barring a few genes that escape silencing (see [43] for a recent review on the subject). Most of what we know about the mechanism of XCI comes from many elegant studies done with mice and has benefited from

http://dx.doi.org/10.5772/63202

the existence of naïve state mESCs with two active X chromosomes that undergo random XCI upon further differentiation. However, there are distinct differences between the XCI process in mice and humans (reviewed in [44]). For example, the imprinted inactivation of the paternal X chromosome seen in the early mouse embryo and maintained in the placenta does not occur in human embryo [45] which only exhibit random XCI. In addition, some of the genes that escape XCI in humans are inactivated in mice [46] and substantial diversity in the timing and regulation of XCI between mammals has been observed [47]. Therefore, having the ability to model XCI using hESCs or hiPSCs is highly desirable and will definitely benefit from optimized methods to derive and maintain female hESCs or hiPSCs in the naïve state with two active X chromosomes.

6. Naïve hPSCs for in vitro modeling of human diseases

Human PSCs can be used to study underlying mechanisms of diseases that cannot be modeled easily in mice. One such example is the repeat-mediated aberrant silencing of the *fragile X mental retardation 1 (FMR1)* gene in fragile X syndrome (FXS). FXS is the most common cause of inherited intellectual disability and autism spectrum disorder [48]. The most common mutation in FXS is the expansion of a CGG-repeat tract in the 5′-untranslated region of the *FMR1* gene that causes its transcriptional silencing by DNA methylation and heterochromatin formation [49, 50]. It is believed that the repeat-mediated *FMR1* gene silencing is developmentally regulated with the gene being active in the early embryo and silenced at a later time during development (see [51] for a recent review). Mouse models for the expanded CGG repeats do not result in *Fmr1* gene silencing [52, 53] and therefore cannot be used for studying the underlying mechanism of gene silencing in FXS or its effect on the developing brain. In the first reported FXS embryonic stem cell line, the *FMR1* gene was active in the ESCs and showed silencing upon differentiation into embryoid bodies [54]. This generated a great interest in developing stem cells as a model to study the mechanisms involved in gene silencing. However, in most of the subsequently generated FXS ESCs, the *FMR1* gene is already silenced [55, 56]. Additionally, iPSCs generated from FXS patient fibroblasts do not show reactivation of the *FMR1* gene [57]. Using a strategy that would favor the generation of naïve iPSCs, Gafni et al. [26] were able to derive naïve state FXS iPSCs that had an active *FMR1* gene. These observations lend support to the idea that the *FMR1* gene is active in the very early embryo but that silencing occurs very soon after that perhaps associated with the transition to the primed state. A better understanding of this process may allow the modeling of developmental silencing of the *FMR1* gene **in vitro**. In addition, these studies will also illuminate the general process by which heterochromatin might be reset during the process of reprogramming cells to pluripotency.

7. Conclusions

Recent research suggests that it might be possible to achieve naïve pluripotency of hESCs and hiPSCs by using specific combinations of small molecule inhibitors and growth factors in the culture medium during their derivation and growth. The ability to generate and maintain naïve pluripotent stem cells will facilitate studies on early human embryonic development and the mechanisms involved in XCI **in vitro**. This will also improve the odds of genetically correcting mutations that cause human diseases and thus help in dissecting the mechanisms of disease pathogenesis. In addition, the ability to stably maintain the epigenetic state of the pluripotent stem cells will be critical for their use in stem cell-based gene therapies.

Acknowledgements

I thank Karen Usdin and Yifan Zhou (Laboratory of Cell and Molecular Biology, National Institute of Diabetes and Digestive and Kidney Diseases, National Institutes of Health), Barbara Mallon (NIH Stem Cell Unit, National Institute of Neurological Disorders and Stroke), and James Kehler (Special volunteer at the National Institutes of Health) for critical reading of the manuscript and helpful discussions.

Author details

Daman Kumari

Address all correspondence to: damank@helix.nih.gov

Laboratory of Cell and Molecular Biology, National Institute of Diabetes and Digestive and Kidney Diseases, National Institutes of Health, Bethesda, Maryland, USA

References

[1] Evans MJ, Kaufman MH. Establishment in culture of pluripotential cells from mouse embryos. Nature. 1981;292(5819):154–6.

[2] Brons IG, Smithers LE, Trotter MW, Rugg-Gunn P, Sun B, Chuva de Sousa Lopes SM et al. Derivation of pluripotent epiblast stem cells from mammalian embryos. Nature. 2007;448(7150):191–5. doi:10.1038/nature05950.

[3] Tesar PJ, Chenoweth JG, Brook FA, Davies TJ, Evans EP, Mack DL et al. New cell lines from mouse epiblast share defining features with human embryonic stem cells. Nature. 2007;448(7150):196–9. doi:10.1038/nature05972.

[4] Nichols J, Smith A. The origin and identity of embryonic stem cells. Development. 2011;138(1):3–8. doi:10.1242/dev.050831.

[5] Huang Y, Osorno R, Tsakiridis A, Wilson V. In vivo differentiation potential of epiblast stem cells revealed by chimeric embryo formation. Cell Rep. 2012;2(6):1571–8. doi: 10.1016/j.celrep.2012.10.022.

[6] Teslaa T, Teitell MA. Pluripotent stem cell energy metabolism: an update. Embo J. 2015;34(2):138–53. doi:10.15252/embj.201490446.

[7] Ficz G, Hore TA, Santos F, Lee HJ, Dean W, Arand J et al. FGF signaling inhibition in ESCs drives rapid genome-wide demethylation to the epigenetic ground state of pluripotency. Cell Stem Cell. 2013;13(3):351–9. doi:10.1016/j.stem.2013.06.004.

[8] Habibi E, Brinkman AB, Arand J, Kroeze LI, Kerstens HH, Matarese F et al. Whole-genome bisulfite sequencing of two distinct interconvertible DNA methylomes of mouse embryonic stem cells. Cell Stem Cell. 2013;13(3):360–9. doi:10.1016/j.stem.2013.06.002.

[9] Marks H, Kalkan T, Menafra R, Denissov S, Jones K, Hofemeister H et al. The transcriptional and epigenomic foundations of ground state pluripotency. Cell. 2012;149(3): 590–604. doi:10.1016/j.cell.2012.03.026.

[10] Niwa H, Ogawa K, Shimosato D, Adachi K. A parallel circuit of LIF signalling pathways maintains pluripotency of mouse ES cells. Nature. 2009;460(7251):118–22. doi:10.1038/nature08113.

[11] Ying QL, Wray J, Nichols J, Batlle-Morera L, Doble B, Woodgett J et al. The ground state of embryonic stem cell self-renewal. Nature. 2008;453(7194):519–23. doi:10.1038/nature06968.

[12] James D, Levine AJ, Besser D, Hemmati-Brivanlou A. TGFbeta/activin/nodal signaling is necessary for the maintenance of pluripotency in human embryonic stem cells. Development. 2005;132(6):1273–82. doi:10.1242/dev.01706.

[13] Levenstein ME, Ludwig TE, Xu RH, Llanas RA, VanDenHeuvel-Kramer K, Manning D et al. Basic fibroblast growth factor support of human embryonic stem cell self-renewal. Stem Cells. 2006;24(3):568–74. doi:10.1634/stemcells.2005-0247.

[14] Betschinger J, Nichols J, Dietmann S, Corrin PD, Paddison PJ, Smith A. Exit from pluripotency is gated by intracellular redistribution of the bHLH transcription factor Tfe3. Cell. 2013;153(2):335–47. doi:10.1016/j.cell.2013.03.012.

[15] Guo G, Yang J, Nichols J, Hall JS, Eyres I, Mansfield W et al. Klf4 reverts developmentally programmed restriction of ground state pluripotency. Development. 2009;136(7): 1063–9. doi:10.1242/dev.030957.

[16] Weinberger L, Ayyash M, Novershtern N, Hanna JH. Dynamic stem cell states: naive to primed pluripotency in rodents and humans. Nat Rev Mol Cell Biol. 2016;17(3):155–69. doi:10.1038/nrm.2015.28.

[17] Buehr M, Meek S, Blair K, Yang J, Ure J, Silva J et al. Capture of authentic embryonic stem cells from rat blastocysts. Cell. 2008;135(7):1287–98. doi:10.1016/j.cell.2008.12.007.

[18] Li P, Tong C, Mehrian-Shai R, Jia L, Wu N, Yan Y et al. Germline competent embryonic stem cells derived from rat blastocysts. Cell. 2008;135(7):1299–310. doi:10.1016/j.cell.2008.12.006.

[19] Fujishiro SH, Nakano K, Mizukami Y, Azami T, Arai Y, Matsunari H et al. Generation of naive-like porcine-induced pluripotent stem cells capable of contributing to embryonic and fetal development. Stem Cells Dev. 2013;22(3):473–82. doi:10.1089/scd.2012.0173.

[20] Fang R, Liu K, Zhao Y, Li H, Zhu D, Du Y et al. Generation of naive induced pluripotent stem cells from rhesus monkey fibroblasts. Cell Stem Cell. 2014;15(4):488–96. doi:10.1016/j.stem.2014.09.004.

[21] Honsho K, Hirose M, Hatori M, Yasmin L, Izu H, Matoba S et al. Naive-like conversion enhances the difference in innate in vitro differentiation capacity between rabbit ES cells and iPS cells. J Reprod Dev. 2015;61(1):13–9. doi:10.1262/jrd.2014-098.

[22] Mak SS, Alev C, Nagai H, Wrabel A, Matsuoka Y, Honda A et al. Characterization of the finch embryo supports evolutionary conservation of the naive stage of development in amniotes. Elife. 2015;4:e07178. doi:10.7554/eLife.07178.

[23] Kawaguchi T, Tsukiyama T, Kimura K, Matsuyama S, Minami N, Yamada M et al. Generation of Naive Bovine Induced Pluripotent Stem Cells Using PiggyBac Transposition of Doxycycline-Inducible Transcription Factors. PLoS One. 2015;10(8):e0135403. doi:10.1371/journal.pone.0135403.

[24] Hanna J, Cheng AW, Saha K, Kim J, Lengner CJ, Soldner F et al. Human embryonic stem cells with biological and epigenetic characteristics similar to those of mouse ESCs. Proc Natl Acad Sci U S A. 2010;107(20):9222–7. doi:10.1073/pnas.1004584107.

[25] Chan YS, Goke J, Ng JH, Lu X, Gonzales KA, Tan CP et al. Induction of a human pluripotent state with distinct regulatory circuitry that resembles preimplantation epiblast. Cell Stem Cell. 2013;13(6):663–75. doi:10.1016/j.stem.2013.11.015.

[26] Gafni O, Weinberger L, Mansour AA, Manor YS, Chomsky E, Ben-Yosef D et al. Derivation of novel human ground state naive pluripotent stem cells. Nature. 2013;504(7479):282–6. doi:10.1038/nature12745.

[27] Takashima Y, Guo G, Loos R, Nichols J, Ficz G, Krueger F et al. Resetting transcription factor control circuitry toward ground-state pluripotency in human. Cell. 2014;158(6):1254–69. doi:10.1016/j.cell.2014.08.029.

[28] Theunissen TW, Powell BE, Wang H, Mitalipova M, Faddah DA, Reddy J et al. Systematic identification of culture conditions for induction and maintenance of naive human pluripotency. Cell Stem Cell. 2014;15(4):471–87. doi:10.1016/j.stem.2014.07.002.

[29] Valamehr B, Robinson M, Abujarour R, Rezner B, Vranceanu F, Le T et al. Platform for induction and maintenance of transgene-free hiPSCs resembling ground state pluripotent stem cells. Stem Cell Reports. 2014;2(3):366–81. doi:10.1016/j.stemcr.2014.01.014.

[30] Ware CB, Nelson AM, Mecham B, Hesson J, Zhou W, Jonlin EC et al. Derivation of naive human embryonic stem cells. Proc Natl Acad Sci U S A. 2014;111(12):4484–9. doi: 10.1073/pnas.1319738111.

[31] Duggal G, Warrier S, Ghimire S, Broekaert D, Van der Jeught M, Lierman S et al. Alternative routes to induce naive pluripotency in human embryonic stem cells. Stem Cells. 2015;33(9):2686–98. doi:10.1002/stem.2071.

[32] Carter MG, Smagghe BJ, Stewart AK, Rapley JA, Lynch E, Bernier KJ et al. A Primitive Growth Factor, NME7AB , Is Sufficient to Induce Stable Naive State Human Pluripotency; Reprogramming in This Novel Growth Factor Confers Superior Differentiation. Stem Cells. 2016;34(4):847–59. doi:10.1002/stem.2261

[33] Buecker C, Chen HH, Polo JM, Daheron L, Bu L, Barakat TS et al. A murine ESC-like state facilitates transgenesis and homologous recombination in human pluripotent stem cells. Cell Stem Cell. 2010;6(6):535–46. doi:10.1016/j.stem.2010.05.003.

[34] Yang Y, Zhang X, Yi L, Hou Z, Chen J, Kou X et al. Naive induced pluripotent stem cells generated from beta-Thalassemia fibroblasts allow efficient gene correction With CRISPR/Cas9. Stem Cells Transl Med. 2016;5(1):267. doi:10.5966/sctm.2015-0157erratum.

[35] Wang J, Xie G, Singh M, Ghanbarian AT, Rasko T, Szvetnik A et al. Primate-specific endogenous retrovirus-driven transcription defines naive-like stem cells. Nature. 2014;516(7531):405–9. doi:10.1038/nature13804.

[36] Osafune K, Caron L, Borowiak M, Martinez RJ, Fitz-Gerald CS, Sato Y et al. Marked differences in differentiation propensity among human embryonic stem cell lines. Nat Biotechnol. 2008;26(3):313–5. doi:10.1038/nbt1383.

[37] Kim K, Zhao R, Doi A, Ng K, Unternaehrer J, Cahan P et al. Donor cell type can influence the epigenome and differentiation potential of human induced pluripotent stem cells. Nat Biotechnol. 2011;29(12):1117–9. doi:10.1038/nbt.2052.

[38] Kyttala A, Moraghebi R, Valensisi C, Kettunen J, Andrus C, Pasumarthy KK et al. Genetic variability overrides the impact of parental cell type and determines iPSC differentiation potential. Stem Cell Reports. 2016;6(2):200–12. doi:10.1016/j.stemcr.2015.12.009.

[39] Lee JH, Lee JB, Shapovalova Z, Fiebig-Comyn A, Mitchell RR, Laronde S et al. Somatic transcriptome priming gates lineage-specific differentiation potential of human-induced pluripotent stem cell states. Nat Commun. 2014;5:5605. doi:10.1038/ncomms6605.

[40] Irie N, Weinberger L, Tang WW, Kobayashi T, Viukov S, Manor YS et al. SOX17 is a critical specifier of human primordial germ cell fate. Cell. 2015;160(1-2):253–68. doi: 10.1016/j.cell.2014.12.013.

[41] Zwaka TP, Thomson JA. Homologous recombination in human embryonic stem cells. Nat Biotechnol. 2003;21(3):319–21. doi:10.1038/nbt788.

[42] Hu Z, Pu J, Jiang H, Zhong P, Qiu J, Li F et al. Generation of Naivetropic Induced Pluripotent Stem Cells from Parkinson's Disease Patients for High-Efficiency Genetic Manipulation and Disease Modeling. Stem Cells Dev. 2015;24(21):2591–604. doi: 10.1089/scd.2015.0079.

[43] Galupa R, Heard E. X-chromosome inactivation: new insights into cis and trans regulation. Curr Opin Genet Dev. 2015;31:57–66. doi:10.1016/j.gde.2015.04.002.

[44] Lessing D, Anguera MC, Lee JT. X chromosome inactivation and epigenetic responses to cellular reprogramming. Annu Rev Genomics Hum Genet. 2013;14:85–110. doi: 10.1146/annurev-genom-091212-153530.

[45] Migeon BR, Wolf SF, Axelman J, Kaslow DC, Schmidt M. Incomplete X chromosome dosage compensation in chorionic villi of human placenta. Proc Natl Acad Sci U S A. 1985;82:3390–4.

[46] Disteche CM. Escape from X inactivation in human and mouse. Trends Genet. 1995;11(1):17–22.

[47] Okamoto I, Patrat C, Thepot D, Peynot N, Fauque P, Daniel N et al. Eutherian mammals use diverse strategies to initiate X-chromosome inactivation during development. Nature. 2011;472(7343):370–4. doi:10.1038/nature09872.

[48] Goldson E, Hagerman RJ. The fragile X syndrome. Dev Med Child Neurol. 1992;34(9): 826–32.

[49] Sutcliffe JS, Nelson DL, Zhang F, Pieretti M, Caskey CT, Saxe D et al. DNA methylation represses FMR-1 transcription in fragile X syndrome. Hum Mol Genet. 1992;1(6):397–400.

[50] Coffee B, Zhang F, Ceman S, Warren ST, Reines D. Histone modifications depict an aberrantly heterochromatinized FMR1 gene in fragile x syndrome. Am J Hum Genet. 2002;71(4):923–32. doi:10.1086/342931.

[51] Usdin K, Kumari D. Repeat-mediated epigenetic dysregulation of the FMR1 gene in the fragile X-related disorders. Front Genet. 2015;6:192. doi:10.3389/fgene.2015.00192.

[52] Brouwer JR, Mientjes EJ, Bakker CE, Nieuwenhuizen IM, Severijnen LA, Van der Linde HC et al. Elevated Fmr1 mRNA levels and reduced protein expression in a mouse model with an unmethylated Fragile X full mutation. Exp Cell Res. 2007;313(2):244–53. doi:10.1016/j.yexcr.2006.10.002.

[53] Entezam A, Biacsi R, Orrison B, Saha T, Hoffman GE, Grabczyk E et al. Regional FMRP deficits and large repeat expansions into the full mutation range in a new Fragile X premutation mouse model. Gene. 2007;395(1-2):125–34. doi:10.1016/j.gene.2007.02.026.

[54] Eiges R, Urbach A, Malcov M, Frumkin T, Schwartz T, Amit A et al. Developmental study of fragile X syndrome using human embryonic stem cells derived from preimplantation genetically diagnosed embryos. Cell Stem Cell. 2007;1(5):568–77. doi: 10.1016/j.stem.2007.09.001.

[55] Avitzour M, Mor-Shaked H, Yanovsky-Dagan S, Aharoni S, Altarescu G, Renbaum P et al. FMR1 epigenetic silencing commonly occurs in undifferentiated fragile X-affected embryonic stem cells. Stem Cell Rep. 2014;3(5):699–706. doi:10.1016/j.stemcr.2014.09.001.

[56] Colak D, Zaninovic N, Cohen MS, Rosenwaks Z, Yang WY, Gerhardt J et al. Promoter-bound trinucleotide repeat mRNA drives epigenetic silencing in fragile X syndrome. Science. 2014;343(6174):1002–5. doi:10.1126/science.1245831.

[57] Urbach A, Bar-Nur O, Daley GQ, Benvenisty N. Differential modeling of fragile X syndrome by human embryonic stem cells and induced pluripotent stem cells. Cell Stem Cell. 2010;6(5):407–11. doi:10.1016/j.stem.2010.04.005.

Chapter 4

Pluripotent Stem Cells and Their Dynamic Niche

Yvonne Reinwald, Jessica Bratt and Alicia El Haj

Additional information is available at the end of the chapter

http://dx.doi.org/10.5772/62671

Abstract

Cell-seeded implants are a regenerative medicine strategy that aims to replace injured tissue and restore tissue function. Pluripotent stem cells are promising cell candidates for the development of regenerative medicine therapies as they have the ability to self-renew and commit towards numerous cell types. *In vivo*, stem cells reside in a dynamic niche, a stem cell-specific microenvironment that possesses chemical, biological and mechanical cues, which drive the stem cell fate and renewal. The connection between stem cells and their niche is a two-way relationship consisting of both cell–cell interaction and cell–extracellular matrix (ECM) interactions. An alternative regenerative medicine approach is the manipulation of the stem cell microenvironment. Hence, novel strategies have been developed including 3D biomaterials and bioreactor technologies providing topographical, chemical and mechanical cues to recreate the stem cell niche. Understanding the mechanisms controlling stem cell fate and the dynamic nature of the stem cell niche will enable researchers to replicate this stem cell-specific microenvironment, and therefore, harness and control the valuable attributes which stem cells possess. This chapter elucidates the importance of pluripotent stem cells and their dynamic niche in regenerative medicine. It further presents novel strategies to replicate chemical, topographical and mechanical stimuli which are essential for the regulation of stem cell fate and hence tissue regeneration.

Keywords: Pluripotent stem cells, Regenerative medicine, Stem cell niche, Mechanical cues, Bioreactor technologies

1. Introduction

Regenerative medicine is a promising field that aims to develop therapies for currently intractable diseases. These approaches include cell–cell and cell–scaffold implants. Self-renewal and the ability of pluripotent stem cells to commit towards particular cell lineages in response to mechanical, chemical and physical stimuli makes them the ideal building block for such therapies. Embryonic stem cells (ESCs) are isolated during embryological development. Their utilisation in regenerative medicine is controversial, and therefore, adult stem cells have been more thoroughly investigated for their potential use for tissue regeneration. Currently, induced pluripotent stem cells (iPSCs) are being investigated as an alternative source for pluripotent stem cells due to their origin and clinical potential [1].

Stem cells in vivo reside in a dynamic, cell type-specific microenvironment, the so-called niche [2–5]. This microenvironment is composed of stem cells, supportive stromal cells and surrounding extracellular matrix (ECM) [5, 6]. The niche provides chemical, mechanical and topographical cues facilitating stem cell renewal and controlling stem cell fate [3, 4]. The ECM is an important component of the niche. It is the biological matrix directly surrounding stem cells *in vivo* and is composed of tissue-specific glycosaminoglycan (GAG), insoluble proteins (e.g. fibronectins, collagens, laminins) and inorganic hydroxyapatites [6, 7]. Research has shown that the biophysical properties of the ECM affect stem cell behaviour. Cells counteract external forces which result from their surroundings, through adjacent cells and ECM stiffness by altering the cytoskeleton tension and through the generation of internal forces [8]. These forces are exerted on and from each individual cell to their environment. The interaction between microenvironment and cells results in the regulation of stem cell behaviour [8–10]. The cellular response to external mechanical cues is defined as mechanotransduction.

An alternative strategy for tissue repair is the manipulation of the stem cell microenvironment to enable tissue repair through endogenous stem cells [11]. To recreate the stem cell niche, it is important to reproduce the physical and mechanical microenvironment stem cells experience *in vivo*. Hence, strategies including novel bioreactor technologies have been developed. By replicating the *in vivo* environment, bioreactors allow the study of mechanical stimuli in combination with chemical and biological signals on cell–cell and cell–biomaterial constructs [12].

This chapter will present the role of pluripotent stem cells and their dynamic niche in regenerative medicine, as well as the importance for the niche replication for the development of novel regenerative therapies. Finally, it will present novel strategies to replicate chemical, biological and mechanical stimuli which are essential for the regulation of stem cell fate and hence tissue regeneration.

2. Stem cell niches

In 1978, the term 'niche' was first described by Schofield [13] who theorised that stem cell self-renewal and character is dependent on their environment. Since then, the theory of the stem cell microenvironment has been expanded [14, 15]. Each stem cell population has a unique and specific environment, but there are features which stem cell niches have in common (**Figure 1**). Tissue-specific (e.g. osteoblasts) and non-specific (stromal cells) heterogeneous cell–cell interactions co-exist in stem cell niches [11]. Secreted-membrane bound factors bind stem cell surface receptors in order to direct stem cell self-renewal and fate (e.g. Wnt, chemokines,

Notch, SCF) [16–22]. During tissue injury and inflammation, immunological cells regulate the niche [23, 24]. The ECM provides structural support and orientation and serves as storage for soluble factors [25]. It further interacts with stem cells through gap junctions, soluble factors and surface receptors [5, 6]. In addition, physical and physiological parameters of the niche such as shape, elasticity, blood flow and oxygen tension influence stem cell differentiation and self-renewal and regulate the metabolic activity [22, 26–32]. The interactions between stem cells and their niche are reciprocal; since stem cells are able to remodel the niche and secrete ECM components in response to the signals they receive from it [33–35].

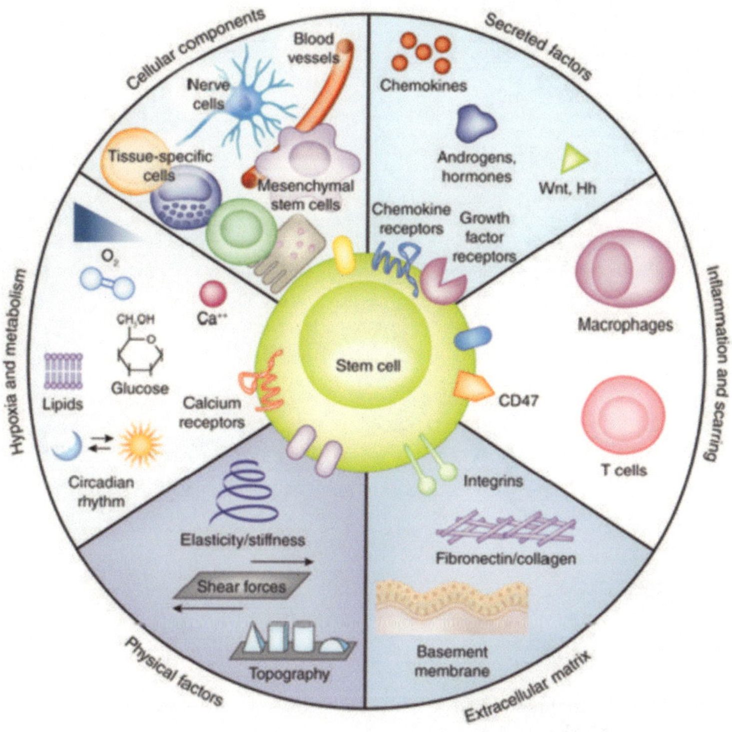

Figure 1. Components of the stem cell niche. Niches are complex dynamic heterogenous microenvironments containing various cell types, extracellular matrix, soluble factors. The stem cell niche is influenced by a variety of factors including physical and metabolic parameters. The ECM and supportive stromal cells interact with stem cells through gap junction, soluble factors and surface receptors. Systematic signals are carried into the niche by blood vessels to facilitate the recruitment of inflammatory cells and neural signals convey distant physiological cues, such as shear stress, tissue stiffness and oxygen tension, to the stem cell niche. Image adapted with the permission from Lane et al. [11].

2.1. Stem cell niches: role

Niches have specific anatomical locations and form unique stem cells surroundings *in vivo*. This microenvironment regulates pathophysiological and physiological processes and directs cellular fate and function (**Figure 2**). The niche provides extracellular signals that maintain a balance between stem cell self-renewal and differentiation enabling stem cells to preserve a dormant and low metabolic state in order to avoid stem cell exhaustion and the accumulation of gene mutations which might result in their transformation into cancer cells [4, 36, 37]. It has been shown that the destabilisation of the stem cell environment is involved in diseases connected to aging, tumorigenesis and degeneration [38]. In adjacent sites within the same tissue, stem cells co-exist in either quiescent or active state [39, 37]. For maintaining stem cell number and to meet the needs for differentiated cells in neighbouring tissues, the balance between asymmetric and symmetric stem cell division is essential [3]. Thus, in order to maintain a healthy stem cell pool, a crosstalk between tissue necessity and state is created by the stem cell niche [40, 41].

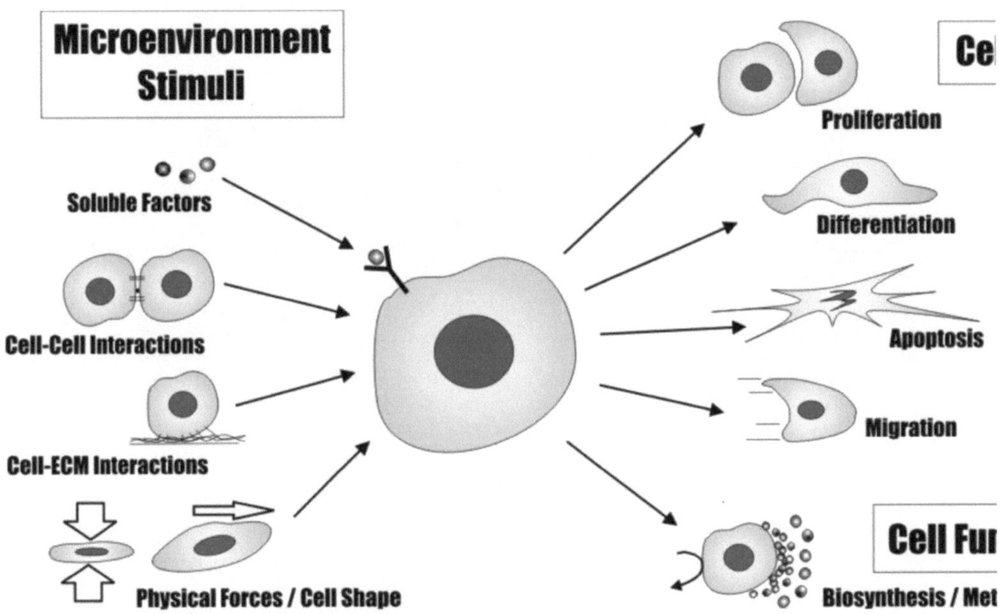

Figure 2. The stem cell microenvironment controls pathophysiological and physiological processes. Image adapted with the permission from Ou et al. [48].

2.2. Extracellular matrix

Due to its dynamic and diverse composition, the ECM provides structural and mechanical support and gives biochemical and physical characteristics to different stem cell niches, which are required for tissue morphogenesis and homeostasis as well as to facilitate stem cell renewal and control stem cell fate [3–5, 7]. It provides scaffolding to cells and stores soluble growth factors [11, 42, 43]. In addition, secreted or cell surface factors, signalling cascades and gradients, as well as physical factors, such as shear stress, oxygen tension and temperature, contribute to control stem cell behaviour in a well-orchestrated manner [40, 41].

The ECM is mainly composed of water, proteins and polysaccharides with each tissue exhibiting its unique composition and topology [42]. This diversity is caused by a combination of specific molecular interactions between ratios, geometries and isoforms of its components [7]. Two main classes of macromolecules are found in the ECM. These are fibrous proteins (e.g. collagens, elastins, fibronectins and laminins) and proteoglycans (e.g. heparin sulphate, chondroitin sulphate, ketaran sulphate) [44, 45].

The most plentiful fibrous protein in the interstitial ECM is collagen. It contributes to approximately 30% of the total protein amount and provides tensile strength it controls chemotaxis, controls cell attachment and migration as well as influences tissue development [46]. In any given tissue, collagen is present as a non-homogenous mixture of collagen fibre types; however, one type is normally dominant (e.g. in bone: collagen type-I, cartilage: collagen type-II) [42]. In close connection to collagen are elastin fibres which recoil when tissues undergo repeated stretching. In fact, the degree of association with collagen is a limiting factor for tissue elasticity [47]. Besides directing the organisation of the interstitial ECM fibronectin also influences cell adhesion and migration [42, 46]. The majority of the extracellular part within tissues is filled by proteoglycans consisting of GAG chains, which are linked to a protein core and are classified according to the GAG and core protein arrangement [45]. These proteoglycans have numerous different purposes that mirror their hydration, binding, buffering and force-resistance properties [42, 48].

Furthermore, the ECM can be divided into two forms, the basal membrane (BM) and the interstitial matrix [49] both of which are composed of a collagen framework with glycoproteins (non-structural proteins) adhering to it to communicate with adjacent cells via integrins [50].

2.3. Niche cells and cell communication

Stem cell niches in adult tissues are populated by a number of different cell types with each having a particular function. This is shown in the adult hematopoietic stem cell (HSC) niche, which is located along the endosteal surface of trabecular bone, in close vicinity of osteoblasts and the endothelial cells of blood vessels [51]. Osteoblasts in the endosteal niche regulate the HSC number [51, 52] and preserve their quiescence by releasing signalling molecules [53–55]. The HSC niche is further inhabited by cell types including stromal cells, bone marrow adipocytes, osteal macrophages, CXCL12-abundant reticular (CAR) cells, nestin-positive mesenchymal stem cells, nestin-positive Schwann cells, endothelial cells [19, 56–59].

Besides cells, which reside permanently in the niche such as nerve cells, endothelial cells and connective tissue fibroblasts, cells of the innate and adaptive immune system and cells that are important for the repair of damaged tissues and promote protection against pathogens are also present [11].

When stem cells undergo cell death or apoptosis neighbouring cells belonging to the niche undergo dedifferentiation in order to replace the lost stem cells [11]. For example, the removal of hair follicle stem cells in mice results in the repopulation of the niche by epithelial cells which then sustain the hair renewal [60]. This reprogramming of endogenous differentiated cells into stem cell controlled by the stem cell niche [16], and the maintenance of stem cell sources [61] might hold important clinical promise [11]. The communication between niche cells and stem cells is facilitated either by indirect contact through the secretion of molecules or directly through physical cell–cell interactions. The direct cell–cell contact is mediated by cell–cell adhesion molecules and receptors with membrane-bound ligands [11]. Indirect communication between cells in the HSC niche is clinically used to modify the HSC niche in vivo. For the treatment of bone marrow failure, genetic disorders and haematological malignancy cytokines (G-CSF, GM-CSF) are administered which result in the activation of hematopoietic stem cells, their expansion as well as diminished adhesion to their niche [11].

2.4. Physical and physiological parameters

Stem cell fate is greatly influenced by physical and physiological parameters (**Figure 2**). Niche topographies modify the stem cell cytoskeletal resulting in the activation of specific signalling pathways and stem cell differentiation [29, 62]. The modification of physical parameters such as substrate stiffness or elasticity, shape and shear forces has been utilised for clinical treatments but also for *in vitro* hematopoietic stem cell culture [26, 27]. Bone tumour and osteoporosis are treated clinically using drugs that change the balance between physical factors, for example rigid (bone) and elastic (arteriolar, dermal connective tissue). Shear forces and drugs which are used to promote blood flow for the development of embryonic HSC *in vivo* [28]. Physiological factors such as oxygen tension are important contributors for cell survival and maintenance. Many cells such as HSC inhabit hypoxic microenvironments [30]. Cells in these niches are carried out the glycolysis and express high levels of hypoxia inducible factor 1α. Growing mammalian cells under hypoxic conditions positively influences cell proliferation, survival and function after engraftment [31].

3. Stem cells in regenerative medicine

Medical conditions such as tissue loss, organ failure, cancer abrasion, congenital structural anomalies can already be treated by clinical procedures such as autologous and allogenic organ transplantation and the use of artificial implants. However, these treatments are limited by organ shortages, impairment of healthy tissue during surgery and immune rejections. Breakthroughs in the field of regenerative medicine may enable the utilisation of stem cells and stem cell-based therapies for the restoration of tissue function [63].

Stem cells can be distinguished by their potency (multipotent versus pluripotent) and through their tissue source, that is ESCs, foetal stem cells (FSC), adult stem cells such as mesenchymal stem cells (MSCs) and induced pluripotent stem cells (iPSC) [64, 65] (**Figure 3**). Stem cells are characterised by their ability to self-renew without senescence for extended culture periods and the potential to differentiate into multiple cell types [66]. Through their pluripotency and multipotency stem cells offer vast cell sources making them ideal for studying degenerative diseases and developing cell-based therapies [64, 65]. They have the capability to generate every tissue type and are essential to human development. However, due to regulatory, ethical and technical considerations involving genetic modification and cell isolation adult stem cells such as MSCs and adipose stem cells have been widely investigated as an alternative cell source to ESCs [63].

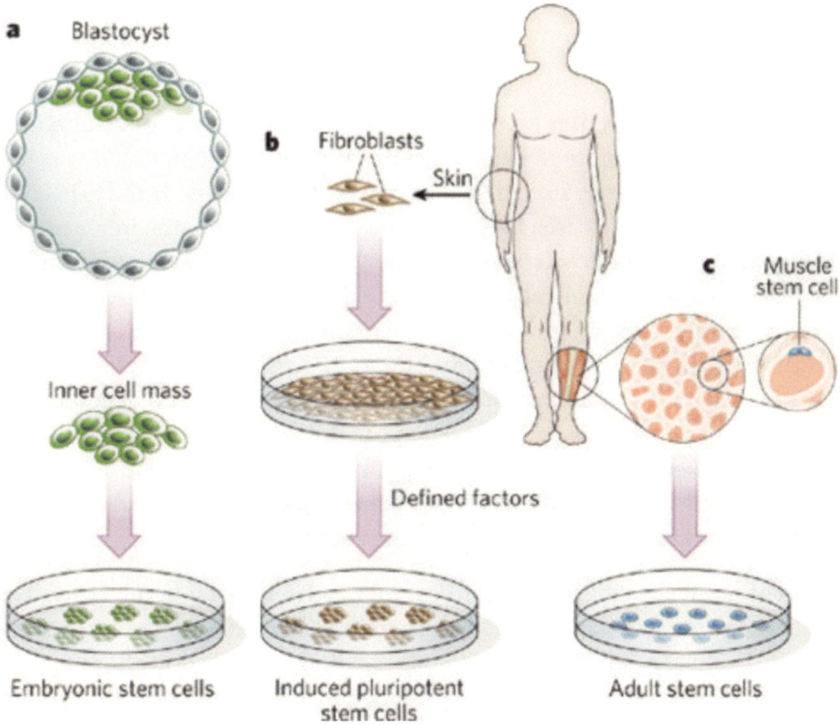

Figure 3. Stem cell types. (a) Embryonic stem cells derived from blastocysts were the first human pluripotent stem cells that could be differentiated to generate various cell types. (b) Induced pluripotent stem cells were first generated by reprogramming somatic adults cells such as skin fibroblasts. (c) Tissue-specific adult stem cells are primed for generating progeny that differentiate into specialized cell types (e.g. stem cells residing in the muscle). Image adapted with the permission from Lutolf et al. [225].

Tissue-specific adult stem cells which can be found in various niches such as the bone marrow or adipose tissue are multipotent and play a crucial role in tissue development, repair and growth [67–70]. Regenerative medicine approaches envision that these somatic stem cells could theoretically be harvested from the patient, be differentiated *in vitro* and injected back into the patient to regenerate impaired tissues without the need to suppress the immune system. Autologous chondrocyte implantation (ACI) for the repair of chondral defects in joints such as the knee is such a regenerative approach. In this two-stage process, the patient's chondrocytes are expanded *in vitro* and the subsequently implanted back into the patient [71–73].

Yet, low cell numbers, complications isolating adult stem cells from healthy tissues well as limited differentiation potential may require other stem cell types [66]. Latest developments indicate that directed differentiation and the development of differentiation protocols could result in the translation of cell therapies for presently incurable diseases [74–77]. The availability of patient-specific cells in required quantities and on demand could revolutionise stem cell therapies and further allow disease modelling and human pluripotent stem cell (hPSC)-based drug discovery [65].

3.1. Pluripotent stem cells

3.1.1. ESCs and FSCs

When ESCs were first isolated, they provided a model for the study of developmental biology but also opened up the possibility to exploit their pluripotency for stem cell-based therapy in order to treat organ damage or dysfunction allowing regenerative medicine to become a reality for the treatment of various diseases [66]. ESCs are harvested from the inner cell mass of the blastocyst-stage embryo, a hollow sphere of cells which is composed of an outer cell layer forming the placenta and an inner cell mass from which the ESCs are derived [66]. ESCs have the ability to proliferate, to maintain an undifferentiated phenotype for extended periods of time in culture and to develop into a large number of somatic cell types [78–80]. Early studies utilising ESCs were aimed at treating traumatic injuries of the CNS and degenerative diseases [81–84]. Implanting early stage and differentiated ESC into laboratory animals has been shown to improve function, behaviour and morphology, but also to cause teratoma formation and hyperproliferation [85]. The application of ESC has also shown great promise for the regeneration of cartilage, cardiac tissue and peripheral nerves [76–89].

The utilisation of tissue-specific stem cell lines isolated from foetal tissue is another possible strategy in regenerative medicine as these cells exhibit higher proliferation potential, more specific differentiation capacity, improved migration and regeneration after implantation [90, 91]. Functional integration of human FSC (hFSC)-derived dopamine neurons in a rodent Parkinson's disease model is one of the most significant examples for the use of hFSC. It became the basis for clinical trials demonstrating similar effects in patients [92–94]. Another study showed that the transplantation of human cortical neuroepithelial stem cells developed from foetal cortical brain did not result in tumour formation and facilitated the recovery of diminished tissue function in a rodent stroke model [95]. Despite associated ethical concerns and the risk for tumour development, these studies have shown great promise for the clinical application of ESC and FSC [64].

3.1.2. Induced pluripotent stem cells

In 2006, Takahashi and Yamanaka [96] discovered that the retroviral expression of pluripotency-specific transcription factors (Oct4, Sox2, Klf4 and c-Myc) reprograms adult somatic cells into a pluripotent state. These iPSCs showed epigenetic and transcriptional similarities to ESCs [97–99]. Over the years, the progress has been made in the generation of virus-free/vector-free reprogramming methodologies to avoid vector-induced tumour development [100–103]. In addition, iPSCs are now being derived from a variety of different cell types such as blood cells, dermal fibroblasts and keratinocytes at higher efficiencies [104–107].

To study the sporadic and genetically inherited diseases, patient-specific iPSCs have been generated (**Table 1**). These *in vitro* studies offer a proof-of-concept for the use of iPSCs for disease modelling with the goal to discover novel drugs and disease-specific pathways aiding their treatment. Since several studies were performed on very limited cell numbers, reproducibility of the observed phenotype still needs to be investigated [108].

Disease	Derived cells	Control cells	Result	References
Parkinson's disease	Dopaminergic neurons	hiPSC	No observed defect, not drug tested	[103]
Parkinson's disease	Dopaminergic neurons	hiPSC	Enhanced chemical sensitivity causes cell death, drug tested	[309]
Spinal muscular atrophy	Motor neurons	hiPSC	Loss of SMN gene expression and neuron formation, drug tested	[310]
RETT syndrome	Neurons	hiPSC	Reduced soma size and spine density, loff of synapse, drug tested	[311]
Familial dysautonomia	Neural crest cells	hiPSC, hESC	Loss of neural crest cells, drug tested	[312]
Long QT 1 syndrome	Cardiomyocytes	hiPSC	Depolarisation of cardiomyocytes, drug tested	[313]
Long QT 2 syndrome	Cardiomyocytes	hiPSC	Depolarisation of cardiomyocytes, drug tested	[314]
A1-antitrypsin deficiency	Hepatocytes	hiPSC	Downregulation of A1-antitrypsin expression, drug tested	[315]
Timothy syndrome	Cardiomyocytes	hiPSC	Depolarisation of cardiomyocytes, drug tested	[316]

Table adapted from Wu and Hochedlinger [108]

Table 1. Patient-specific disease models utilising human iPSC-derived cells.

The generation of autologous cells for cell therapy is another possible application for the iPSC technology as it minimises the challenges associated with human ESC-based therapies [108]. Researchers have described that iPSCs facilitate the reduction of the blood cell phenotype in a sickle cell anaemia mouse model [109]. iPSCs were derived from a transgenic mouse exhibiting a mutation in the human haemoglobin sequence. These iPSCs were genetically corrected and differentiated into haematopoietic progenitor cells. Subsequent implantation of these cells into the mouse model resulted in a normal haemoglobin level and restored phenotype [109]. Despite these remarkable results, haematopoietic stem cells with the ability for multilineage differentiation have not yet been generated. In addition, this study utilised retrovirus-derived iPSCs and it remains unclear if similar results can be achieved in a retrovirus-free approach. Comparable studies implanting iPSC-derived progenitors for mesodermal and ectodermal cells into animal models have been described (e.g. neurons, cardiomyocytes, blood) [110–112].

The conversion of somatic cells into iPSCs and the advances that have been made in their generation has enabled researchers to utilise disease-specific cells for disease modelling as well as drug screening. Moreover, it opens up opportunities for the generation of custom-made iPSCs for cell therapies [108] (**Figure 4**). Patient-specific iPSCs are interesting autologous cell sources as they would eliminate the need to suppress the patient's immune system and could be generated in larger quantities. However, future studies are required to investigate the effect of genetic modifications of host and donor cells when injecting iPSC into patients [64].

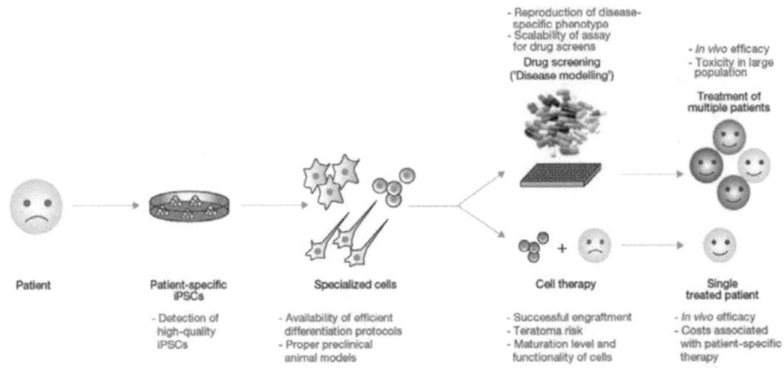

Figure 4. The potential of iPSC technology in regenerative medicine. Reprogramming a patient's somatic cells, for example blood cells or skin cells generates patient-specific iPSCs. These iPSCs can be differentiated into specific cell types which are subsequently used for either cell therapy or disease modelling. Cell therapy aims to fabricate autologous differentiated cells for implantation into a single patient. Disease modelling on the other hand is based on reproducing a cell phenotype from iPSC-derived specialist cells as present in the patient and utilising these cells for large-scale drug screening for the treatment of any patient with the same disease. Image adapted with the permission from Wu and Hochedlinger [108].

3.1.3. Comparison of ESCs and iPSCs

ESC and iPSC cell lines display biological differences between one another and how comparable the both cell types are might affect their functionality and safety. Researchers have shown that hiPSC and hESCs are highly similar [113–116]. However, variations in gene expression [116], DNA methylation [117, 118], differentiation potential [119, 120] and teratoma-forming propensity [121] have been reported. Genetic background [122], passage number [116, 123], lab-to-lab differences and the use of vectors in their generation have extensive influence on the function and gene expression of PSC. Moreover, hiPSC and hESC have been shown to carry copy number variations (deletions and duplications) [124–126] and point mutations [127], which are consequences of their culture. Some mutations might also occur during reprogramming [108]. Since differences in transcriptional and DNA methylation profiles between 12 hiPSC and 20 hESC lines have been described, large sample numbers are required to robustly investigate potential differences and their influence on pluripotent stem cells [128].

3.2. Clinical potential of pluripotent stem cells

Translation of pluripotent stem cell-based therapies into the clinic will depend on several factors such as cell purification, the efficiency of cell-lineage-specific differentiation leading to functional cells, eliminating tumour generation and finally the generation of novel organs [108]. Even though iPSCs have been differentiated into a number of cell types including cardiomyocytes [129–133], neurons [134, 135], blood [133, 136, 137] and pancreatic cells [137, 138], purities over 95% have not been reported and isolating these cells from a heterogeneous cell population is difficult. Furthermore, ESC-/iPSC-derived cells are mostly immature and whether their stage will affect their clinical performance remains to be investigated in a cell- and disease-specific manner [108].

Another important issue for the application of PSC in regenerative medicine is their integration into the host tissue. Organs exhibit a balance between the numbers of each cell type, their geometrical arrangement and their developmental stages. Whether the injection of cell suspension will auto-regulate cell type numbers to generate an endogenous tissue and whether the transplanted cells will function in synergy with existing cells still needs to be investigated [108]. In addition, regulatory requirements for pluripotent stem cell–derived therapies remain high as shown by a clinical trial performed by Geron, a biopharmaceutical company that tested human ESC–derived oligodendrocytes for spinal cord injury [139]. Microscopic cysts found in cell-transplanted mice resulted in extensive studies for the batch-to-batch assessment of cyst formation as well as follow-up safety study on these cyst containing grafts were requested by the FDA [139].

Concerns over the use of hiPSCs for cell therapy are also arsing due to studies reporting on genetic and epigenetic modifications during the reprogramming process such as protein coding and DNA methylation [127, 140]. To-date, progress in the development of iPSC protocols has eliminated the need for vectors reducing the risk of tumour formation [113, 141, 142]. Furthermore, whether iPSC-derived cell types retain their epigenetic 'memory' remains under investigation. Studies have shown that both mouse and human iPSCs preserve an epigenetic profile which is related to the donor cell source and this may affect subsequent differentiation [143–146]. Drawbacks for the application of patient-specific iPSCs also include regulatory, cost and time requirements for the generation of patient-specific iPSC-based treatments as well as scalability and good manufacturing practice (GMP)-compliant cell therapies.

Despite all this, in recent years, the focus has been on translating hESC research into the clinic such as the trial sponsored by Geron which has highlighted the obstacles but also promises of pluripotent stem cell therapy. Furthermore, Japanese researchers were applying for regulatory approval to use patient-specific iPSC-derived RPE cells to treat macular degeneration [147]. Pluripotent stem cell research has evolved from the isolation of hPSC to the development of differentiation protocols and early clinical trials. Less than two decades into hPSC research, the utilisation of human pluripotent stem cell–derived treatments are under clinical investigation [65].

3.3. Multipotent stem cells as alternative cell sources

Limited availability of embryonic and foetal tissue as well as ongoing research into the development of improved protocols yielding matured differentiated cells may be limiting factors for the use of ESC, FSC and iPSC. Alternatives to pluripotent stem cells are multipotent stem cells such as MSCs and adipose-derived stem cells which have been extensively studied for regenerative medicine therapies.

3.3.1. Mesenchymal stem cells

Non-hematopoietic multipotent MSCs are derived from the mesoderm and are present in a variety of connective tissues and postnatal organs [63]. Their discovery was reported over 130 years ago when research suggested that bone marrow is the source collagen depositing cells that can develop into various cell types including bone and cartilage [148]. Caplan [149] then named these marrow cells MSCs after their capability to differentiate into cells of mesenchyme origin such as bone, cartilage, tendon, muscle and adipose tissue. The clinical routine to treat haematological diseases with bone marrow makes MSCs interesting contenders for cell-based therapies. MSCs isolated from bone marrow are probably among the most characterised and clinical utilised stem cell types [150]. The multipotency of MSC allows them to differentiate into bone, cartilage, muscle and neural cells [151–153]. They offer the possibility for autologous cell transplantation and at the same time eliminating the risk for graft-versus-host diseases. MSCs are non-immunogenic, affect the maturation and response of immune cells and do not cause tumour formation when implanted in allogeneic hosts [64]. Finally, the injection of MSC leads to the secretion of growth factors by host cells including brain-derived neurotrophic factor (BDNF), vascular endothelial growth factor (VEGF) and nerve growth factor (NGF) among others. In summary, the capacity of MSC to change their default state and their potential to promote tissue regeneration may surpass their application for haematopoietic diseases and might make them a suitable tool to treat degenerative diseases such as neurological and neurodegenerative disorders [154]. Employing MSC in combination with biomaterials or alone resulted in functional regeneration of paralysed limbs, diminished cavity formation in the spinal cord and axonal regrowth [155–158]. MSCs have also been utilised in first clinical trials to treat myocardial infarcts, stroke and diseases of the CNS [159, 160]. These studies have demonstrated that bone marrow–derived MSCs are a promising stem cell type for clinical application and stem cell therapies [64]. However, limited integration into the host tissue was observed in clinical trials indicating that the detected benefit in these trials were most likely due to the secretion of cytokines and soluble factors [161].

MSCs with similar properties to bone marrow–derived MSC have been isolated from trabecular bone [162], periosteum [163], synovial membrane [164], skeletal muscle [165], skin [166], pericytes [167], peripheral blood [168] and umbilical cord [169, 170]. Due to their low number upon isolation, adult stem cells require in vitro expansion and modification before being employed therapeutically [63].

3.3.2. Adipose-Derived stem cells

Adipose-derived stem cells (ASCs) are a promising cell type as they can be easily isolated in large numbers from adipose tissue without causing severe donor site morbidity and discomfort [171, 172]. Over 80% of adipose tissue is composed of mature adipocytes and the stromal vascular fraction consisting of vascular smooth muscle cells, fibroblasts, preadipocytes, endothelial cells, resident immune cells and ASCs [173, 174]. Many of the ASC properties and density differ according to the harvest location of the adipose tissue [63]. Studies have shown that the ASCs proliferation rate is dependent on age, tissue location and type, the culture conditions and isolation procedure, but is in general higher compared to the doubling rate of bone marrow–derived MSCs [175–178]. Recent studies have shown that their application is not restricted to mesodermal tissues instead they can be employed for endodermal and ectodermal tissue regeneration [63]. ASCs have been successfully differentiated into adipogenic [179–181], osteogenic [182], cardiomyogenic [183, 184], chondrogenic [182, 185, 186], angiogenic [187, 188], tenogenic [189], hepatogenic [190, 191] lineages. Moreover, studies utilising scaffolds in combination with growth factors have been carried out to assess their tissue regenerative potential [192, 193]. In summary, ASCs are a valuable cell source for the development of cell-based therapies and have been shown to be safe and effective in clinical and preclinical studies [171, 172]. However, since ASCs are characterised as multipotent adult stem cells, their differentiation potential is restricted compared to ESC and iPSC. Even though limited numbers of clinical studies have investigated the therapeutic potential of ASCs and their lineage-specific differentiation depends on the site of harvest, gender and age of donor [194, 195], there are practical advantages of using ASCs in regenerative medicine [63].

4. Regenerative medicine strategies: engineering artificial niches for the control of stem cell fate

Adams and Scadden [196] elucidated the concept that the stem cell niche is 'dynamic' and its properties change during development and with varying physiological conditions. These changes affect stem cell fate, but could also be utilised as potential therapeutic tool in regenerative medicine. Consequently, the control of nanotopography, mechanical and chemical properties of the ECM among others in engineered constructs as well as their mechanical loading is essential for directing stem cell fate in bioartificial systems and for the development of regenerative therapies (**Figure 5**) [197].

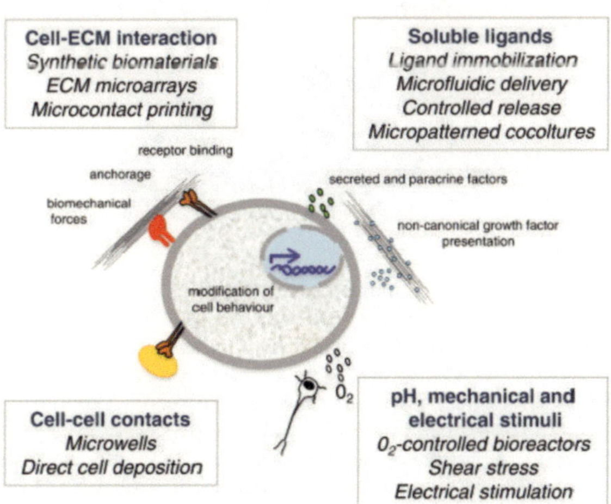

Figure 5. Engineering stem cell niches for the development of regenerative medicine strategies. To replicate niches in order to control stem cell fate chemical, topographical and mechanical properties are being mimicked using engineering techniques. Image adapted with the permission from Gazzatto et al. 2014 [5].

4.1. Biochemical signals as stem cell fate regulators

Chemical signals that cells are exposed to in the ECM activate signalling cascades determining cell proliferation, differentiation, migration and apoptosis [198, 199]. Hence, it is essential to mimic chemical cues of the ECM to control stem cell behaviour. Growth factors, which regulate cell adhesion, proliferation and lineage development, are an important type of chemical cue [200]. Growth factors are either added to the culture medium or secreted by niche cells and stem cells. They are crucial for stem cell fate decision and are spatiotemporally regulated during embryonic development [201]. A chemically defined protocol for the directed differ-

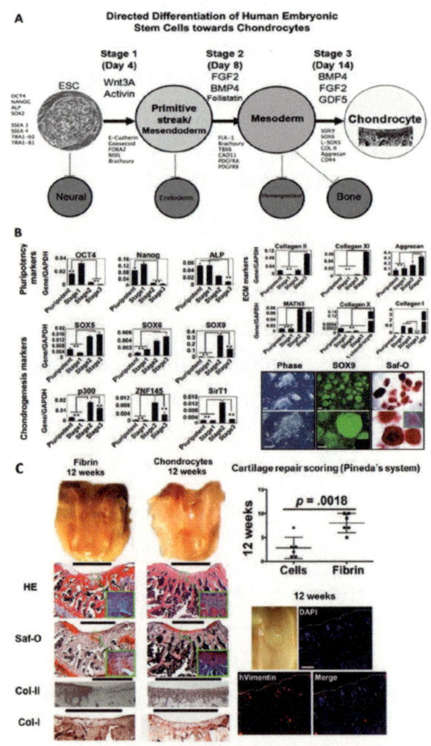

Figure 6. Differentiation of hESC and *in vivo* cartilage formation of hESC-derived chondroprogenitors. (A) hESC underwent a 14-day defined protocol for promoting chondrogenic differentiation. (B) Characterisation of chondroprogenitor cells by quantifying the upregulation of pluripotency, chondrogenic and cartilage ECM markers using qPCR. Results are shown as means ±SD (n – 3). Chondrogenic cells derived from MAN7 showed high expression of SOX9 (inset shows IgG control) and Safranin-O staining (inset shows pre-treated with chondroitinase ABC) at the end of the protocol. Scale bars = 100 mm. (C) Assessment of *in vivo* cartilage repair capacity of pre-differentiated hESC after 12 weeks by macroscopic observation of the gross appearance of the RNU rat patella groove. Histological sections through the knee were stained for HE, Saf-O immunological assessment through Col-I and COL-II staining. Cartilage repair was scored using Pineda's system (0-worst to 14-best). Human cells were detected 12 weeks after implantation by immunohistochemistry staining. Scale bars black = defect area; white = 500 mm. Abbreviations: alkaline phosphatase (ALP); extracellular matrix (ECM); fibronectin (FN); glyceraldehyde-3-phosphate dehydrogenase (GAPDH); gelatine (GEL); Safranin-O (Saf-O); collagen type-I (Col-I); collagen type-II (Col-II); haematoxylin (HE). Figures were modified with the permission from Cheng et al. [202] and Oldershaw et al. [203].

entiation of hESC towards chondrocytes in 2D culture was developed where the pluripotent stem cells undergo intermediate developmental stages by supplementing the culture medium with exogenous growth factors and culturing cells on substrates of known matrix proteins over a culture period of 14 days [202, 203] (**Figure 6A**). The cartilage repair capacity of these chondroprogenitors was assessed by incorporation into fibrin hydrogels and implantation using an osteochondral defect model in the patellar groove of athymic RNU rats. Results showed that chondrogenic cells derived from hESC using a chemically defined differentiation protocol promoted cartilage repair [202] (**Figure 6B**).

Furthermore, the importance of cell–cell interactions in soluble factor signalling has revealed that certain cell types will respond only to locally secreted growth factors when in direct contact with adjacent cells [204]. Growth factors affect secreting cells (autocrine) and other cells (paracrine) *in vivo*. Soluble factors are often bound to the ECM limiting their diffusion and enhancing their efficiency. This can be replicated by tethering these to a biomaterial substrate [205]. Kuhl et al. [200] demonstrated that growth factors in their matrix-bound state were more effective than their soluble form. When proteins are incorporated or linked to biomaterials, they are commonly denatured or degraded. Consequently, short peptide sequences mimicking growth factors and chemokines are integrated to control stem cell fate [206]. The RGD sequence present in ECM proteins has been widely linked to biomaterial surfaces demonstrating enhanced osteogenesis and chondrogenesis compared to non-modified surfaces [206–208]. Bone morphogenic proteins have been shown to play a key role in stem cell activity and osteoblastic differentiation [209, 210]. Another approach is a surface modification of natural and synthetic biomaterials with specific functional chemical groups [211]. Examples include the functionalisation of PEG hydrogels with phosphate groups for bone mineralisation, carboxyl groups resembling GAG for cartilage and tert-butyl groups to mimic lipids for adipose tissue [212]. Moreover, self-assembled monolayers (SAM) functionalised with methyl-, hydroxyl-, amino- and carboxyl groups have been shown to promote osteogenic differentiation of MSC with the amino group being the most effective chemical group [213].

Growth factors and reactive chemical groups have been successfully used to guide stem cell differentiation. It was shown that immobilised chemical cues are more effective than soluble factors. However, improved control over bioactivity, spacing and orientation of the immobilised growth factors is required to guide-/direct-specific stem cell differentiation [214]. Further studies investigating the effect of chemical signals on stem cell fate are summarised in **Table 2**.

Cell type	Material	Chemical group	Result	References
Soluble factors				
hMSC	PLGA	BMP-2	Enhanced osteogenic differentiation	[317]
hMSC	PLLA	BMP-2	Enhanced osteogenic differentiation	[318]
hMSC	Chitosan/collagen IV	VEGF	Endothelia differentiation	[319–321]

Cell type	Material	Chemical group	Result	References
hESC	TCP coated with FN, gelatine and both	Sequential addition of Wnt3a, Activin-A, BMP-4, Follistatin, GDF5, FGF2, NT4	Chondrogenic differentiation	[202, 203]
Peptides				
hMSC	Alginate	Osteopontin peptide	Osteogenic differentiation	[322]
hMSC	PLGA	Osteocalcin peptide	Osteogenic differentiation	[323]
hMSC	BCP/PLA	RGD	osteogenic differentiation	[324]
rMSC	HA-PLG	BMP-2 peptide	Increased ALP expression, ectopic bone formation in vivo	[325]
Chemically reactive groups				
hMSC	Silk fibroin	–COO–, =C=O, SO_3H, NH_2, CH_3	Enhanced osteogenic differentiation	[326]
hMSC	PEG	PO_3	Increase of osteogenic markers at protein and gene level	[212, 327, 328]
hMSC	Glass	COOH, CH_3, OH, NH_2, SO_3H, SH	–NH2 and –SH group promoted and maintained osteogenesis, –OH and –COOH promoted chondrogenesis	[329–331]

Table adapted with the permission from Griffin et al. [214]
hMSC: human mesenchymal stem cells; hESC: human embryonic stem cells; PLGA: poly (lactic-*co*-glycolic acid); PLLA: poly(L-lactic acid); PEG: polyethylene glycol; HA-PLG: hydroxyapatite (HA)/poly(lactic-*co*-glycolic acid); BMP: bone morphogenetic protein; PMMA-g-PEG: poly(methyl methacrylate)-graft-poly(ethylene glycol; BCP: biphasic calcium phosphate; TCP: tissue culture plastic; EGF: epidermal growth factor; PDGF-AA: VEGF: vascular endothelial growth factor.

Table 2. Effect of chemical signals on stem cell differentiation.

4.2. Topographical signals as regulators of stem cell fate

Cells in their native in vivo environment engage with a variety of macro (tissues)-, micro (cells)- and nano-sized (proteins and ligands) topographical features. The basement membrane with its nanoscale pores, ridges and fibres is the most crucial ECM structure providing tissue organisation and support [215]. Micro- and nanopatterning techniques including soft lithography, electrospinning, layer-by-layer microfluidic patterning, three-dimensional printing, reactive ion etching and ion milling have resulted in the fabrication of scaffolds with controlled porosity, geometry and rigidity and texture [216–218]. The production of specific surface topography scales (nano, micro), types (ridges, pit, pillar, grooves) and distributions (random, regular) has enabled researchers to study the influence of topographical signals on stem cell differentiation [214]. Structural cues greatly affect gene expression as they determine cell shape, elongation, positioning of focal adhesion and cell–cell interactions [29, 219–221]. Cell shape is a crucial regulator for cell physiology and function as well as dictator for stem cell differentiation [222]. Manasek et al. [223] reported that cell shape regulates myocardial development, whereas ECM-induced cell shape changes are responsible for the proliferation and differentiation of capillary endothelial cells [224]. Numerous studies have demonstrated that stem cell fate can be guided through controlling their shape by artificial extracellular matrices [197]. Traditionally, stem cells are expanded two-dimensionally on rigid, flat-coated or non-coated tissue culture plastic and are exposed to soluble factors in the liquid growth medium. These plastic culture substrates are usually coated with collagen or laminin, feeder cell layers or with hydrogels such as Matrigel [225]. However, this culture approach is very different from the native microenvironment cells are experiencing in vivo, where they are residing in the stem cell niche anchored to ECM though adhesion molecules [225]. Two-dimensional culture approaches provide simplified methods to expand stem cells and to study individual cues influencing their fate decisions in vitro. This has allowed the identification, patterning and concentration of soluble to tethered ECM molecules regulating stem cell niche and its inhabitants. ECM arrays have been utilised to screen for molecules individually or in combination that induce fate changes [226, 227]. The development of 3D-culture systems compared to traditional 2D culture systems results in a more rounded cell morphology. Culturing chondrocytes in 2D, for example resulted in a dedifferentiation and change from chondrogenic to fibroblastic phenotype [228], whereas pellet culture or incorporation of chondrocytes in hydrogels maintained their native phenotype [229, 230]. It was further demonstrated that for bone marrow–derived MSCs to undergo chondrogenic differentiation cells would have to be cultured with a spherical shape either as spheroid culture or encapsulated in hydrogels [221, 231–233]. However, when MSCs were cultured in hydrogels facilitating their adhesion and spreading stem cells exhibited a fibroblastic phenotype [234]. **Table 3** provides a summary of studies examining the correlation between surface topography and stem cell differentiation. Oh et al. [235] found that MSCs cultured on ~100-nm nanotubes differentiated towards the osteogenic lineage, whereas MSCs seeded on 30-nm nanotubes did not differentiate [235]. Moreover, surface topography can also cause cell elongation on material surfaces resulting in the distortion of the nuclei shape as the nuclei are mechanically integrated into the cells, which has been shown to influence osteoblastic stem cell differentiation [237]. When ESCs are aggregated into embryoid bodies (EB), which mimic early stages of embryonic development, it becomes clear that geometrical signals also affect cell–cell interaction and hence guide stem cell differentiation [66]. The investigation of hexagonal patterns, random nanopits, disordered and ordered squares utilising on hMSCs electron beam lithography (EBL) showed that highly asymmetric structures resulted in highest osteogenic expression in the absence of growth factor supplemented medium [237] indicating that nano-sized topographical induction of stem cell differentiation is as effective as chemical induction [238]. Micropatterns have also been shown to affect cell shape. Changes from rounded (on small islands) to flattened (on large islands) shapes result in modifications of the actin cytoskeleton and focal adhesions and hence controlling the lineage commitment of MSC into a variety of cell phenotypes [29].

Cell type	Material	Topographical feature	Result	References
hMSC	TiO_2	Nanotubes (15, 100 nm)	Cells grown on 15 nm exhibited enhanced integrin clustering, cell spreading, osteogenesis	[332]
hMSC	TiO_2	Nanotubes (30, 50, 70 and 100 nm)	Osteogenic differentiation promoted on 70–100 nm nanotubes	[235]
hMSC	PDMS	Islands	1000 μm^2 facilitated osteogenesis	[29]
hMSC	PDMS	Grafting (350 nm)	Neurogenesis	[333]
hMSC	PDMS	Micropattern, stripped grooves collagen type-I coated	Neuronal differentiation enhanced	[334]
hMSC	PMMA	Hexagonal pattern, nanopits, disordered and ordered squares	Enhanced osteogenesis on disordered squares	[237]
hBMSC	Hydrogenated amorphous carbon	Grooves (80/40, 40/30, 30/20 μm-width/ spacing; 24 nm depth)	Neurogenesis, absence of growth factor supplemented medium	[335]
hESC	PDMS	Square shaped fibronectin surrounded by pluronic-F127, micropattern	Myogenesis and chondrogenesis	[336]
hESC	PDMS	Grooves	Neuronal	[337]
mESC	PLLA	Fibrous grating (50–500 nm), TCP	Enhanced osteogenesis on fibrous gratings	[338]

Table adapted with the permission from Griffin et al. [214]
hMCS: human mesenchymal stem cell; hBMSC: human bone marrow–derived stem cells; hESC: human embryonic stem cell; mESC: mouse embryonic stem cell; PDMS: polydimethylsiloxane; PMMA: polymethyl methacrylate; PLLA: poly(L-lactide); TiO_2: titanium dioxide, TCP: tissue culture plastic.

Table 3. Effect of surface nanotopography and stem cell fate.

4.3. Biomaterials affecting stem cell fate

The control of stem cell fate through biomaterial nanotopography is promising as these approaches are not subjected to short-term degradation as are chemical cues. In addition, an individual biomaterial could exhibit various nanotopographies and gradients which would allow for further clinical applications [214]. Biomaterials are being designed and utilised to mimic the microenvironment and its cues stem cells experience *in vivo*. Nano- and microtopography replicating the ECM signals have been shown to direct stem cell differentiation. ECM components linked to the biomaterials surface in form of small peptides or through the incorporation of growth factors have been successful approaches for guided stem cell differentiation and provide advanced stem cell-based clinical approaches [214, 225] To overcome limiting factors for the clinical use of pluripotent stem cells, new culture systems based on advanced biomaterials are required which more closely mimic the native *in vivo* milieu and support application related stem cell fate decisions [225]. In summary, simplifying a complex three-dimensional stem cell niche into a two-dimensional biomaterial approach is a potent tool to study control machineries regulating stem cell biology. In order to reconstruct the complexity and interplay of the stem cell niche and its components it is, however, necessary to utilise 3D culture approaches [225]. Hence, there have been numerous strategies to create 3D biomaterial matrices with a variety of structural, chemical and mechanical properties as artificial growth environments for cells each with its advantages and disadvantages [239]. Since the investigation of niche components and their effect on stem cell fate is a complex undertaking, 3D high-throughput approaches similar to 2D ECM protein arrays are being developed [240–242]. These could be produced by printing or liquid-dispensing technologies [263]. For example, 3D PEG gel arrays were prepared to study the effect of gel degradation and cell adhesion ligand concentration [264]. Once ideal biomaterial candidates have been identified, it is essential to examine their *in vivo* performance by implantation into hosts, some of which have been reviewed elsewhere [239].

Biomaterials have also been utilised to study the interplay between stem cells and support cells such as vascular cells, neural cells and stromal cells [225]. Electropatterning of living cells within PEG hydrogels resulted in the deposition of multicellular aggregates of known size and shape, which was shown to regulate biosynthesis of chondrocytes by increase of sulphated GAG in larger cell aggregates [245]. Moreover, 3D angiogenesis was studied combining microfluidics with gel patterning [246]. Gradients are known to be crucial for the regulation of dynamic processes during development and during tissue regeneration and homeostasis. Hence, the development of biomaterial gradients has found wide interest in regenerative medicine [225, 247–249]. It has been shown for bone/marrow [52, 250], prostate [251] and breast [252] that cell–cell interactions in those stem cell niches are influenced by paracrine hormone signalling [197].

Furthermore, cell–cell interactions have been studied in co-culture approaches, which on the other hand, do not allow for the identification of the particular function of soluble or immobilised molecules. In vivo, cytokines and growth factors are mostly immobilised to the ECM suggesting that tethered chemical factors are more stable and their signalling is longstanding. So has for example, the immobilisation of FGF2 to a synthetic polymer resulted in an increase in its potency and stability and subsequent enhancement in ECS proliferation and ERKI activation [253]. Additionally, the tethering of EGF to a biomaterial scaffold has demonstrated an increased effectiveness compared to its soluble equivalent [254]. These studies have demonstrated that investigating chemical cues individually is informative, but testing growth factors in a high-throughput manner on polymer arrays is more desirable for identifying pluripotent stem cell regulators [255].

4.4. ECM elasticity as stem cell fate regulator

ECM subjects cells to multiple physical signals including mechanical signals due to its stiffness [256–258]. Cells adhering to their surrounding employ contractile forces resulting in tensile stresses within their cytoskeleton [259]. The connection between these forces and ECM stiffness has a huge influence on cell migration [260, 261], proliferation [262] and apoptosis [263] (**Table 4**). Stem cells have shown to respond to mechanical properties of their surrounding microenvironment [214, 264]. Engler et al. [264] observed that when MSCs were cultured on collagen-coated polyacrylamide hydrogels with varying stiffness, cells differentiation without the use of supplemented medium. On soft gels mimicking brain tissue (0.1–1.0 kPa), MSCs underwent neurogenic differentiation. Stiffer gels resembling muscle tissue enabled myogenic development and very stiff gels (25–40 kPa) replicating the bone properties resulted in osteogenic differentiation of MSC [264]. Similar observations were made by other researchers for two-dimensional and three-dimensional culture [265–269]. MSC incorporated into collagen-GAG scaffolds with stiffness of 0.5, 1 and 1.5 kPa exhibited different chondrogenic differentiation. Softer gels allowed for chondrogenic and stiffer gels for osteogenic differentiation [269]. It is assumed that the substrate stiffness induces alterations in the activity of focal adhesion and causes remodelling, which triggers a cascade if signalling pathways enabling cell differentiation [270, 271]. In addition, integrins are thought to be a central cell structure for sensing mechanical stimulation [270, 271]. A limitation to these studies is the fact that different tissues may have similar stiffness, and hence, it might not be possible to direct stem cell fate by a single mechanical property of the surrounding. Instead, it is necessary to consider a more complex interplay of extrinsic and intrinsic factors influencing stem cell differentiation [197].

Cell type	Material	Mechanical cues	Result	References
ECM stiffness				
MSC	Polyacrylamide gel, collagen coated	Stiffness-0.1–1.0, 25–40 kPa	Lineage commitment according to substrate stiffness; softer gel-neuronal, stiffer gel-osteoblastic differentiation	[264]
ANSC	Interfacial hydrogel	Substrate moduli 0.01–10 kPa	Self-renewal, cell spreading and differentiation inhibited on soft gels (0.01 kPa); cell proliferation and neuronal differentiation maintained on ≥0.1 kPa; glial differentiation on gels 1–10 kPa	[339]
hMSC	Polyacrylamide	Marrow and	Cells were quiescent but	[35]

Cell type	Material	Mechanical cues	Result	References
	gel, collagen and fibronectin coated	adipose tissue 0.25 kPa	maintained multilineage potential,	
MSC	Collagen-glycosaminoglycan	0.5, 1 and 1.5 kPa	Softer gels triggered chondrogenic differentiation; stiffer gels resulted in osteogenic differentiation	[269]
Mechanical stimulation				
MSC	Protein-coated membranes	1, 5, 10, 15% cyclic uniaxial stretch	Myogenic differentiation for 5 and 10%; 1 and 15% failure of myogenic lineage	[340–342]
AMSC		10% uniaxial cyclic strain at 1 Hz for 7 days	Decreased expression of myogenic markers	[343]
MSC		Uniform biaxial strain	Enhanced osteogenesis and calcium deposition	[344–347]
AMSC		Pulsatile fluid flow	Enhanced osteogenesis	[348]
mESC		4–12% strain, 1 Hz, 24 h	Differentiation into vascular smooth muscle cells; increase in proliferation	[349]
hESC		Cyclic strain	promoting of self-renewal, inhibition of differentiation,	[350]
MSC	Agarose	Cyclic confined compression	Chondrogenic differentiation	[351]
MSC	Spheroids	IHP	Chondrogenesis	[352]
hBMSC	PCL	IHP, 270 kPa, 1 Hz, over 21 days	Enhanced metabolic activity, upregulation of osteogenic markers and calcium deposition	[295]
FCSC	*Ex vivo* cultured chick femurs	IHP, 270 kPa, 1 Hz, over 21 days	Enhanced mineralisation	[287]

MSC: mesenchymal stem cells; ANSC: adult neuronal stem cells, AMSC: adipose-derived mesenchymal stem cells, hBMSC: human bone marrow–derived mesenchymal stem cells, FCSC: fetal chick stem cells, kPa: kilo Pascal, Hz: hertz, IHP: intermittent hydrostatic pressure.

Table 4. Effect of ECM stiffness and mechanical stimulation on stem cell differentiation.

4.5. Biomechanical regulation of stem cell fate

The body's cells are constantly subjected to a variety of mechanical stimuli through muscle action, blood flow, gravity and other physical and physiological processes [197].To ensure tissue health and function, the interplay between cells and mechanical cues is essential. It is assumed that these mechanical factors are involved in diseases including osteoarthritis, osteoporosis and atherosclerosis [277]. Furthermore, there is mounting evidence that mechanical forces are crucial for development [278, 279] stem cell lineage commitment and fate decisions [197]. Despite early *in vivo* studies by Glückmann et al. [280], Fang and Hall [281], Hall and Herring [282] and Murray and Drachman [283] on the effect of mechanical factors on development little was known on how biomechanical cues affect gene expression and stem cell fate [280–283]. One difficulty in studying these interactions is the complexity by which cells sense mechanical signals. Mechanical loading of tissue can result in a variety of stresses, strains, pressures, fluid flows, osmotic pressures and electric charges on biological molecules [284]. These modifications of the cell's microenvironment result in structural changes of ECM proteins, as well as the activity of immobilised or soluble growth factors. For this reason it is challenging to differentiate the effect of direct mechanical forces *in vivo* from indirect mechanically driven effects in adhesive and paracrine signals and resulting changes such as cell shape [197, 285]. Nevertheless, mechanical forces are influencing cellular processes directly and the mechanism by which cells perceive external mechanical stimuli have been described by Liedtke and Kim [285]. Cellular responses to physical stimuli are not simply a reaction to the input stimuli, instead they are also linked to cytoskeletal changes, ECM interactions and the production of cellular forces [29, 264]. Stem cell response to controlled physical forces as well as the biomechanical mechanism and signalling pathways that direct stem cell lineage commitment are being investigated [286]. Several research groups have attempted to isolate the effect of applied mechanical stimuli such as fluid shear, strain and compression. **Table 4** summarises studies on the effect of externally applied mechanical stimuli and ECM stiffness on stem cell fate. These studies demonstrate that the effect of physical stimulation is dependent on stem cell source, type and state of pre-differentiation. Dynamic mechanical compression, for example increases chondrogenic marker expression of BhMSC. However, ESC-derived EB exhibit downregulation of chondrogenic markers when subjected to the same stimulation regime [197]. It is also under investigation whether mechanical cues provided by bioreactors alone or in combination with other cues can induce lineage commitment in PSC [287, 288].

5. Bioreactor technologies promoting mechanical stimuli

As elucidated, earlier physical forces *in vivo* are essential cues during development and for regulating stem cell fate. Furthermore, it has been widely acknowledged that *in vitro* mechanical cues (e.g. stress, strain, shear, compression, hydrostatic pressure) are greatly influencing the cell morphology, cell adhesion, proliferation and gene regulation [214, 289]. In order to provide physiological relevant growth environments for cells and tissue-engineered constructs, various types of bioreactors have been developed [287, 290]. Bioreactors are devices that utilise mechanical forces to influence biological processes under closely controlled and monitored conditions [291, 292]. They offer biophysical cues encouraging cells to differentiate or/and produce ECM prior to implantation *in vivo* [293]. In general, bioreactors meet the following applications in regenerative medicine-1: they provide spatially uniform cell distribution, 2: they deliver physiological relevant concentrations of oxygen and carbon dioxide as well as nutrients in the culture medium, 3: they support mass transport to the core of the tissue engineered construct facilitating cell survival throughout tissue-engineered constructs, 4: they provide physical stimuli to regulate stem cell differentiation and proliferation, 5: they facilitate tissue development and accelerate construct maturation [287, 294–297]. Studies utilising mechanical forces for the direction of stem cell fate and tissue development are summarised in **Table 4.**

5.1. Bioreactor designs and their application for tissue regeneration

Many attempts in regenerative medicine employ traditional cell culture approaches. These, however, are associated with numerous disadvantages. To overcome limitations in cell culture techniques and to provide mechanical stimulus for tissue development, various types of custom-made and commercially available bioreactors have been manufactured [293] (**Figure 7**).

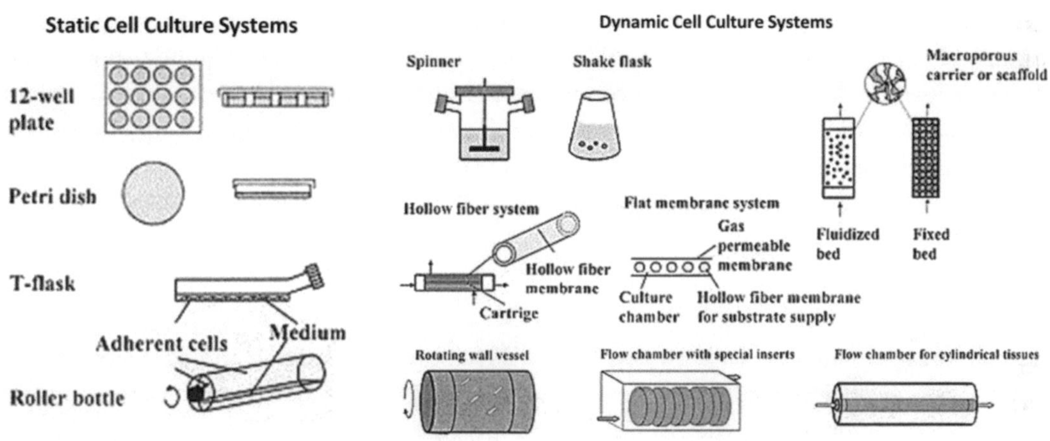

Figure 7. Static and dynamic cell culture systems. Traditional cell culture systems include well plate formats, petri dishes, T-flasks and roller bottles (left). To overcome disadvantages of static culture dynamic culture systems have been developed such as spinner flasks, rotating vessels, hollow fibre systems (right). Image adapted with the permission from Pörtner et al. 2005.

5.1.1. Rotating wall bioreactors

The rotating wall vessel bioreactor was originally developed by Schwarz and colleagues at Nasa's Johnson Space Centre to protect cell culture experiments from high forces during space shuttle take-off and landing [290, 298]. This bioreactor consists of a cylindrical chamber in which the scaffolds are free to move within the cell culture media. A permeable membrane is inserted in the vessel wall to allow for sufficient gas exchange. The vessel is rotated so that the downward gravitational force and the upward hydrodynamic force are balanced so that the scaffolds remain suspended in the media experiencing microgravity. This bioreactor has been utilised for regenerative studies on bone, cartilage, human ovarian surface epithelial cells among others [299–301].

5.1.2. Flow perfusion

A fluid pump is used to pump media through a sample chamber which is designed so that fluid flow is directed through the centre of cellular scaffolds [293]. Superior fluid transport was obtained for flow perfusion bioreactors when comparing spinner flasks, rotating wall bioreactors and flow perfusion bioreactors. For the comparison of the three bioreactor types, consistent flow rate, scaffold type and cell densities were utilised resulting in homogenous cell distribution throughout the scaffold only for the flow perfusion bioreactor [302, 303]. In order to obtain optimum experimental outcome, fluid flow rate needs to be adjusted to facilitate a balance between mass transfer of waste products and nutrients and to retain newly produced ECM [293].

5.1.3. Compression bioreactors

Compression bioreactors are commonly used for cartilage and bone tissue engineering and can be adjusted for both dynamic and static loading as studies have exhibited shown that dynamic loading is more suitable for certain tissues such as cartilage, but negative for others [291, 292]. A compression bioreactor commonly consists of a motor, a controlling mechanism providing various magnitudes of displacements and frequencies as well as a tool to provide linear motion [304]. Evenly distributed load is applied to cellular scaffolds via flat plates [305].

5.1.4. Strain bioreactors

Strain bioreactors are normally utilised for the development of tendon and ligament tissues, but also for bone, cartilage and cardiovascular tissues [293]. The bioreactor design is similar to compression bioreactors whereas instead of applying force through flat plates, scaffolds are hold in place via clamps when tensile force is employed.

5.1.5. Hydrostatic pressure bioreactor

Hydrostatic pressure bioreactors have been utilised for the development of bone and cartilage tissues among others. Professor El Haj's group has developed a novel hydrostatic bioreactor allowing for ease of handling and scale up of sample numbers that can be mechanically stimulated simultaneously [287, 288] (**Figure 8**). This bioreactor system is composed of a sealed aluminium chamber suitable for standard tissue culture well plates. By compressing incubator air and subsequently pumping it into the bioreactor chamber, hydrostatic pressure is applied to samples creating a gas–liquid interface between cell culture medium and the air. A temperature controller maintains the temperature of the compressed air at 37°C. The gas phase is removed from the bioreactor chamber, pumped back into the incubator and reused for the next stimulation cycle. A system accompanying software controls continuous and sinusoidal waveforms at various pressures (0–280 kPa) and frequencies (0.0001–2 Hz).

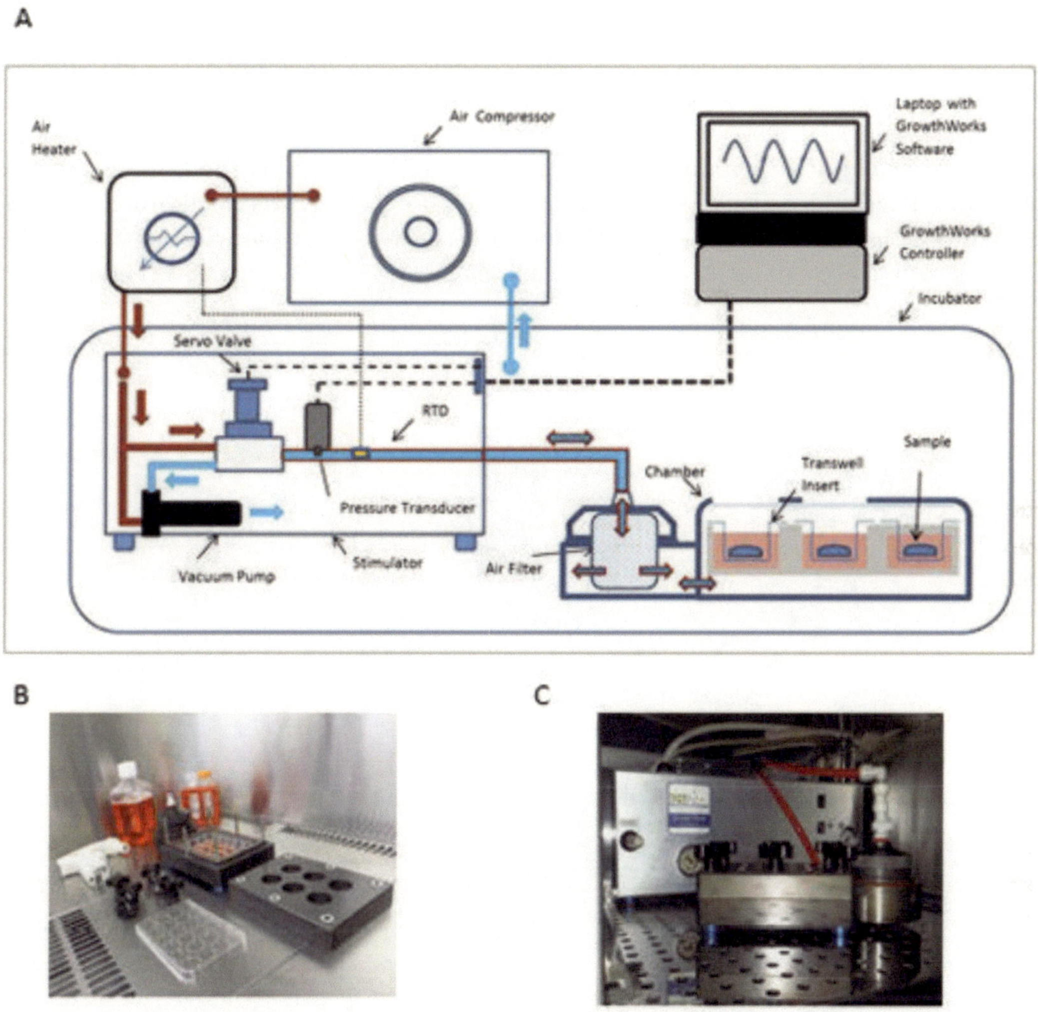

Figure 8. Hydrostatic force bioreactor. The bioreactor system (A) consists of a compressor, computer for monitoring of applied pressures and a bioreactor chamber (B) fitting standard tissue culture well plates, which is connected to a stimulator within a standard cell culture incubator (C). Image was adapted with the permission from Reinwald et al. 2015 [287].

Pluripotent stem cells are interesting cell types due to their capability to differentiate into numerous cell types. A limitation, however, is that hPSCs develop their functionality at later stages in development. At present, protocols yield in differentiated cells that match the embryonic stages of development. Protocols are needed that result in cells with adult-like functionality [65]. The investigation of signalling pathways controlling cell development has led to successful strategies to direct stem cell fate [306], and small molecules have been utilised for the creation of differentiation protocols [202, 203, 307, 308]. In addition, bioreactors have been successfully employed to direct stem cell fate [288, 295]. Recently, a collaborative approach between Manchester University and Keele University started to investigate the effect of hydrostatic pressure on the maturation of hESC-derived chondroprogenitors and hMSC. Chondroprogenitors were obtained following a defined 2 weeks of differentiation protocol (**Figure 6A**) [202, 203]. To enhance maturation of these cells and to differentiate hMSC towards chondrogenic lineage, they were cultured in growth factor supplemented medium as either spheroids or embedded in fibrin hydrogels and subjected to intermittent hydrostatic pressure at 270 kPa, 1Hz for 1 h daily. Results suggested that a combination of mechanical and chemical cues resulted in the production of matrix proteins collagen and GAG as well as the upregulation and maintained expression of chondrogenic markers Aggrecan and SOX9 (**Figure 9**).

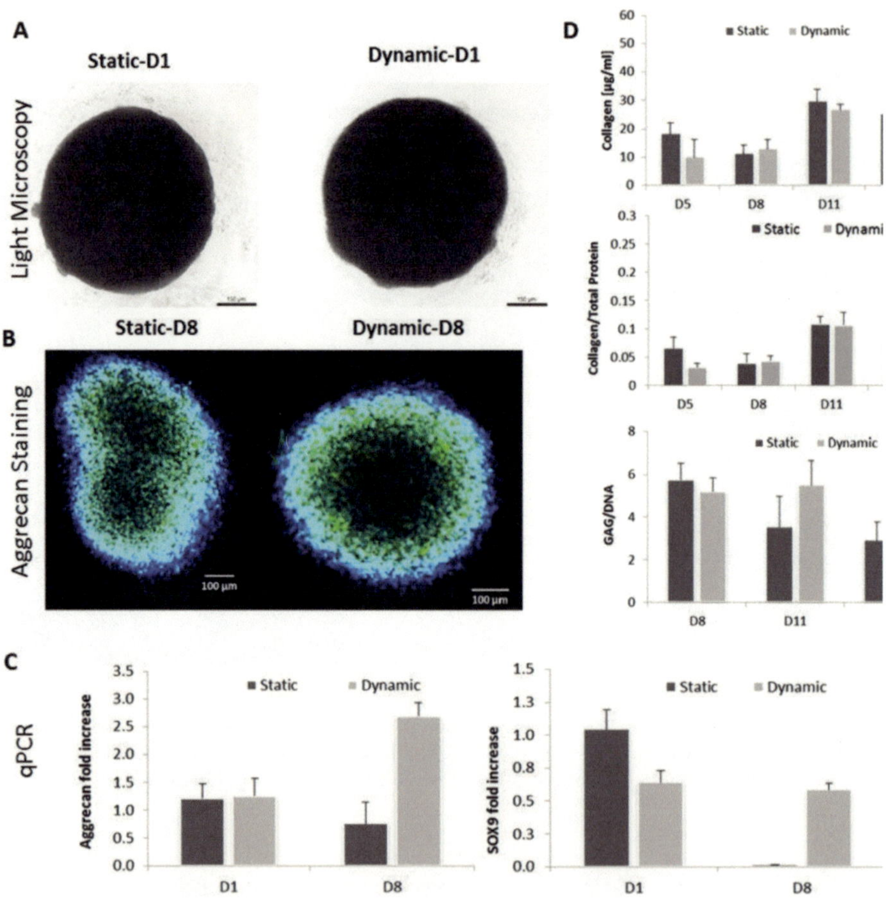

Figure 9. Stem cell differentiation was promoted by mechanical and chemical cues. hMSC spheroids (A) were cultured in growth factor supplemented medium either under static conditions or when subjected to hydrostatic pressure at 270 kPa, 1 Hz and 1 h daily over a period of 14 days. Positive immunocytochemistry staining for Aggrecan was observed for both culture conditions (B). To quantify changes in the expression of chondrogenic markers, Aggrecan and SOX9 quantitative polymerase chain reactor was performed revealing 2.5-fold Aggrecan upregulation and maintained SOX9 expression compared to static controls (n = 5). Increased levels of matrix proteins collagen and GAG and total protein were detected from biochemical assays (D) (n = 5). Scale bars = 150 µm (light microscopic images) and 100 µm (immunocytochemistry stains).

6. Conclusion

This chapter highlights the valuable role of multipotent and pluripotent stem cells and the importance of replicating their niche *in vitro* to develop novel regenerative therapies. Despite the ethical and safety concerns associated with PSC, they offer valuable differentiation and proliferation potential and could possibly offer a valuable cell source for clinical applications.

The stem cell niche influences stem cell fate in many ways, including mechanical support, elasticity, topography, biochemical signals, oxygen tension and cell communication. By artificially engineering the dynamic stem cell niche, stem cell fate can be directed. Advanced 3D biomaterials are being harnessed to mimic the *in vivo* environment by providing physical and chemical support and by allowing for cell–cell interactions. Bioreactor technologies are able to replicate the mechanical stimuli and mimic the physiologically relevant environment.

Whilst promising advances have been made in regenerative medicine, significant obstacles have been identified and these must be overcome before novel cell and tissue engineering therapies are clinically established.

Acknowledgements

This work was supported by the UKRMP Hub. The authors would also like to thank Mr. Diogo Mosqueira-Alves-Moreira-Silva, Mr. Luis Costa-Marques and Mr. Shah Mijaan Ali.

Author details

Yvonne Reinwald*, Jessica Bratt and Alicia El Haj

*Address all correspondence to: y.reinwald@keele.ac.uk

Institute for Science and Technology in Medicine, Keele University, Medical School, Guy Hilton Research Centre, UHNM, Stoke-on-Trent, United Kingdom

References

[1] Shohreh Mashayekhan, Maryam Hajiabbas and Ali Fallah (2013). Stem Cells in Tissue Engineering, Pluripotent Stem Cells, Dr. Deepa Bhartiya (Ed.), InTech, DOI: 10.5772/54371. Available from: http://www.intechopen.com/books/pluripotent-stem-cells/stem-cells-in-tissue-engineering

[2] Moore K.A.,Lemischka I.R. Stem cells and their niches. Science. 2006; 311:1880–1885.

[3] Fuchs E.,Tumbar T., Guasch G. Socializing with the neighbors: stem cells and their niche. Cell. 2004; 116:769–778.

[4] Morrison S.J.,Spradling A.C. Stem cells and niches: mechanisms that promote stem cell maintenance throughout life. Cell.2008; 132(4)598–611.

[5] Gattazzo F.,Urciuolo A., Bonaldo P. Extracellular matrix: a dynamic microenvironment for stem cell niche. Biochim Biophys Acta. 2014; 1840(8):2506–2519.

[6] Place E.S.,Evans N.D., Stevens M.M. Complexity in biomaterials for tissue engineering. Nat. Mater. 2009; 8:457–470.

[7] Stevens M.M.,George J.H. Exploring and engineering the cell surface interface. Science. 2005; 310:1135–8.

[8] Halder G.,Dupont S., Piccolo S. Transduction of mechanical and cytoskeletal cues by YAP and TAZ. Nat. Rev. Mol. Cell Biol. 2012; 13:591–600.

[9] Mammoto A.,Mammoto T., Ingber D.E. Mechanosensitive mechanisms in transcriptional regulation. J. Cell Sci. 2012; 125:3061–3073.

[10] Mammoto T.,Ingber D.E. Mechanical control of tissue and organ development. Development. 2010; 137:1407–1420.

[11] Lane S.W.,Williams D.A., Watt F.M. Modulating the stem cell niche for tissue regeneration. Nat. Biotechnol. 2014; 32:795–803.

[12] Reinwald Y.,Leonard K.H.L., Henstock J.R.,. Evaluation of the growth environment of a hydrostatic force bioreactor for preconditioning of tissue-engineered constructs. Tissue Eng. Part C Methods. 2015; 21(1):1–14.

[13] Schofield R. The relationship between the spleen colony-forming cell and the haemopoietic stem cell. Blood Cells. 1978; 4:7–25.

[14] Hall P.A.,Watt F.M. Stem cells: the generation and maintenance of cellular diversity. Development. 1989; 106:619–633.

[15] Watt F.M.,Hogan B.L. Out of Eden: stem cells and their niches. Science. 2000; 287:1427–1430.

[16] van Es J.H., et al. Dll1+ secretory progenitor cells revert to stem cells upon crypt damage. Nat. Cell Biol. 2012; 14:1099–1104.

[17] Katayama Y., et al. Signals from the sympathetic nervous system regulate hematopoietic stem cell regress from bone marrow. Cell. 2006; 124:407–421.

[18] Barker N., et al. Identification of stem cells in small intestine and colon by marker gene Lgr5. Nature. 2007; 449:1003–1007.

[19] Greenbaum A., et al. CXCL12 in early mesenchymal progenitors is required for haematopoietic stem-cell maintenance. Nature. 2013; 495:227–230.

[20] Ding L.,Morrison S.J. Haematopoietic stem cells and early lymphoid progenitors occupy distinct bone marrow niches. Nature. 2013; 495:231–235.

[21] Reya T.,Clevers H. Wnt signalling in stem cells and cancer. Nature. 2005; 434:843–850.

[22] Méndez-Ferrer S.,Lucas D., Battista M., Frenette P.S. Haematopoietic stem cell release is regulated by circadian oscillations. Nature. 2008; 452:442–447.

[23] Fujisaki J., et al. In vivo imaging of Treg cells providing immune privilege to the haematopoietic stem-cell niche. Nature. 2011; 474:216–219.

[24] Jaiswal S., et al. CD47 is upregulated on circulating hematopoietic stem cells and leukemia cells to avoid phagocytosis. Cell. 2009; 138:271–285.

[25] Williams D.A.,Rios M., Stephens C., Patel V.P. Fibronectin and VLA-4 in haematopoietic stem cell-microenvironment interactions. Nature. 1991; 352:438–441.

[26] Holst J., et al. Substrate elasticity provides mechanical signals for the expansion of hemopoietic stem and progenitor cells. Nat. Biotechnol. 2010; 28:1123–1128.

[27] Gilbert P.M., et al. Substrate elasticity regulates skeletal muscle stem cell self-renewal in culture. Science. 2010; 329:1078–1081.

[28] North T.E., et al. Hematopoietic stem cell development is dependent on blood flow. Cell. 2009; 137:736–748.

[29] McBeath R.,Pirone D.M., Nelson C.M., Bhadriraju K., Chen C.S. Cell shape, cytoskeletal tension, and RhoA regulate stem cell lineage commitment. Dev. Cell. 2004; 6:483–495.

[30] Kimura W.,Sadek H.A. The cardiac hypoxic niche: emerging role of hypoxic microenvironment in cardiac progenitors. Cardiovasc. Diagn. Ther. 2012; 2:278–289.

[31] Muscari C., et al. Priming adult stem cells by hypoxic pre-treatments for applications in regenerative medicine. J. Biomed. Sci. 2013; 20:63.

[32] Forristal C.E., et al. Pharmacologic stabilization of HIF-1alpha increases hematopoietic stem cell quiescence in vivo and accelerates blood recovery after severe irradiation. Blood. 2013; 121:759–769.

[33] Watt F.M., Driskell R.R. The therapeutic potential of stem cells. Philos. Trans. R. Soc. Lond. B Biol. Sci. 2010; 365:155–163.

[34] Trumpp A., Essers M., Wilson A. Awakening dormant haematopoietic stem cells. Nat. Rev. Immunol. 2010; 10:201–209.

[35] Kurtz A., Oh S. Age related changes of the extracellular matrix and stem cell maintenance. Prev. Med. Baltimore. 2012; 54:S50–S56 (Suppl.).

[36] Orford K.W., Scadden D.T. Deconstructing stem cell self-renewal: genetic insights into cell-cycle regulation. Nat. Rev. Genet. 2008; 9:115–128.

[37] Greco V., Guo S. Compartmentalized organization: a common and required feature of stem cell niches? Development. 2010; 137:1586–1594.

[38] Jones D.L., Wagers A.J. No place like home: anatomy and function of the stem cell niche. Nat. Rev. Mol. Cell Biol. 2008; 9:11–21.

[39] Li L., Clevers H. Coexistence of quiescent and active adult stem cells in mammals. Science. 2010; 327:542–545.

[40] Lander A.D., Kimble J., Clevers H., Fuchs E., Montarras D., Buckingham M., Calof A.L., Trumpp A., Oskarsson T. What does the concept of the stem cell niche really mean today? BMC Biol. 2012; 10:19.

[41] Wagers A.J. The stem cell niche in regenerative medicine. Cell Stem Cell. 2012; 10:362–369.

[42] Frantz C., Stewart K.M., Weaver V.M. The extracellular matrix at a glance. J. Cell Sci. 2011; 123:4195–4200.

[43] Brizzi M.F., Tarone G., Defilippi P. Extracellular matrix, integrins, and growth factors as tailors of the stem cell niche. Curr. Opin. Cell Biol. 2012; 24:645–651.

[44] Jarvelainen H., Sainio A., Koulu M., Wight T.N., Penttinen R. Extracellular matrix molecules: potential targets in pharmacotherapy. Pharmacol. Rev. 2009; 61:198–223.

[45] Schaefer L., Schaefer R.M. Proteoglycans: from structural compounds to signaling molecules. Cell Tissue Res. 2010; 339:237–246.

[46] Rozario T., DeSimone D.W. The extracellular matrix in development and orphogenesis: a dynamic view. Dev. Biol. 2010; 341:126–140.

[47] Wise S.G., Weiss A.S. Tropoelastin. Int. J. Biochem. Cell Biol. 2009; 41:494–497.

[48] Ou K.L., Hosseinkhani H. Development of 3D in vitro technology for medical applications. Int. J. Mol. Sci. 2014; 15:17938–17962.

[49] Kular J.K., Basu S., Sharma R.I. The extracellular matrix: Structure, composition, age-related differences, tools for analysis and applications for tissue engineering. J. Tissue Eng. 2014; 5:1–17.

[50] Bosman F.T., Stamenkovic I. Functional structure and composition of the extracellular matrix. J. Pathol. 2003; 200:423–428.

[51] Zhang J., Niu C., Ye L., Huang H., He X., Tong W.G., Ross J., Haug J., Johnson T., Feng J.Q., Harris S., Wiedemann L.M., Mishina Y., Li L. Identification of the haematopoietic stem cell niche and control of the niche size. Nature. 2003; 425:836–841.

[52] Calvi L.M., Adams G.B., Weibrecht K.W., Weber J.M., Olson D.P., Knight M.C., Martin R.P., Schipani E., Divieti P., Bringhurst F.R., Milner L.A., Kronenberg H.M., Scadden D.T. Osteoblastic cells regulate the haematopoietic stem cell niche. Nature. 2003; 425:841–846.

[53] Yoshihara H., Arai F., Hosokawa K., Hagiwara T., Takubo K., Nakamura Y., Gomei Y., Iwasaki H., Matsuoka S., Miyamoto K., Miyazaki H., Takahashi T., Suda T. Thrombopoietin/MPL signaling regulates hematopoietic stem cell quiescence and interaction with the osteoblastic niche. Cell Stem Cell. 2007; 1:685–697.

[54] Arai F., Hirao A., Ohmura M., Sato H. Tie2/angiopoietin-1 signaling regulates hematopoietic stem cell quiescence in the bone marrow niche. Cell. 2004; 118:149–161.

[55] Weber J.M., Calvi L.M. Notch signaling and the bone marrow hematopoietic stem cell niche. Bone. 2010; 46:281–285.

[56] Nakamura-Ishizu A., Suda T., Hematopoietic stem cell niche: an interplay among a repertoire of multiple functional niches. Biochim. Biophys. Acta. 2013; 1830:2404–2409.

[57] Kunisaki Y., Frenette P.S. The secrets of the bone marrow niche: enigmatic niche brings challenge for HSC expansion. Nat. Med. 2012; 18:864–865.

[58] Shen Y., Nilsson S.K. Bone, microenvironment and hematopoiesis. Curr. Opin. Hematol. 2012; 19:250–255.

[59] Smith J.N.P., Calvi L.M. Concise review: current concepts in bone marrow microenvironmental regulation of hematopoietic stem and progenitor cells. Stem Cells. 2013; 31:1044–1050.

[60] Rompolas P., Mesa K.R., Greco V. Spatial organization within a niche as a determinant of stem-cell fate. Nature. 2013; 502:513–518.

[61] Tian H., Biehs B., Warming S., Leong K.G., Rangell L., Klein O.D., de Sauvage F.J. A reserve stem cell population in small intestine renders Lgr5-positive cells dispensable. Nature. 2011; 478:255–259.

[62] Kilian K.A., Bugarija B., Lahn B.T., Mrksich M. Geometric cues for directing the differentiation of mesenchymal stem cells. Proc. Natl. Acad. Sci. USA. 2010; 107:4872–4877.

[63] Mizuno H., Tobita M., Uysal A.C. Concise review: adipose-derived stem cells as a novel tool for future regenerative medicine. Stem Cells. 2012; 30:804–810.

[64] Sykova E., Forostyak S. Stem cells in regenerative medicine. Laser Ther. 2013; 22(2):87–92.

[65] Tabar V., Studer L. Pluripotent stem cells in regenerative medicine: challenges and recent progress. Nat. Rev. Genet. 2014; 15(2):82–92.

[66] Rippon H.J., Bishop A.E. Embryonic stem cells. Cell Prolif. 2004; 37:23–34.

[67] Lavker R.R., Sun T.T. Epidermal stem cells: properties, markers and location. Proc. Natl. Acad. Sci. USA. 2000; 97:13473.

[68] Uchida N., Buck D.W., He D., Reitsma M.J., Masek M., Phan T.V., Tsukamoto A.S., Gage F.H., Weissman I.L. Direct isolation of human central nervous system stem cells. Proc. Natl. Acad. Sci. USA. 2000; 97:14720.

[69] Vessey C.J., de la Hall P.M. Hepatic stem cells: a review. Pathology (Phila). 2001; 33:130.

[70] Wagers A.J., Christensen J.L., Weissman I.L. Cell fate determination from stem cells. Gene Ther. 2002; 9:606.

[71] McCarthy H.S., Roberts S. A histological comparison of the repair tissue formed when using either Chondrogide® or periosteum during autologous chondrocyte implantation. Osteoarthr. Cartil. 2013; 21:2048–2057.

[72] Bailey A., Goodstone N., Roberts S., Hughes J., Roberts S., van Niekerk L., Richardson J., Rees D. Rehabilitation after oswestry autologous-chondrocyte implantation: the OsCell protocol. J. Sport Rehabil. 2003; 12:104–118.

[73] Roberts S., Menage J., Sandell L.J., Evans E.H., Richardson J.B. Immunohistochemical study of collagen types I and II and procollagen IIA in human cartilage repair tissue following autologous chondrocyte implantation. Knee. 2009; 16:398–404.

[74] Kriks S., Shim J.W., Piao J., Ganat Y.M., Wakeman D.R., Xie Z., Carrillo-Reid L., Auyeung G., Antonacci C, Buch A., Yang L., Beal M.F., Surmeier D.J., Kordower J.H., Tabar V., Studer L., . Dopamine neurons derived from human ES cells efficiently engraft in animal models of Parkinson's disease. Nature. 2011; 480:547–551.

[75] Ma L., Hu B., Liu Y., Vermilyea S.C., Liu H., Gao L., Sun Y., Zhang X., Zhang S.C. Human embryonic stem cell-derived GABA neurons correct locomotion deficits in quinolinic acid-lesioned mice. Cell Stem Cell. 2012; 10:455–464.

[76] Wang S., Bates J., Li X., Schanz S., Chandler-Militello D., Levine C., Maherali N., Studer L., Hochedlinger K., Windrem M., Goldman S.A.Human iPSC-derived oligodendrocyte progenitor cells can myelinate and rescue a mouse model of congenital hypomyelination. Cell Stem Cell. 2013; 12:252–264.

[77] Shiba Y., Fernandes S., Zhu W., Filice D., Veronica Muskheli V., Kim J., Palpant N.J., Gantz J., White Moyes K., Reinecke H., Van Biber B., Dardas T., Mignone J.L., Izawa A., Hanna R., Viswanathan M., Gold J.D., Kotlikoff M.I., Sarvazyan N., Kay M.W., Murry C.E., Laflamme M.A. Human ES-cell-derived cardiomyocytes electrically couple and suppress arrhythmias in injured hearts. Nature. 2012; 489:322–325.

[78] Smith A., Heath J.K., Donaldson D.D., Wong G.G., Moreau J., Stahl M., Rogers, D. Inhibition of pluripotential embryonic stem cell differentiation by purified polypeptides. Nature. 1988; 336:688.

[79] Williams R.L., Hilton D.J., Pease S., Willson T.A., Stewart C.I., Gearing D.P., Wagner E.F., Metcalf D., Nicola N.A., Gough N.M. Myeloid leukaemia inhibitory factor maintains the developmental potential of embryonic stem cells. Nature. 1988; 36:684.

[80] Thomson J.A., Itskovitz-Eldor J., Shapiro S.S., Waknitz M.A., Swiergiel J.J., Marshall V.S., Jones J.M. Embryonic stem cell lines derived from human blastocysts. Science. 1998; 282(5391):1145–1147.

[81] Hofstetter C.P., Schwarz E.J., Hess D., Widenfalk J., El Manira A., Prockop D.J., Olson L. Marrow stromal cells form guiding strands in injured spinal cord and promote recovery. Proc. Natl. Acad. Sci. USA. 2002; 99(4):2199–2204.

[82] Xu L., Yan J., Chen D., Welsh A.M., Hazel T., Johe K., Hatfield G., Koliatsos V.E. Human neural stem cell grafts ameliorate motor neuron disease in SOD-1 transgenic rats. Transplantation. 2006; 82(7):865–875.

[83] McDonald J.W., Liu X.Z., Qu Y., Liu S., Mickey S.K., Turetsky D., Gottlieb D.I., Choi D.W. Transplanted embryonic stem cells survive, differentiate and promote recovery in injured rat spinal cord. Nat. Med. 1999; 5(12):1410–1412.

[84] Nistor G.I., Totoiu M.O., Haque N., Carpenter M.K., Keirstead H.S. Human embryonic stem cells differentiate into oligodendrocytes in high purity and myelinate after spinal cord transplantation. Glia. 2005; 49(3):385–396.

[85] Seminatore C., Polentes J., Ellman D., Kozubenko N., Itier V., Tine S., Tritschler L., Brenot M., Guidou E., Blondeau J., Lhuillier M., Bugi A., Aubry L., Jendelova P., Sykova E., Perrier A.L., Finsen B., Onteniente B. The postischemic environment differentially impacts teratoma or tumor formation after transplantation of human embryonic stem cell-derived neural progenitors. Stroke J. Cereb. Circ. 2010; 41 (1), 153–159.

[86] Toh W.S., Lee, E.H, Guo X.M., Chan J.K., Yeow C.H., Choo A.B., Cao T. Cartilage repair using hyaluronan hydrogel-encapsulated human embryonic stem cell-derived chondrogenic cells. Biomaterials. 2010; 31(27):6968–6980.

[87] Kehat I., Khimovich L., Caspi O., Gepstein A., Shofti R., Arbel G., Huber I., Satin J., Itskovitz-Eldor J., Gepstein L. Electromechanical integration of cardiomyocytes derived from human embryonic stem cells. Nat. Biotechnol. 2004; 22(10):1282–1289 .

[88] Hwang N.S., Varghese S., Lee H.J., Zhang Z., Ye Z., Bae J., Cheng L., Elisseeff J. In vivo commitment and functional tissue regeneration using human embryonic stem cell-derived mesenchymal cells. Proc. Natl. Acad. Sci. USA. 2008; 105(52):20641–20646.

[89] Lee E.J., Xu L., Kim G.H., Kang S.K., Lee S.W., Park S.H., Kim S., Choi T.H., Kim H.S. Regeneration of peripheral nerves by transplanted sphere of human mesenchymal stem cells derived from embryonic stem cells. Biomaterials. 2012; 33:7039–7046.

[90] Horiguchi S., Takahashi J., Kishi Y., Morizane A., Okamoto Y., Koyanagi M., Tsuji M., Tashiro K., Honjo T., Fujii S., Hashimoto N. Neural precursor cells derived from human embryonic brain retain regional specificity. J. Neurosci. Res. 2004; 75(6):817–824.

[91] Burnstein R.M., Foltynie T., He X., Menon D.K., Svendsen C.N., Caldwell M.A. Differentiation and migration of long term expanded human neural progenitors in a

partial lesion model of Parkinson's disease. Intl. J. Biochem. Cell Biol. 2004; 36(4): 702–713.

[92] Brundin P., Strecker R.E., Gage F.H., Lindvall O., Bjorklund A. Intracerebral transplantation of dopamine neurons: understanding the functional role of the mesolimbocortical dopamine system and developing a therapy for Parkinson's disease. Ann. NY Acad. Sci. 1988; 537:148–160.

[93] Clarke D.J., Brundin P., Strecker R.E., Nilsson O.G., Bjorklund A., Lindvall O. Experimental brain research. Exp. Cereb. 1988; 73(1):115–126.

[94] Lindvall O., Brundin P., Widner H., Rehncrona S., Gustavii B., Frackowiak R., Leenders, K.L, Sawle G., Rothwell J.C., Marsden C.D., Björklund A. Grafts of fetal dopamine neurons survive and improve motor function in Parkinson's disease. Science. 1990; 247(4942):574–577.

[95] Pollock K., Stroemer P., Patel S., Stevanato L., Hope A., Miljan E., Dong Z., Hodges H., Price J., Sinden J.D. A conditionally immortal clonal stem cell line from human cortical neuroepithelium for the treatment of ischemic stroke. Exp. Neurol. 2006; 199(1):143–155.

[96] Takahashi K., Yamanaka S. Induction of pluripotent stem cells from mouse embryonic and adult fibroblast cultures by defined factors. Cell. 2006; 126:663–676.

[97] Maherali N., Sridharan R., Xie W., Utikal J., Eminli S., Arnold K., Stadtfeld M., Yachechko R., Tchieu J., Jaenisch R., Plath K., Hochedlinger K. Directly reprogrammed fibroblasts show global epigenetic remodeling and widespread tissue contribution. Cell Stem Cell. 2007; 1:55–70.

[98] Okita K., Ichisaka T., Yamanaka S. Generation of germline-competent induced pluripotent stem cells. Nature. 2007; 448:313–317.

[99] Wernig M., vMeissner, A., Foreman, R., Brambrink, T., Ku, M., Hochedlinger, K., Bradley E., Bernstein, B.E., Jaenisch, R. In vitro reprogramming of fibroblasts into a pluripotent ES-cell-like state. Nature. 2007; 448:318–324.

[100] Okita K., Nakagawa M., Hyenjong H., Ichisaka T., Yamanaka S. Generation of mouse induced pluripotent stem cells without viral vectors. Science. 2008; 322:949–953.

[101] Yu J., Hu K., Smuga-Otto K., Tian S., Stewart R., Slukvin I.I., Thomson J.A. Human induced pluripotent stem cells free of vector and transgene sequences. Science. 2009; 324:797–801.

[102] Stadtfeld M., Nagaya M., Utikal J., Weir G., Hochedlinger K. Induced pluripotent stem cells generated without viral integration. Science. 2008; 322:945–949.

[103] Soldner F., Hockemeyer D., Beard C., Gao Q., Bell G.W., Cook E.G., Hargus G., Blak A., Cooper O., Mitalipova M., Isacson O., Jaenisch R. Parkinson's disease patient-

derived induced pluripotent stem cells free of viral reprogramming factors. Cell. 2009; 136:964–977.

[104] Lowry W.E., Richter L., Yachechko R., Pyle A.D., Tchieu J., Sridharan R., Clark A.T., Plath K. Generation of human induced pluripotent stem cells from dermal fibroblasts. Proc. Natl. Acad. Sci. USA. 2008; 105:2883–2888.

[105] Loh Y.H., Agarwal S., Park I., Urbach A., Huo H., Heffner G.C., Kim K., Miller J.D., Ng K., Daley G.Q. Generation of induced pluripotent stem cells from human blood. Blood. 2009; 113:5476–5479.

[106] Kunisato A., Wakatsuki M., Shinba H., Ota T., Ishida I., Nagao K. Direct generation of induced pluripotent stem cells from human non-mobilized blood. Stem Cells Dev. 2011; 20:159–168.

[107] Aasen T., Raya A., Barrero M.J., Garreta E., Consiglio A., Gonzalez F., Vassena R., Bilić J., Pekarik V., Tiscornia G., Edel M., Boué S., Izpisúa Belmonte J.C. Efficient and rapid generation of induced pluripotent stem cells from human keratinocytes. Nat. Biotechnol. 2008; 26:1276–1284.

[108] Wu S.M., Hochedlinger K. Harnessing the potential of induced pluripotent stem cells for regenerative medicine. Nat. Cell Biol. 2011; 13(5):497–505.

[109] Hanna J., Wernig M., Markoulaki S., Sun C.W., Meissner A., Cassady J.P., Beard C., Brambrink T., Wu L.C., Townes T.M., Jaenisch R. Treatment of sickle cell anemia mouse model with iPS cells generated from autologous skin. Science. 2007; 318:1920–1923.

[110] Wernig M., Zhao J.P., Pruszak J., Hedlund E., Fu D., Soldner F., Broccoli V., Constantine-Paton M., Isacson O., Jaenisch R. Neurons derived from reprogrammed fibroblasts functionally integrate into the fetal brain and improve symptoms of rats with Parkinson's disease. Proc. Natl. Acad. Sci. USA. 2008; 105:5856–5861.

[111] Zhang J., Wilson G.F., Soerens A.G., Koonce C.H., Yu J., Palecek S.P., Thomson J.A., Kamp T.J. Functional cardiomyocytes derived from human induced pluripotent stem cells. Circ. Res. 2009; 104:e30–e41.

[112] Nelson T.J., Martinez-Fernandez A., Yamada S., Perez-Terzic C., Ikeda Y., Terzic A. Repair of acute myocardial infarction by human stemness factors induced pluripotent stem cells. Circulation. 2009; 120:408–416.

[113] Yu J., Vodyanik M.A., Smuga-Otto K., Antosiewicz-Bourget J., Frane J.L., Tian S., Nie J., Jonsdottir G.A., Ruotti V., Stewart R., Slukvin I.I., Thomson J.A. Induced pluripotent stem cell lines derived from human somatic cells. Science. 2007; 318:1917–1920.

[114] Park I.H., Zhao R., West J.A., Yabuuchi A., Huo H., Ince T.A., Lerou P.H., Lensch M.W., Daley G.Q. Reprogramming of human somatic cells to pluripotency with defined factors. Nature. 2008; 451:141–146.

[115] Guenther M.G., Frmapton G.M., Soldner F., Hockemeyer D., Mitalipova M., Jaenisch R., Young R.A. Chromatin structure and gene expression programs of human embryonic and induced pluripotent stem cells. Cell Stem Cell. 2010; 7:249–257.

[116] Chin M.H., Mason M.J., Xie W., Volinia S., Singer M., Peterson C., Ambartsumyan G., Aimiuwu O., Richter L., Zhang J., Khvorostov I., Ott V., Grunstein M., Lavon N., Benvenisty N., Croce C.M., Clark A.T., Baxter T., Pyle A.D., Teitell M.A., Pelegrini M., Plath K., Lowry W.E. Induced pluripotent stem cells and embryonic stem cells are distinguished by gene expression signatures. Cell Stem Cell. 2009; 5:111–123.

[117] Doi A., Park I.H., Wen B., Murakami P., Aryee M.J., Irizarry R., Herb B., Ladd-Acosta C., Rho J., Loewer S., Miller J., Schlaeger T., Daley G.Q., Feinberg A.P. Differential methylation of tissue- and cancer-specific CpG island shores distinguishes human induced pluripotent stem cells, embryonic stem cells and fibroblasts. Nat. Genet. 2009; 41:1350–1353.

[118] Deng J., Shoemaker R., Xie B., Gore A., LeProust E.M., Antosiewicz-Bourget J., Egli D., Maherali N., Park I.H., Yu J., Daley G.Q., Eggan K., Hochedlinger K., Thomson J., Wang W., Gao Y., Zhang K. Targeted bisulfite sequencing reveals changes in DNA methylation associated with nuclear reprogramming. Nat. Biotechnol. 2009; 27:353–360.

[119] Hu B.Y., Weick J.P., Yu J., Ma L.X., Zhang X.Q., Thomson J.A., Zhang S.C. Neural differentiation of human induced pluripotent stem cells follows developmental principles but with variable potency. Proc. Natl. Acad. Sci. USA. 2010; 107:4335–4340.

[120] Feng Q., Lu S.J., Klimanskaya I., Gomes I., Kim D., Chung Y., Honig G.R., Kim K.S., Lanza R. Hemangioblastic derivatives from human induced pluripotent stem cells exhibit limited expansion and early senescence. Stem Cells. 2010; 28:704–712.

[121] Miura K., Okada Y., Aoi T., Okada A., Takahashi K., Okita K., Nakagawa M., Koyanagi M., Tanabe K., Ohnuki M., Ogawa D., Ikeda E., Okano H., Yamanaka S. Variation in the safety of induced pluripotent stem cell lines. Nat. Biotechnol. 2009; 27:743–745.

[122] Stadtfeld M., Apostolou E., Akutsu H., Fukuda A., Follett P., Natesan S., Kono T., Shioda T., Hochedlinger K. Aberrant silencing of imprinted genes on chromosome 12qF1 in mouse induced pluripotent stem cells. Nature. 2010; 465:175–181.

[123] Polo J.M., Liu S., Figueroa M.E., Kulalert W., Eminli S., Tan K.Y., Apostolou E., Stadtfeld M., Li Y., Shioda T., Natesan S., Wagers A.J., Melnick A., Evans T., Hochedlinger K. Cell type of origin influences the molecular and functional properties of mouse induced pluripotent stem cells. Nat. Biotechnol. 2010; 28:848–855.

[124] Laurent, L.C., Ulitsky, I., Slavin, I., Tran, H., Schork, A., Morey, R., Lynch, C., Harness, J.V., Lee, S., Barrero, M.J., Ku, S., Martynova, M., Semechkin, R., Galat, V., Gottesfeld, J., Izpisua Belmonte, J.C., Murry, C., Keirstead, H.S., Park, H.S., Schmidt, U., Laslett, A.L., Muller, F.J., Nievergelt, C.M., Shamir, R., Loring, J.F. Dynamic changes in the copy number of pluripotency and cell proliferation genes in human ESCs and iPSCs during reprogramming and time in culture. Cell Stem Cell. 2011; 8:106–118.

[125] Hussein S.M., Batada N.N., Vuoristo S., Ching R.W., Autio R., Närvä E., Ng S., Sourour M., Hämäläinen R., Olsson C., Lundin K., Mikkola M., Trokovic R., Peitz M., Brüstle O., Bazett-Jones D.P., Alitalo K., Lahesmaa R., Nagy A., Otonkoski T. Copy number variation and selection during reprogramming to pluripotency. Nature. 2011; 471:58–62.

[126] Mayshar Y., Ben-David U., Lavon N., Biancotti J.C., Yakir B., Clark A.T., Plath K., Lowry W.E., Benvenisty N. Identification and classification of chromosomal aberrations in human induced pluripotent stem cells. Cell Stem Cell. 2010; 7:521–531.

[127] Gore A., Li Z., Fung H.L., Young J.E., Agarwal S., Antosiewicz-Bourget J., Canto I., Giorgetti A., Israel M.A., Kiskinis E., Lee J.H., Loh Y.H., Manos P.D., Montserrat N., Panopoulos A.D., Ruiz S., Wilbert M.L., Yu J., Kirkness E.F., Izpisua Belmonte J.C., Rossi D.J., Thomson J.A., Eggan K., Daley G.Q., Goldstein L.S., Zhang K. Somatic coding mutations in human induced pluripotent stem cells. Nature. 2011; 471:63–67.

[128] Bock C., Kiskinis E., Verstappen G., Gu H., Boulting G., Smith Z.D., Ziller M., Croft G.F., Amoroso M.W., Oakley D.H., Gnirke A., Eggan K., Meissner A. Reference Maps of human ES and iPS cell variation enable high-throughput characterization of pluripotent cell lines. Cell. 2011; 144:439–452.

[129] Yang L., Soonpaa M.H., Adler E.D., Roepke T.K., Kattman S.J., Kennedy M., Henckaerts E., Bonham K., Abbott G.W., Linden R.M., Field L.J., Keller G.M. Human cardiovascular progenitor cells develop from a KDR+ embryonic-stem-cell derived population. Nature. 2008; 453:524–528.

[130] van Laake L.W., Passier R., Monshouwer-Kloots J., Verkleij A.J., Lips D.J., Freund C., den Ouden K., Ward-van Oostwaard D., Korving J., Tertoolen L.G., van Echteld C.J., Doevendans P.A., Mummery C.L. Human embryonic stem cell-derived cardiomyocytes survive and mature in the mouse heart and transiently improve function after myocardial infarction. Stem Cell Res. 2007; 1:9–24.

[131] Laflamme M.A., Chen K., Naumova A.V., Muskheli V., Fugate J.A., Dupras S.K., Hassanipour M., Police S., O'Sullivan, C., Collins L., Chen Y., Minami E., Gill E.A., Ueno S., Yuan C., Gold J., Murry C.E. Cardiomyocytes derived from human embryonic stem cells in pro-survival factors enhance function of infarcted rat hearts. Nat. Biotechnol. 2007; 25:1015–1024.

[132] Kattman S.J., Witty A.D., Gagliardi M., Dubois N.C., Niapour M., Hotta A., Ellis J., Keller G. Stage-specific optimization of activin/nodal and bmp signaling promotes cardiac differentiation of mouse and human pluripotent stem cell lines. Cell Stem Cell. 2011; 8:228–240.

[133] Irion S., Clarke R.L., Luche H., Kim I., Morrison S.J., Fehling H.J., Keller G.M. Temporal specification of blood progenitors from mouse embryonic stem cells and induced pluripotent stem cells. Development. 2010; 137:2829–2839.

[134] Swistowski A., Peng J., Liu Q., Mali P., Rao M.S., Cheng L., Zeng, X. Efficient generation of functional dopaminergic neurons from human induced pluripotent stem cells under defined conditions. Stem Cells. 2010; 28:1893904.

[135] Nizzardo M., Simone C., Falcone M., Locatelli F., Riboldi G., Comi G.P., Corti S. Human motor neuron generation from embryonic stem cells and induced pluripotent stem cells. Cell. Mol. Life Sci. 2010; 67:3837–3847.

[136] Ma YD, Lugus JJ, Park C, Choi K. Differentiation of Mouse Embryonic Stem Cells into Blood. *Current protocols in stem cell biology*. 2008;CHAPTER:Unit-1F.4. doi: 10.1002/9780470151808.sc01f04s6.

[137] D'Amour, K.A, Agulnick A.D., Eliazer S., Kelly O.G., Kroon E., Baetge E.E. Efficient differentiation of human embryonic stem cells to definitive endoderm. Nat. Biotechnol. 2005; 23:1534–1541.

[138] Kroon E., Martinson L.A., Kadoya K., Bang A.G., Kelly O.G., Eliazer S., Young H., Richardson M., Smart N.G., Cunningham J., Agulnick A.D., D'Amour K.A., Carpenter M.K., Baetge E.E. Pancreatic endoderm derived from human embryonic stem cells generates glucose responsive insulin-secreting cells in vivo. Nat. Biotechnol. 2008; 26:443–452.

[139] Strauss S. Geron trial resumes, but standards for stem cell trials remain elusive. Nat. Biotechnol. 2010; 28:989–990.

[140] Lister R., Kida Y.S., Hawkins R.D., Nery J.R., Hon G., Antosiewicz-Bourget J., O'Malley, R., Castanon R., Klugman S., Downes M., Yu., R., Stewart, R., Ren, B., Thomson, J.A., Evans, R.M., Ecker, J.R. Hotspots of aberrant epigenomic reprogramming in human induced pluripotent stem cells. Nature. 2011; 471:68–73.

[141] Warren L., Manos P.D., Ahfeldt T., Loh Y.H., Li H., Lau F., Ebina W., Mandal P.K., Smith Z.D., Meissner A., Daley G.Q., Brack A.S., Collins J.J., Cowan C., Schlaeger T.M., Rossi D.J. Highly efficient reprogramming to pluripotency and directed differentiation of human cells with synthetic modified mRNA. Cell Stem Cell. 2010; 7:618–630.

[142] Fusaki N., Ban H., Nishiyama A., Saeki K., Hasegawa M. Efficient induction of transgene-free human pluripotent stem cells using a vector based on Sendai virus, an RNA virus that does not integrate into the host genome. Proc. Jpn. Acad. Ser. B Phys. Biol. Sci. 2009; 85:348–362.

[143] Polo J.M., Liu S., Figueroa M.E., Kulalert W., Eminli S., Tan K.Y., Apostolou E., Stadtfeld M., Li Y., Shioda T., Natesan S., Wagers A.J., Melnick A., Evans T., Hochedlinger K. Cell type of origin influences the molecular and functional properties of mouse induced pluripotent stem cells. Nat. Biotechnol. 2010; 28:848–855.

[144] Kim K., Zhao R., Doi A., Ng K., Unternaehrer J., Cahan P., Huo H., Loh Y.H., Aryee M.J., Lensch M.W., Li H., Collins J.J., Feinberg A.P., Daley G.Q. Donor cell type can

influence the epigenome and differentiation potential of human induced pluripotent stem cells. Nat. Biotechnol. 2011; 29:1117–1119.

[145] Kim K., Doi A., Wen B., Ng K., Zhao R., Cahan P., Kim J., Aryee M.J., Ji H., Ehrlich L.I.R., Yabuuchi A., Takeuchi A., Cunniff K.C., Hongguang H., Mckinney-Freeman S., Naveiras O., Yoon T.J., Irizarry R.A., Jung N., Seita J., Hanna J., Murakami P., Jaenisch R., Weissleder R., Orkin S.H. Epigenetic memory in induced pluripotent stem cells. Nature. 2010; 467:285–290.

[146] Bar-Nur O., Russ H.A., Efrat S., Benvenisty N. Epigenetic memory and preferential lineage-specific differentiation in induced pluripotent stem cells derived from human pancreatic islet β cells. Cell Stem Cell. 2011; 9:17–23.

[147] Cyranoski D. Stem cells cruise to clinic. Nature. 2013; 494:413.

[148] Friedenstein A.J., Gorskaja J.F., Kulagina N.N. Fibroblast precursors in normal and irradiated mouse hematopoietic organs. Exp. Hematol. 1976; 4:267–274.

[149] Caplan A.I. Mesenchymal stem cells. J. Orthop. Res. 1991; 9(5):641–50.

[150] Arboleda D., Forostyak S., Jendelova P., Marekova D., Amemori T., Pivonkova H., Masinova K., Sykova E. Transplantation of predifferentiated adipose-derived stromal cells for the treatment of spinal cord injury. Cell. Mol. Neurobiol. 2011; 31(7):1113–1122.

[151] Krause D.S., Plasticity of marrow-derived stem cells. Gene Ther. 2002; 9(11):754–758.

[152] Mezey E., Chandross K.J., Harta G., Maki R.A., McKercher S.R. Turning blood into brain: cells neuronal antigens generated in vivo from bone marrow. Science. 2000; 290(5497):1779–1782.

[153] Prockop D.J. Marrow stromal cells as stem cells for nonhematopoietic tissues. Science. 1997; 276(5309):71–74.

[154] La Spada A., Ranum L.P. Molecular genetic advances in neurological disease:special review issue. Hum. Mol. Genet. 2010; 19(R1):R1–3.

[155] Urdzikova L., Jendelova P., Glogarova K., Burian M., Hajek M., Sykova E. Transplantation of bone marrow stem cells as well as mobilization by granulocyte-colony stimulating factor promotes recovery after spinal cord injury in rats. J. Neurotrauma. 2006; 23(9):1379–1391.

[156] Sykova E., Jendelova P. Migration, fate and in vivo imaging of adult stem cells in the CNS. Cell Death Differ. 2007; 14(7):1336–1342.

[157] Hejcl A., Sedy J., Kapcalova M., Toro D.A., Amemori T., Lesny P., Likavcanova-Masinova K., Krumbholcova E., Pradny M., Michalek J., Burian M., Hajek M., Jendelova P., Sykova E. HPMA-RGD hydrogels seeded with mesenchymal stem cells improve functional outcome in chronic spinal cord injury. Stem cells and development 2010; 19(10):1535–1546.

[158] Ohta M., Suzuki Y., Noda T., Ejiri Y., Dezawa M., Kataoka K., Chou H., Ishikawa N., Matsumoto N., Iwashita Y., Mizuta E., Kuno S., Ide C. Bone marrow stromal cells infused into the cerebrospinal fluid promote functional recovery of the injured rat spinal cord with reduced cavity formation. Exp. Neurol. 2004; 187 (2), 266–278.

[159] Daley G.Q. The promise and perils of stem cell therapeutics. Cell Stem Cell. 2012; 10:740–749.

[160] Bang O.Y., Lee J.S., Lee P.H., Lee G. Autologous mesenchymal stem cell transplantation in stroke patients. Ann. Neurol. 2005; 57 (6), 874–882.

[161] McNiece I. Stem cells and regenerative medicine. J. Regen. Med. 2012; 1:1.

[162] Song L., Young N.J., Webb N.E., Tuan R.S. Origin and characterization of multipotential mesenchymal stem cells derived from adult human trabecular bone. Stem Cells Dev. 2005; 14:712–721.

[163] Choi Y.S., Noh S.E., Lim S.M., Lee C.W., Kim C.S., Im M.W., Lee M.H., Kim D.I. Multipotency and growth characteristic of periosteum-derived progenitor cells for chondrogenic, osteogenic, and adipogenic differentiation. Biotechnol. Lett. 2008; 30:593–601.

[164] De Bari C., Dell'Accio F., Tylzanowski P., Luyten F.P. Multipotent mesenchymal stem cells from adult human synovial membrane. Arthritis Rheum. 2001; 44:1928–1942.

[165] Dodson M.V., Hausman G.J., Guan L., Du M., Rasmussen T.P., Poulos S.P., Mir P., Bergen W.G., Fernyhough M.E., McFarland D.C., Rhoads R.P., Soret B., Reecy J.M., Velleman S.G., Jiang Z. Skeletal muscle stem cells from animals I. Basic cell biology. Int. J. Biol. Sci. 2010; 6:465–474.

[166] Belicchi M., Pisati F., Lopa R., Porretti L., Fortunato F., Sironi M., Scalamogna M., Parati E.A., Bresolin N., Torrente Y. Human skin-derived stem cells migrate throughout forebrain and differentiate into astrocytes after injection into adult mouse brain. J. Neurosci. Res. 2004; 77:475–486.

[167] Feng J., Mantesso A., Sharpe P.T. Perivascular cells as mesenchymal stem cells. Exp. Opin. Biol. Ther. 2010; 10:1441–1451.

[168] Shi M., Ishikawa M., Kamei N., Nakasa T., Adachi N., Deie M., Asahara T., Ochi M. Acceleration of skeletal muscle regeneration in a rat skeletal muscle injury model by local injection of human peripheral blood-derived cd133-positive cells. Stem Cells. 2009; 27:949–960.

[169] Baksh D., Yao R., Tuan R.S. Comparison of proliferative and multilineage differentiation potential of human mesenchymal stem cells derived from umbilical cord and bone marrow. Stem Cells. 2007; 25:1384–1392.

[170] Musina R.A., Bekchanova E.S., Sukhikh G.T. Comparison of mesenchymal stem cells obtained from different human tissues. Bull. Exp. Biol. Med. 2005; 139:504–509.

[171] Gimble J.M., Katz A.J., Bunnell B.A. Adipose-derived stem cells for regenerative medicine. Circ. Res. 2007; 100:1249–1260.

[172] Tobita M., Orbay H., Mizuno H. Adipose-derived stem cells: Current findings and future perspectives. Discov. Med. 2011; 11:160–170.

[173] Weisberg S.P., McCann D., Desai M., Rosenbaum M., Leibel R.L., Ferrante A.W., Jr. Obesity is associated with macrophage accumulation in adipose tissue. J. Clin. Investig. 2003; 112:1796–1808.

[174] Xu H., Barnes G.T., Yang Q., Tan G., Yang D., Chou C.J., Sole J., Nichols A., Ross J.S., Tartaglia L.A., Chen H. Chronic inflammation in fat plays a crucial role in the development of obesity-related insulin resistance. J. Clin. Investig. 2003; 112:1821–1830.

[175] Zuk P.A., Zhu M., Mizuno H., Huang J., Futrell J.W., Katz A.J., Benhaim P., Lorenz H.P., Hedrick M.H. Multilineage cells from human adipose tissue: implications for cell-based therapies. Tissue Eng. 2001; 7:211–228.

[176] De Ugarte D.A., Morizono K., Elbarbary A., Alfonso Z., Zuk P.A., Zhu M., Dragoo J.L., Ashjian P., Thomas B., Benhaim P., Chen I., Fraser J., Hedrick M.H. Comparison of multi-lineage cells from human adipose tissue and bone marrow. Cells Tissues Organs. 2003; 174:101–109.

[177] Izadpanah R., Trygg C., Patel B., Kriedt C., Dufour J., Gimble J.M., Bunnell B.A. Biologic properties of mesenchymal stem cells derived from bone marrow and adipose tissue. J. Cell Biochem. 2006; 99:1285–1297.

[178] Mitchell J.B., McIntosh K., Zvonic S., Garrett S., Floyd Z.E., Kloster A., Di Halvorsen Y., Storms R.W., Goh B., Kilroy G., Wu X., Gimble J.M. Immunophenotype of human adipose-derived cells: temporal changes in stromal-associated and stem cell-associated markers. Stem Cells. 2006; 24:376–385.

[179] Brayfield C.A., Marra K.G., Rubin J.P. Adipose tissue regeneration. Curr. Stem Cell Res. Ther. 2010; 5:116–121.

[180] Cherubino M., Marra K.G. Adipose-derived stem cells for soft tissue reconstruction. Regen. Med. 2009; 4:109–117.

[181] Rubin J.P., Marra K.G. Soft tissue reconstruction. Methods Mol. Biol. 2011; 702:395–400.

[182] Dragoo JL, Samimi B, Zhu M . Tissue-engineered cartilage and bone using stem cells from human infrapatellar fat pads. J. Bone Jt. Surg. Br. 2003; 85:740–747.

[183] Lee W.C., Sepulveda J.L., Rubin J.P., Marra K.G. Cardiomyogenic differentiation potential of human adipose precursor cells. Int. J. Cardiol. 2009; 133:399–401.

[184] Planat-Bénard V., Menard C., André M., Puceat M., Perez A., Garcia-Verdugo J.M., Pénicaud L., Casteilla L. Spontaneous cardiomyocyte differentiation from adipose tissue stroma cells. Circ. Res. 2004; 94:223–229.

[185] Estes B.T., Diekman B.O., Gimble J.M., Guilak F. Isolation of adipose derived stem cells and their induction to a chondrogenic phenotype. Nat. Protoc. 2010; 5:1294–1311.

[186] Estes B.T., Guilak F. Three-dimensional culture systems to induce chondrogenesis of adipose-derived stem cells. Methods Mol. Biol. 2011; 702:201–217.

[187] Rehman J., Traktuev D., Li J., Merfeld-Clauss S., Temm-Grove C.J., Bovenkerk J.E., Pell C.L., Johnstone B.H., Considine R.V., March K.L. Secretion of angiogenic and anti-apoptotic factors by human adipose stromal cells. Circulation. 2004; 109:1292–1298.

[188] Cherubino M., Rubin J.P., Miljkovic N., Kelmendi-Doko A., Marra K.G. Adipose-derived stem cells for wound healing applications. Ann. Plast. Surg. 2011; 66:210–215.

[189] Uysal A.C., Mizuno H. Tendon regeneration and repair with adipose derived stem cells. Curr. Stem Cell Res. Ther. 2010; 5:161–167.

[190] Aurich H., Sgodda M., Kaltwasser P., Vetter M., Weise A., Liehr T., Brulport M., Hengstler J.G., Dollinger M.M., Fleig W.E., Christ B. Hepatocyte differentiation of mesenchymal stem cells from human adipose tissue in vitro promotes hepatic integration in vivo. Gut. 2009; 58:570–581.

[191] Banas A., Teratani T., Yamamoto Y., Tokuhara M., Takeshita F., Osaki M., Kato T., Okochi H., Ochiya T. Rapid hepatic fate specification of adipose-derived stem cells and their therapeutic potential for liver failure. J. Gastroenterol. Hepatol. 2009; 24:70–77.

[192] Mesimäki K., Lindroos B., Törnwall J., Mauno J., Lindqvist C., Kontio R., Miettinen S., Suuronen R. Novel maxillary reconstruction with ectopic bone formation by gmp adipose stem cells. Int. J. Oral. Maxillofac. Surg. 2009; 38:201–209.

[193] Lendeckel S., Jödicke A., Christophis P., Heidinger K., Wolff J., Fraser J.K., Hedrick M.H., Berthold L., Howaldt H.P. Autologous stem cells (adipose) and fibrin glue used to treat widespread traumatic calvarial defects: case report. J. Craniomaxillofac. Surg. 2004; 32:370–373.

[194] Gimble J.M., Guilak F., Bunnell B.A. Clinical and preclinical translation of cell-based therapies using adipose tissue-derived cells. Stem Cell Res. Ther. 2010; 1:19.

[195] Bailey A.M., Kapur S., Katz A.J. Characterization of adipose-derived stem cells: an update. Curr. Stem Cell Res. Ther. 2010; 5:95–102.

[196] Adams G.B., Scadden D.T. A niche opportunity for stem cell therapeutics. Gene Ther. 2008; 15:96–99.

[197] Guilak F., Cohen D.M., Estes B.T., Gimble J.M., Liedtke W., Chen C.S. Control of stem cell fate by physical interactions with the extracellular matrix. Cell Stem Cell. 2009;5:17–26.

[198] Ratne B.D. Biomaterials Science: an Introduction to Materials in Medicine, 2nd ed. Amsterdam, The Netherlands: Elsevier Academic Press, 2004.

[199] Lafrenie, R.M, Yamada K.M. Integrin-dependent signal transduction. J Cell. Biochem. 1996; 61:543–553.

[200] Kuhl P.R., Griffith-Cima L.G. Tethered epidermal growth factor as a paradigm for growth factor-induced stimulation from the solid phase. Nat. Med. 1996; 2:1022e7.

[201] Kelly P.N., Dakic A., Adams J.M., Nutt S.L., Strasser, A. Tumor growth need not be driven by rare cancer stem cells. Science. 2007; 317:337.

[202] Cheng A., Kapacee Z., Peng J., Lucas R.J., Hardingham T.E., Kimber S.J. Cartilage repair using human embryonic stem cell-derived chondroprogenitors. Stem Cells Transl. Med. 2014; 3:1287–1294.

[203] Oldershaw R.A., Baxter M.A., Lowe E.T., Bates N., Grady L.M., Soncin F., Brison D.R., Hardingham T.E., Kimber S.J. Directed differentiation of human embryonic stem cells toward chondrocytes. Nat. Biotechnol. 2010; 28:1187–1194.

[204] Guo W., Lasky J.L., Chang C.J., Mosessian S., Lewis X., Xiao Y., Yeh J.E., Chen J.Y., Iruela-Arispe M.L., Varella-Garcia M., Wu H. Nature. 2008; 453:529.

[205] Neering S.J., Bushnell T., Sozer S., Ashton J., Rossi R.M., Wang P.Y., Bell D.R., Heinrich D., Bottaro A., Jordan C.T. Leukemia stem cells in a genetically defined murine model of blast-crisis CML. Blood. 2007; 110:2578.

[206] Hern D.L., Hubbell J.A. Incorporation of adhesion peptides into non-adhesive hydrogels useful for tissue resurfacing. J. Biomed. Mater. Res. 1998; 39:266–276.

[207] Chien C.Y., Tsai W.B. Poly(dopamine)-assisted immobilization of Arg-Gly-Asp peptides, hydroxyapatite, and bone morphogenic protein-2 on titanium to improve the osteogenesis of bone marrow stem cells. ACS Appl. Mater. Interfaces. 2013; 5:6975–6983.

[208] Hsiong S.X., Huebsch N., Fischbach C., Kong H.J., Mooney D.J. Integrin-adhesion ligand bond formation of pre-osteoblasts and stem cells in three-dimensional RGD presenting matrices. Biomacromolecules. 2008; 9:1843–1851.

[209] Luu H.H., Song W.X., Luo X., Manning D., Luo J., Deng Z.L., Sharff K.A., Montag A.G., Haydon R.C., He T.C. Distinct roles of bone morphogenetic proteins in osteogenic differentiation of mesenchymal stem cells. J Orthop. Res. 2007; 25:665–677.

[210] Zouani O.F., Kalisky J., Ibarboure E., Durrieu M.C. Effect of BMP-2 from matrices of different stiffnesses for the modulation of stem cell fate. Biomaterials. 2013; 34:2157–2166.

[211] Shin H., Zygourakis K., Farach-Carson M.C., Yaszemski M.J., Mikos A.G. Attachment, proliferation, and migration of marrow stromal osteoblasts cultured on biomimetic hydrogels modified with an osteopontin-derived peptide. Biomaterials. 2004; 25:895–906.

[212] Benoit D.S., Schwartz M.P., Durney A.R., Anseth K.S. Small functional groups for controlled differentiation of hydrogel encapsulated human mesenchymal stem cells. Nat. Mater. 2008; 7:816–823.

[213] Phillips J.E., Petrie T.A., Creighton F.P., García A.J. Human mesenchymal stem cell differentiation on self-assembled monolayers presenting different surface chemistries. Acta Biomater. 2010; 6:12–20.

[214] Griffin M.F., Butler P.E., Seifalian A.M., Kalaskar D.M. Control of stem cell fate by engineering their micro and nanoenvironment. World J Stem Cells. 2015; 7(1):37–50.

[215] Abrams G.A., Goodman S.L., Nealey P.F., Franco M., Murphy C.J. Nanoscale topography of the basement membrane underlying the corneal epithelium of the rhesus macaque. Cell Tissue Res. 2000; 299:39–46.

[216] van Dorp W.F., Zhang X., Feringa B.L., Hansen T.W., Wagner J.B., De Hosson J.T. Molecule-by-molecule writing using a focused electron beam. ACS Nano. 2012; 6:10076–10081.

[217] Lai G.J., Shalumon K.T., Chen S.H., Chen J.P. Composite chitosan/silk fibroin nanofibers for modulation of osteogenic differentiation and proliferation of human mesenchymal stem cells. Carbohydr. Polym. 2014; 111:288–297.

[218] Xu T., Zhao W., Zhu J.M., Albanna M.Z., Yoo J.J., Atala A. Complex heterogeneous tissue constructs containing multiple cell types prepared by inkjet printing technology. Biomaterials. 2013; 34:130–139.

[219] Dalby M.J., McCloy D., Robertson M., Wilkinson C.D., Oreffo R.O. Osteoprogenitor response to defined topographies with nanoscale depths. Biomaterials. 2006; 27:1306–1315.

[220] Zanetti N.C., Solursh M. Induction of chondrogenesis in limb mesenchymal cultures by disruption of the actin cytoskeleton. J. Cell Biol. 1984; 99(1 Pt 1):115–123.

[221] McBride S.H., Knothe Tate M.L. Modulation of stem cell shape and fate A: the role of density and seeding protocol on nucleus shape and gene expression. Tissue Eng. Part A. 2008; 14:1561–1572.

[222] Folkman J., Moscona A. Role of cell shape in growth control. Nature. 1978; 273:345–349.

[223] Manasek F.J., Burnside M.B., Waterman R.E. Myocardial cell shape change as a mechanism of embryonic heart looping. Dev. Biol. 1972; 29:349–371.

[224] Ingber D. Extracellular matrix and cell shape: potential control points for inhibition of angiogenesis. J. Cell. Biochem. 1991; 47:236–241.

[225] Lutolf M.P., Gilbert P.M., Blau H.M. Designing materiasl to direct stem-cell fate. Nature. 2009; 46(26):433–441.

[226] Flaim C.J., Chien S., Bhatia S.N. An extracellular matrix microarray for probing cellular differentiation. Nat. Methods. 2005; 2:119–125.

[227] Soen Y., Mori A., Palmer T.D., Brown P.O. Exploring the regulation of human neural precursor cell differentiation using arrays of signaling microenvironments. Mol. Syst. Biol. 2006; 2:37.

[228] Holtzer H., Abbott J., Lash J., Holtzer S. The loss of phenotypic traits by differentiated cells in vitro, I. Dedifferentiation of cartilage cells. Proc. Natl. Acad. Sci. USA. 1960; 46:1533–1542.

[229] Abbott J., Holtzer H. The loss of phenotypic traits by differentiated cells. 3. The reversible behavior of chondrocytes in primary cultures. J. Cell Biol. 1966; 28:473–487.

[230] Benya P.D., Shaffer J.D. Dedifferentiated chondrocytes re-express the differentiated collagen phenotype when cultured in agarose gels. Cell. 1982; 30:215–224.

[231] Erickson G.R., Gimble J.M., Franklin D.M., Rice H.E., Awad H., Guilak F. Chondrogenic potential of adipose tissue-derived stromal cells in vitro and in vivo. Biochem. Biophys. Res. Commun. 2002; 290:763–769.

[232] Hoben G.M., Koay E.J., Athanasiou K.A. Fibrochondrogenesis in two embryonic stem cell lines: effects of differentiation timelines. Stem Cells. 2008; 26:422–430.

[233] Johnstone B., Hering T.M., Caplan A.I., Goldberg V.M., Yoo J.U. In vitro chondrogenesis of bone marrow-derived mesenchymal progenitor cells. Exp. Cell Res. 1998; 238:265–272.

[234] Awad H.A., Wickham M.Q., Leddy H.A., Gimble J.M., Guilak F. Chondrogenic differentiation of adipose-derived adult stem cells in agarose, alginate, and gelatin scaffolds. Biomaterials. 2004; 25:3211–3222.

[235] Oh S., Brammer K.S., Li Y.S., Teng D., Engler A.J., Chien S., Jin S. Stem cell fate dictated solely by altered nanotube dimension. Proc. Natl. Acad. Sci. USA. 2009; 106:2130–2135.

[236] Brammer K.S., Oh S., Cobb C.J., Bjursten L.M., van der Heyde H., Jin S. Improved bone-forming functionality on diametercontrolled TiO(2) nanotube surface. Acta Biomater. 2009; 5:3215–3223.

[237] Dalby M.J., Gadegaard N., Tare R., Andar A., Riehle M.O., Herzyk P., Wilkinson C.D., Oreffo R.O. The control of human mesenchymal cell differentiation using nanoscale symmetry and disorder. Nat. Mater. 2007; 6:997–1003.

[238] Zouani O.F., Chanseau C., Brouillaud B., Bareille R., Deliane F., Foulc M.P., Mehdi A., Durrieu M.C. Altered nanofeature size dictates stem cell differentiation. J. Cell Sci. 2012; 125:1217–1224.

[239] Reinwald, Y., Shakesheff, K. and Howdle, S. (2011) Biomedical Devices, in Porous Polymers (eds M. S. Silverstein, N. R. Cameron and M. A. Hillmyer), John Wiley & Sons, Inc., Hoboken, NJ, USA. doi: 10.1002/9780470929445.ch9

[240] Underhill G.H., Bhatia S.N. High-throughput analysis of signals regulating stem cell fate and function. Curr. Opin. Chem. Biol. 2007; 11:357–366.

[241] Gidrol X., Fouqué B., Ghenim L., Haguet V., Picollet-D'hahan N., Schaack B. 2D and 3D cell microarrays in pharmacology. Curr. Opin. Pharmacol. 2009; 9:664–668.

[242] Fernandes T.G., Diogo M.M., Clark D.S., Dordick J.S., Cabral J.M.S. High-throughput cellular microarray platforms: applications in drug discovery, toxicology and stem cell research. Trends Biotechnol. 2009; 27:342–349.

[243] Lee, M.Y., Kumar, R.A., Sukumaran, S.M., Hogg, M.G., Clark, D.S., Dordick, J.S. Three-dimensional cellular microarray for high-throughput toxicology assays. Proc. Natl. Acad. Sci. USA. 2008; 105:59–63.

[244] Jongpaiboonkit L., King W.J., Murphy W.L. Screening for 3D environments that support human mesenchymal stem cell viability using hydrogel arrays. Tissue Eng. A. 2009; 15:343–353.

[245] Albrecht D.R., Underhill G.H., Wassermann T.B., Sah R.L., Bhatia S.N. Probing the role of multicellular organization in three-dimensional microenvironments. Nat. Methods. 2006; 3:369–375.

[246] Sudo R., Chung S., Zervantonakis I.K., Vickerman V., Toshimitsu Y., Griffith L.G., Kamm R.D. Transport-mediated angiogenesis in 3D epithelial coculture. FASEB J. 2009; 23:2155–2164.

[247] Whitesides G.M. The origins and the future of microfluidics. Nature. 2006; 442:368–373.

[248] Chung B.G., Flanagan L.A., Rhee S.W., Schwartz P.H., Lee A.P., Monuki E.S., Jeon N.L. Human neural stem cell growth and differentiation in a gradient generating microfluidic device. Lab Chip. 2005; 5:401–406.

[249] Choi N.W., Cabodi M., Held B., Gleghorn J.P., Bonassar L.J., Stroock A.D. Microfluidic scaffolds for tissue engineering. Nat. Mater. 2007; 6:908–915.

[250] Kronenberg H.M. PTH regulates the hematopoietic stem cell niche in bone. Adv. Exp. Med. Biol. 2007; 602:57–60.

[251] Kasper S. Exploring the origins of the normal prostate and prostate cancer stem cell. Stem Cell Rev. 2008; 4:193–201.

[252] Brisken C., Duss S. Stem cells and the stem cell niche in the breast:an integrated hormonal and developmental perspective. Stem Cell Rev. 2007; 3:147–156.

[253] Nur-E-Kamal A., Ahmed I., Kamal J., Babu A.N., Schindler M., Meiners S. Covalently attached FGF-2 to three-dimensional polyamide nanofibrillar surfaces demonstrates enhanced biological stability and activity. Mol. Cell. Biochem. 2008; 309:157–166.

[254] Fan V.H., Tamama K., Au A., Littrell R., Richardson L.B., Wright J.W., Wells A., Griffith L.G. Tethered epidermal growth factor provides a survival advantage to mesenchymal stem cells. Stem Cells. 2007; 25:1241–1251.

[255] Anderson D.G., Levenberg S., Langer R. Nanoliter-scale synthesis of arrayed biomaterials and application to human embryonic stem cells. Nat. Biotechnol. 2004; 22:863–866.

[256] Kshitiz D.M.G., Hubbi M.E., Ahn E.H., Downey J., Afzal J., Kim D.H., Rey S., Chang C., Kundu A., Semenza G.L., Abraham, R.M., Levchenko, A. Matrix rigidity controls endothelial differentiation and morphogenesis of cardiac precursors. Sci. Signal. 2012; 5:ra41.

[257] Dela Paz N.G., Walshe T.E., Leach L.L., Saint-Geniez M., D'Amore, P.A. Role of shear-stress-induced VEGF expression in endothelial cell survival. J. Cell Sci. 2012; 125:831–843.

[258] Bakeine G.J., Ban J., Grenci G., Pozzato A., Zilio S.D., Prasciolu M., Businaro L., Tormen M., Ruaro M.E. Design, fabrication and evaluation of nanoscale surface topography as a tool in directing differentiation and organisation of embryonic stem-cell-derived neural precursors. Microelect. Eng. 2009; 86:1435–1438.

[259] Ingber D.E. The mechanochemical basis of cell and tissue regulation. Mech. Chem. Biosyst. 2004; 1:53–68.

[260] Guo W.H., Frey M.T., Burnham N.A., Wang Y.L. Substrate rigidity regulates the formation and maintenance of tissues. Biophys. J. 2006; 90:2213–2220.

[261] Pelham R.J. Jr., Wang Y. Cell locomotion and focal adhesions are regulated by substrate flexibility. Proc. Natl. Acad. Sci. USA. 1997; 94:13661–13665.

[262] Hadjipanayi E., Mudera V., Brown R.A. Close dependence of fibroblast proliferation on collagen scaffold matrix stiffness. J. Tissue Eng. Regen. Med. 2008; 3:77–84.

[263] Wang H.B., Dembo M., Wang Y.L. Substrate flexibility regulates growth and apoptosis of normal but not transformed cells. Am. J. Physiol. Cell Physiol. 2000; 279:C1345–C1350.

[264] Engler A.J., Sen S., Sweeney H.L., Discher D.E. Matrix elasticity directs stem cell lineage specification. Cell. 2006; 126:677–689.

[265] Wingate K., Bonani W., Tan Y., Bryant S.J., Tan W. Compressive elasticity of three-dimensional nanofiber matrix directs mesenchymal stem cell differentiation to vascular cells with endothelial or smooth muscle cell markers. Acta Biomater. 2012; 8:1440–1449.

[266] Nieponice A., Soletti L., Guan J., Deasy B.M., Huard J., Wagner W.R., Vorp D.A. Development of a tissue-engineered vascular graft combining a biodegradable scaffold, muscle-derived stem cells and a rotational vacuum seeding technique. Biomaterials. 2008; 29:825–833.

[267] Zhang G., Drinnan C.T., Geuss L.R., Suggs L.J. Vascular differentiation of bone marrow stem cells is directed by a tunable three-dimensional matrix. Acta Biomater. 2010; 6:3395–3403.

[268] Banerjee A., Arha M., Choudhary S., Ashton R.S., Bhatia S.R., Schaffer D.V., Kane R.S. The influence of hydrogel modulus on the proliferation and differentiation of encapsulated neural stem cells. Biomaterials. 2009; 30:4695–4699.

[269] Murphy C.M., Matsiko A., Haugh M.G., Gleeson J.P., O'Brien, F.J. Mesenchymal stem cell fate is regulated by the composition and mechanical properties of collagen-glycosaminoglycan scaffolds. J. Mech. Behav. Biomed. Mater. 2012; 11:53–62.

[270] Janmey P.A., McCulloch C.A. Cell mechanics: integrating cell responses to mechanical stimuli. Annu. Rev. Biomed. Eng. 2007; 9:1–34.

[271] Chen W., Villa-Diaz L.G., Sun Y., Weng S., Kim J.K., Lam R.H., Han L., Fan R., Krebsbach P.H., Fu J. Nanotopography influences adhesion, spreading, and self-renewal of human embryonic stem cells. ACS Nano. 2012; 6:4094–4103.

[272] Chuang C.K., Sung L.Y., Hwang S.M., Lo W.H., Chen H.C., Hu Y.C. Baculovirus as a new gene delivery vector for stem cell engineering and bone tissue engineering. Gene Ther. 2007; 14:1417–1424.

[273] Lee S.J., Kang S.W., Do H.J., Han I., Shin D.A., Kim J.H., Lee S.H. Enhancement of bone regeneration by gene delivery of BMP2/Runx2 bicistronic vector into adipose-derived stromal cells. Biomaterials. 2010; 31:5652–5659.

[274] Morito A., Kida Y., Suzuki K., Inoue K., Kuroda N., Gomi K., Arai T., Sato T. Effects of basic fibroblast growth factor on the development of the stem cell properties of human dental pulp cells. Arch. Histol. Cytol. 2009; 72:51–64.

[275] Zhang J., Wilson G.F., Soerens A.G., Koonce C.H., Yu J., Palecek S.P., Thomson J.A., Kamp T.J. Functional cardiomyocytes derived from human induced pluripotent stem cells. Circ. Res. 2009; 104:e30–e41.

[276] Levenberg S., Ferreira L.S., Chen-Konak L., Kraehenbuehl T.P., Langer R. Isolation, differentiation and characterization of vascular cells derived from human embryonic stem cells. Nat. Protocol. 2010; 5:1115–1126.

[277] Ingber D.E. Mechanobiology and diseases of mechanotransduction. Ann. Med. 2003; 35:564–577.

[278] Estes B.T., Gimble J.M., Guilak F. Mechanical signals as regulators of stem cell fate. Curr. Top. Dev. Biol. 2004; 60:91–126.

[279] Kshitiz Park, J., Kim, P., Helen, W., Engler, A.J., Levchenko, A., Kim, D.H. Control of stem cell fate and function by engineering physical microenvironments. Integr. Biol. 2012; 4(9):1008–1018.

[280] Glücksmann A. The role of mechanical stresses in bone formation in vitro. J. Anat. 1942; 76:132–139.

[281] Fang J., Hall B.K. Differential expression of neural cell adhesion molecule (NCAM) during osteogenesis and secondary chondrogenesis in the embryonic chick. Int. J. Dev. Biol. 1995; 39:519–528.

[282] Hall B.K., Herring S.W. Paralysis and growth of the musculoskeletal system in the embryonic chick. J. Morphol. 1990; 206:45–56.

[283] Murray P.D.F., Drachman B.D. The role of movement in the development of joints and related structures: the head and neck in the chick embryo. J. Embryol. Exp. Morphol. 1969; 22:349–371.

[284] Guilak F., Sah R.L., Setton L.A. Physical regulation of cartilage metabolism. In: Mow V.C., Hayes W.C. (eds.). Basic Orthopaedic Biomechanics. (Philadelphia: Lippincott-Raven), 1997; pp. 179–207.

[285] Liedtke W., Kim C. Functionality of the TRPV subfamily of TRP ion channels: add mechano-TRP and osmo-TRP to the lexicon! Cell. Mol. Life Sci. 2005; 62:2985–3001.

[286] Wang J.H., Thampatty B.P. Mechanobiology of adult and stem cells. Int. Rev. Cell Mol. Biol. 2008; 271:301–346.

[287] Reinwald Y., Leonard K.H.L., Henstock J.R., Whiteley J.P., Osborne J.M., Waters S.L., Levesque P., El Haj A.J. Evaluation of the growth environment of a hydrostatic force bioreactor for preconditioning of tissue-engineered constructs. Tissue Eng. Part C. 2015; 21(1):1–14.

[288] Henstock J.R., Rotherham M., Rose J.B., El Haj A.J. Cyclic hydrostatic pressure stimulates enhanced bone development in the foetal chick femur in vitro. Bone. 2013; 53:468.

[289] Luo W., Jones S.R., Yousaf M.N. Geometric control of stem cell differentiation rate on surfaces. Langmuir. 2008; 24:12129–12133.

[290] El Haj A.J., Cartmell S.H. Bioreactors for bone tissue engineering. Proc. Inst. Mech. Eng. 2010H; 224:1523.

[291] Darling E.M., Athanasiou K.A. Biomechanical strategies for articular cartilage regeneration. Ann. Biomed. Eng. 2003; 31:1114–1124.

[292] Darling E.M., Athanasiou K.A. Articular cartilage bioreactors and bioprocesses. Tissue Eng. 2003; 9:9–26.

[293] Partap S., Plunkett N.A., O'Brien, F.J. Bioreactors in Tissue Engineering. In: Eberli D. (ed.).Tissue Engineering. 2010; ISBN:978-953-307-079-7.

[294] Freed L.E., Guilak F., Guo X.E., Gray M.L., Tranquillo R., Holmes J.W., Radisic M., Sefton M.V., Kaplan D., Vunjak-Novakovic G. Advanced tools for tissue engineering: scaffolds, bioreactors, and signaling. Tissue Eng. 2006; 12:3285–3305.

[295] Reinwald Y., El Haj A.J. Mechanical and topographical cues affecting the mesenchymal stem cell fate for bone tissue regeneration. 2016 (manuscript under revision).

[296] Vunjak-Novakovic G., Obradovic B., Martin I., Bursac P.M., Langer R., Freed L.E. Dynamic cell seeding of polymer scaffolds for cartilage tissue engineering. Biotechnol. Prog. 1998; 14:193–202.

[297] Bancroft G.N., Sikavitsas V.I., Mikos A.G. Design of a flow perfusion bioreactor system for bone tissue-engineering applications. Tissue Eng. 2003; 9:549–554.

[298] Schwarz R.P., Goodwin T.J., Wolf D.A. Cell culture for three-dimensional modeling in rotating-wall vessels: an application of simulated microgravity. J. Tissue Cult. Methods. 1992; 14:51–57.

[299] Li W.J., Jiang Y.J., Tuan R.S., Cell-nanofiber-based cartilage tissue engineering using improved cell seeding, growth factor, and bioreactor technologies. Tissue Eng. Part A. 2008; 14(5):69–48.

[300] Lawrenson K., Benjamin E., Turmaine M., Jacobs I., Gayther S., Dafou D. In vitro three-dimensional modelling of human ovarian surface epithelial cells. Cell Prolif 2009; 42(3): 385–93.

[301] Consolo F., Bariani C., Mantalaris F., Redaelli A., Morbiducci U., Computional modelling for the optimization of a cardiogenic 3D bioprocess of encapsulated embryonic stem cells. Biomech. Model. Mechanobiol. 2012; 11(1–2):261–77.

[302] Goldstein A.S., Juarez T.M., Helmke C.D., Gustin M.C., Mikos A.G. Effect of convection on osteoblastic cell growth and function in biodegradable polymer foam scaffolds. Biomaterials. 2001; 22:1279–1288.

[303] Yu X., Botchwey E.A., Levine E.M., Pollack S.R., Laurencin C.T. Bioreactor-based bone tissue engineering: the influence of dynamic flow on osteoblast phenotypic expression and matrix mineralization. Proc. Natl. Acad. Sci. USA. 2004; 101:11203–11208.

[304] Huang C.Y., Hagar K.L., Frost L.E., Sun Y., Cheung H.S. Effects of cyclic compressive loading on chondrogenesis of rabbit bone-marrow derived mesenchymal stem cells. Stem Cells. 2004; 22:313–323.

[305] Thorpe S.D., Buckley C.T., Vinardell T., O'Brien F.J., Campbell V.A., Kelly D.J. Dynamic compression can inhibit chondrogenesis of mesenchymal stem cells. Biochem. Biophys. Res. Commun. 2008; 377:458–462.

[306] Barberi T., Bradbury M., Dincer Z., Panagiotakos G., Socci N.D., Studer L. Derivation of engraftable skeletal myoblasts from human embryonic stem cells. Nat. Med. 2007; 13:642–648.

[307] Li W., Li K., Wei W., Ding S. Chemical approaches to stem cell biology and therapeutics. Cell Stem Cell. 2013; 13:270–283.

[308] Li W., Ding S. Small molecules that modulate embryonic stem cell fate and somatic cell reprogramming. Trends Pharmacol. Sci. 2010; 31:36–45.

[309] Nguyen H.N., Byers B., Cord B., Shcheglovitov A., Byrne J., Gujar P., Kee K., Schüle B., Dolmetsch R.E., Langston W., Palmer T.D., Pera R.R. LRRK2 mutant iPSC-derived DA neurons demonstrate increased susceptibility to oxidative stress. Cell Stem Cell. 2011; 8:267–280.

[310] Ebert A.D., Yu J., Rose F.F.,Jr., Mattis V.B., Lorson C.L., Thomson J.A., Svendsen C.N. Induced pluripotent stem cells from a spinal muscular atrophy patient. Nature. 2009; 457:277–280.

[311] Marchetto M.C., Carromeu C., Acab A., Yu D., Yeo G.W., Mu Y., Chen G., Gage F.H., Muotri A.R. A model for neural development and treatment of Rett syndrome using human induced pluripotent stem cells. Cell. 2010; 143:527–539.

[312] Lee G., Papapetrou E.P., Kim H., Chambers S.M., Tomishima M.J., Fasano C.A., Ganat Y.M., Menon J., Shimizu F., Viale A., Tabar V., Sadelain M., Studer L. Modelling pathogenesis and treatment of familial dysautonomia using patient-specific iPSCs. Nature. 2009; 461:402–406.

[313] Moretti A., Bellin M., Welling A., Jung C.B., Lam J.T., Bott-Flügel L., Dorn T., Goedel A., Höhnke C., Hofmann F., Seyfarth M., Sinnecker D., Schömig A., Laugwitz K.L. Patient-specific induced pluripotent stem-cell models for long-QT syndrome. N. Engl. J. Med. 2010; 363:1397–1409.

[314] Itzhaki I., Maizels L., Huber I., Zwi-Dantsis L., Caspi O., Winterstern A., Feldman O., Gepstein A., Arbel G., Hammerman H., Boulos M., Gepstein L. Modelling the long QT syndrome with induced pluripotent stem cells. Nature. 2011; 471:225–229.

[315] Rashid S.T., Corbineau S., Hannan N., Marciniak S.J., Miranda E., Alexander G., Huang-Doran I., Griffin J., Ahrlund-Richter L., Skepper J., Semple R., Weber A., Lomas D.A., Vallier L. Modeling inherited metabolic disorders of the liver using human induced pluripotent stem cells. J. Clin. Investig. 2010; 120:3127–3136.

[316] Yazawa M., Hsueh B., Jia X., Pasca A.M., Bernstein J.A., Hallmayer J., Dolmetsch R.E. Using induced pluripotent stem cells to investigate cardiac phenotypes in Timothy syndrome. Nature. 2011; 471:230–234.

[317] Ko E., Yang K., Shin J., Cho S.W. Polydopamine-assisted osteoinductive peptide immobilization of polymer scaffolds for enhanced bone regeneration by human adipose-derived stem cells. Biomacromolecules. 2013; 14:3202–3213.

[318] Beazley K.E., Nurminskaya M. BMP2 cross-linked by transglutaminase 2 to collagen-plla scaffold promotes osteogenic differentiation in mesenchymal stem cells. Biotechnol. Lett. 2014; 36:1901–1907.

[319] Chiang C.K., Chowdhury M.F., Iyer R.K., Stanford W.L., Radisic M. Engineering surfaces for site-specific vascular differentiation of mouse embryonic stem cells. Acta Biomater. 2010; 6:1904–1916.

[320] Poh C.K., Shi Z., Lim T.Y., Neoh K.G., Wang W. The effect of VEGF functionalization of titanium on endothelial cells in vitro. Biomaterials. 2010; 31:1578–85.

[321] Rahman N., Purpura K.A., Wylie R.G., Zandstra P.W., Shoichet M.S. The use of vascular endothelial growth factor functionalized agarose to guide pluripotent stem cell aggregates toward blood progenitor cells. Biomaterials. 2010; 31:8262–8270.

[322] Lee J.Y., Choo J.E., Park H.J., Park J.B., Lee S.C., Jo I., Lee S.J., Chung C.P., Park Y.J. Injectable gel with synthetic collagen-binding peptide for enhanced osteogenesis in vitro and in vivo. Biochem. Biophys. Res. Commun. 2007; 357:68–74.

[323] Lee J.S., Lee J.S., Murphy W.L. Modular peptides promote human mesenchymal stem cell differentiation on biomaterial surfaces. Acta Biomater. 2010; 6:21–28.

[324] Shin Y.M., Jo S.Y., Park J.S., Gwon H.J., Jeong S.I., Lim Y.M. Synergistic effect of dual-functionalized fibrous scaffold with BCP and RGD containing peptide for improved osteogenic differentiation. Macromol. Biosci. 2014; 14:1190–1198.

[325] Lin Z.Y., Duan Z.X., Guo X.D., Li J.F., Lu H.W., Zheng Q.X., Quan D.P., Yang S.H. Bone induction by biomimetic PLGA-(PEGASP) n copolymer loaded with a novel synthetic BMP-2-related peptide in vitro and in vivo. J. Control Release. 2010; 144:190–195.

[326] Murphy A.R., St John P., Kaplan D.L. Modification of silk fibroin using diazonium coupling chemistry and the effects on hMSC proliferation and differentiation. Biomaterials. 2008; 29:2829–2838.

[327] Nuttelman C.R., Tripodi M.C., Anseth K.S. Synthetic hydrogel niches that promote hMSC viability. Matrix Biol. 2005; 24:208–218.

[328] Nuttelman C.R., Benoit D.S., Tripodi M.C., Anseth K.S. The effect of ethylene glycol methacrylate phosphate in PEG hydrogels on mineralization and viability of encapsulated hMSCs. Biomaterials. 2006; 27:1377–1386.

[329] Liu X., Feng Q., Bachhuka A., Vasilev K. Surface modification by allylamine plasma polymerization promotes osteogenic differentiation of human adipose-derived stem cells. ACS Appl. Mater. Interfaces. 2014; 6:9733–9741.

[330] Ren Y.J., Zhang H., Huang H., Wang X.M., Zhou Z.Y., Cui F.Z., An Y.H. In vitro behavior of neural stem cells in response to different chemical functional groups. Biomaterials. 2009; 30:1036–1044.

[331] Curran J.M., Chen R., Hunt J.A. The guidance of human mesenchymal stem cell differentiation in vitro by controlled modifications to the cell substrate. Biomaterials. 2006; 27:4783–4793.

[332] Park J., Bauer S., von der Mark K., Schmuki P. Nanosize and vitality: TiO_2 nanotube diameter directs cell fate. Nano Lett. 2007; 7:1686–1691.

[333] Yim E.K., Pang S.W., Leong K.W. Synthetic nanostructures inducing differentiation of human mesenchymal stem cells into neuronal lineage. Exp. Cell Res. 2007; 313:1820–1829.

[334] Biehl J.K., Yamanaka S., Desai T.A., Boheler K.R., Russell B. Proliferation of mouse embryonic stem cell progeny and the spontaneous contractile activity of cardiomyocytes are affected by microtopography. Dev. Dyn. 2009; 238:1964–1973.

[335] D'Angelo F., Armentano I., Mattioli S., Crispoltoni L., Tiribuzi R., Cerulli G.G., Palmerini C.A., Kenny J.M., Martino S., Orlacchio A. Micropatterned hydrogenated amorphous carbon guides mesenchymal stem cells towards neuronal differentiation. Eur. Cell Mater. 2010; 20:231–244.

[336] Gao L., McBeath R., Chen C.S. Stem cell shape regulates a chondrogenic versus myogenic fate through Rac1 and N-cadherin. Stem Cells. 2010; 28:564–572.

[337] Lu D., Chen C.S., Lai C.S., Soni S., Lam T., Le C., Chen E.Y., Nguyen T., Chin W.C. Microgrooved surface modulates neuron differentiation in human embryonic stem cells. Methods Mol. Biol. 2016; 1307:281–7.

[338] Smith L.A., Liu X., Hu J., Wang P., Ma P.X. Enhancing osteogenic differentiation of mouse embryonic stem cells by nanofibers. Tissue Eng. Part A. 2009; 15:1855–1864.

[339] Saha K., Keung A.J., Irwin E.F., Li Y., Little L., Schaffer D.V., Healy K.E. Substrate modulus directs neural stem cell behavior. Biophys. J. 2008; 95:4426–4438.

[340] Yang Y., Beqaj S., Kemp P., Ariel I., Schuger L. Stretch-induced alternative splicing of serum response factor promotes bronchial myogenesis and is defective in lung hypoplasia. J. Clin. Investig. 2000; 106:1321–1330.

[341] Hamilton D.W., Maul T.M., Vorp D.A. Characterization of the response of bone marrow-derived progenitor cells to cyclic strain: implications for vascular tissue-engineering applications. Tissue Eng. 2004; 10:361–369.

[342] Gong Z., Niklason L.E. Small-diameter human vessel wall engineered from bone marrow-derived mesenchymal stem cells (hMSCs). FASEB J. 2008; 22:1635–1648.

[343] Lee W.C., Maul T.M., Vorp D.A., Rubin J.P., Marra K.G. Effects of uniaxial cyclic strain on adipose-derived stem cell morphology, proliferation, and differentiation. Biomech. Model. Mechanobiol. 2007; 6:265–273.

[344] Sen B., Xie Z., Case N., Ma M., Rubin C., Rubin J. Mechanical strain inhibits adipogenesis in mesenchymal stem cells by stimulating a durable beta-catenin signal. Endocrinology. 2008; 149:6065–6075.

[345] Simmons C.A., Matlis S., Thornton A.J., Chen S., Wang C.Y., Mooney D.J. Cyclic strain enhances matrix mineralization by adult human mesenchymal stem cells via the

extracellular signal-regulated kinase (ERK1/2) signaling pathway. J. Biomech. 2003; 36:1087–1096.

[346] Thomas G.P., El Haj A.J. Bone marrow stromal cells are load responsive in vitro. Calcif. Tissue Int. 1996; 58:101–108.

[347] Yoshikawa T., Peel S.A., Gladstone J.R., Davies J.E. Biochemical analysis of the response in rat bone marrow cell cultures to mechanical stimulation. Biomed. Mater. Eng. 1997; 7:369–377.

[348] Knippenberg M., Helder M.N., Doulabi B.Z., Semeins C.M., Wuisman P.I., Klein-Nulend J. Adipose tissue-derived mesenchymal stem cells acquire bone cell-like responsiveness to fluid shear stress on osteogenic stimulation. Tissue Eng. 2005; 11:1780–1788.

[349] Shimizu N., Yamamoto K., Obi S., Kumagaya S., Masumura T., Shimano Y., Naruse K., Yamashita J.K., Igarashi T., Ando J. Cyclic strain induces mouse embryonic stem cell differentiation into vascular smooth muscle cells by activating PDGF receptor beta. J. Appl. Physiol. 2008; 104:766–772.

[350] Saha S., Ji L., de Pablo J.J., Palecek S.P. Inhibition of human embryonic stem cell differentiation by mechanical strain. J. Cell. Physiol. 2006; 206:126–137.

[351] Elder S.H., Kimura J.H., Soslowsky L.J., Lavagnino M., Goldstein S.A. Effect of compressive loading on chondrocyte differentiation in agarose cultures of chick limb-bud cells. J. Orthop. Res. 2000; 18:78–86.

[352] Angele P., Yoo J.U., Smith C., Mansour J., Jepsen K.J., Nerlich M., Johnstone B. Cyclic hydrostatic pressure enhances the chondrogenic phenotype of human mesenchymal progenitor cells differentiated in vitro. J. Orthop. Res. 2003; 21:451–457.

Chapter 5

Rejuvenation on the Road to Pluripotency

Tapash Jay Sarkar and Vittorio Sebastiano

Additional information is available at the end of the chapter

http://dx.doi.org/10.5772/63219

Abstract

The technology of reprogramming differentiated cells into a pluripotent state, which can be used to derive virtually any cell type in vitro, has ignited the field of regenerative medicine. An equally revolutionary, but yet to be harnessed phenomenon, is the reset of age that occurs en route to pluripotency. This rejuvenation is clearly evident during reproduction, resulting in a young offspring from aged parental cells. Artificial reprogramming techniques built off this process, such as somatic cell nuclear transfer (SCNT) and induced pluripotent stem cell (iPSC) reprogramming techniques, are showing growing evidence for rejuvenation at the cellular level. These findings all points to an intimate relationship between reprogramming to pluripotency and the reset of age, and iPSC technology, especially, offers the possibility of a man-made intervention in the aging process. Though in vitro cell reprogramming has been studied arguably for the last three decades, this application of specifically developing a protocol to rejuvenate cells, tissues, even whole organs has only just begun to be explored. There are still many challenges to realization but this technology has already famously shown that cell differentiation is more than a one-way street, and, maybe, so is aging.

Keywords: iPSC reprogramming, aging, rejuvenation, organoids, epigenetics

In their recent *Overview on Chronic Disease*, the Center for Disease and Control (CDC) cited multiple studies in the last few years highlighting the predominance (86% in 2010) of chronic diseases in the US national health care spending. Each of these diseases has their own treatments directed by their individualized fields of research. Yet the key risk factor and fundamental correlations behind many conditions like arthritis ($128 billion in 2003), heart disease and stroke ($315.4 billion in 2010), and type II diabetes ($245 billion in 2012) are of course middle and advanced age [1]. To further quantify the core role of aging, a 2013 study did a prospective study and estimated that the economic value of delaying aging in Americans by only about 2 years would save $7.1 trillion over 50 years [2]. This study was prompted by the growing field of aging research and longevity technologies, but now an even bolder direction is emerging. Researchers are venturing beyond just slowing age or promoting healthy aging and actually attempting to reverse the manifestations of age, from the cellular to the tissue and organ levels.

In another field of research, stem cell science has rapidly grown in the last few decades as a means of studying tissue development and maintenance and to develop methods to artificially produce cells that are either absent or dysfunctional in patients. Again the applications here are pathology specific, conditions like blood and immunological diseases, brain and spinal cord injury, and type I diabetes, as well as some of the more aging-correlated disease like Alzheimer's and heart disease [3]. A unifying approach in the field is the use of pluripotent stem cells, the crown jewel state that can produce any cell in the body. Natural and artificial processes all reach this state by starting with differentiated cell types and "reprogramming" them back to pluripotency. Clearly, this process of reprogramming represents a drastic change and powerful technology, whose potential is just beginning to be explored.

The focus of this chapter is convergence of these two fields. We will see how rejuvenation is intimately linked to reprogramming. Furthermore, we discuss how technologies that induce pluripotency may hold the key to a wholescale and stable reset of cellular age, with tissue- and organ-level consequences.

1. Natural reprogramming and rejuvenation

Rejuvenation is actually not at all an esoteric idea. It happens consistently and is crucial to the survival of virtually every species. Most would immediately think of relatively simple organisms like the hydra and the jellyfish *Turritopsis dohrnii*, which have virtually unlimited regenerative potential [4]. Yet there is a rejuvenation mechanism in much more "complicated" animals, and most importantly in humans, which hides in plain sight: the process of reproduction. The most popular and widely investigated feature of this process is attaining pluripotency, the cellular state that can differentiate into the full diversity of cells in a new organism, starting from two highly specialized and differentiated cells (i.e., the germ cells). However, rejuvenation is an equally exciting phenomenon that occurs during reproduction and is only beginning to be explored. A simple input output analysis verifies this: The inputs are the sperm and oocyte cells of aged parents, typically in their second or third decade of life, while the outputs are the myriad of different cells that make up a new, young organism, retaining none of aged aspects of the parent cells. While one could argue that germ cells have evolved special mechanisms that, unlike somatic cells, prevent them from aging, multiple studies have shown that germ cells do indeed age. Sperm show progressively accumulated DNA damage, elevated reactive ion species levels, and loss of chromatin integrity with age [5]. Oocytes show altered expression of genes implicated in DNA methylation and histone acetylation and mitochondrial dysfunction with age, as well as microniche-imposed aging, through nutrition and hormonal pathways [6]. If these manifestations of age are not erased during reproduction, each generation would progressively accumulate age and the species would no longer have the capability to produce viable offspring after just a few generations.

The mechanisms specifically behind the reversal of age during reproduction are largely unknown. However, there are two main categories of changes that drive embryogenesis as a whole: genetic and epigenetic. In terms of genetic, the main change is the genetic recombination of the parental chromosomes, which are gathered during zygote formation, right after fertilization, and fused during the first mitotic division into blastomeres [7]. The result is a new genome derived from the parental genomes. There are also more active DNA repair process but they can be tied to upregulation of genes for double-strand break repair and cell cycle checkpoint control that result from the epigenetic changes we will cover next [8]. The reactivated repair processes would definitely have an age-reversing effect, but there is no evidence that the recombination to produce a new genome, alone, should rejuvenate the cell. Epigenetics also undergo large shifts in the initial stage of zygote formation, characterized by a genome-wide loss of DNA methylation except in certain centromeric and parentally imprinted regions and retrovirally derived repetitive elements. In addition, there are widespread alterations to methylation and acetylation of H3 and H4 histones from the maternal genome, while the paternal genome abandons its own protamines for these modified histones as well. Increased activity of methlytransferase has also been observed during this stage; these types of enzymes preserve methylation patterns during cell division for both DNA and histones. Together, these changes further regulate overall heterochromatin organization [9]. Subsequent stages in pre-implantation development up until blastocyst formation show a more passive loss of DNA methylation with cell division, primarily due to reduced levels of methlytransferase in the cell nucleus. The blastocyst then marks the beginning of lineage specification, with an increase in methylation and additional histone modifications that are primarily correlated to differentiation rather than rejuvenation [10]. Ultimately, the bulk erasure of parental epigenetic markers, for cell type and most likely for age, is during the reprogramming of the parental germ cells into a totipotent zygote; totipotency here refers to the zygote's ability to differentiate into all embryonic and extraembryonic (supporting) tissue.

Artificial reprogramming to totipotency was further pursued in the last few decades, with the primary objective of attaining cells that could differentiate into any desired cell type. The first key deviation from the natural process was the technology of somatic cell nuclear transfer (SCNT). In fertilization, only the sperm's nuclei is transferred and it is the oocyte that provides the reprogramming environment. SCNT replaces the sperm nucleus with a somatic cell nucleus for fertilization, but the overall reprogramming process is fundamentally the same, again with large-scale genetic and epigenetic shifts to yield a viable, young offspring. [11] This method, however, established that somatic cells, in addition to germ cells, could be reprogrammed, and thus rejuvenated. The next key breakthrough, which is our main focus, is the technology of induced pluripotent stem cell (iPSC) reprogramming. This technology focuses on reprogramming somatic cells to pluripotent cells, which can specifically form only the embryonic tissues and thus is more relevant than totipotency. More importantly, the strategy here is substantially different. Instead of the reprogramming driven by the oocyte, iPSC technologies introduce exogenous pluripotency genes into somatic cells to induce reprogramming to pluripotency [12]. This process does not include a genetic recombination of two different genomes, like reproduction or SCNT. Thus, the success of this technology established that a global shift in the gene expression profile, induced by epigenetic remodeling, without substantial genetic alterations is sufficient to execute reprogramming and, as we will see in the next section, can also implement multiple changes toward rejuvenation.

2. iPSC reprogramming: signs of rejuvenation

iPSC reprogramming was developed to artificially attain the embryonic stem cell (ESC) state, a derivative of the totipotent zygote state established by natural fertilization and SCNT reprogramming. The principle behind this technology is that forced overexpression of genes that normally maintain the pluripotent state in ESCs is sufficient to induce reprogramming in somatic cells to revert back to an ESC-like state. The most widely utilized strategy, which most of the result we discuss have applied, is the exogenous introduction of the genes Oct4, Sox2, Klf4, and cMyc, typically achieved by viral or episomal vectors [12]. In evaluating the final product of these procedures, it has been noted that the transition to pluripotency is accompanied by multiple indications of rejuvenation that, in most cases, are retained upon subsequent redifferentiation of the iPSC. As iPSC reprogramming is a cellular technology, it requires a set of cellular biomarkers by which to measure age. The search is still ongoing for a comprehensive and precise list, but the "Hallmarks of Aging" presented by Lopez-Otin et al. in 2013 provide one most generally accepted categorization of the known biomarkers [13]. We will review the multiple signs of rejuvenation through iPSC reprogramming in the context of these hallmarks.

2.1. Epigenetic alterations

Epigenetics include all the changes to the structure of DNA, but not the DNA sequence itself, which control transcriptional potential and ultimately gene expression. Epigenetics most notably establishes cell type and functionality, but also includes many markers of age. DNA methylation is a key aspect of epigenetics and aging correlates with a global trend toward hypomethlyation in numerous cells types across tissues [14]. This is linked to the decline in mRNA transcription for DNA methyltransferase (DNMT), which controls the transfer of methylation patters to daughter cells [15]. Histones, which act as the "spools" around which DNA strands are wound, are another aspect of epigenetics which show age-related modifications, for instance, trimethlyation levels increase for H3K27 and H3K9 but decrease for H4K20, which are all involved in heterochromatin formation, the most densely packed form of chromatin [16]. A key driver of these alterations to histone methylations is the decreased transcription of methyltransferase proteins with age, like those of the Polycomb and trithorax groups [17]. There is an overall loss of heterochromatin with age, driven by the histone modifications but also lost during cell division itself, as recondensation processes are not perfectly efficient. These global trends lead to a loss in gene silencing with age and overall increase in transcriptional noise. iPSC reprogramming reverses of many of these changes, with a youthful restoration of the levels of HP1γ (a key marker for heterochomatin) and trimethlyated H3K9, increased heterochromatin, but there is still much yet to be explored [18]. Changes in the levels of endogenous Polycomb proteins have not been studied, but the artificial knockdown or their genes yields decreased reprogramming efficiency—hinting at a correlation between the two [19]. There is also an overall decrease in DNA methylation but this is primarily because global hypomethlyation is a characteristic of the pluripotent state, as seen in ESCs. A better comparison, that has yet to be investigated, is between the methylation levels of the original source cells and iPSC-derived cells of the same cell type. A correlation between artificially increased DNMT and reprogramming efficiency has also been made, but no further study has been done on whether reprogramming itself stimulates DNMT transcription [20]. Ultimately, reprogramming itself is characterized by a large-scale epigenetic remodeling, most infamously for dedifferentiation but also for rejuvenation.

2.2. DNA and nuclear damage

Direct damage to DNA from natural metabolic activities, stochastic chemical (endogenous and environmental), and radiative interactions regularly occurs throughout life at rate of 10^5 total molecular lesions per cell per day [21]. There are number of enzymatic repair mechanisms in place to fix these aberrations, but these processes are not perfectly efficient and some processes, like homologous recombination (HR) and non-homologous end joining (NHEJ), which occur during cell division, also decline in efficiency with age [22, 23]. This leads to the accumulated DNA damage, which impairs cell functionality, and decreases the number of viable progeny for proliferative cells; this is the central idea in the DNA damage theory of aging that leads to tissue wide consequences in organs like brain, liver muscle, and kidney. In addition, the nuclear lamina degrades with age with conformational defects like folding and blebbing; this impairs the lamina's function as the overall structural support for the nuclear material, especially heterochromatin in the epigenetics hallmark. These nuclear lamina defects result from age-related decreases in production of laminar proteins, like lamina-associated protein 2α (LAP2α), and altered distribution of these proteins, like the localization of the laminar matrix protein lamin A/C to the nuclear rim but not the nucleoplasm [24]. iPSCs show evidence of reducing accumulated DNA damage through a marked upregulation in genes and activity in both the HR and NHEJ pathways. Further evidence for reinvigorated repair mechanism are the restored levels of phosphorylated histone H2AX which is a key biomarker for the repair of double-strand breaks, in the iPSC state and after redifferentiation. In addition, lamina structure is shown to be restored in the iPSC state which replenished levels of LAP2α. In general, reprogrammed cells also show as an increased sensitivity to extreme irradiation damage noted by a greater propensity to undergo apoptosis, to eliminate mutated cells; this is ultimately healthier for the tissue by clearing cells that may instigate cancerous growth [25].

2.3. Telomere attrition

During cell division, helicase "unzips" DNA in one direction and DNA polymerases must build off of the two template strands but can only move in the $5' \rightarrow 3'$ direction. Thus, one polymerase can only move oppositely to the helicase, meaning it must repeatedly attach, synthesize, detach, and catch up to the helicase. Attachment sites are directed by RNA primers, which bind to DNA behind the helicase. Synthesis can only start after the primer, so the DNA by covered primer is not synthesized until the subsequent synthesis when it detaches and the polymerase starts further upstream. The problem arises at the end of the chromosome, where it is no further upstream site for polymerase to begin at and thus the end is simply not copied.

Nature's solution is to pad the end of the chromosomes with telomeres, repeats of noncoding junk DNA (TTAGGG), which are lost instead of coding segments. Telomere erosion is characteristic of cell division and thus aging itself. A number of studies have shown an elongation of telomeres, around 40%, upon reprogramming to iPSC [26–28]. This elongation of telomeres is mechanistically tied to the reactivated transcription of telomerase, a reverse transcriptase enzyme which synthesizes additional telomere repeats. Genes for telomerase are only expressed in stem cells as well as some other highly proliferative somatic cells, but they are silenced upon differentiation. Upon redifferentiation, all lines have been shown to have shorter telomeres than the original iPSC but some who have shown still have longer telomeres than the original source cell [27, 28]. This subsequent loss of telomere length is not necessarily due to accelerated telomere attrition, but just due to the lengthy redifferentiation protocols and variations in each cell line's reprogramming and redifferentiation efficiency. If these protocols can be made shorter and more efficient, the rejuvenation effects could be better retained. This is another example of how a genetic hallmark is addressed by alterations in gene expression. It has also yet to be assessed exactly when telomerase is reactivated and when it is lost during redifferentiation; this knowledge may help to further optimize the retention of this rejuvenation effect.

2.4. Mitochondrial dysfunction

Older cells exhibit a greater percentage of mitochondria in the less efficient "condensed" and "ultra-condensed" configurations than in the healthier "orthodox" configuration; the former have their inner and outer mitochondrial membranes further apart, promoting a smaller matrix and larger cristae. This condensed structure leads to a lower transmembrane potential, which is correlated with higher, less regulated respiratory activity. iPSC's and their redifferentiated lineages exhibit a reversal of this trend, with a higher proportion of orthodox mitochondria as well as higher membrane potentials when compared with the aged source cells. In addition, reprogramming and redifferentiation studies have also shown decreased mitochondrial mass, increased levels of adenosine triphosphate (ATP), and decreased reactive oxygen species (ROS) levels, which are all characteristic of a more youthful phenotype [29]. The lowered ROS levels are especially promising, as elevated levels are a key player in the free radical theory of aging. This idea holds that the electrically active by-products of metabolic reactions, like ROS, can aberrantly ionize the biomolecular machinery of cells, ultimately impairing the functionality of mitochondria and other cell compartments. These youthful changes can be fundamentally tied to upregulated mitochondrial biogenesis genes like PGC-1 alpha and antioxidant genes like GPX1 during reprogramming [30]. This upregulated biogenesis may directly explain the noted increase in youthful mitochondrial morphology and functionality, while the antioxidant genes explain the loss of ROS and further damage to mitochondria. The time course of these changes during reprogramming has not been documented, with the exception of mitochondrial mass, which has been noted to decrease as early as 3 days after the induction of reprogramming [31]. Interestingly, during SCNT reprogramming, cells began exhibiting higher membrane potential in the first week of reprogramming and were linked to increased expression of Glut1, Pfkm, Hxk2, and Ldha which drive glycolytic activity over oxidative phosphorylation [32]. This transition would explain the decreased ROS levels despite the observed increase in ATP, as glycolysis also produces ATP but without the production of ROS. This initial week, however, does not show the onset of pluripotency, as marked by the expression of pluripotency genes like Nanog and Oct4, suggesting a temporal segregation of the gene expression changes driving rejuvenation from those that drive differentiation.

2.5. Loss of proteostasis

Proteostasis is regulated by modulating levels of both protein transcription/stabilization, to ensure the cell has the proteins it needs, and proteolytic activity, to prevent proteotoxicity. The former, of course, is a direct result of gene expression dynamics. The latter is implemented through the ubiquitin-proteasome and autophagy-lysosome systems; the activity of both notably diminishes with age. This leads to an aggregation of macromolecules as well as degraded organelles that contribute to other aging hallmarks—like mitochondrial dysfunction and senescence [32, 34]. Reprogramming studies have explored relatively little in regard to rejuvenated proteolytic activity. The key finding so far has been an elevated level of autophagy upon induction of reprogramming, peaking on day 2 then relaxing to basal levels from day 3 on until pluripotency is reached [35, 36]. This transient activity has been linked to increased mitophagy, a key role of autophagy, which is responsible for the clearing of degraded mitochondria. This works in tandem with the aforementioned increased mitochondrial biogenesis to create the noted skew in the mitochondria distribution to the younger, orthodox configuration [26]. It also postulated that the additional autophagy is responsible for eliminating cell type-specific protein complexes early on, thus promoting dedifferentiation. This effect may be transient, but the additional degraded proteins and organelles that are removed during the interim definitely contribute to cell rejuvenation. It is crucial to note that these studies were on murine cells, so translation of these effects to human systems also needs to be investigated. Proteasomal activity has yet to be studied in the context of iPSC reprogramming, but increased activities have been noted in human ESCs, linked to elevated expression of the PSMD11 gene [38].

2.6. Senescence

Senescence is defined as the state of full mitotic arrest, relevant primarily for proliferative cells. In vitro, this is linked to the Hayflick limit—the point at which cells have undergone enough cell divisions to deplete their telomeres. In vivo, senescence is also linked to the accumulated DNA damage with age and triggered by epigenetic (chromatin) remodeling at the INK4a/ARF locus. Senescence is likely a programmed biological mechanism for preventing the propagation of cells with damaged DNA, which is more prevalent in aged cells as characterized by shortened telomeres. This prevents mutation-induced cancerous growth and aberrant functionality, but it also arrests tissue growth and repair in general. iPSC reprogramming has been achieved on completely senescent populations, propagated for 51 population doublings then maintained for 2 months, and show revived proliferation in the pluripotent state as well as upon redifferentiation [39]. In addition, expressions of cell cycle-inhibiting proteins like p16 and p21, which initiate the INK4a/ARF remodeling, are known to increase with age but diminish after reprogramming and redifferentiation. Crucially, the redifferntiated progeny can once again be passaged into senescence, thus verifying that no cancerous growth was induced during the entire process.

2.7. Deregulated nutrient sensing

This hallmark concerns poorer regulation in gene expression pathways that regulated cell metabolism based on the availability of nutrients. The mTOR family of genes is a key glucose- and amino acid-sensing pathway, while the AMPK pathway is route triggered by adenosine monophosphate (AMP) levels, a by-product of ATP synthesis. Artificial downregulation of mTOR and upregulation of AMPK genes have both led to increased longevity in model organisms, while the inverse (as occurs naturally) has been linked with many age-related pathologies like diabetes, hinting at epigenetic alterations to these pathways with age [40, 41]. During reprogramming, mTOR was downregulated in the same pattern that autophagy increased and AMPK activity was upregulated in the first 6 days; both pathways are drivers for autophagy [31, 36]. In addition, downregulation of mTOR is also involved in mitochondrial biogenesis [36]. Thus, the deregulated nutrient-sensing hallmark further links the loss of proteostasis and mitochondrial dysfunction hallmarks. The two remaining FOXO and Sirtuin pathway have yet to be studied for their role in reprogramming.

There are two remaining hallmarks. The hallmark of *stem cell exhaustion* is defined by age-associated losses in both proliferative capacity and stem cells reserves, like in hippocampal neural stem cell populations, as well as altered epigenetics for differentiation potential, for instance, aged hematopoietic stem cells bias toward myeloid lineages over lymphoid lineages [42, 43]. The hallmark of *altered intracellular communication* results from increased expression and secretion of pro-inflammatory cytokines by senescent and pre-senescent cells as well as proteotoxicity from the loss of proteostasis. This leads to a larger, immune system phenotype of "inflamaging," an age-induced propensity to sustain a chronic low-grade innate immune response in multiple tissues, which ultimately interferes with single cell function as well as cell-to-cell interactions [44]. To our knowledge, no iPSC reprogramming studies have really examined rejuvenation of these hallmarks, primarily because both truly need be assessed within aged tissue. The organoid technology we will discuss in the next section may offer the platform to evaluate the rejuvenation of these hallmarks as well through iPSC reprogramming.

These rejuvenated biomarkers are summarized pictorially in **Figure 1**, but there are still many other known biomarkers within these hallmarks that must be assessed for iPSC- and iPSC-derived cells and, as new ones are being discovered, they must also be evaluated. In addition, the results are primarily in fibroblasts, the most common source material for iPSC reprogramming, so the generalization of the rejuvenation effect to other cell types still needs to be assessed. Still, these studies provide initial evidence that iPSC reprogramming can reset age to some degree. In the next section, we will see where this technology stands as a possible rejuvenation procedure.

http://dx.doi.org/10.5772/63219

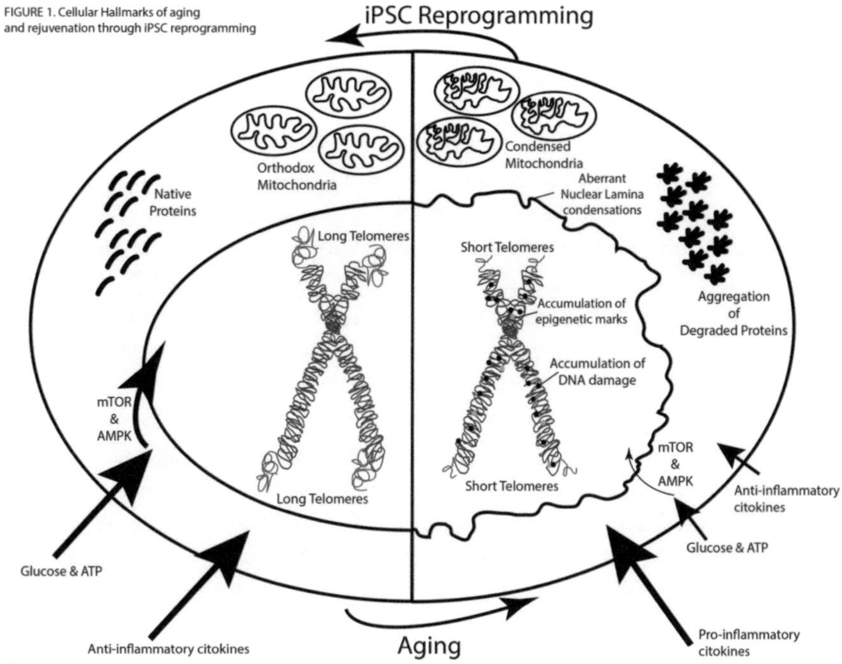

Figure 1. Cellular Hallmarks of aging and rejuvenation through iPSC reprogramming.

3. iPSC rejuvenation: mechanism, comparisons and outlook

The field of iPSC reprogramming was developed to attain and harness the power of pluripotency. The rejuvenation effects that accompany the process were secondary discovery, though one with great potential as well. As we now move to capitalize on this effect, we must evaluate it in the broader field of aging research and technologies. In this section, we will first explore how the mechanistic principle behind this iPSC rejuvenation fits in with unifying theories of aging. Then we will assess the reprogramming strategy in the context of other technologies for healthy aging and rejuvenation. Finally, we will discuss the future perspectives and outlook of this technology in terms of scalability and clinical application.

3.1. iPSC reprogramming and programmed aging

The common theme we note from all the hallmarks are that multifaceted changes in gene expression that are linked to and propagate the aging phenotype. We similarly saw that the rejuvenation effects of iPSC reprogramming could be traced to alterations in gene expression. We noted a few examples, yet these are but a small sample of the full gamut of changes that drive rejuvenation along with dedifferentiation. The core driver of these changes in iPSC reprogramming is the massive overhaul of the epigenetic landscape, which as previously noted, alters the transcription potential throughout the genome. Additionally, the noted changes in nuclear lamina influence epigenetics and further regulate the distribution of transcription factors as well as the transport of mRNA for translation. Furthest downstream, the increased proteolytic activity, though transient, also clears out gene products—another form of gene regulation. The key concept this strategy has tapped into is the so-called programmed aging hypothesis [45]. This idea holds that the aging phenotype is driven in large part by deterministic and programmed changes, and thus is reversible. At the cellular level, this directly points to changes in gene expression, which is reversible especially through the modes of epigenetic and nuclear lamina modifications. So if age is programmed, iPSC technology could possibly reprogram age—truly apt naming in hindsight. The dual theory is the damage-induced aging hypothesis, which holds that aging is driven by the stochastic degradation of multiple cellular components. This damage is driven by environmental interactions as well as internal degradation as a result of metabolic processes [46]. We discussed manifestations of this in previous section, such as DNA damage, ROS damage, and proteotoxicity from accumulated macromolecules. Unlike programmed aging, the results of this damage are random and thus are not inherently reversible. iPSC reprogramming cannot directly oppose damage but it can help to mitigate some of this damage. As we have seen, it can boost expression for natural repair mechanisms, like homologous repair of DNA damage; it can promote the synthesis of new organelles. From an evolutionary perspective, the two hypotheses may be fundamentally linked. The continued accumulation of age-related damage and the resulting loss in functionality may make retaining older individual less beneficial from a species-level perspective. Older individuals would be less fit and less capable of performing their role in a communal society, more susceptible to and may further transmit pathogens, more likely to produce mutated or dysfunctional offspring, and still take up resources that could go to the younger generations. Thus, species may have evolved programmed mechanisms to further the decay with age and thus increase the mortality and clearing of the older individuals. This could explain why a natural rejuvenation exists but only occurs in the production of the next generation, instead of somehow being applied to retain the youthful phenotype [6], like in the case of mitochondrial biogenesis, and it can boost the clearing of damaged components by transiently increasing proteolytic activity. At the same time, there are age-induced damages that iPSC reprogramming cannot counter (for instance, genetic mutations or damage) without an intact reference template. So in regard to damage-induced aging, there are definitely inherent limitations to this gene expression level technology.

3.2. iPSC reprogramming vs other longevity/rejuvenation technologies

The predominant strategy in the field of aging is to identify individual gene expression pathways that maintain the youthful phenotype or advance aging and artificially control their expression to promote healthy aging. Treatment involves either controlling the expression of endogenous genes through cues like DNA-binding agents and small molecules or introducing exogenous genetic material into the cells through vectors (primarily viral and episomal). The strategies have been applied to genes for telomerase, autophagy, mTOR, AMPK, p16, and p21 to name a few of the aging biomarkers we discussed [47]. Additionally, many pharmacological strategies have been developed to induce these pathways with more exogenous control, based on administration and dosage of the drug. Notable examples, specifically in regard to nutrient sensing, include rapamycin, an mTOR inhibitor, metformin, an AMPK activator, and fisetin and resveratrol, both are situin activators [48]. Both the genetic and pharmacologic approaches are fundamentally lacking in two respects. First, these approaches are too narrow in scope. These strategies target one or a handful of genes, but aging is a complex phenotype propagated by a full array of genetic pathways. iPSC reprogramming, in contrast, is more comprehensive with a whole scale epigenetic remodeling that instigates global changes in gene expression which encompass many of the candidate pathways for gene/pharmacological therapies as well as others for aspects like mitochondrial function and gene repair. Even though neither the pathways of aging nor all the gene expression changes in reprogramming are known, the signs of iPSC rejuvenation already discussed and the intuition that iPSC reprogramming mimics natural reprogramming during reproduction both suggest that iPSC reprogramming may offer a more holistic reset of age. The second failing of the traditional genetic/pharmacological approaches is that they are too simplistic in application. These approaches are primarily static over/under expressions of their target genes, but outside of the direct rejuvenation effects, these changes may have detrimental consequences on other cellular processes. For instance, genetic knockout or knockdown of p16 genes may prevent senescence and promote tissue repair, but they increase the chance of spontaneous cancer development [49, 50]. Another example is rapamycin's induction of immunosuppressant pathways, which would outweigh its benefit in inhibiting mTOR [51]. A more viable solution would be a dynamic sequence of changes to gene expression which can respond and compensate for each other to produce a stable, youthful homeostasis. Reprogramming also meets this criterion. iPSCs are very similar to the stable, naturally occurring pluripotent state of ESCs, in terms of both epigenetics and gene expression [52]. In addition, redifferentiation of iPSCs yields viable somatic cells which do not show signs of aberrant functionality or advanced aging. In addition, they regain the aged and senescent phenotype after extensive passaging, having avoided cancerous mutations, as noted earlier.

Another direction in the aging field has gained attention in recent years: heterochronic parabiosis, the surgical joining of circulatory system between old and young mice. This promotes more than just longevity or healthy aging; it rejuvenates whole tissue in multiple organs, like muscles, heart, and brain, in the aged mouse. The mechanism behind this is the replacement of old organism's aged extracellular milieu—distributed through the bloodstream—with that of a youthful organism [53]. This milieu contains cytokines, hormones, and other signaling molecules which influence and bias gene expression at the single cell level expression through the intracellular communication and nutrient-sensing hallmarks. Programmed changes to the immune and neuroendocrine systems with age drive the production of this old milieu, and thus this technology also hinges on the programmed aging hypothesis, specifically its extracellular manifestation. The drawback of parabiosis, and immune and hormonal therapies in general, is that the "youthful" milieu is a dynamic and complex mixture, again a homeostatically maintained youthful state. This would require continuous transplantation of youthful blood from another patient, which is not feasible. The stand in would be artificial synthesis and constant administration of the youthful factors in the blood, which is not completely known and would be extremely difficult to develop and maintain. The same is true for immune and hormonal therapies, which also have detrimental side effects; a clear example is the use immunosuppressors would reduce chronic inflamaging but also weaken the response to pathogens. Ultimately, this extracellular rejuvenation approach would require rejuvenation of the entire immune and neuroendocrine systems to regulate and maintain. iPSC reprogramming is still primarily a single cell technology, and an entire rejuvenated system is still far beyond its scope. Yet the technology can still have extracellular effects since every somatic cell secretes signaling molecules that influence the cells in their local microniche. Thus, rejuvenated iPSC-derived cells, with sufficient numbers, can start to propagate the youthful phenotype in the tissue in which they are transplanted. In addition, multiple groups are working on producing rejuvenated iPSC-derived lymphoid lineages and T cells. As part of immunoaging, hematopoietic stem cells tend to differentiate more to the myeloid rather than the lymphoid lineage, thus artificial iPSC derivation of the latter can help compensate for this shift [54]. In addition, adaptive immunity to counter specific pathogens diminishes with age due to weakened response by T cells. In vitro production of rejuvenated T cells from iPSCs and transplantation could restore this functionality [55]. The key target that is being explored in iPSC endocrine rejuvenation is the production of insulin producing β-islet cells, which deplete with age [56]. If iPSC-derived cells can be used to holistically rejuvenate just these two systems, then it may capture or surpass the effects seen in parabiosis and subsequently rejuvenate the rest of the somatic cells.

3.3. Future perspectives

Outside of single rejuvenated cell production, iPSC technology is showing some progress in developing rudimentary tissue through the use protocols for the derivation of organoids. These protocols are methods of guided differentiation that direct the iPSC to a somatic stem cell fate, instead of a fully differentiated cell. By mimicking the environmental signals during embryo development and providing 3-D development infrastructure (air-liquid interface or Matrigel embedding), pluripotent cells can be directed to form a rudimentary organ bud. The cells at this state can be further proliferated or instructed to differentiate into multiple cell types by switching to a differentiation-promoting media. The final product resembles a fragment of in vivo tissue in its diversity and arrangement of the different cell types; successful examples include intestine, liver, and thyroid [57]. This is still a very nascent technology, with the primary focus of using organoids to study normal and cancerous development. There have not been any studies assessing to what degree the rejuvenation effects of iPSC reprogramming are retained in the final product, especially since there are now two phases of differentiation to reach the initial organ bud and then to attain the fully differentiated tissue. One encouraging sign is that these organoids can be maintained up to and some times longer than 6 months in culture, despite most of the primary cell donors being of advanced age (50s–70s). This has been noted to surpass the Hayflick limit, so at least telomere attrition and senescence may be reversed in the derivation of these organoids. Future studies can investigate the same aging hallmarks in the cells of organoids, but also explore the two additional hallmarks yet untouched in previous studies. They would be able to assess the intercellular communication—how the interactions of different cell types maintain or degrade the youthful phenotype. In addition, the differentiated organoid also retains some somatic stem cells, so these studies can also look at the stem cell exhaustion in maintaining and regenerating the organoid. The organoid platform also provides the opportunity to look at an additional dimension of aging in addressing the rejuvenation of tissue functionality. This moves beyond the Hallmarks of Aging, which are fundamentally at the cellular level, to focus on the emergent function of specific tissues which degrades due to the hallmarks. So far, very little work has been done on assessing organoid functionality. One notable example is a 2015 study that show albumin and bile acid production, ammonia elimination, CYP3A4 activity, and midazolam metabolism in liver organoids, all of which decline with age [58]. These organoids were formed directly from somatic stem cells of the patient, another method for organoid development, and thus show levels of functionality in these areas similar to in vivo. Future studies can use the iPSC derivation and compare this functionality with the in vivo counterpart for a possible rejuvenation effect.

Even if iPSC technology progresses to the point where generating whole rejuvenated tissue or organs becomes feasible, transplantation still presents a major hurdle. iPSC-derived transplant do have the benefit of being autologous, thus avoiding issues like rejection and graft-vs-host disease. However, they still require extensive surgical procedures, which would be especially detrimental to aged patients. In addition, heterochronic transplant studies, specifically transplanting muscle from a young organism into an aged one, show that the young tissue succumb to the aged environment within a few months [59]. Multiple aged tissues and organs must be replaced together, possibly even entire systems (like immune and neuroendocrine mentioned earlier), to change the aged environment and get a lasting rejuvenation effect. These obstacles necessitate a procedure that can scale and rejuvenate multiple tissues simultaneously (which even organoids could not eventually do). Ideally, all of this would be in vivo itself. In vivo reprogramming to iPSC has been achieved in mice, using transgenic specimen but also through the injection of exogenous transcription factors. The immediate drawback is that the subsequent differentiation of the iPSCs cannot be guided because they invariably form teratomas. This actually captures the core problem behind both scalability and in vivo use: the undesired dedifferentiation. The technology iPSC reprogramming was developed to replace cells that were depleted or deficient in a patient. The rejuvenation effect of this process was discovered, essentially, as a side effect. Now with the explicit goal of rejuvenation, we can assume that the cells are already present in the patient but they be made younger, so dedifferentiation is unnecessary. This dedifferentiation necessitates complicated, lengthy, and cell type-specific redifferentiation protocols that can at best generate very rudimentary tissue. They cannot generate multiple tissue types simultaneously, cannot be carried out in vivo, and may erase some of the rejuvenation effects. Reprogramming, itself, is just a series of epigenetic changes so a subset of these must be for dedifferentiation and another subset (possibly overlapping) must be for rejuvenation. The optimal solution, now, would be to identify and execute the epigenetics for rejuvenation without those for dedifferentiation. This may be the best hope for one day utilizing the rejuvenation potential of iPSC reprogramming as a clinical treatment.

Ultimately, rejuvenation is realized on the road to pluripotency. Harnessing these mechanisms to reset the age of cells in an organism is indeed an audacious goal, but not more audacious than artificially achieving pluripotency itself. One can argue that the true scientific leap was developing the iPSC technology, that made reprogramming controllable and generalizable to any cell type, and this new direction is just building upon, optimizing, and repurposing the technology for the ulterior motive of rejuvenation—comparably a less daunting task. We believe that as our understanding of aging become clearer, and hard evidence for age reversal becomes more prevalent, the dogma that aging is an immutable, irreversible process will be shattered. The field of medicine is fundamentally about challenging these limitations and revolutionizing the human condition. We believe the next revolution is upon us, as rejuvenation goes from mythology to gerontology.

Author details

Tapash Jay Sarkar[1,2] and Vittorio Sebastiano[2*]

*Address all correspondence to: vsebast@stanford.edu

1 Department of Applied Physics 348 Via Pueblo Mall Stanford University, Stanford, CA, USA

2 Department of Obstetrics and Gynecology, Institute for Stem Cell Biology and Regenerative Medicine, Stanford University, Stanford, CA, USA

References

[1] Centers for Disease Control and Prevention. Overview of Chronic Disease. National Center for Chronic Disease Prevention and Health Promotion; 2015. http://www.cdc.gov/chronicdisease/overview/index.htm. Accessed on January 12, 2016.

[2] Goldman, D. P., et al. "Substantial health and economic returns from delayed aging may warrant a new focus for medical research." *Health Affairs* 32.10 (2013): 1698–1705.

[3] Mimeault, M., R. Hauke, and S. K. Batra. "Stem cells: a revolution in therapeutics—recent advances in stem cell biology and their therapeutic applications in regenerative medicine and cancer therapies." *Clinical Pharmacology & Therapeutics* 82.3 (2007): 252–264.

[4] Petralia, R. S., M. P. Mattson, and P. J. Yao. "Aging and longevity in the simplest animals and the quest for immortality." *Ageing Research Reviews* 16 (2014): 66–82.

[5] Paul, C., and B. Robaire. "Ageing of the male germ line." *Nature Reviews Urology* 10.4 (2013): 227–234.

[6] Ge, Z-J., et al. "Oocyte ageing and epigenetics." *Reproduction* 149.3 (2015): R103–R114.

[7] Longo, V. D., J. Mitteldorf, and V. P. Skulachev. "Programmed and altruistic ageing." *Nature Reviews Genetics* 6.11 (2005): 866–872.

[8] Jaroudi, S., and S. SenGupta. "DNA repair in mammalian embryos." *Mutation Research/Reviews in Mutation Research* 635.1 (2007): 53–77.

[9] Santos, F., and W. Dean. "Epigenetic reprogramming during early development in mammals." *Reproduction* 127.6 (2004): 643–651.

[10] Morgan, H. D., et al. "Epigenetic reprogramming in mammals." *Human Molecular Genetics* 14(Suppl. 1) (2005): R47–R58.

[11] Willadsen, S. M. "Nuclear transplantation in sheep embryos." *Nature* 320.6057 (1986): 63–65.

[12] Takahashi, K., et al. "Induction of pluripotent stem cells from adult human fibroblasts by defined factors." *Cell* 131.5 (2007): 861–872.

[13] López-Otín, C., et al. "The hallmarks of aging." *Cell* 153.6 (2013): 1194–1217.

[14] Wilson, V. L., and P. A. Jones. "DNA methylation decreases in aging but not in immortal cells." *Science* 220.4601 (1983): 1055–1057.

[15] Fraga, M. F., and M. Esteller. "Epigenetics and aging: the targets and the marks." *Trends in Genetics* 23.8 (2007): 413–418.

[16] Munoz-Najar, U., and J. M. Sedivy. "Epigenetic control of aging."*Antioxidants & Redox Signaling* 14.2 (2011): 241–259.

[17] Pollina, E. A., and A. Brunet. "Epigenetic regulation of aging stem cells." *Oncogene* 30 (2011): 3105–3126.

[18] Miller, J. D., et al. "Human iPSC-based modeling of late-onset disease via progerin-induced aging." *Cell Stem Cell* 13.6 (2013): 691–705.

[19] Onder, T. T., et al. "Chromatin-modifying enzymes as modulators of reprogramming." *Nature* 483.7391 (2012): 598–602.

[20] De Carvalho, D. D., J. S. You, and P. A. Jones. "DNA methylation and cellular reprogramming." *Trends in Cell Biology* 20.10 (2010): 609–617.

[21] Caldecott, K. W. "Single-strand break repair and genetic disease." *Nature Reviews Genetics* 9.8 (2008): 619–631.

[22] Freitas, A. A., and J. P. de Magalhaes. "A review and appraisal of the DNA damage theory of ageing." *Mutation Research/Reviews in Mutation Research* 728.1 (2011): 12–22.

[23] Lindahl T. "Instability and decay of the primary structure of DNA." *Nature* 362 (1993): 709–715.

[24] Mao, Z., et al. "Sirtuin 6 (SIRT6) rescues the decline of homologous recombination repair during replicative senescence." *Proceedings of the National Academy of Sciences* 109.29 (2012): 11800–11805.

[25] Scaffidi, P., and M. T. Lamin. "A-dependent nuclear defects in human aging." *Science* 312 (2006): 1059–1063.

[26] Fan, J., et al. "Human induced pluripotent cells resemble embryonic stem cells demonstrating enhanced levels of DNA repair and efficacy of nonhomologous end-joining." *Mutation Research/Fundamental and Molecular Mechanisms of Mutagenesis* 713.1 (2011): 8–17.

[27] Terai, M., et al. "Investigation of telomere length dynamics in induced pluripotent stem cells using quantitative fluorescence in situ hybridization."*Tissue and Cell* 45.6 (2013): 407–413.

[28] Suhr, Steven T., et al. "Telomere dynamics in human cells reprogrammed to pluripotency." *PloS one* 4.12 (2009): e8124.

[29] Vaziri, H., et al. "Spontaneous reversal of the developmental aging of normal human cells following transcriptional reprogramming". *Regenerative Medicine* 5 (2010): 345–363.

[30] Lee, H-C., et al. "Increase in mitochondrial mass in human fibroblasts under oxidative stress and during replicative cell senescence." *Journal of Biomedical Science* 9.6 (2002): 517–526.

[31] Prigione, Alessandro, et al. "The senescence–related mitochondrial/oxidative stress pathway is repressed in human induced pluripotent stem cells." *Stem cells* 28.4 (2010): 721-733.

[32] Ma, T., et al. "Atg5-independent autophagy regulates mitochondrial clearance and is essential for iPSC reprogramming." *Nature Cell Biology* (2015).

[33] Folmes, C. D. L., et al. "Somatic oxidative bioenergetics transitions into pluripotency-dependent glycolysis to facilitate nuclear reprogramming." *Cell Metabolism* 14.2 (2011): 264–271.

[34] Rubinsztein, D. C., Marinõ, G., and Kroemer, G. Autophagy and aging. *Cell* 146 (2011): 682–695.

[35] Tomaru, U., et al. "Decreased proteasomal activity causes age-related phenotypes and promotes the development of metabolic abnormalities." *American Journal Pathology* 180 (2012): 963–972.

[36] Wang, S., et al. "Transient activation of autophagy via Sox2-mediated suppression of mTOR is an important early step in reprogramming to pluripotency." *Cell Stem Cell* 13.5 (2013): 617–625.

[37] Wu, Y., et al. "Autophagy and mTORC1 regulate the stochastic phase of somatic cell reprogramming." *Nature Cell Biology* 17.6 (2015): 715–725.

[38] Vilchez, D., et al. "Increased proteasome activity in human embryonic stem cells is regulated by PSMD11." *Nature* 489.7415 (2012): 304–308.

[39] Lapasset, L., et al. "Rejuvenating senescent and centenarian human cells by reprogramming through the pluripotent state." *Genes & Development* 25.21 (2011): 2248–2253.

[40] Passtoors, W. M., et al. "Gene expression analysis of mTOR pathway: association with human longevity." *Aging Cell* 12.1 (2013): 24–31.

[41] Mair, W., et al. "Lifespan extension induced by AMPK and calcineurin is mediated by CRTC-1 and CREB." *Nature* 470.7334 (2011): 404–408.

[42] Perkins, G. A., and M. H. Ellisman. "Mitochondrial configurations in peripheral nerve suggest differential ATP production." *Journal of Structural Biology* 173.1 (2011): 117–127.

[43] Encinas, J. M., et al. "Division-coupled astrocytic differentiation and age-related depletion of neural stem cells in the adult hippocampus." *Cell Stem Cell* 8.5 (2011): 566–579.

[44] Pang, W. W., et al. "Human bone marrow hematopoietic stem cells are increased in frequency and myeloid-biased with age." *Proceedings of the National Academy of Sciences* 108.50 (2011): 20012–20017.

[45] Salminen, A., K. Kaarniranta, and A. Kauppinen. "Inflammaging: disturbed interplay between autophagy and inflammasomes." *Aging (Albany NY)* 4.3 (2012): 166.

[46] Jin, K. "Modern biological theories of aging." *Aging and Disease* 1.2 (2010): 72.

[47] Hipkiss, Alan R. "Aging, proteotoxicity, mitochondria, glycation, NAD and carnosine: possible inter-relationships and resolution of the oxygen paradox."*Front Aging Neurosci* 2.10 (2010):1-4

[48] de Jesus, B. B., et al. "Telomerase gene therapy in adult and old mice delays aging and increases longevity without increasing cancer."*EMBO Molecular Medicine* 4.8 (2012): 691–704.

[49] Chen, T., et al. "Rapamycin and other longevity-promoting compounds enhance the generation of mouse induced pluripotent stem cells." *Aging Cell* 10.5 (2011): 908–911.

[50] Rayess, H., M. B. Wang, and E. S. Srivatsan. "Cellular senescence and tumor suppressor gene p16." *International Journal of Cancer* 130.8 (2012): 1715–1725.

[51] Kim, W. Y., and N. E. Sharpless. "The regulation of INK4/ARF in cancer and aging." *Cell* 127.2 (2006): 265–275.

[52] Lamming, D. W., et al. "Rapalogs and mTOR inhibitors as anti-aging therapeutics." *The Journal of Clinical Investigation* 123.3 (2013): 980–989.

[53] Robinton, D. A., and G. Q. Daley. "The promise of induced pluripotent stem cells in research and therapy." *Nature* 481.7381 (2012): 295–305.

[54] Conboy, I. M., and T. A. Rando. "Heterochronic parabiosis for the study of the effects of aging on stem cells and their niches." *Cell Cycle* 11.12 (2012): 2260–2267.

[55] Slukvin, I. I. "Hematopoietic specification from human pluripotent stem cells: current advances and challenges toward de novo generation of hematopoietic stem cells." *Blood* 122.25 (2013): 4035–4046.

[56] Nishimura, T., et al. "Generation of rejuvenated antigen-specific T cells by reprogramming to pluripotency and redifferentiation." *Cell Stem Cell* 12.1 (2013): 114–126.

[57] Quiskamp, N., J. E. Bruin, and T. J. Kieffer. "Differentiation of human pluripotent stem cells into β-cells: Potential and challenges." *Best Practice & Research Clinical Endocrinology & Metabolism* 29.6 (2015): 833–847.

[58] Huch, M., and B-K. Koo. "Modeling mouse and human development using organoid cultures." *Development* 142.18 (2015): 3113–3125.

[59] Huch, M., et al. "Long-term culture of genome-stable bipotent stem cells from adult human liver." *Cell* 160.1 (2015): 299–312.

[60] Carlson, B. M., and J. A. Faulkner. "Muscle transplantation between young and old rats: age of host determines recovery." *American Journal of Physiology – Cell Physiology* 256.6 (1989): C1262–C1266.

Chapter 6

Genomic Instability of Pluripotent Stem Cells: Origin and Consequences

Elena Lo Furno, Siem van der Laan and
Domenico Maiorano

Additional information is available at the end of the chapter

http://dx.doi.org/10.5772/62906

Abstract

Maintenance of genomic stability is crucial in ensuring cellular homeostasis and perpetuation of life. Perpetuation of the genetic information relies upon faithful replication of the genome. Mutations, generated during DNA synthesis and/or cell division and induced by exposure to external chemical agents, are drivers of genetic and associated genomic instability believed to fuel malignant transformation. Curiously, pluripotent stem cells (PSCs) are characterized by a high degree of genomic instability of unknown origin, which resembles that observed in cancer cells. This peculiar feature of PSCs raises the questions of the reasons responsible for this apparent aberrant regulation and of how genome integrity is kept under control. Genomic instability of PSCs also raises important concerns about their use in regenerative medicine, which sets severe limitations in clinical applications. The aim of this chapter is to review current knowledge about the molecular grounds of genomic instability of PSCs of diverse origin, such as embryonic (ESCs), induced pluripotent (iPSCs), and adult (ASCs) stem cells. We will also review how these features undermine the use of PSCs in clinical applications and discuss new emerging perspectives aimed at reducing genomic instability so to improve their use in clinical applications.

Keywords: DNA damage, checkpoints, replication stress, oncogenesis, nucleus, chromatin

1. Introduction

Maintenance of genome stability is primordial for stem cells, given their potential to generate multiple distinct cell lineages. Mutations may lead to the inheritance of DNA discontinuities in differentiated cells with potentially catastrophic consequences such as chromosomal rearrangements and deletions [1]. Genetic aberrancies can affect the stem cell pool or increase the chances of malignant transformation since these can lead to oncogenes activation and/or tumor suppressors silencing [2, 3]. Paradoxically, embryonic stem cells (ESCs) and induced pluripotent stem cells (iPSCs) display signs of genomic instability, to a level comparable to that observed in cancer cells [4]. In contrast, adult stem cells (ASCs), which have a strongly reduced pluripotency, appear to have more stable genomes. The occurrence of genomic instability also undermines the use of PSCs in regenerative medicine since these cells are known to induce tumors once injected in the organism.

2. Molecular basis of genomic instability

Genomic instability consists in the tendency of cells to accumulate mutations that directly or indirectly affect the structure of the genome, such as deletions, translocations, variation in the chromosomes copy number (CNVs) [5]. Maintenance of genome stability depends upon cellular processes that regulate DNA metabolism, such as DNA replication, transcription, repair, chromatin remodeling and their coordination with the cell cycle. Such coordination is orchestrated by cell cycle checkpoints [6]. Once activated, these signaling pathways slow down the cell cycle, activate DNA repair, and promote recovery of proliferation so to ensure that genetic information is faithfully transmitted to the daughter cells. For instance, the S-phase checkpoint restrains the onset of M-phase so to ensure that all DNA has been replicated before cells enter division. On the other hand, M-phase checkpoint delays anaphase so to ensure that condensed chromosomes are faithfully transmitted to the daughter cell. Importantly, checkpoints also preserve tissues homeostasis, since they can trigger cell death to avoid propagation of cells with unstable and/or highly damaged genomes [5, 7] (see also **Figure 1**).

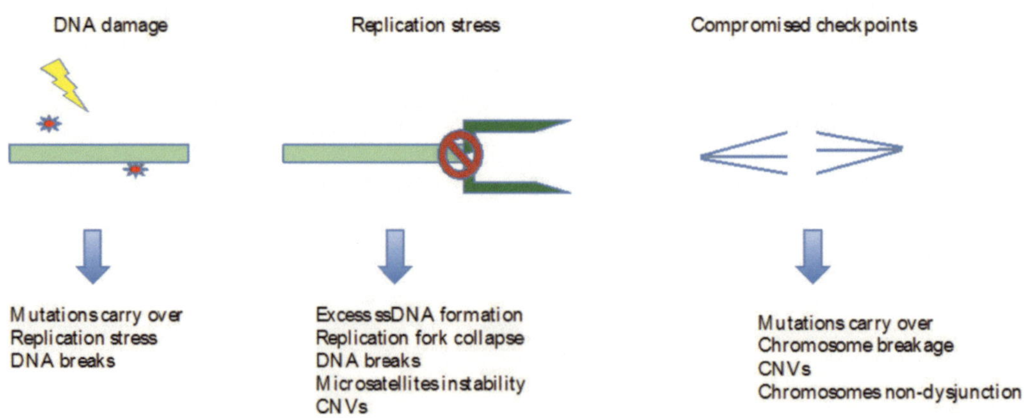

Figure 1. Main causes of genomic instability. Schematic representation of the main causes of genomic instability observed in cells. See text for more details.

2.1. DNA damage

Exogenous cues, such as chemical and radiations, for instance, but equally the metabolism of the DNA itself, generate DNA damage that threatens genome integrity. DNA damage elicits a DNA damage response (DDR) by activating cell cycle checkpoints [8]. Efficient DNA repair mechanisms ensure that DNA lesions are fixed to minimize loss or modification of the genetic information; among these are nucleotide excision repair (NER), base excision repair (BER), mismatch repair (MMR), post-replication repair (PRR), interstrand crosslinks (ICL) repair, homologous recombination (HR), and non-homologous end joining (NHEJ). A defect in any of these DNA repair pathways can compromise genome stability directly, by affecting the structure of the chromosome, or indirectly by generating mutations in genes important for the maintenance of genomic stability. Hence, occurrence of mutations in genes controlling cell cycle checkpoints, promote strong chromosome imbalance [9]. Notwithstanding, not all of these pathways allow faithfully repair of DNA lesions. For instance, repair of double-strand breaks (DSBs) by NHEJ, which involves fusion of broken DNA ends after processing without template assistance, is error-prone [10], in contrast to HR which requires a DNA template for repair and is more error-free. Equally, ICL repair and PRR, both involving DNA translesion synthesis, are also error-prone.

2.2. DDR

The DDR involves the activation of apical PI3KK protein kinases DNA–PK*cs*, ATR, and ATM. ATR is most closely related to ATM, a protein kinase encoded by the gene mutated in the ataxia telangiectasia syndrome. This disorder is characterized by a greatly reduced ability to repair radiation-induced DSBs and increased risk of developing cancer [11]. Activation of PI3KK follows a phosphorylation cascade that leads to the activation of a large number of substrates [12], among which the tumor suppressor proteins p53, BRCA1, and CHK1. These proteins also gather at sites of DNA damage and inhibit DNA replication and cell division apart from promoting DNA repair, recombination, or apoptosis. For example, activated CHK1 (its phosphorylated form) delays cell cycle progression, stabilizes stalled replication forks, and induces the S-phase checkpoint [13]. ATR is activated following several forms of DNA damage, including damaged nucleotides, stalled replication forks, and indirectly by DSBs [14]. ATM instead is more specialized in the response to DSBs and in sensing modifications of the chromatin state. DNA–PK*cs* is involved in the repair of DSBs by non-homologous recombination, and more recently, it has also been implicated in signaling DNA damage synergistically with ATR [15–17]. CHK1 and/or CHK2 phosphorylation mediates cell cycle slow down or arrest by affecting the stability and post-translational state of master cell cycle regulators, such as CDC25 proteins (A, B, and C) and CDKs.

In S-phase, ATR is chromatin-bound to monitor replication fork progression [18] and is activated following generation of excess single-stranded (ss) DNA as a result of replication forks delay or stalling at damaged sites [19]. ATR activation requires synthesis of replication intermediates onto ssDNA followed by the recruitment of specific proteins that recognize this substrate, such as Rad17 and the 9-1-1 checkpoint clamp [20–24]. ATR-dependent phosphorylation of the histone variant H2AX (γH2AX) constitutes a widely used marker of replication stress (RS) and genomic instability [25]. ATR can be also recruited and activated at DSBs following generation of ssDNA by resection [14].

ATM assembles at DSBs immediately after their formation. Its recruitment depends upon the MRN trimeric complex (made of Mre11, Rad50, and Nbs1 proteins), which holds two DNA ends together, by interacting with Nbs1. DNA damage results in ATM conversion from an inactive homodimer into an active monomer with protein kinase activity [26], which phosphorylates effector molecules that carry out the DDR including H2AX, p53, BRCA1, CHK2, RAD17, RAD9, NBS1 to form repair foci. The MDC1 protein is recruited by γH2AX *via* its BRCT domains and is phosphorylated by ATM, mediating the localization of ubiquitin ligase RNF8 that triggers monoubiquitination of H2AX. RNF168, a second ubiquitin ligase, is recruited and amplifies the ubiquitination response resulting in γH2AX polyubiquitination, which leads to the recruitment of Rad18, p53-binding protein 1 (53BP1), and BRCA1, among other proteins thus promoting DBSs repair by either HR or NHEJ [27].

2.3. RS

RS, defined as a more or less pronounced slow down or arrest of the DNA replication process, is a major source of genomic instability in proliferating cells [28]. Many obstacles can interfere with DNA synthesis. These can be specialized DNA or chromatin structures, or DNA damage (see below). The metabolism of the cell can also induce RS by affecting the availability of nucleotides and/or proteins required for DNA synthesis [29], as well as by production of reactive oxygen species (ROS) that generate a large amount of DNA lesions (about 100,000 per cell per day in an organism) [30]. RS is also generated by interference between DNA synthesis and DNA transcription induced, for instance, by unscheduled re-entry into the cell cycle, a situation observed during malignant transformation [31] and during reprogramming of somatic cells into iPSCs (see Section 3.2). Conflicts between DNA replication and transcription may lead to under replication of the genome [32, 33] as a result of DNA synthesis arrest, or to over replication as a result of aberrant reinitiation of DNA synthesis induced by certain oncogenes [32]. RS induces DNA damage (whose molecular bases are not completely understood) and thus generates a cellular response similar to that observed when cells are challenged with DNA damaging agents.

At the molecular level, the consequence of RS can be: (a) generation of excess ssDNA if the progression of the DNA polymerases and not that of the replicative DNA helicase is perturbed. In this situation, the ssDNA binding protein RPA is recruited and the replication fork can undergo remodeling in a process known as fork regression, dependent upon the Rad51 protein, to limit the extent of ssDNA; (*b*) a pause or a permanent arrest of the replication fork with no excess ssDNA formation due to an impediment to both DNA polymerases and helicase activities ([34] for review). In this situation, stalled replication forks can restart through generation of DSBs followed by resection and HR mediated by the PARP-1 enzyme [28]. (c) Generation of extra copies of the DNA as a result of over replication of the genome leading to collision between replication forks [32]. Recent evidence highlights the presence of regressed replication forks in G_2/M phases generated by unscheduled activation of Cyclin E and CDC25A [35]. At this stage, the endonuclease Mus81 can cleave the DNA and replication can occur to minimize the lost of genetic information. Nevertheless, DNA damage that could not be fixed before entering mitosis persists in the next cell cycle leading to the formation of nuclear bodies containing 53BP1 in G_1 phase [36]. These nuclear bodies appear symmetrical in the two daughter cells suggesting that they probably mark sister loci from the previous S-phase, where unresolved replication intermediates are still present. This ATM-dependent process hints to the possibility that ATM activation by RS is necessary to preserve genome integrity into following cell cycles.

RS-induced genome instability is a feature of almost all human cancers which can arise from mutations in DNA repair genes as stated by the mutator hypothesis [37]. According to this model, genomic instability is present in precancerous lesions and causes tumor development by increasing the spontaneous mutation rate. Then, mutations occurring in genes controlling the DNA damage checkpoint would allow anarchic proliferation of cells having collapsed replication forks and unstable genomes [38, 39]. The main instability found in cancer cells is chromosomal instability (CIN) or CNVs, where chromosome structure and number varies significantly in comparison with normal cells. Cancer cells can show other forms of genomic instability, including microsatellite instability (MSI or MIN), in which the number of DNA repeats present in microsatellite sequences increases or decreases [40, 41] in addition to increased frequencies of base pair mutations [42, 43].

3. Genomic instability of stem cells

Genomic instability has been extensively reported for ESCs, while ASCs appear to have a much more stable genome. Chromosomes 8 and 11 trisomy in ESCs [44, 45] and trisomy of chromosomes 12 and 17 in hESCs [46] with the amplification of the chromosome arm 20p in these latter have been observed [47]. These changes were reported to confer proliferation advantage. In addition, hESCs were reported to have the tendency to become aneuploid [48]. Very recent data now suggest that aneuploidy in hESCs arises as consequence of RS and chromosome condensation defects [49]. Detection of several markers of RS has been reported in ESCs (though without full activating the DDR) compared with somatic cells, suggesting that ESCs have a strong predisposition to genetic instability. One explanation for this feature may be that ESCs exhibit a contracted cell cycle structure, consisting of a short G_1- and G_2-phase and a high proportion of cells in S-phase [50–53]. These cells are also marked by open heterochromatic structure and an abundance of chromatin-remodeling factors [54, 55].

3.1. ESCs

Due to a highly contracted cell cycle, mESCs have an inefficient G_1/S checkpoint which does not allow them to arrest in G_1 in the presence of DNA damage [56–58], while the S-phase checkpoint is normally activated [53]. The consequence of this regulation is that lesions generated in G_1 are not sensed and therefore cannot be efficiently repaired, so they will persist in S-phase. For instance, unrepaired ssDNA breaks generated in G_1 may be replicated during S-phase, thus generating DSBs that in turn can induce genomic rearrangements. Curiously, the situation seems to be inversed in hESCs, where the G_1/S checkpoint has been suggested to be functional [59], while the S-phase checkpoint appears to be inefficient [60]. This difference can be explained as possible differences in the molecular circuits that regulate pluripotency between mouse and human stem cells. Absence of a G_1/S checkpoint in mESCs was originally suggested to be due to inefficient p53 function [57]. However, it has been shown that p53 can transactivate target genes in these cells [53, 56, 61]. The molecular grounds of inefficient G1/S checkpoint in mESCs have been more recently explained by the presence of high levels of the CDC25A phosphatase [58] due to its stabilization by the ubiquitin hydrolase DUB3 which is expressed at high levels in mESCs [53]. This results in constitutive dephosphorylation of the CDK2 kinase, which pushes cells into S-phase even in the presence of DNA damage, similar to the phenotype observed in cancer cells overexpressing CDC25A [62]. Interestingly, DUB3 has been shown to regulate the ubiquitination of both H2AX and γH2AX in somatic cells [63]. If this is also the case in mESCs, then it may explain why these cells repair inefficiently DSBs [64], aside from expressing low levels of DNA–PK*cs* [65]. Suppression of the G_1/S checkpoint is untimely linked to pluripotency. The expression of the DUB3 gene in mESCs [53] and that of the CDC25A gene in hESCs [66] are under control of pluripotency factors. Indeed, down-regulation of CDC25A induces a G_1/S delay upon DNA damage and cells spontaneously differentiate [53]. Consistent with this observation, DUB3 is more rapidly downregulated than OCT4 upon onset of differentiation (starting from day 1), making this gene a novel and highly specific marker of pluripotency in mESCs. Another work has shown that the contraction of the G_1 phase is crucial to suppress differentiation of mESCs [67]. Collectively these observations suggest that cell cycle contraction is an essential feature of pluripotency in mESCs.

mESCs exhibit spontaneous formation of γH2AX, RPA, and Rad51 foci but do not appear to display DSBs accumulation consistent with the absence of 53BP1 foci [68, 69], although activation of downstream DDR transducers (CHK1/2, CDC25A) does not seem to be affected [53]. It is possible to envisage the presence of multiple levels of regulation of the S-phase checkpoint by various factors, such as effectors of signaling pathways, unique to stem cells. One example is provided by the observation that the CHK2 kinase appears to be sequestered at the centrosome in mESCs so that it is not activated following induction of DSBs [70]. New evidence suggests that H2AX phosphorylation in cultured ESCs is neither DNA–PK*cs*- nor ATM-dependent but is in part ATR-dependent. This is associated with ssDNA gaps accumulation, reduced fork speed, and frequent fork reversal. All these features are lost upon onset of differentiation [71]. Why is ATR spontaneously activated in mESCs? Ahuja et al. [71] show that hypoxia, DNA methylation, and transcription do not seem to be the main cause of RS in mESCs. RS appears to be linked to the maintenance of self-renewal of embryonic stem cells. Turinetto et al. [68] demonstrated that γH2AX level decrease during mESCs differentiation, while it increases upon treatment with self-renewal-enhancing small molecules such as GSK and MEK inhibitor, which correlates with increased OCT4 and NANOG expression. Further, a pluripotent state-specific gene, named FILIA, has been recently shown to be important for genomic stability in mESCs [72]. This protein is constitutively localized to the centrosomes, is recruited to DNA damage sites, where it stimulates PARP1 enzymatic activity, and contributes to CHK2 activation independently of ATM.

The main kind of spontaneous mutations observed in mESCs are loss of heterozygosity as a consequence of chromosome loss/reduplication. However, the mutation rate of mESCs has been found be to 100 times lower than that of isogenic somatic cells [73]. This is surprising given the high level of RS observed in mESCs and may suggest that these cells could counterbalance genetic instability by increased DNA repair efficiency [74]. One of this could be reduced efficiency of mitotic recombination in ESCs compared to somatic cells as observed ([73] and references therein). In addition, because mESCs have an inefficient G_1/S checkpoint, they activate DNA damage- and p53-dependent differentiation if injured, as a way to enter apoptosis more easily since this checkpoint is restored in differentiated cells [75]. This latter control mechanism then avoids that damaged cells would be part of the pool of differentiated cells.

3.2. iPSCs

Somatic cells expressing defined pluripotency factors can be reprogrammed into iPSCs [76]. These cells share several similarities with ESCs such as a similar contracted cell cycle [77], the ability to undergo self-renewal and differentiation, as well as expression of pluripotency markers such as NANOG, OCT4, SOX2, and SSE-4 amongst others. Reprogramming increases γH2AX levels [78–80] and induces accumulation of genomic aberrancies ranging from whole chromosome aneuploidies, CNVs to point mutations [81], as well as epigenetic abnormalities [82]. The mutation frequency of iPSCs is also increased and has been estimated to be 10 times higher than that of ESCs [83–84]. This is in some way not surprising since the reprogramming protocol involves overexpression of oncogenes, such as *c-myc*, that introduces RS. Decreased genomic instability can be achieved by overexpression of the CHK1 kinase or by nucleosides supplementation during reprogramming [80]. ATM is also important for reprogramming. It has been reported that iPSCs deleted of ATM reprogram less efficiently and have increased genomic instability. Interestingly, these cells display gene expression profiles similar to wild-type ESCs and maintain the ability to differentiate into all three germ layers [85]. In line with this data, iPSCs exhibit G2/M cell cycle arrest and efficient DSB repair if ATM-dependent checkpoint activation signaling cascade is activated by ionizing radiation. iPSCs arrest the cell cycle in G2-phase and repair DSB by HR probably by overexpressing DNA repair genes [86]. Altogether these observations point out to a general requirement for the DNA damage checkpoint in sustaining reprogramming, suggesting that forced induction of proliferation induces RS and cells need a functional DDR to cope with this.

3.3. ASCs

ASCs are characterized by a narrower differentiation potential than ESCs. These cells self-renew to preserve both specific tissue and organ homeostasis throughout the life of an individual. Although ASCs show much less signs of genomic instability than ESCs, they deteriorate with age [87]. It is likely that the accumulation of lesions and mutations observed during ageing of stem cells is caused by acquired defects in DNA repair pathways that reduce stem cell potential. Interestingly, defective DNA repair is tightly linked to regeneration failure in certain tissues. Fanconi anemia patients, who are deficient in ICL repair, are characterized by a premature failure of bone marrow hematopoiesis. This event is triggered by the accumulation of DNA lesions leading to excessive DDR activation in hematopoietic stem cells (HSCs) and their progenitors [88]. In addition, NER is required for the maintenance of HSCs and prevention of premature ageing [89]. NHEJ is critical for the maintenance of skeletal muscle and muscle stem cells, since decreased Ku80 expression (a subunit of the heterodimeric complex made of Ku70 and 80 proteins that that functions with DNA–PK*cs* in NHEJ) causes accelerated exhaustion of stem cell pool and ageing [90]. In HSCs and their progenitors, ROS accumulation can be provoked by loss of ATM, affecting cell cycle progression. Conditional depletion of ATR or its downstream effector CHK1 is responsible for premature ageing phenotypes in skin, bones, small intestine, and the hematopoietic system [91, 92], resulting in apoptosis and cell cycle arrest because of rapid accumulation of DNA lesions [93, 94].

Interestingly, it has been reported that aging HSCs have a higher rate of genomic instability than young HSCs, fuelled by a high level of RS generated by the reduced expression of components of the MCM2-7 replicative helicase [95]. Reduced expression of the MCM3 gene was also recently shown to be sufficient to impair hematopoietic progenitor cells due to RS [94]. A recently identified protein, NUCLEOSTEMIN, rules a primary function in maintaining the genomic stability of neural stem cells. This protein promotes recruitment of RAD51 to replication-induced DNA damage foci and activates growth arrest independently of p53 [96]. Analysis of the transcriptional program of ESCs compared to ASCs (i.e., neural and hematopoietic) showed unexpected high similarities of gene expression profiles and identified a core set of about 200 genes expressed in all three cell types, accordingly coined as the "*stemness*" factors [97].

Cell type	Causes of genomic instability of stem cells
ESCs	Short cell cycle
	Inefficient checkpoints
	RS
iPSCs	Short cell cycle
ASCs	Inefficient checkpoints
	RS
	Reprogramming-induced DNA damage
	Mutations carry over
	High mutation frequency
	Aging-induced RS*

* Observed in HSCs [100].

Table 1. Summary of main causes of genomic instability of pluripotent stem cells.

The cell cycle of ASCs is remarkably different from that of ESCs. ASCs are mostly quiescent (being mostly in the G_0 state) and display a very slow cell cycle. For example, about 75% of HSCs reside in G_0 [98], whereas ESCs grown in culture display less than 20% of cells in G_1. It has been suggested that HSCs may divide once every 145 days (about five times during a mouse lifetime [99]), while ESCs divide every 11–12 h. These features make ASCs able to activate checkpoints and allow efficient repair. Since NHEJ can also act during G_0/G_1 because of its template independency, HSCs make an attempt to avoid DNA lesions by maintaining a hypoxic status [100] and decreasing the generation of ROS. In these conditions, ATP is generated mainly through glycolysis rather than mitochondrial respiration. Since this latter is activated only following cell cycle entry [101, 102], it explains why the first process is mainly employed by HSCs that are usually quiescent (**Table 1**).

4. Implications of PSCs genomic instability in regenerative medicine

PSCs are of great interest for their use in cell-based therapy. Current protocols involve PSCs differentiation into a specific cell type and then injection into an organ in the aim of replacing existing faulty cells. From a clinical point of view, this is a major concern due to the threat of transplanting immature cells with instable genome. Indeed, when injected in mice, immature PSCs induce teratoma. Hence, a greater understanding of the factors that regulate genomic stability in PSCs is critical to address this issue.

4.1. Hurdles in translating iPSCs technology into the clinic: problems and perspectives

Genetic instability and a high mutation rate constitute the dark side of iPSCs when taken into the clinic [4]. Hence, current efforts are made to generate iPSCs with reduced mutation load and having more stable genomes. Ji et al. [103] have shown that antioxydants reduce the level of γH2AX and *de novo* formation of CNVs in iPSCs suggesting that excessive ROS production in iPSCs increases their genetic instability. Indeed, a very recent report that analyzed the "metabolome" of naïve ESCs compared to that of primed ESCs show significant differences between these two cell types, which in turn may impinge on the level of ROS [104]. Further, recent work suggests that the use of non-integrative vectors to induce reprogramming significantly reduces the number of CNVs in the resulting iPSCs [105, 106]. Furthermore, a recent report that analyzed the mutational load of three distinct pluripotency induction methods shows that a non-integrative approach results in lower mutation load than either retrovirus or Sendai virus-based reprogramming methods [104]. Because integrative vectors induce DNA damage by generating DSBs, this suggests that the manipulation of the DDR can be a useful tool to reduce the genetic instability of iPSCs. It is then conceivable to think that DNA damage generated during reprogramming may be not well taken care, one reason being that iPSCs have inefficient checkpoints [86]. Indeed, a recent report indicates that manipulating the DDR can decrease the genomic instability of iPSCs [80]. This work shows that increasing the cellular levels of the CHK1 protein kinase decreases the level of γH2AX in these cells. In sum, in order to reduce undesired genetic burden arising during reprogramming of somatic cells, supplementing medium with both antioxidants and nucleoside should be combined to significantly reduce RS and CNVs in iPSCs.

In an effort to reduce genetic manipulation and consequent DNA damage, it has been shown that several transcription factors needed for iPSCs generation (except OCT4) can be replaced with a cocktail of chemical compounds [107]. More recently, generating iPSCs with a mix of small molecules inhibitors that can also replace OCT4 appears to strongly suppress the level of γH2AX, suggesting a reduction in spontaneous DNA damage, while keeping these cells pluripotent [108, 109]. As an example for the downside of genetic manipulation, NANOG expression has been shown to be cell cycle-regulated in human and mouse ESCs [53, 66, 110], whereas during reprogramming, this transcription factor is under constitutive expression. Altogether these data suggest that the use of an optimized set of chemical compounds may not alter natural gene expression during reprogramming and therefore would likely reduce unwanted off-target effects otherwise generated using genetic manipulation for reprogramming. Hence, chemical reprogramming remains a potentially more appropriate method since standardization of the approach is foreseeable and paves a new way of keeping genomic instability of iPSCs under control using pharmacological inhibitors.

Interestingly, mutation in the cell surface protein Glypican4 (Gpc4), implicated in the control of the Wnt/ß-catenin signaling pathway, has been reported to strongly reduce formation of teratoma upon implantation of mESCs in nude mice without affecting pluripotency [111]. Gpc4 mutant ESCs appear to be able to differentiate in all three embryonic layers when injected into developing blastocysts, although with faster kinetics compared to wild-type ESCs. These data propose Gpc4 as a promising target to modulate the teratogenic potential of ESCs. Indeed, more recent data show that ESCs bearing a hypomorphic Gpc4 allele improve recovery of motoneuron defects in a rat model for Parkinson disease without generating teratoma [112]. It is not yet known whether Gpc4 mutations have a similar effect also on human ESCs or iPSCs, and whether spontaneous DNA damage and/or genomic instability are affected.

5. Conclusions and perspectives

Differentiation of iPSCs has been successfully achieved to generate hematopoietic cells, neurons, pancreatic β-islet, and cardiomyocytes; however, production of other cell types is still challenging. One major hurdle is the efficiency of differentiation that still remains very low. In addition, PSCs show several signs of genetic instability, not only in culture but also *in vivo* [71, 106], yet embryos manage to keep this instability under control by generating viable and healthy organisms. Hence, the question arises of how this control is achieved. First, cells with unstable genomes can be eliminated by apoptosis during differentiation, which is actually what it is observed during *in vitro* differentiation. However, γH2AX detection in blastocysts shows that most of the cells stain positive for this marker [71, 106], which makes unlikely that most of them bear indeed highly unstable genomes. Another possibility is that the γH2AX observed in these cells is not only a mark of genetic instability but perhaps also a marker of other DNA transactions, including chromatin remodeling. Chromatin remodeling is known to change dramatically during differentiation; hence, the decrease of γH2AX observed upon differentiation onset may be also due to changes in chromatin structure. If this is the case, the chromatin structure and epigenetic marks responsible for constitutive γH2AX in ESCs remain to be discovered. Furthermore, recent data show the presence of a high proportion of ssDNA gaps in mESCs; however, the link between these gaps and the level of γH2AX is unclear. Despite the high level of genetic instability, a highly contracted cell cycle and an inefficient G1/S checkpoint, the mutation rate of mESCs has been surprisingly reported to be lower than that of isogenic somatic cells. The significance of this discrepancy needs to be further understood.

How our current knowledge on PSCs can be translated into improving their genetic stability so to foster the development of PSCs with lower mutation load that can be used with success in regenerative medicine? In principle, identifying the molecular basis of genomic instability of PSCs opens the perspective of manipulating the genes implicated, in the aim to decrease their tendency to introduce mutations and so doing, reduce their teratogenicity. For instance, being able to manipulate the structure of the cell cycle of PSCs so to decrease RS and still maintain pluripotency may be of value. Further work in this direction is expected to generate novel insights and hopes into this rather difficult though exciting task. Clinically-compatible quantitative methods to comprehensively analyze the genetic stability of iPSCs would greatly facilitate the selection process of most appropriate iPSCs clones. Recent efforts have shifted the practice and proposed pathway signaling as readout to compare for functionality [113]. Ease of application and selection would guarantee large-scale testing in clinics.

Author details

Elena Lo Furno[1], Siem van der Laan[2] and Domenico Maiorano[1*]

*Address all correspondence to: domenico.maiorano@igh.cnrs.fr

1 Institute of Human Genetics, CNRS-UPR1142, Genome Surveillance and Stability Laboratory, Department of Molecular Bases of Human Diseases, University of Montpellier, Montpellier, France

2 CNRS FRE3690, Parc Euromedicine Cap Delta, Montpellier, France

References

[1] Wood RD (1996) DNA repair in eukaryotes. Annual Review of Biochemistry 65: 135–167.

[2] Kenyon J, Gerson SL (2007) The role of DNA damage repair in aging of adult stem cells. Nucleic Acids Research 35: 7557–7565.

[3] Hakem R (2008) DNA-damage repair; the good, the bad, and the ugly. The EMBO Journal 27: 589–605.

[4] Pera MF (2011) Stem cells: The dark side of induced pluripotency. Nature 471: 46–47.

[5] Shen Z (2011) Genomic instability and cancer: an introduction. Journal of Molecular Cell Biology 3: 1–3.

[6] Hartwell L (1992) Defects in a cell cycle checkpoint may be responsible for the genomic instability of cancer cells. Cell 71: 543–546.

[7] Matt S, Hofmann TG (2016) The DNA damage-induced cell death response: a roadmap to kill cancer cells. Cellular and molecular life sciences: CMLS.

[8] Goldstein M, Kastan MB (2015) The DNA damage response: implications for tumor responses to radiation and chemotherapy. Annual Review of Medicine 66: 129–143.

[9] Schuyler SC, Wu YF, Kuan VJ (2012) The Mad1-Mad2 balancing act—a damaged spindle checkpoint in chromosome instability and cancer. Journal of Cell Science 125: 4197–4206.

[10] Waters CA, Strande NT, Pryor JM, Strom CN, Mieczkowski P, et al. (2014) The fidelity of the ligation step determines how ends are resolved during nonhomologous end joining. Nature Communications 5: 4286.

[11] Savitsky K, Bar-Shira A, Gilad S, Rotman G, Ziv Y, et al. (1995) A single ataxia telangiectasia gene with a product similar to PI-3 kinase. Science 268: 1749–1753.

[12] Matsuoka S, Ballif BA, Smogorzewska A, McDonald ER, 3rd, Hurov KE, et al. (2007) ATM and ATR substrate analysis reveals extensive protein networks responsive to DNA damage. Science 316: 1160–1166.

[13] Gonzalez Besteiro MA, Gottifredi V (2015) The fork and the kinase: a DNA replication tale from a CHK1 perspective. Mutation Research Reviews in Mutation Research 763: 168–180.

[14] Cimprich KA, Cortez D (2008) ATR: an essential regulator of genome integrity. Nature Reviews Molecular Cell Biology 9: 616–627.

[15] Vidal-Eychenie S, Decaillet C, Basbous J, Constantinou A (2013) DNA structure-specific priming of ATR activation by DNA–PKcs. The Journal of Cell Biology 202: 421–429.

[16] Lin YF, Shih HY, Shang Z, Matsunaga S, Chen BP (2014) DNA–PKcs is required to maintain stability of Chk1 and Claspin for optimal replication stress response. Nucleic Acids Research 42: 4463–4473.

[17] Ashley AK, Shrivastav M, Nie J, Amerin C, Troksa K, et al. (2014) DNA–PK phosphorylation of RPA32 Ser4/Ser8 regulates replication stress checkpoint activation, fork restart, homologous recombination and mitotic catastrophe. DNA Repair 21: 131–139.

[18] Hekmat-Nejad M, You Z, Yee MC, Newport JW, Cimprich KA (2000) Xenopus ATR is a replication-dependent chromatin-binding protein required for the DNA replication checkpoint. Current Biology 10: 1565–1573.

[19] Byun TS, Pacek M, Yee MC, Walter JC, Cimprich KA (2005) Functional uncoupling of MCM helicase and DNA polymerase activities activates the ATR-dependent checkpoint. Genes & Development 19: 1040–1052.

[20] Van C, Yan S, Michael WM, Waga S, Cimprich KA (2010) Continued primer synthesis at stalled replication forks contributes to checkpoint activation. The Journal of Cell Biology 189: 233–246.

[21] Recolin B, Van der Laan S, Maiorano D (2012) Role of replication protein A as sensor in activation of the S-phase checkpoint in Xenopus egg extracts. Nucleic Acids Research 40: 3431–3442.

[22] Betous R, Pillaire MJ, Pierini L, van der Laan S, Recolin B, et al. (2013) DNA polymerase kappa-dependent DNA synthesis at stalled replication forks is important for CHK1 activation. The EMBO Journal 32: 2172–2185.

[23] Duursma AM, Driscoll R, Elias JE, Cimprich KA (2013) A role for the MRN complex in ATR activation via TOPBP1 recruitment. Molecular Cell 50: 116–122.

[24] Lee J, Dunphy WG (2013) The Mre11-Rad50-Nbs1 (MRN) complex has a specific role in the activation of Chk1 in response to stalled replication forks. Molecular Biology of the Cell 24: 1343–1353.

[25] Ward IM, Chen J (2001) Histone H2AX is phosphorylated in an ATR-dependent manner in response to replicational stress. Journal of Biological Chemistry 276: 47759–47762.

[26] Bakkenist CJ, Kastan MB (2003) DNA damage activates ATM through intermolecular autophosphorylation and dimer dissociation. Nature 421: 499–506.

[27] Lukas J, Lukas C, Bartek J (2011) More than just a focus: The chromatin response to DNA damage and its role in genome integrity maintenance. Nature Cell Biology 13: 1161–1169.

[28] Lambert S, Carr AM (2013) Impediments to replication fork movement: stabilisation, reactivation and genome instability. Chromosoma 122: 33–45.

[29] Zeman MK, Cimprich KA (2014) Causes and consequences of replication stress. Nature Cell Biology 16: 2–9.

[30] Fraga CG, Shigenaga MK, Park JW, Degan P, Ames BN (1990) Oxidative damage to DNA during aging: 8-hydroxy-2′-deoxyguanosine in rat organ DNA and urine. Proceedings of the National Academy of Sciences of the United States of America 87: 4533–4537.

[31] Gorgoulis VG, Vassiliou LV, Karakaidos P, Zacharatos P, Kotsinas A, et al. (2005) Activation of the DNA damage checkpoint and genomic instability in human precancerous lesions. Nature 434: 907–913.

[32] Neelsen KJ, Zanini IM, Mijic S, Herrador R, Zellweger R, et al. (2013) Deregulated origin licensing leads to chromosomal breaks by rereplication of a gapped DNA template. Genes & Development 27: 2537–2542.

[33] Jones RM, Mortusewicz O, Afzal I, Lorvellec M, Garcia P, et al. (2012) Increased replication initiation and conflicts with transcription underlie Cyclin E-induced replication stress. Oncogene.

[34] Recolin B, van der Laan S, Tsanov N, Maiorano D (2014) Molecular mechanisms of DNA replication checkpoint activation. Genes 5: 147–175.

[35] Neelsen KJ, Zanini IM, Herrador R, Lopes M (2013) Oncogenes induce genotoxic stress by mitotic processing of unusual replication intermediates. The Journal of Cell Biology 200: 699–708.

[36] Lukas C, Savic V, Bekker-Jensen S, Doil C, Neumann B, et al. (2011) 53BP1 nuclear bodies form around DNA lesions generated by mitotic transmission of chromosomes under replication stress. Nature Cell Biology 13: 243–253.

[37] Loeb LA (1991) Mutator phenotype may be required for multistage carcinogenesis. Cancer Research 51: 3075–3079.

[38] Halazonetis TD, Gorgoulis VG, Bartek J (2008) An oncogene-induced DNA damage model for cancer development. Science 319: 1352–1355.

[39] Kinzler KW, Vogelstein B (1997) Cancer-susceptibility genes. Gatekeepers and caretakers. Nature 386: 761, 763.

[40] Fishel R, Lescoe MK, Rao MR, Copeland NG, Jenkins NA, et al. (1993) The human mutator gene homolog MSH2 and its association with hereditary nonpolyposis colon cancer. Cell 75: 1027–1038.

[41] Leach FS, Nicolaides NC, Papadopoulos N, Liu B, Jen J, et al. (1993) Mutations of a mutS homolog in hereditary nonpolyposis colorectal cancer. Cell 75: 1215–1225.

[42] Al-Tassan N, Chmiel NH, Maynard J, Fleming N, Livingston AL, et al. (2002) Inherited variants of MYH associated with somatic G:C→T:A mutations in colorectal tumors. Nature Genetics 30: 227–232.

[43] Nowell PC (1976) The clonal evolution of tumor cell populations. Science 194: 23–28.

[44] Longo L, Bygrave A, Grosveld FG, Pandolfi PP (1997) The chromosome make-up of mouse embryonic stem cells is predictive of somatic and germ cell chimaerism. Transgenic Research 6: 321–328.

[45] Liu X, Wu H, Loring J, Hormuzdi S, Disteche CM, et al. (1997) Trisomy eight in ES cells is a common potential problem in gene targeting and interferes with germ line transmission. Developmental Dynamics 209: 85–91.

[46] Maitra A, Arking DE, Shivapurkar N, Ikeda M, Stastny V, et al. (2005) Genomic alterations in cultured human embryonic stem cells. Nature Genetics 37: 1099–1103.

[47] Lefort N, Feyeux M, Bas C, Feraud O, Bennaceur-Griscelli A, et al. (2008) Human embryonic stem cells reveal recurrent genomic instability at 20q11.21. Nature Biotechnology 26: 1364–1366.

[48] Baker DE, Harrison NJ, Maltby E, Smith K, Moore HD, et al. (2007) Adaptation to culture of human embryonic stem cells and oncogenesis in vivo. Nature Biotechnology 25: 207–215.

[49] Lamm N, Ben-David U, Golan-Lev T, Storchova Z, Benvenisty N, et al. (2016) Genomic instability in human pluripotent stem cells arises from replicative stress and chromosome condensation defects. Cell Stem Cell 18: 253–261.

[50] Becker KA, Ghule PN, Therrien JA, Lian JB, Stein JL, et al. (2006) Self-renewal of human embryonic stem cells is supported by a shortened G1 cell cycle phase. Journal of Cellular Physiology 209: 883–893.

[51] Savatier P, Lapillonne H, Jirmanova L, Vitelli L, Samarut J (2002) Analysis of the cell cycle in mouse embryonic stem cells. Methods in Molecular Biology 185: 27–33.

[52] Ballabeni A, Park IH, Zhao R, Wang W, Lerou PH, et al. (2011) Cell cycle adaptations of embryonic stem cells. Proceedings of the National Academy of Sciences of the United States of America 108: 19252–19257.

[53] van der Laan S, Tsanov N, Crozet C, Maiorano D (2013) High Dub3 expression in mouse ESCs couples the G1/S checkpoint to pluripotency. Molecular Cell 52: 366–379.

[54] Meshorer E, Misteli T (2006) Chromatin in pluripotent embryonic stem cells and differentiation. Nature Reviews Molecular Cell biology 7: 540–546.

[55] Pajerowski JD, Dahl KN, Zhong FL, Sammak PJ, Discher DE (2007) Physical plasticity of the nucleus in stem cell differentiation. Proceedings of the National Academy of Sciences of the United States of America 104: 15619–15624.

[56] Prost S, Bellamy CO, Clarke AR, Wyllie AH, Harrison DJ (1998) p53-independent DNA repair and cell cycle arrest in embryonic stem cells. FEBS Letters 425: 499–504.

[57] Aladjem MI, Spike BT, Rodewald LW, Hope TJ, Klemm M, et al. (1998) ES cells do not activate p53-dependent stress responses and undergo p53-independent apoptosis in response to DNA damage. Current Biology 8: 145–155.

[58] Koledova Z, Kafkova LR, Kramer A, Divoky V (2010) DNA damage-induced degradation of Cdc25A does not lead to inhibition of Cdk2 activity in mouse embryonic stem cells. Stem Cells 28: 450–461.

[59] Barta T, Vinarsky V, Holubcova Z, Dolezalova D, Verner J, et al. (2010) Human embryonic stem cells are capable of executing G1/S checkpoint activation. Stem Cells 28: 1143–1152.

[60] Desmarais JA, Hoffmann MJ, Bingham G, Gagou ME, Meuth M, et al. (2012) Human embryonic stem cells fail to activate CHK1 and commit to apoptosis in response to DNA replication stress. Stem Cells 30: 1385–1393.

[61] Sabapathy K, Klemm M, Jaenisch R, Wagner EF (1997) Regulation of ES cell differentiation by functional and conformational modulation of p53. EMBO Journal 16: 6217–6229.

[62] Falck J, Mailand N, Syljuasen RG, Bartek J, Lukas J (2001) The ATM-Chk2-Cdc25A checkpoint pathway guards against radioresistant DNA synthesis. Nature 410: 842–847.

[63] Delgado-Diaz MR, Martin Y, Berg A, Freire R, Smits VA (2014) Dub3 controls DNA damage signalling by direct deubiquitination of H2AX. Molecular oncology 8: 884–893.

[64] Momcilovic O, Navara C, Schatten G (2011) Cell cycle adaptations and maintenance of genomic integrity in embryonic stem cells and induced pluripotent stem cells. Results and Problems in Cell Differentiation 53: 415–458.

[65] Banuelos CA, Banath JP, MacPhail SH, Zhao J, Eaves CA, et al. (2008) Mouse but not human embryonic stem cells are deficient in rejoining of ionizing radiation-induced DNA double-strand breaks. DNA Repair 7: 1471–1483.

[66] Zhang X, Neganova I, Przyborski S, Yang C, Cooke M, et al. (2009) A role for NANOG in G1 to S transition in human embryonic stem cells through direct binding of CDK6 and CDC25A. Journal of Cell Biology 184: 67–82.

[67] Pauklin S, Vallier L (2013) The cell-cycle state of stem cells determines cell fate propensity. Cell 155: 135–147.

[68] Turinetto V, Orlando L, Sanchez-Ripoll Y, Kumpfmueller B, Storm MP, et al. (2012) High basal gammaH2AX levels sustain self-renewal of mouse embryonic and induced pluripotent stem cells. Stem Cells 30: 1414–1423.

[69] Banath JP, Banuelos CA, Klokov D, MacPhail SM, Lansdorp PM, et al. (2009) Explanation for excessive DNA single-strand breaks and endogenous repair foci in pluripotent mouse embryonic stem cells. Experimental Cell Research 315: 1505–1520.

[70] Hong Y, Stambrook PJ (2004) Restoration of an absent G1 arrest and protection from apoptosis in embryonic stem cells after ionizing radiation. Proceedings of the National Academy of Sciences of the United States of America 101: 14443–14448.

[71] Ahuja AK, Jodkowska K, Teloni F, Bizard AH, Zellweger R, et al. (2016) A short G1 phase imposes constitutive replication stress and fork remodelling in mouse embryonic stem cells. Nature Communications 7: 10660.

[72] Zhao B, Zhang WD, Duan YL, Lu YQ, Cun YX, et al. (2015) Filia is an ESC-specific regulator of DNA damage response and safeguards genomic stability. Cell Stem Cell 16: 684–698.

[73] Hong Y, Cervantes RB, Tichy E, Tischfield JA, Stambrook PJ (2007) Protecting genomic integrity in somatic cells and embryonic stem cells. Mutation Research 614: 48–55.

[74] Maynard S, Swistowska AM, Lee JW, Liu Y, Liu ST, et al. (2008) Human embryonic stem cells have enhanced repair of multiple forms of DNA damage. Stem Cells 26: 2266–2274.

[75] Lin T, Chao C, Saito S, Mazur SJ, Murphy ME, et al. (2005) p53 induces differentiation of mouse embryonic stem cells by suppressing Nanog expression. Nature Cell Biology 7: 165–171.

[76] Yamanaka S (2008) Induction of pluripotent stem cells from mouse fibroblasts by four transcription factors. Cell Proliferation 41(Suppl 1): 51–56.

[77] Ruiz S, Panopoulos AD, Herrerias A, Bissig KD, Lutz M, et al. (2011) A high proliferation rate is required for cell reprogramming and maintenance of human embryonic stem cell identity. Current Biology: CB 21: 45–52.

[78] Blasco MA, Serrano M, Fernandez-Capetillo O (2011) Genomic instability in iPS: time for a break. The EMBO Journal 30: 991–993.

[79] Marion RM, Strati K, Li H, Murga M, Blanco R, et al. (2009) A p53-mediated DNA damage response limits reprogramming to ensure iPS cell genomic integrity. Nature 460: 1149–1153.

[80] Ruiz S, Lopez-Contreras AJ, Gabut M, Marion RM, Gutierrez-Martinez P, et al. (2015) Limiting replication stress during somatic cell reprogramming reduces genomic instability in induced pluripotent stem cells. Nature Communications 6: 8036.

[81] Weissbein U, Benvenisty N, Ben-David U (2014) Quality control: Genome maintenance in pluripotent stem cells. The Journal of Cell Biology 204: 153–163.

[82] Lister R, Pelizzola M, Kida YS, Hawkins RD, Nery JR, et al. (2011) Hotspots of aberrant epigenomic reprogramming in human induced pluripotent stem cells. Nature 471: 68–73.

[83] Ji J, Ng SH, Sharma V, Neculai D, Hussein S, et al. (2012) Elevated coding mutation rate during the reprogramming of human somatic cells into induced pluripotent stem cells. Stem Cells 30: 435–440.

[84] Gore A, Li Z, Fung HL, Young JE, Agarwal S, et al. (2011) Somatic coding mutations in human induced pluripotent stem cells. Nature 471: 63–67.

[85] Kinoshita T, Nagamatsu G, Kosaka T, Takubo K, Hotta A, et al. (2011) Ataxia-telangiectasia mutated (ATM) deficiency decreases reprogramming efficiency and leads to

genomic instability in iPS cells. Biochemical and Biophysical Research Communications 407: 321–326.

[86] Momcilovic O, Knobloch L, Fornsaglio J, Varum S, Easley C, et al. (2010) DNA damage responses in human induced pluripotent stem cells and embryonic stem cells. PLos One 5: e13410.

[87] Liu L, Rando TA (2011) Manifestations and mechanisms of stem cell aging. Journal Cell Biology 193: 257–266.

[88] Ceccaldi R, Parmar K, Mouly E, Delord M, Kim JM, et al. (2012) Bone marrow failure in Fanconi anemia is triggered by an exacerbated p53/p21 DNA damage response that impairs hematopoietic stem and progenitor cells. Cell Stem Cell 11: 36–49.

[89] Rossi DJ, Bryder D, Seita J, Nussenzweig A, Hoeijmakers J, et al. (2007) Deficiencies in DNA damage repair limit the function of haematopoietic stem cells with age. Nature 447: 725–729.

[90] Didier N, Hourde C, Amthor H, Marazzi G, Sassoon D (2012) Loss of a single allele for Ku80 leads to progenitor dysfunction and accelerated aging in skeletal muscle. EMBO Molecular Medicine 4: 910–923.

[91] Ruzankina Y, Pinzon-Guzman C, Asare A, Ong T, Pontano L, et al. (2007) Deletion of the developmentally essential gene ATR in adult mice leads to age-related phenotypes and stem cell loss. Cell Stem Cell 1: 113–126.

[92] Greenow KR, Clarke AR, Jones RH (2009) Chk1 deficiency in the mouse small intestine results in p53-independent crypt death and subsequent intestinal compensation. Oncogene 28: 1443–1453.

[93] Ruzankina Y, Schoppy DW, Asare A, Clark CE, Vonderheide RH, et al. (2009) Tissue regenerative delays and synthetic lethality in adult mice after combined deletion of Atr and Trp53. Nature Genetics 41: 1144–1149.

[94] Alvarez S, Diaz M, Flach J, Rodriguez-Acebes S, Lopez-Contreras AJ, et al. (2015) Replication stress caused by low MCM expression limits fetal erythropoiesis and hematopoietic stem cell functionality. Nature Communications 6: 8548.

[95] Flach J, Bakker ST, Mohrin M, Conroy PC, Pietras EM, et al. (2014) Replication stress is a potent driver of functional decline in ageing haematopoietic stem cells. Nature 512: 198–202.

[96] Meng L, Lin T, Peng G, Hsu JK, Lee S, et al. (2013) Nucleostemin deletion reveals an essential mechanism that maintains the genomic stability of stem and progenitor cells. Proceedings of the National Academy of Sciences of the United States of America 110: 11415–11420.

[97] Ramalho-Santos M, Yoon S, Matsuzaki Y, Mulligan RC, Melton DA (2002) "Stemness": transcriptional profiling of embryonic and adult stem cells. Science 298: 597–600.

[98] Cheshier SH, Morrison SJ, Liao X, Weissman IL (1999) In vivo proliferation and cell cycle kinetics of long-term self-renewing hematopoietic stem cells. Proceedings of the National Academy of Sciences of the United States of America 96: 3120–3125.

[99] Wilson A, Laurenti E, Oser G, van der Wath RC, Blanco-Bose W, et al. (2008) Hematopoietic stem cells reversibly switch from dormancy to self-renewal during homeostasis and repair. Cell 135: 1118–1129.

[100] Nombela-Arrieta C, Pivarnik G, Winkel B, Canty KJ, Harley B, et al. (2013) Quantitative imaging of haematopoietic stem and progenitor cell localization and hypoxic status in the bone marrow microenvironment. Nature Cell Biology 15: 533–543.

[101] Takubo K, Nagamatsu G, Kobayashi CI, Nakamura-Ishizu A, Kobayashi H, et al. (2013) Regulation of glycolysis by Pdk functions as a metabolic checkpoint for cell cycle quiescence in hematopoietic stem cells. Cell Stem Cell 12: 49–61.

[102] Yu WM, Liu X, Shen J, Jovanovic O, Pohl EE, et al. (2013) Metabolic regulation by the mitochondrial phosphatase PTPMT1 is required for hematopoietic stem cell differentiation. Cell Stem Cell 12: 62–74.

[103] Ji J, Sharma V, Qi S, Guarch ME, Zhao P, et al. (2014) Antioxidant supplementation reduces genomic aberrations in human induced pluripotent stem cells. Stem Cell Reports 2: 44–51.

[104] Sperber H, Mathieu J, Wang Y, Ferreccio A, Hesson J, et al. (2015) The metabolome regulates the epigenetic landscape during naive-to-primed human embryonic stem cell transition. Nature Cell Biology 17: 1523–1535.

[105] Kang X, Yu Q, Huang Y, Song B, Chen Y, et al. (2015) Effects of integrating and non-integrating reprogramming methods on copy number variation and genomic stability of human induced pluripotent stem cells. PLos One 10: e0131128.

[106] Sobol M, Raykova D, Cavelier L, Khalfallah A, Schuster J, et al. (2015) Methods of reprogramming to induced pluripotent stem cell associated with chromosomal integrity and delineation of a chromosome 5q candidate region for growth advantage. Stem Cells and Development 24: 2032–2040.

[107] Zhu S, Li W, Zhou H, Wei W, Ambasudhan R, et al. (2010) Reprogramming of human primary somatic cells by OCT4 and chemical compounds. Cell Stem Cell 7: 651–655.

[108] Xu Y, Shi Y, Ding S (2008) A chemical approach to stem-cell biology and regenerative medicine. Nature 453: 338–344.

[109] Park HS, Hwang I, Choi KA, Jeong H, Lee JY, et al. (2015) Generation of induced pluripotent stem cells without genetic defects by small molecules. Biomaterials 39: 47–58.

[110] Singh AM, Chappell J, Trost R, Lin L, Wang T, et al. (2013) Cell-cycle control of developmentally regulated transcription factors accounts for heterogeneity in human pluripotent cells. Stem Cell Reports 1: 532–544.

[111] Fico A, De Chevigny A, Egea J, Bosl MR, Cremer H, et al. (2012) Modulating Glypican4 suppresses tumorigenicity of embryonic stem cells while preserving self-renewal and pluripotency. Stem Cells 30: 1863–1874.

[112] Fico A, de Chevigny A, Melon C, Bohic M, Kerkerian-Le Goff L, et al. (2014) Reducing glypican-4 in ES cells improves recovery in a rat model of Parkinson's disease by increasing the production of dopaminergic neurons and decreasing teratoma formation. The Journal of Neuroscience 34: 8318–8323.

[113] Makarev E, Fortney K, Litovchenko M, Braunewell KH, Zhavoronkov A, et al. (2015) Quantifying signaling pathway activation to monitor the quality of induced pluripotent stem cells. Oncotarget 6: 23204–23212.

Section 3

Culture methods

Chapter 7

Moving toward Xeno-free Culture of Human Pluripotent Stem Cells

Kamthorn Pruksananonda and
Ruttachuk Rungsiwiwut

Additional information is available at the end of the chapter

http://dx.doi.org/10.5772/62770

Abstract

Human pluripotent stem cells (hPSCs) were conventionally cultured on feeder cells that are isolated from mouse embryonic fibroblast (MEF). However, these culture components could contaminate the hPSCs and can limit the application of hPSCs for clinical use. On the other hand, we demonstrated that exogenous basic fibroblast growth factor (bFGF) could be omitted from the hPSC culture media if we used the suitable feeder cells. We also showed that although hPSCs can proliferate on the feeder-free culture system, however, genetic instability of hPSCs has been reported in such environment. Feeder cells enable hPSCs to maintain their pluripotency. The feeder cells are usually grown in a culture medium containing fetal bovine serum (FBS) prior to coculture with hPSCs. The use of FBS might limit the clinical application of hPSCs. We proposed the use of human cord blood-derived serum (hUCS) and showed a positive effect on culture of mesenchymal stem cells. The results showed that human foreskin fibroblasts (HFFs) cultured in hUCS-containing medium (HFF-hUCS) displayed fibroblastic features, short population doubling times, high proliferation rates, and normal karyotypes after prolonged culture. These studies of hPSC xeno-free culture have been growing in both basic research and clinical trial. The data regarding the current clinical trials of using hPSCs convince the researchers not only about the possibility of application of hPSCs for cell-based therapy, but also the quality of established hPSC lines. Most of the hPSC lines that were published in the literature and registered in the National Institute of Health (NIH), hPSCreg of the European Union are not Good Manufacturing Practice (GMP) grade cell lines. Since one of the goals of using hPSCs is therapeutic purpose, GMP for derivation, cultivation, and handling the hPSCs are required. This chapter also reviews the state-of-the-art xeno-free culture system of hPSCs in the respect of future clinical applications.

Keywords: xeno-free, pluripotent stem cells, cell culture, cell therapy, regenerative medicine

1. Introduction

The ultimate goal of pluripotent stem cell research is to improve quality of life and patient treatment. The human pluripotent stem cells (hPSCs) can be classified as human embryonic stem cells (hESCs) and human-induced pluripotent stem cells (hiPSCs). hESCs can be derived from the inner cell mass (ICM) of the preimplantation embryos [1]. hiPSCs can be generated by reprogramming the somatic cells using the exogenous transcriptional factors and resulting in the pluripotent cells [2]. The successful derivation of hESCs and hiPSCs holds the great hope of treatment of incurable diseases, and the hPSC-related studies have been growing with regard to the potential of hPSCs.

hPSCs was firstly cultured in the medium containing the fetal bovine serum (FBS) and cocultured with mouse embryonic fibroblast (MEF) feeders [1, 2]. However, such conditions might introduce the contamination of animal pathogens to the hPSCs and make hPSCs unsuitable for clinical applications. Thus, elimination of all animal products during the derivation, long-term culture as well as differentiation of hPSCs is necessary prior to application of hPSCs in clinical cell therapy. Because the ultimate aim of pluripotent stem cell research is to improve quality of life and patient outcome, the more defined culture systems of hPSCs were progressively developed. The researchers firstly developed xeno-free culture system that consisted of a growth medium containing Knockout serum replacement (KO-SR) and basic fibroblast growth factor (bFGF) instead of FBS [3–5]. MEF feeder cells were replaced by the humanized feeder cells derived from human tissues. Later, the researchers have developed various xeno-free culture media, and xeno-free extracellular matrices which support the growth and pluripotency of hPSCs [6–10]. Recently, commercial xeno-free, defined culture media and extracellular matrix has been developed and available for the researchers.

2. Human pluripotent stem cells properties

hPSCs are widely accepted as the valuable source for cell-based therapy due to their ability of differentiation toward three embryonic germ layers and process cell division indefinitely [11]. Although the origin of hESCs and hiPSCs is different, they share similarities in the morphological appearances and the molecular levels. The unique morphological appearances of hPSCs can be identified by the tightly pack colonies with defined borders. The clear cytoplasm and distinctive nuclei of the individual cells can be observed. hPSCs expressed the pluripotent-related transcriptional factors including OCT-4, Nanog, Sox2, Nodal, hTERT, Rex1 as well as positive for stage-specific-embryonic antigen (SSEA)-3, SSEA −4, tumor-recognition antigen (TRA)-1-60 and −1-81, expressed high level of alkaline phosphatase and telomerase activity [1–

10]. Mellon et al. 12] and Guenther et al. [13] reported that there were no significant gene expression differences between hESCs and hiPSCs. However, some studies reported the differences in the methylation profile between hESCs and hiPSCs [14–16]. This can reflect the genetic and epigenetic abnormalities during the reprogramming process or even the differences of cell culture condition in different laboratories [17]. Although the similarities and differences between hESCs and hiPSCs are contradictory, the researchers around the world tried to standardize the culture conditions and characterization of hPSCs with the ultimate goal to achieve the development of the hPSC culture system for the future clinical application [18, 19].

3. Good Manufacturing Practice for human Pluripotent Stem Cells

Since one of the goals of using hPSCs is therapeutic purpose, Good Manufacturing Practice (GMP) for derivation, cultivation, and handling the hPSCs are required. Most of the hPSC lines that were registered in the National Institute of Health (NIH), hPSCreg of the European Union as well as the data published in the literature are not GMP grade cell lines. Non-GMP grade cell lines are usually contacts to the animal products especially through the feeder cells, serum, and extracellular matrix [20, 21]. The researchers found that culture of hESCs on the feeder cells derived from the mouse embryonic tissues may contaminate nonhuman sialic acid (Neu5Gc) molecules, which can stimulate the immune rejection after transplantation [22]. Moreover, FBS that was supplemented in the culture medium of feeder cells might contain bovine pathogens such as bovine spongiform encephalitis (BSE) and limits the use of hPSCs for clinical applications [23]. Therefore, the materials for hPSC culture system that compliant for clinical application should be verified that the materials are GMP-grade and animal-free products. Importantly, the clinical-grade hPSCs should be generated under the ethical consideration guidelines that strictly followed the national law, from the traceable and healthy donors and the culture system should follow the standard operating procedures (SOPs) of GMP.

4. Derivation of human pluripotent stem cells

The recent advances in GMP derivation of hPSC lines are using xeno-free reagents/materials and replacing the mouse feeders with GMP-qualified human feeders or recombinant human proteins [24–28]. Regarding the xeno-free hPSC derivation, there are several steps that should be considered in order to meet the GMP or criteria for clinical grade hPSCs.

For hESC derivation, the zona pellucida of the blastocyst embryos has to be removed prior to isolation of the ICM. The zona pellucida could be digested by the acid Tyrode's solution or pronase [3–5, 29]. To avoid the destructive effect of the acid or enzyme, the laser-assisted zona pellucida removal and the mechanical cut using surgical blade or needle were successfully applied [30–32]. Enzymatic removal, laser assisted, and mechanical dissection of zona pellucida are the xeno-free procedures due to these methods has no contact to the animal materials. However, further step for derivation of hESCs is separation of the ICM from the trophectoderm (TE) which the embryo or ICM might contact to the animal products. Traditionally, the TE is separated from the ICM by the procedure called immunosurgery. The immunosurgery technique involves the selective lyses of the TE but not the ICM, resulting to the TE-free ICM. The use of anti-human serum combined with complement mediated lysis using guinea pig serum made the immunosurgery technique unsuitable for derivation of xeno-free hESCs. Therefore, other means such as microdissection of blastocysts using fine needle, laser-assisted biopsy, and whole blastocyst culture are the promising techniques for xeno-free ICM isolation [30–32]. The isolated ICM is subsequently plated and cultured for generation of hESC lines. The following procedures including culture conditions, cell propagation, and cryopreservation have to manage under the xeno-free conditions.

For hPSC derivation, the early step for derivation of xeno-free hPSCs is isolation and culture of the somatic cells prior to reprogramming. The somatic cells should be isolated under the aseptic condition with the minimal invasive technique. It has been reported that hPSCs can be generated by using several somatic cell types such as skin fibroblasts, peripheral blood mononuclear cells and even the cells isolated from urine [2, 33, 34]. Replacement of the animal products by using human grade or recombinant synthetic products during the isolation and culture of somatic cells is necessary. For example, human dermal fibroblasts can be isolated and propagated in the culture medium supplemented with human serum. These fibroblast cells can multiply under the xeno-free condition and suitable for using as the starting cells for reprogramming. We proposed the use of human cord blood-derived serum (hUCS) and showed a positive effect on culture of mesenchymal stem cells. The results showed that human foreskin fibroblasts (HFFs) cultured in hUCS-containing medium (HFF-hUCS) displayed fibroblastic features, high proliferation rates, short population doubling times, and normal karyotypes after prolonged culture. Inactivated HFF-hUCS expressed important genes, including Activin A, FGF2, and transforming growth factor beta-1 (TGFβ1), which have been implicated in the maintenance of hPSC pluripotency. Moreover, hPSC lines maintained pluripotency, differentiation capacities, and karyotypic stability after being cocultured for extended period with inactivated HFF-hUCS. Therefore, the results demonstrated the benefit of hUCS for hPSC culture system.

However, one of the bottle-neck of derivation of xeno-free hPSCs is the method for reprogramming of the somatic cells. Originally, the exogenous genes were delivered to the somatic cells by means of virus such as lentivirus or retrovirus [2, 35]. Because the integration of lentiviral- or retroviral genome to the host (somatic cells) genome might occur during the reprogramming process, thus the hPSCs generated by lentivirus- or retrovirus are not the ideal hPSCs for the clinical use. The clinical applications will require hiPSCs that are free of exogenous DNA and that can be manufactured through GMP. Recently, reprogramming by protein, microRNA, episomal vectors, or Sendai virus, a nonintegrating virus is considered suitable for delivery of the exogenous genes into the somatic cells [36]. We demonstrated recently that the transgene-free hPSCs can be generated by the temperature-sensitive strain of Sendai virus and the viral particles were clearly eliminated from the established hPSCs by the heat treatment [37]. Thus, derivation of transgene-free hPSCs is a very important step for generation of xeno-free hPSCs that match the future clinical applications.

5. Culture of human pluripotent stem cells

The culture conditions that allow hPSCs to maintain their pluripotency is the complex environment and still under investigation. The researchers proved that the pluripotency of hPSCs is controlled by the TGF pathway which involves the cooperation of Activin and Nodal signaling pathway [38]. The pathways that related to the maintenance of the pluripotency of hPSCs might be provided by the secretion of cytokines, growth factors, extracellular matrix from the feeder cells, synthetic substrates, or culture medium. The improvement of culture system is not only necessary for maintenance of hPSC pluripotency but also for large-scale propagation of hPSCs prior to cell transplantation. The aims of our study were to determine (i) the effect of exogenous bFGF supplementation in the hPSC culture media on the morphology and gene expression of inactivated human cesarean scar fibroblasts (HSFs) and (ii) the feasibility of using the inactivated HSFs as the feeder cells for culturing the hPSCs. Our results showed that the cells shrunk and an increase in gap between the cells were observed in the inactivated HSFs that were cultured in 4 and 8 ng/ml but not in 0 ng/ml bFGF. Expression of Activin A, bFGF, TGF-β, and BMP4 was similar between inactivated HSFs cultured in 0, 4, and 8 ng/ml bFGF. After two hPSC lines including hESC line (Chula2.hES) and hiPSC line (PFX12) were cocultured with three conditions of inactivated HSFs for more than ten passages, the hPSCs lines were subjected to characterization. The results showed that the hESCs and hiPSCs cultured in 0, 4, and 8 ng/ml bFGF could maintain their undifferentiation state, differentiate in vitro into three embryonic germ layers, and maintain their normal karyotype. In conclusion, exogenous bFGF supplementation in the culture medium can be omitted when using HSFs as the feeder cells for culturing the hPSCs.

5.1. Feeder-dependent culture system

Feeder-dependent culture system referred to the culture system that the hPSCs grown by the support of feeder cells. The feeder-dependent culture system is the traditional system that was developed for generation and propagation of hPSCs [1–6]. Feeder cells secreted cytokines, extracellular matrices and provided the niches that support the growth and maintenance of pluripotency of hPSCs. Generally, MEFs are used as the feeder for culture of hPSCs. Due to, MEFs might stimulate the immune response of the patients after transplantation of hPSCs [22]. Therefore, the feeder cells that isolated from human tissues are the better choice for culture of the clinical-grade hPSCs. Several studies demonstrated that the supportive feeder cells of hPSCs could be derived from the human tissues. Hovatta et al. [3] firstly demonstrated that HFFs supported the derivation and culture of hESCs. HFFs supported the pluripotency of hPSCs by expression of the key genes such as TFG, FGF2, and Activin A which have been implicated in the maintenance of hPSC pluripotency [39]. Ma et al. [40], the HFFs produced interleukin-6 and can be used for coculture with mouse embryonic stem cells. However, not all the HFF lines exhibited the characteristics of supportive feeder cells. The data confirmed that the foreskin fibroblasts from the different donor secreted different amount of growth factors, cytokines, and specific genes related to pluripotency [39, 41]. Therefore, it is necessary to explore the new supportive feeder cell types in order to develop the xeno-free feeder dependent culture system of hPSCs. Up to date, feeder cells derived from other sources of human tissues, for example, human adult fallopian tubal fibroblasts, human fetal skin, human adult skin, human fetal muscle, human adult muscle, human adult marrow cells, human umbilical cord mesenchymal stem cells, human amniotic epithelial cells, and human iPS cell derived fibroblast-like cells had been reported as the supportive feeder cells for culture of the hPSCs [42–45]. In addition, genetic modification of human feeder cells such as immortal HFFs or bFGF-secreting HFFs has been developed and proved to be useful for culture of hPSCs [46, 47]. Although replacement of MEFs by human feeder cells can reduce the risk of contamination of mouse pathogens to the hPSCs, the use of FBS in the feeder cell culture medium does not meet the xeno-free condition of hPSCs. We recently demonstrated that HFFs cultured in the medium containing human umbilical cord blood serum (hUCBS) retain their supportive feeder characteristics for maintenance of hPSC lines [48]. In addition, we used hUCBS for isolation and culture of cesarean scar-derived fibroblasts (**Figure 1**). The xeno-free fibroblasts will be used for the feeder cells and the starting cells for generation of transgene-free hPSCs.

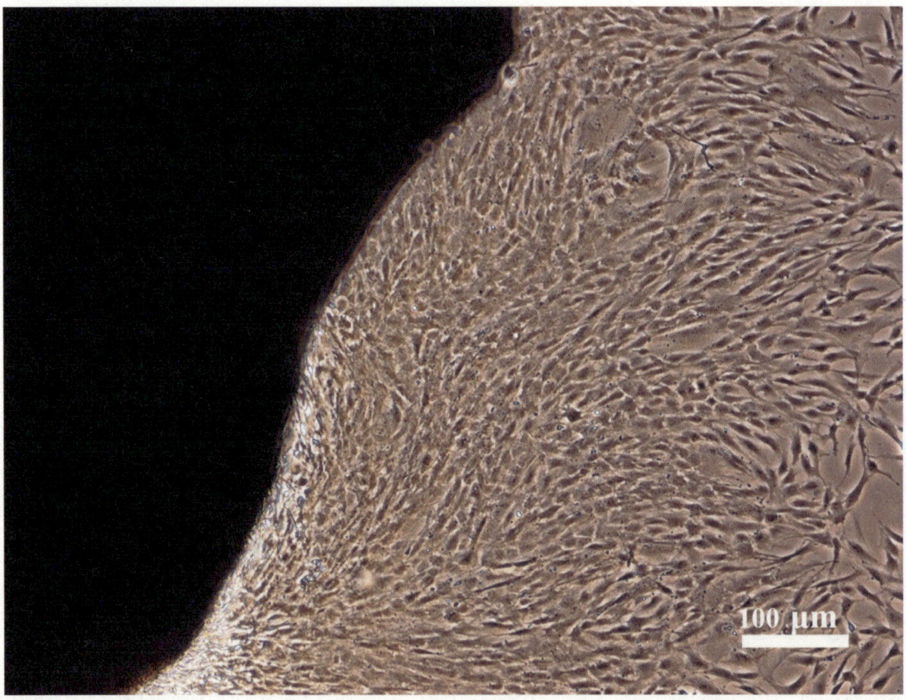

Figure 1. Isolation of fibroblast cells from human cesarean scar tissue. Fibroblast cells grew out from the tissue after culture for 7 days in the culture medium containing human umbilical cord blood serum (hUCBS). Scale bar = 100 μm.

To generate the clinical-grade human feeder cells, the cells should be derived under the culture system containing humanized or recombinant GMP materials and reagents. Interestingly,

Prathalingam et al. [27] demonstrated that the clinical-grade feeder cells can be derived and cultured in the medium containing the FBS that was derived under the GMP condition. Prior to use of the feeder cells for derivation of the new hPSC lines or routinely coculture with the existing hPSC lines, the clinical-grade human feeder cells should be tested for the potential of the cells to support the proliferation of hPSCs. Although the standard protocol for characterization of the feeder cells is lacking, several methods have been adopted. Inactivation of feeder cells by irradiation or Mitomycin C treatment is the method for inactivation of feeder cell proliferation prior to use for coculture with hPSCs. After being inactivated, the feeder cells should be able to secrete the enough amounts of cytokines and growth factors into the culture medium. Moreover, the extracellular matrices provided by the feeder play important roles for maintenance of the pluripotency of hPSCs [7]. The cells that easy to apoptosis after the inactivation process might probably not be a good supportive feeder cells.

One major concern of feeder-dependent culture system is coating the culture vessel with the xeno-free matrix in order to enhance the attachment of feeder cells to the surface of the culture vessel. Typically, porcine-skin-derived gelatin was used for coating the culture vessel prior of seeding the inactivated feeder cells. To meet the criteria of xeno-free condition, porcine-skin-derived gelatin should be replaced by recombinant gelatin. Tannenbaum et al. [26] developed the platform of derivation of xeno-free and GMP-grade hESCs by using recombinant gelatin and they found no adverse effect of recombinant gelatin on the pluripotency of hESCs. In addition, Ding et al. [49] showed that human cord blood serum can be used for coating the culture vessels instead of the porcine-skin-derived gelatin for derivation and maintenance of hPSCs.

5.2. Feeder-free culture system

It has been proposed that coculture of hPSCs with the feeder layer might contaminate the harmful substance and cause the risk of graft rejection, viral or bacterial infection, or zoonoses [50]. To culture hPSCs under the feeder-free condition, several media formulations have been developed that eliminate the use of fetal bovine serum and reduce or eliminate all animal-derived components. Therefore, development of feeder-free and xeno-free culture system for hPSCs by the use of xeno-free reagents and matrices in every cell-handling process is important for obtaining clinical-grade hPSCs.

Feeder-free culture system is applied for culturing hPSCs without the feeder layers but the matrices such as Matrigel or recombinant human proteins, together with the commercially defined hPSC culture medium [51, 52]. Although the genetic and epigenetic instability could be detected in hPSCs grown under feeder-free condition [53–55], the feeder-free condition together with defined culture medium is still important because it allows the hPSCs to proliferate robustly.

In order to propagate and maintain the pluripotency of hPSCs under the feeder-free condition, it is necessary to provide the cell adhesion coating that support the proliferation of hPSCs. Matrigel, the solubilized basement matrix extracted from mouse Engelbreth-Holm-Swarm sarcoma is the widely used matrix. The major components of Matrigel include laminin, collagen IV, heparan sulfate proteoglycan, and entactin. In addition, Matrigel contains growth factors such as TGF-beta, fibroblast growth factor, TPA, and insulin-like growth factor [56, 57]. Matrigel provides the complex extracellular matrices that support the cell communication and the suitable microenvironment for the hPSCs to grow as well as maintain their pluripotency. Although hPSCs proliferated robustly on Matrigel, the origin of Matrigel makes this matrix unsuitable for maintaining the clinical-grade hPSCs. Recently, Ding et al. [49] developed the humanized ECM by using umbilical cord blood serum (UCBS) as the matrix for derivation and culture of hPSCs. The best result was obtained when UCBS was used in combination with the basal medium supplemented with bFGF, fibronectin, and Y-27632. Other humanized matrices, for example, human placental-derived ECM and H9-hESC embryoid body–derived ECM have been proved to support the pluripotency of hPSCs [58, 59]. The hPSC lines cultured with the ECM derived from both types of human tissues or cells in the defined serum and xeno-free culture media such as TeSR2, sustained their pluripotency after prolonged culture. However, lot-to-lot variations in the individual serum donor or tissue-derived ECM can affect the effectiveness of the serum or ECM in maintaining the pluripotent state of hPSCs. In order to develop the defined-feeder-free cell adhesion coating, individual extracellular matrix (ECM) proteins have been used instead of Matrigel. Several ECM including laminin, fibronectin, vitronectin, and collagen support the proliferation of hPSCs. Those ECM can be used as single or in various combinations. Interestingly, the studies have demonstrated that recombinant laminin-511 and recombinant vitronectin enable to support the growth and maintain the pluripotency of hPSCs [60, 61]. To improve feeder-free culture system, cell adhesion proteins can be added to ECM component. E-cadherin, the cell adhesion protein improved the attachment and expansion of hPSCs when the cells were cultured in defined media such as mTeSR1 and xeno-free media, TeSR2 [62]. Moreover, recombinant E-cadherin fusion protein supported the proliferation of the other pluripotent cells [63]. Additional developments in defined culture surfaces include synthetic peptide coatings such as Synthemax, Peptide acrylate surfaces (PAS), or synthetic polymers have been proved to support the growth and pluripotency of hPSCs after prolonged culture [64]. The combination of human ECM proteins or cell adhesion molecules and synthetic biomaterials with well-designed surfaces and/or structures in the presence of a chemically defined medium containing recombinant growth factors would offer a xeno-free alternative to feeder cells for culturing hPSCs and maintaining their pluripotency. Because the xeno-free and feeder-free culture system is suitable for generation of hPSCs for clinical application, we reprogrammed $CD34^+$ cells by the temperature-sensitive Sendai viral vectors carries OCT-4, SOX2, KLF4 and c-MYC, on Pronectin F plus (Sanyo Chemical Industries, Kyoto, Japan) coating matrix and cultured the cells in ReproFF (ReproCELL, Yokohama, Japan) culture medium (**Figure 2**). We will use this newly established hPSCs for further development of xeno-free culture system. In our study [48], the hPSC lines were derived and cultured on HFF feeder layer. The hPSC lines were continuously maintained on either mitomycin-C inactivated HFF-FBS or mitomycin-C inactivated HFF-hUCS in serum-free hPSC culture medium. The serum-free hPSC culture medium comprising 80% Knockout Dulbecco's modified Eagles' medium (KO-DMEM), 20% KO-SR, 1% nonessential amino acid, 1% Glutamax, 1% penicillin-streptomycin, 0.1 mM β-mercaptoethanol (all from Invitrogen), and 8 ng/ml bFGF (R&D Systems, Minneapolis, MN, USA).

http://dx.doi.org/10.5772/62770

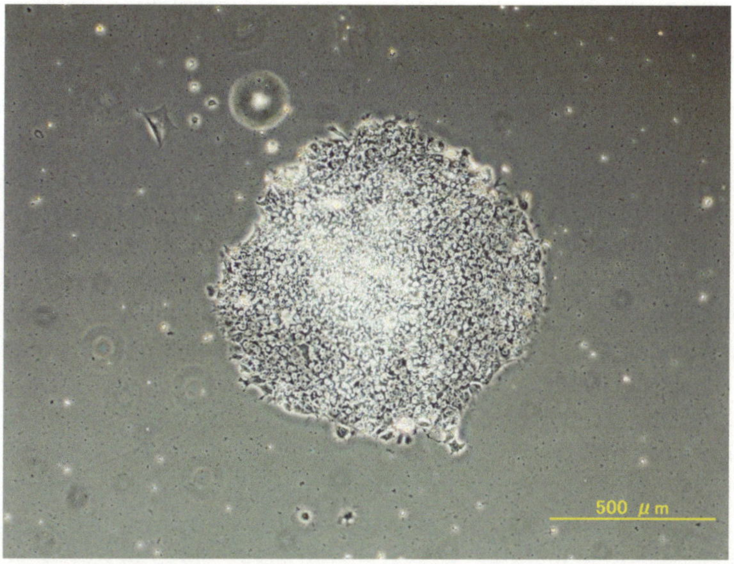

Figure 2. Human-induced pluripotent stem cell culture on feeder-free condition. The transgene-free human-induced pluripotent stem cells generated from CD34$^+$ cells and culture in feeder-free culture system. Scale bar = 500 μm.

6. Contamination concern

Prior to clinical application, xeno-free hPSC lines should be tested and evaluated for their biological safety. The evaluation should follow the guidelines that developed for clinical trials or therapies. The hPSC lines should be free from serious pathogenic microorganisms such as human immunodeficiency virus (HIV) and human papilloma virus (HPV). On the other hand, hPSC lines should also be free from mycoplasma contamination. Although Romorini [65] demonstrated the successful elimination of the mycoplasma contamination in their newly established hPSC lines but the mycoplasma-free newly established cell line is the ideal cell line for therapeutic application. Moreover, the level of endotoxin in the culture medium of hPSCs should meet the requirement of the national or international standard. Other biological test, for example, injection of the hiPSCs into the chorioallantoic and yolk sac of the chicken for testing unknown pathogenic microorganisms can be performed [25].

7. Clinical trial and therapeutic application

The clinical trial and therapeutic application requires hPSCs that were derived, cultured, and differentiated under the xeno-free and GMP conditions. Despite controversies and difficulties, the clinical trials using hESCs started in 2010 and continue ongoing. Geron Corporation, the biotechnology company based in California, USA began a clinical trial in patients with spinal cord injuries using hESCs in 2010. The patient was transplanted with oligodendrocyte precursor cells derived from hESCs. Unfortunately, the project was discontinued a year later because the company changed their business strategy. There is no official data that has been published from the study of Geron. However, the project has been started and continued again by Advance Cell Technology, the biotechnology company based in Massachusetts. The company just recently published their preclinical safety data [66]. In 2012, the clinical trials using hESC-derivative cells were initiated and conducted in the USA. Advance Cell Technology, performed the clinical trials in the patients with Stargardt's macular degeneration and age-related dry macular degeneration, which caused the loss of photoreceptor and results in blindness. Retinal-pigmented epithelial (RPE) cells derived from hESCs were transplanted to replace the degenerated cells. Interestingly, the visual improvement of the patients had been preliminarily reported [67, 68]. ViaCyte, the company based in California, USA reported the successful surgical transplantation of pancreatic precursor cells derived from hESCs for treatment of Type I diabetes [69]. Recently, the Japanese researchers from RIKEN institute performed the first clinical trial using hiPSC-derived cells for treatment of age-related macular degeneration. Autologous hPSCs were differentiated to retinal-pigmented epithelial (RPE) sheet and later transplanted to the patient. However, the second trial was cancelled due to the occurring of genetic instability of the cells prior to transplantation [70]. Besides the clinical trials mentioned above, the hPSC-based products for the treatment of Parkinson's disease and others diseases are currently in the pipeline [71].

8. Perspective of collaboration

Although clinically compliant hPSC has been derived, limitations in scale of production and high costs for culture of hPSCs remain significant challenges. Besides efficiency and safety, the development of hPSC-based cell therapies are expected to be expensive, time consuming and might face the clinical failure. Therefore, the collaboration between academy, SME and large pharmaceutical company can take the advantage of opportunities to tap into various sources of support. Funding for basic research carried out in the academic setting is available from national research grant agencies and institutes. In the private sector, SMEs have the possibility to attract venture capital to fund efforts in regenerative medicine. In addition, the pharmaceutical companies have the possibility to directly support the project using their internal resources. Besides individually applying for grants or securing other type of funding, the partners can also join together and obtain consortium grants or support based on collaborative networks from international initiatives such as the Framework Programs funded by the European Union. We had developed the similar model by carried out the project with the two European partners, The University of Copenhagen, Denmark and BioTalentum, Ltd, Hungary. The project was funded by the European Union under the Framework 7 through the Marie Curies Actions. The results of the project were not only the publications [72] but also transferring the researchers and knowledge between the academy and industry. We believed that having research partners cover both non-profit and for-profit organizations is of great advantage when developing hPSC-based cell therapy.

9. Conclusion

To enable a xeno-free culture system capable of clinical use, it is important to examine the sources of components used in hPSC derivation, culture, including the reagents used for handling cells during passaging and the reagents used for cryopreservation. The sufficient protocols and documentations, development of hPSC derivation and culture conditions under the strict cleanroom and the use of GMP-grade reagents and materials should be prepared. In addition to the technical issues involved, establishment of the donor consent documentation appropriately and utilize forms that specifically states that the cells will be used for cell therapy purposes have to be prepared. The clinical grade hPSC lines that derived completely under the GMP conditions will be a valuable source for the future clinical therapies.

Acknowledgements

This work was supported by Chulalongkorn University Research Grant (GRB_APS_03_57_30_01).

Author details

Kamthorn Pruksananonda* and Ruttachuk Rungsiwiwut

*Address all correspondence to: pkamthorn@yahoo.com

Human Embryonic Stem Cells Research Center, Department of Obstetrics and Gynecology, Faculty of Medicine, Chulalongkorn University, Bangkok, Thailand

References

[1] Thomson JA, Itskovitz-Eldor S, Shapiro SS, Waknitz MA, Swiergiel JJ, Marshall VS, et al. (1998). Embryonic stem cells lines derived from human blastocysts. *Science* 282: 1145–1147.

[2] Takahashi K, Tanabe K, Ohnuki M, Narita M, Ichisaka T, Tomoda K, et al. (2007). Induction of pluripotent stem cells from adult human fibroblasts by defined factors. *Cell* 131: 861–872.

[3] Hovatta O, Mikkola M, Gertow K, Stromberg AM, Inzunza J, Hreinsson J, et al. (2003). A culture system using human foreskin fibroblasts as feeder cells allows production of human embryonic stem cells. *Hum. Reprod.* 18(7): 1404–1409.

[4] Inzunza J, Gertow J, Stromberg MA, Matilainen E, Blennow E, Skottman H, et al. (2005). Derivation of human embryonic stem cell lines in serum replacement medium using postnatal human fibroblasts as feeder cells. *Stem Cell.* 23(4): 544–549.

[5] Pruksananonda K, Rungsiwiwut R, Numchaisrika P, Ahnonkitpanit V, Virutamasen P. (2009). Development of human embryonic stem cell derivation. *J. Med. Assoc. Thai.* 92: 443–450.

[6] Ellerstorm C, Strehl R, Moya K, Andersson K, Bergh C, Lundin K, et al. (2006). Derivation of a xeno-free human embryonic stem cell line. *Stem Cells* 24(10): 2170–2176.

[7] Meng G, Liu S, Li X, Krawetz R, Rancourt DE. (2010). Extracellular matrix isolated from foreskin fibroblasts supports long-term xeno-free human embryonic stem cell culture. *Stem Cells Dev.* 19(4): 547–556.

[8] Vuoristo S, Toivonen S, Weltner J, Mikkola M, Ustinov J, Trokovic R, et al. (2013). A novel feeder-free culture system for human pluripotent stem cell culture and induced pluripotent stem cell derivation. *PLoS One* 8(10): e76205.

[9] Desai N, Judgin J, Goldberg J, Falcone T. (2013). Development of a xeno-free non-contact co-culture system for derivation and maintenance of embryonic stem cells using a novel human endometrial cell line. *J. Assist. Reprod. Genet.* 30(5): 609–615.

[10] Montes R, Ligero G, Sanchez L, Catalina P, de la Cueva T, Nieto A, Melen GJ, Rubio R, Garcia-Castro J, Bueno C, Menedez P. (2009). Feeder-free maintenance of hESCs in mesenchymal stem cell-conditioned media: distinct requirements of TGF-beta and IGF-II. *Cell Res.* 19(6): 698–709.

[11] Philonenko ES, Shutova MV, Chestkov IV, Lagarkova MA, Kiselev SL. (2011). Current progress and potential practical application for human pluripotent stem cells. *Int. Rev. Cell Mol. Biol.* 292: 153–169.

[12] Mallon BS, Hamilton RS, Kozhich OA, Johnson KR, Fann YC, Rao MS, et al. (2014). Comparison of the molecular profiles of human embryonic and induced pluripotent stem cells of isogenic origin. *Stem Cell Res.* 12(2): 376–385.

[13] Guenther MG, Frampton GM, Soldner F, Hockemeyer D, Mitalipova M, Jaenisch R, et al. (2010). Chromatin structure and gene expression programs of human embryonic and induced pluripotent stem cells. *Cell Stem Cell* 7(2): 249–257.

[14] Bock C, Kiskinis E, Verstappen G, Gu H, Boulting G, Smith ZD, et al. (2011). Reference maps of human ES and iPS cell variation enable high-throughput characterization of pluripotent cell lines. *Cell* 144(3): 439–452.

[15] Deng J, Shoemaker R, Xie B, Gore A, LeProust EM, Antosiewicz-Bourget J, et al. (2009). Targets bisulfite sequencing reveals changes in DNA methylation associated with nuclear reprogramming. *Nat. Biotechnol.* 27(4): 353–360.

[16] Nashun B, Hill PW, Hajkova P. (2015). Reprogramming of cell fate: epigenetic memory and the erasure of memories past. *EMBO J.* 34(10): 1296–1308.

[17] Wutz A. (2012). Epigenetic alterations in human pluripotent stem cells: a tale of two cultures. *Cell Stem Cell* 11(1): 9–15.

[18] International Stem Cell Initiative, Adewumi O, Aflatoonian B, Ahrlund-Richter L, Amit M, Andrews PW, et al. (2007). Characterization of human embryonic stem cells by the International Stem Cell Initiative. *Nat. Biotechnol.* 25(7): 803–816.

[19] Andrews PW, Baker D, Benvinisty N, Miranda B, Bruce K, Brüstle O, et al. (2015). Points to consider in the development of seed stocks of pluripotent stem cells for clinical applications: International Stem Cell Banking Initiative (ISCBI). *Regen. Med.* 10(2 Suppl): 1–44.

[20] Shetty P, Bharucha K, Tanavde V. (2007). Human umbilical cord blood serum can replace fetal bovine serum in the culture of mesenchymal stem cells. *Cell Biol. Int.* 31: 293–298.

[21] Kocaoemer A, Kern S, Kluter H, Bieback K. (2005). Human AB serum and thrombin-activated platelet-rich plasma are suitable alternatives to fetal calf serum for expansion of mesenchymal stem cells from adipose tissue. *Stem Cells* 25(5): 1270–1278.

[22] Martin MJ, Muotri A, Gage F, Varkj A. (2005). Human embryonic stem cells express an immunogenic nonhuman sialic acid. *Nat. Med.* 11(2): 228–232.

[23] Chou ML, Bailey A, Avory T, Tanimoto J, Burnouf T. (2015). Removal of transmissible spongiform encephalopathy prion from large volume of cell culture media supplemented with fetal bovine serum by using hollow fiber anion-exchange membrane chromatography. *PLoS One* 10(4): e0122300.

[24] Durruthy-Durruthy J, Briggs SF, Awe J, Ramathal CY, Karumbayaram S, Lee PC, Heidmann JD, Clark A, Karakikes I, Loh KM, Wu JC, Hoffman AR, Byrne J, Reijo Pera RA, Sebastiano V. (2014). Rapid and efficient conversion of integration-free human induced pluripotent stem cells to GMP-grade culture conditions. *PLoS One* 9(4): e94231.

[25] Wang J, Hao J, Bai D, Gu Q, Han W, Wang L, et al. (2015). Generation of clinical-grade human induced pluripotent stem cells in xeno-free conditions. *Stem Cell Res. Ther.* 6(1): 223.

[26] Tannenbaum SE, Turetsky TT, Singer O, Aizenman E, Kirshberg S, Ilouz N, et al. (2012). Derivation of xeno-free and GMP-grade human embryonic stem cells—platforms for the future clinical applications. *PLoS One* 7(6): e35325.

[27] Prathalingam N, Ferguson L, Young L, Lietz G, Oldershaw R, Healy L, et al. (2012). Production and validation of a good manufacturing practice grade human fibroblast like for supporting human embryonic stem cells derivation and culture. *Stem Cell Res Ther.* 3(2): 12.

[28] Chen VC, Couture SM, Ye J, Lin Z, Hua G, Huang HI, et al. (2012). Scalable GMP compliant suspension culture system of human ES cells. *Stem Cell Res.* 8(3): 388–402.

[29] Pruksananonda K, Rungsiwiwut R, Numchaisrika P, Ahnonkitpanit V, Isarasena N, Virutamasen P. (2012). Eighteen-year cryopreservation does not negatively affect the pluripotency of human embryos: evidence from embryonic stem cell derivation. *BioRes Open Access.* 1: 166–173.

[30] Kim HS, Oh SK, Park YB, Ahn HJ, Sung KC, Kang MJ, Lee LA, Suh CS, Kim SH, Kim DW, Moon SY. (2005). Methods for derivation of human embryonic stem cells. *Stem Cells* 23: 1228–1233.

[31] Strom S, Rodriguez-Wallberg K, Holm F, Bergstrom R, Eklund L, et al. (2010). No relationship between embryo morphology and successful derivation of human embryonic stem cell lines. *PLoS One* 5: e15329.

[32] Turetsky T, Aizenman E, Gil Y, Weinberg N, ShufaroY, Revel A, et al. (2008). Laser-assisted derivation of human embryonic stem cell lines from IVF embryos after preimplantation genetic diagnosis. *Hum. Reprod.* 23(1): 46–53.

[33] Trokovic R, Weltner J, Nishimura K, Ohtaka M, Nakanishi M, Salomaa V, et al. (2014). Advanced feeder-free generation of induced pluripotent stem cells directly from blood cells. *Stem Cells Transl. Med.* 3(12): 1402–1409.

[34] Zhang SZ, Li HF, Ma LX, Qian WJ, Wang ZF. (2015). Urine-derived induced pluripotent stem cells as a modeling tool for paroxysmal kinesigenic dyskinesia. *Biol. Open.* 4(12): 1744–1752.

[35] Li W, Zhou H, Abujarour R, Zhu S, Young Joo J, Lin T, et al. (2009). Generation of human-induced pluripotent stem cells in the absence of exogenous Sox2. *Cell* 27(12): 2992–3000.

[36] Zhou YY, Zeng F. (2013) Integration-free methods for generating induced pluripotent stem cells. *Genomics, Proteomics Bioinf.* 11(5): 284–287.

[37] Rungsiwiwut R, Pavarajarn W, Numchaisrika P, Virutamasen P, Pruksananonda K. (2016). Transgene-free human induced pluripotent stem cell line (HS5-SV.hiPS) generated from caesarean scar-derived fibroblasts. *Stem Cell Res..* 16(1): 10–13.

[38] James D, Levine AJ, Besser D, Hemmati-Brivanlou A. (2005). TGFbeta/activin/nodal signaling is necessary for the maintenance of pluripotency in human embryonic stem cells. *Development* 132(6): 1273–1282.

[39] Eiselleova L, Peterkova I, Neradil J, Slaninova I, Hampl A, Dvorak P. (2008). Comparative study of mouse and human feeder cells for human embryonic stem cells. *Int. J. Dev. Biol.* 52(4): 353–363.

[40] Ma Y, Gu J, Li C, Wei X, Tang F, Shi G, et al. (2012). Human foreskin fibroblast produces interleukin-6 to support derivation and self-renewal of mouse embryonic stem cells. *Stem Cell Res. Ther.* 3(4): 29.

[41] Mamidi MK, Pal R, Mori NA, Arumugam G, Thrichelvam ST, Noor PJ, et al. (2011). Co-culture of mesenchymal-like stromal cells derived from human foreskin permits long term propagation and differentiation of human embryonic stem cells. *J. Cell. Biochem.* 112(5): 1353–1363.

[42] Cheng L, Hammond H, Ye Z, Zhan X, Dravid G. (2003). Human adult marrow cells support prolonged expansion of human embryonic stem cells in culture. *Stem Cells* 21(2): 131–142.

[43] Ding DC, Shyu WC, Lin SZ, Liu HW, Chiou SH, Chu TY. (2012). Human umbilical cord mesenchymal stem cells support nontumorigenic expansion of human embryonic stem cells. *Cell Transplant.* 21(7): 1515–1527.

[44] Ávila-Gonzalez D, Vega-Hernandez E, Regalado-Hernandez JC, De la Jara-Diaz JF, Garcia-Castro IL, Molina-Hernandez A, et al. (2015). Human amniotic epithelial cells as feeder layer to derived and maintain human embryonic stem cells from poor-quality embryos. *Stem Cell Res.* 15(2): 322–324.

[45] Du SH, Tay JC, Chen C, Tay FC, Tan WK, Li ZD, et al. (2015). Human iPS cell-derived fibroblast-like cells as feeder layers for iPS cell derivation and expansion. *J. Biosci. Bioeng.* 120(2): 210–217.

[46] Saxena S, Hanwate M, Deb K, Sharma V, Totey S. (2008). FGF2 secreting human fibroblast feeder cells: a novel culture system for human embryonic stem cells. *Mol. Reprod. Dev.* 75(10): 1523–1532.

[47] Unger C, Gao S, Cohen M, Jaconi M, Bergstrom R, Holm F, et al. (2009). Immotalized human skin fibroblast feeder cells support growth and maintenance of both human embryonic and induced pluripotent stem cells. *Hum. Reprod.* 24(10): 2567–2581.

[48] Rungsiwiwut R, Ingrungruanglert P, Numchaisrika P, Virutamasen P, Phermthai T, Pruksananonda K. (2016). Human umbilical cord blood-derived serum for culturing the supportive feeder cells of human pluripotent stem cell lines. *Stem Cells Int.* 2016: 4626048.

[49] Ding Y, Yang H, Yu L, Xu CL, Zeng Y, Qiu Y, et al. (2015). Feeder-free and xeno-free culture of human pluripotent stem cells using UCBS matrix. *Cell Biol. Int.* 39: 1111–1119.

[50] Skottman H, Hovatta O. (2006). Culture conditions for human embryonic stem cells. *Reproduction* 132(5): 691–698.

[51] Amit M, Shariki C, Margulets V, Itskovitz-Eldor J. (2004). Feeder layer- and serum-free culture of human embryonic stem cells. *Biol. Reprod.* 70(3): 837–845.

[52] Xu C, Inokuma MS, Denham J, Golds K, Kundu P, Gold JD, et al. (2001). Feeder-free growth of undifferentiated human embryonic stem cells. *Nat. Biotechnol.* 19(10): 971–974.

[53] Catalina P, Montes R, Ligero G, Sanchez L, de la Cueva T, Bueno C, et al. (2008). Human ESCs predisposition to karyotypic instability: Is a matter of culture adaptation or differential vulnerability among hESC lines due to inherent properties? *Mol. Cancer* 7: 76.

[54] Draper JS, Smith K, Gokhale P, Moore HD, Maltby E, Johnson J, et al. (2004). Recurrent gain of chromosomes 17q and 12 in cultured human embryonic stem cells. *Nat. Biotechnol.* 22(1): 53–54.

[55] Garitaonandia I, Amir H, Boscolo FS, Wambua GK, Schultheisz HL, Sabatini K, et al. (2015). Increased risk of genetic and epigenetic instability in human embryonic stem cells associated with specific culture conditions. *PLoS One* 10(2): e0118307.

[56] Kleinman HK, Martin GR. (2005). Matrigel: basement membrane matrix with biological activity. *Semin. Cancer Biol.* 15(5): 378–385.

[57] Ludwig TE, Bergendahl V, Levenstein ME, Yu J, Probasco MD, Thomson JA. (2006). Feeder-independent culture of human embryonic stem cells. *Nat. Methods* 3(8): 637–646.

[58] Fu X, Toh WS, Liu H, Lu K, Li M, Hande MP, et al. (2010). Autologous feeder cells from embryoid body outgrowth support the long-term growth of human embryonic stem cells more effectively than those from direct differentiation. *Tissue Eng. Part C Methods* 16(4): 719–733.

[59] Fukusumi H, Shofuda T, Kanematsu D, Yamamoto A, Suemizu H, Nakamura M, et al. (2013). Feeder-free generation and long-term culture of human induced pluripotent stem cells using pericellular matrix of decidual derived mesenchymal cells. *PLoS One* 8(1): e55226.

[60] Chen. (2011). Chemically defined conditions for human iPSC derivation and culture. *Nat. Methods* 8(5): 424–429.

[61] Rodin. (2010). Long-term self-renewal of human pluripotent stem cells on human recombinant laminin-511. *Nat. Biotechnol.* 28(6): 611–615.

[62] Braam SR. (2008). Recombinant vitronectin is a functionally defined substrate that supports human embryonic stem cell self-renewal via alphavbeta5 integrin. *Stem Cells* 26(9): 2257–2265.

[63] Nagaoka M, Si-Taved K, Akaike T, Duncan SA. (2010). Culture of human pluripotent stem cells using completely defined conditions on a recombinant E-cadherin substratum. *BMC Dev. Biol.* 2(10): 60.

[64] Enam S, Jin S. (2015). Substrate for clinical application of stem cells. *World J. Stem Cells.* 7(2): 243–252.

http://dx.doi.org/10.5772/62770

[65] Romorini L, Riva DA, Bluguermann C, Videla Richardson GA, Scassa ME, Sevlever GE, et al. (2013). Effect of antibiotics against Mycoplasma sp. On human embryonic stem cells undifferentiated status, pluripotency, cell viability and growth. *PLoS One.* 8(7): e70267.

[66] Priest CA, Manley NC, Denham J, Wirth ED 3rd, Lebkowski JS. (2015). Preclinical safety of human embryonic stem cell-derived oligodendrocyte progenitors supporting clinical trials in spinal cord injury. *Regen. Med.* 10(8): 939–958.

[67] Schwart SD, Hubschman JP, Heilwell G, Franco-Cardenas V, Pan CK, et al. (2012). Embryonic stem cell trials for macular degeneration: a preliminary report. *Lancet* 379(9817): 713–720.

[68] Schwart SD, Regillo CD, Lam BL, Eliott D, Rosenfeld PJ, Gregori NZ, et al. (2015). Human embryonic stem cell-derived retinal pigmented epithelium in patients with age-related macular degeneration and Stargardt's macular dystrophy: follow-up of two open-label phase 1/2 studies. *Lancet* 385(9967): 509–516.

[69] Schulz TC. (2015). Concise review: manufacturing of pancreatic endoderm cells for clinical trials in type 1 diabetes. *Stem Cell Transl. Med.* 4(8): 927–931.

[70] Garber K. (2015). RIKEN suspends first clinical trial involving induced pluripotent stem cells. *Nat. Biotechnol.* 33: 890–891.

[71] Hunsberger JG, Rao M, Kurtzberg J, Bulte JW, Atala A, LaFerla FM, et al. (2016). Accelerating stem cell trials for Alzheimer's disease. *Lancet Neurol.* 15: 219–230.

[72] Rungsiwiwut R, Manolertthewan C, Numchaisrika P, Ahnonkitpanit V, Virutamasen P, Techakumphu M, Pruksananonda K. (2013). The ROCK inhibitor Y-26732 enhances the survival and proliferation of human embryonic stem cell-derived neural progenitor cells upon dissociation. *Cells Tissues Organs* 198(2): 127–138.

Chapter 8

Affordable Clean Air Facility for (Stem) Cell Therapy in an Academic Setting

Una Chen

Additional information is available at the end of the chapter

http://dx.doi.org/10.5772/62757

Abstract

Together with her colleagues, the author has worked extensively on the differentiation potential of mouse embryonic stem cells (ESCs) to blood islands containing embryoid bodies (EBs) in culture. Such EBs contain hematopoietic stem- precursor cells (HSCs) committed to lymphoid, myeloid lineages and cells possessing Natural Killer (NK) cell marker. Further differentiation of such EBs to mature B-lymphoid cells can be demonstrated in a second stage of tissue culture, by co-culturing separated EBs with a mitogen, Lipopolysaccharide (LPS). The demonstration of differentiation of such cells to mature T- and B-lymphoid cells was performed in vivo, i.e., by implanting EBs subcutaneously in nude mice. The differentiation of ESCs to EBs needed to be scaled up to obtain enough HSC-containing EBs for a second stage of differentiation, in vitro and in vivo. Selection of HSC-containing EBs must be performed manually under a microscope outside the tissue culture incubator. For research purposes, the cell culture medium contains antibiotics to inhibit the growth of bacteria. However for clinical application, the medium should be free of antibiotics. A conventional research laboratory environment is not sufficient for such an application. A GMP (Good Manufacturing Production) facility with clean air became necessary. Commercial clean air facilities used in the operation room and/or by industry were over budget. The attention of the author has shifted to developing an affordable GMP facility for clinical research in academia. These efforts were successful and the price for such a GMP facility is 10% of a comparable industrial model. After that experience, "a GMP facility for cell therapy on wheels" was further developed. This chapter will describe and share with academic colleagues our experience of constructing such affordable GMP facilities from scratch, starting with the purchase of high-efficiency particulate arrestance (HEPA) filters for the clean air facility.

Keywords: cGMP, clinical application, mobile HEPA filter, mobile clean air facility, mini-clean air box

1. Introduction

cGMP (current Good Manufacturing Production) for (stem) cell therapy is the making (growth, propagation, expansion, scaled up production) of (stem) cells for clinical purposes. The procedures and requirements include two parts: hardware and software.

Here, I shall discuss and share some experiences that we have conducted and performed by converting a regular research tissue culture room-facility, especially the hardware part, into

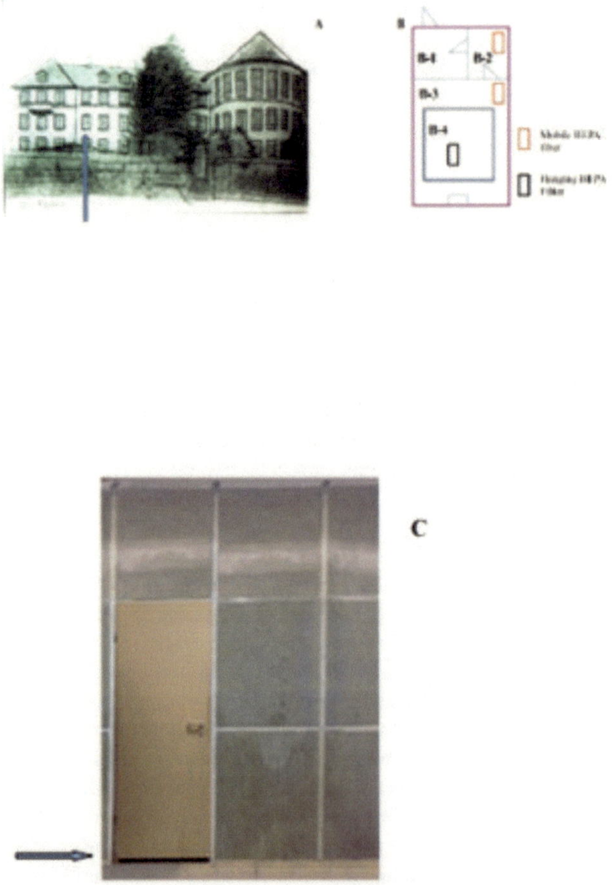

Figure 1: GMP facility. A: Biochemistry Institute Building in Uni. Clinic Giessen, cold needle graphic by U.C. Blue arrow points to the location of GMP facility. B: Planning of the GMP facility: Space 1-B-1: A front room; Space 1-B-2: Air lock; Space 1-B-3: main space of the GMP facility; Space 1-B-4: Soft wall where clean bench was located. Red long square: Mobile HEPA filter, Black long square: Hanging HEPA filter.**C. Washable wall and doors for the facility.** The existing wall of the space (1-B, purple color) was coated with transparent lacquer. The newly constructed wall and doors were these between entrance preparation room (1-B-1) and airlock (1-B-2); airlock and the main room (1-B-3) of the facility. The material was iron frame coated with white color paint and acryl glass (shown in 1-C). A row of brushes (black color, indicated by a blue arrow) at the bottom of the door (blue triangles in 1-B-1 and 1-B-2) for directing the outward air flowing.

http://dx.doi.org/10.5772/62757

an open system clean air facility inside an old building (**Figure 1**), and then transferring the facility in a container (**Figure 5**) to become a "GMP facility on wheels".

2. Procedures and results

2.1. Software

The software includes: Documentation, Standard Operational Procedures (S.O.P.), QA (Quality Assurance), QC (Quality Control) tests.

The criteria and requirements can vary from country to country and be updated periodically. One must be in contact with the relevant regulatory authority regularly to achieve the standard required. There are guidelines constantly updated from regulatory authorities such as FDA [1]. The information provided by the UK stem cell bank could also be of interest [2]. The appointment of a QA person, possibly from industry working in the field, who can guide the production, give advice and suggestions, deal with update of the guidelines, and communicate with regulatory authorities, is highly recommended.

Standard Operation Procedure (S.O.P.)-1 (**Figure 7**) and S.O.P.-2 provide examples on how the clean air facility operation procedures were written, practiced, and controlled by the staff members. Our S.O.P. were updated whenever necessary and as often as possible to describe the real steps of practical exercise so that new/any staff members could hold a printout sheet in their hands, read, and follow the instructions, step by step, to meet the requirements and to operate properly.

"S.O.P. 1: Into the clean air room" (**Figure 7**) instructs the staff on what to do in the entrance preparation room, including how to change clothes, pin up hair, wear protective glasses; on how to take air shower when entering the airlock, on how to operate first in the clean air facility such as to measure particles counts, to use particle collectors, to place agar plates for collection of particles for microbiological sterile tests, before starting the cell culture experiments. Any leak and/or contamination of the air particles from outside will be detected by these procedures.

"S.O.P. 2: Out of the clean air room" (**Figure 8**) describes the reverse steps of "Into the clean air room", stating how to operate before leaving the facility, what to take out and where to place the samples collected, how to document the operation and to be double checked by a supervisor of the operation..

S.O.P. 3 (**Figure 9**) provides the regular and routine control of the clean air in the facility and describes, step by step, how and where (in clean bench, HEPA Filter 1, HEPA Filter 2, air lock) agar plates should be placed, when to be removed, where and how long they should be tested. The procedure describes how to clean the benches and what should be done before and after the cleaning, namely, agar plate tests. It describes when and where to measure particles using particle counter and particle collector. Any inappropriate operation in the entire procedure or contamination is documented and double-triple controlled by the operators and supervisor.

2.2. Hardware

The hardware includes: standardization, inspection, and routinely periodic control of all equipment used for tissue culture, including temperature and cleanness control of refrigerator, accuracy control of pipette-aids, temperature and CO_2 content of CO_2 incubator, Laminar Flow-clean bench, etc. One goes through such inspection routinely once a week, and documents the results on the so-called Monday S.O.P. (S.O.P. 3).

The equipment control is essential. The equipment do not have to be purchased new, though new equipments are under guarantee by the manufacturing companies and require less work, but the existing equipments must be inspected, (re)-standardized, and controlled routinely. Tissue culture working area should have no (bacterial–fungal) contamination (S.O.P. 3), i.e., clean air 100/ISO 6 (see below under criteria and in **Table 1**).

Class	Maximum particles/m³						FED STD 209D
	≥0.1 μm	≥0.2 μm	≥0.3 μm	≥0.5 μm	≥1 μm	≥5 μm	equivalent
ISO 1	10	2.37	1.02	0.35	0.083	0.0029	
ISO 2	100	23.7	10.2	3.5	0.83	0.029	
ISO 3	1000	237	102	35	8.3	0.29	Class 1
ISO 4	10,000	2370	1020	352	83	2.9	Class 10
ISO 5	100,000	23,700	10,200	3520	832	29	Class 100
ISO 6	1.0×10^6	237,000	102,000	35,200	8320	293	Class 1000
ISO 7	1.0×10^7	2.37×10^6	1,020,000	352,000	83,200	2930	Class 10,000
ISO 8	1.0×10^8	2.37×10^7	1.02×10^7	3,520,000	832,000	29,300	Class 100,000
ISO 9	1.0×10^9	2.37×10^8	1.02×10^8	35,200,000	8,320,000	293,000	Room air

Table 1. ISO 14644-1 cleanroom standards vs US FED STD 209D standard.

2.3. Clean air facility: closed vs open systems

Two systems for cell culture work at the clean air bench: closed- system vs. open- system. Companies such as Phoenix in N.Y. sell such closed clean air benches. For our purpose of growing cells and selecting ES-cell derived embryoid bodies under a microscope, we could only use the open clean air bench located inside a clean air room.

Clean air rooms are classified according to the number and size of particles permitted per volume of air. There are basic codes of classification for cleanrooms:

1. US FED STD 209D and 209E cleanroom standards [3]:

It denotes the number of particles of size 0.5 μm or larger permitted per cubic foot (ft^3) of air. For example "Class 100/A" is 100 particles/ft^3, "class 1000/B" is 1000 particles/ft^3, "class 10,000/C" is the normal air environment. For a tissue culture bench, it is recommended to operate under condition A, i.e., 100 particles or less/ft^3. One m^3 is 35 ft^3. This criteria of clean room classification was cancelled by the General Services Administration in 2001 [4, 5]. However, this system is still commonly used by many organizations, including us.

2. ISO 14644-1 standards [6]:

International Organization for Standardization (ISO) codes ISO 14644-1 cleanroom standards are the most commonly used. ISO classifies cleanrooms on a scale from 1 to 9, with 1 being the "cleanest" air. ISO classification 1 cleanrooms have a maximum of 10 particles smaller than 0.1 micrometers in a cubic meter of airspace. Class 9 is used to describe normal room air.

3. BS 5295 cleanroom standards and standards of other countries:

There are many other standards, such as Britain BS 5295, Australia AS 1386, France AFNOR X44101, and Germany VD I.2083. Table 6 in [7] describes the details.

Table 1 lists the comparison of US and ISO systems (1 and 2, above). Small numbers refer to ISO 14644-1 standards, which specify the decimal logarithm of the number of particles 0.1 μm or larger permitted per m^3 of air. The table provides an immediate comparison and understanding of the criteria, such that an ISO class 5 clean room has at most 10^5 particles/m^3. In our facility, we succeeded in achieving (marked in red color) class 1000/ISO 6 for clean air room, class 100/ISO 5 for clean bench open working area.

2.4. A mini-clean air box

For the purpose of growing cells, flasks equipped with mini-filters are available. However, for picking ES-derived EBs, Petri dishes are preferable for the practical reason that an access to the open dish under the microscope is absolutely required. Therefore, we have developed and made **"mini-tissue culture boxes"**. They are self-made, from material such as ready-made plastic boxes or acryl glass derived from Mishell-Dutton cell incubation system, and equipped with an air in-let and out-let system for CO_2 gassing before they are placed into the CO_2 incubator (**Figure 6**).

2.5. Open clean air facility

The clean air facility was located on the second floor of an old building where the Biochemistry Institute of University clinic of Giessen is situated (**Figure 1-A**), having one window (blue arrow). The three windows to the right of this facility are those of my research laboratory.

2.6. Washable wall and doors for the facility

Due to financial considerations, the option to have a commercial hardwall facility inside the existing building wall was not chosen. The existing wall of the space (**Figure 1-B**, purple color) was coated with a transparent vinyl lacquer/paint to reduce the release of particles from plaster.

The newly constructed wall and doors were those between the entrance preparation room (1-B-1) and air shower-airlock (1-B-2); air shower-air lock and the main clean air room (1-B-3) of the facility. The material of the newly constructed wall was white-painted iron frame and acryl glass (shown in **Figure 1-C**). The surface was smooth, washable, and inert, so that it will not attract and accumulate dust. A row of brushes (black color in **Figure 1-C**, indicated by a blue arrow) was fixed to the bottom of the door (blue triangles in 1-B-1 and 1-B-2) to direct the air flowing outward. Above the window was an air-conditioning apparatus (blue box).

The facility (**Figure 1-B**) was constructed and divided into the following spaces: an entrance preparation room for changing clothes and a storage shelf (Space 1-B-1), an air shower-airlock with a mobile high efficiency particulate air (HEPA) filter for taking air shower (Space 1-B-2), the main workspace (1-B-3) with a softwall cleanroom (Space B-4) with hanging HEPA filter from the ceiling, and a mobile HEPA filter moving inside and outside the softwall clean room. Inside the softwall clean room, a two-person-laminar-air-flow cabin/clean bench, and two chairs were placed for doing cell culture. The CO_2 incubator could also be placed inside the softwall cleanroom. The red long box indicated the place where the mobile HEPA filter was located. The black long box indicated the place where the hanging HEPA filter was located.

2.7. HEPA filter on wheels

The construction of the mobile HEPA filter is explained in this figure (**Figure 2**). It requires access to a metal workshop so that one can do the metal work by oneself, and/or contracting the design to a skilful mechanic who knows how to work with metal. A frame-in metal carrier with four wheels was designed with the precise size of the outer measurement of the HEPA filter. In this case, the HEPA filter from Zander (EU 3 and H 14, Zander, [8]) was purchased (**Figure 2-A** left image) and the designed construct was made by a mechanic (**Figure 2-A** right lower corner image). The HEPA filter could be lifted with the help of two persons and loaded to fit into/onto this carrier (**Figure 2-B**). The mounted HEPA filter could move around the space freely and securely, no extra screw was needed to fasten these two pieces together for this construct.

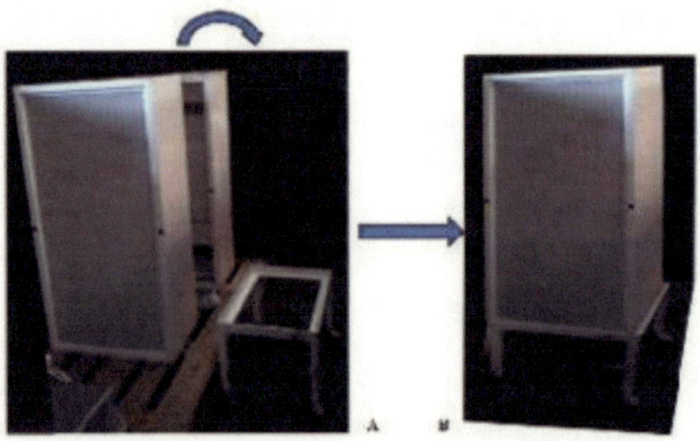

Figure 2. HEPA Filter on the wheels. The left image (A) contains two HEPA filters from Zander (No. EU 3 and H 14). To the right lower corner of these filters is the image of a self-made metal carrier with four wheels. The right image (B) shows a HEPA filter loaded on this wheeled metal carrier to become mobile.

http://dx.doi.org/10.5772/62757

2.8. Soft-wall clean room self-construction

2.8.1. HEPA filter to hang from the ceiling

The same type of HEPA filter from Zander (EU 3 and H 14) was used for this construction (**Figure 3-A**). The HEPA filter was lifted onto self-constructed metal in-frames and suspended with additional metal support from the ceiling of the room (**Figure 3-B**).

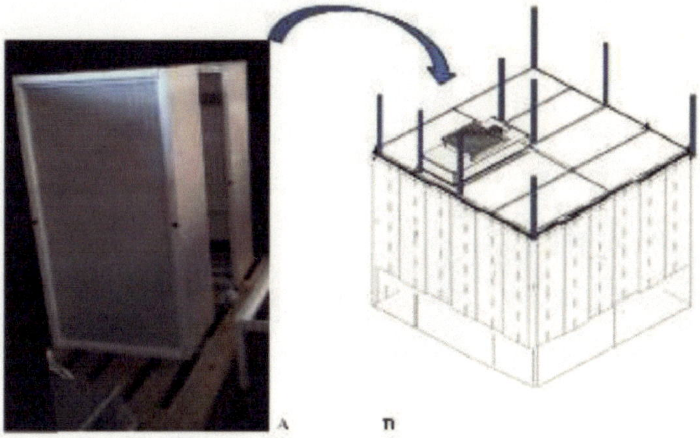

Figure 3. HEPA filter hanging from the ceiling. The HEPA filter from Zander (A) was lifted onto metal in-frames (B). For construction of soft wall, the additional material such as metal frame, ceiling metal hangers, plastic strips were locally available.

2.8.2. Self-constructed metal frame and heavy plastic hanging strips

If there is sufficient budget, one could purchase and install the commercially available softwall clean room, such as the one from the company, Cleanairproducts [9]. However, we have chosen a lower budget option: to purchase the HEPA filter and to have a skilful worker perform the rest of the construction.

For the construction of the softwall clean room, the additional material such as metal frame, ceiling metal hangers for suspension, heavy plastic strips, were purchased and constructed. The design of metal frame for the softwall clean room depends on the size of the room and the type of metal pieces available locally. **Figure 3-B** shows one option. Around the metal frames were heavy plastic sheets with free flip strips a few centimeters away from the floor to allow the air to escape.

2.8.3. Outfit for working in the clean air facility

Among all the items to enter the clean room, human beings carry the most dust. Thus, it is essential to remove as many dust particles as possible from the workers in the air lock before entering the clean room.

As shown in **Figure 4**, the complete impermeable synthetic suit, either of tightly woven polyester or a nonwoven material, such as Tyvek, is preferable: Examples are polyester suit (Countdown clean systems, Derby, England), Jumpsuit (Tyvek Pro-Tech Classic, DuPontTM, US) was the recommended gown for working in the clean air room [10, 11]. In addition, hair cap, eye glasses (for non-glasses wearing workers), mouth mask, hand gloves, and washable plastic shoes were worn by the worker in the entrance preparation room before taking an air shower in the airlock. Most items were used once and disposed. Exceptions to this general rule were the protecting glasses and plastic shoes. The gown might be recycled a few times, if an X-ray machine is accessible to allow irradiation for sterilization.

A

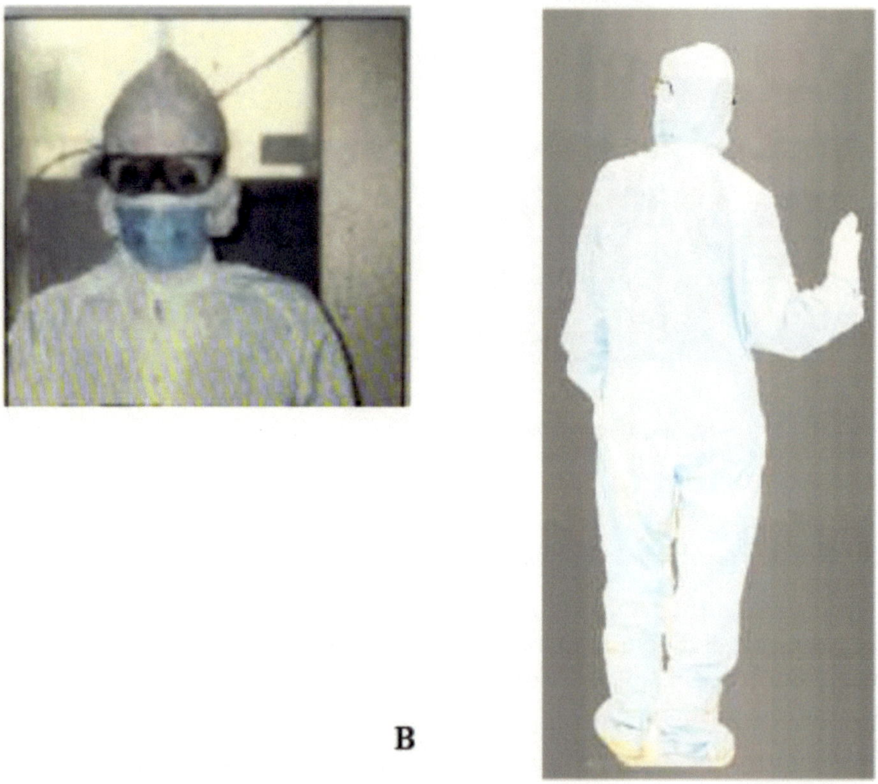

B

Figure 4. Outfits for working in the clean air facility. A: A staff member wearing glasses, mouth mask, hair hood to cover the hair and skin in the entrance preparationroom before entering the airlock. **Tyvek® Coveralls** gown was the clothes of choice [11]. B: A backview of the clean air gown with shoes (derived from [10]).

2.9. Mobile clean air facility/A clean air facility on wheels

Two containers, Variante 6 from ALHO [12] (**Figure 5-A**), were rented to place in a parking lot of Europaviertal, an industrial area outside the city of Giessen. The enlarged image (**Figure 5-B**) shows the location of HEPA filters (in red) and other items in the facility. The container was built from smooth surface metal material so that no extra coating process to repel dust was performed. The space was divided into the following working areas: The container to the left was the preparation space for storage and for changing clothes (Space B-1); an airlock with a mobile HEPA filter for taking air shower (Space B-2); the container to the right was the main clean air working space (B-3). The ceiling of the container was not high enough to accommodate any hanging HEPA filter. Instead, two mobile HEPA filters were placed inside this container. The two-person-clean bench (in black) and two chairs were placed there for cell culture. A CO_2 incubator was placed next to the clean bench for growing cells.

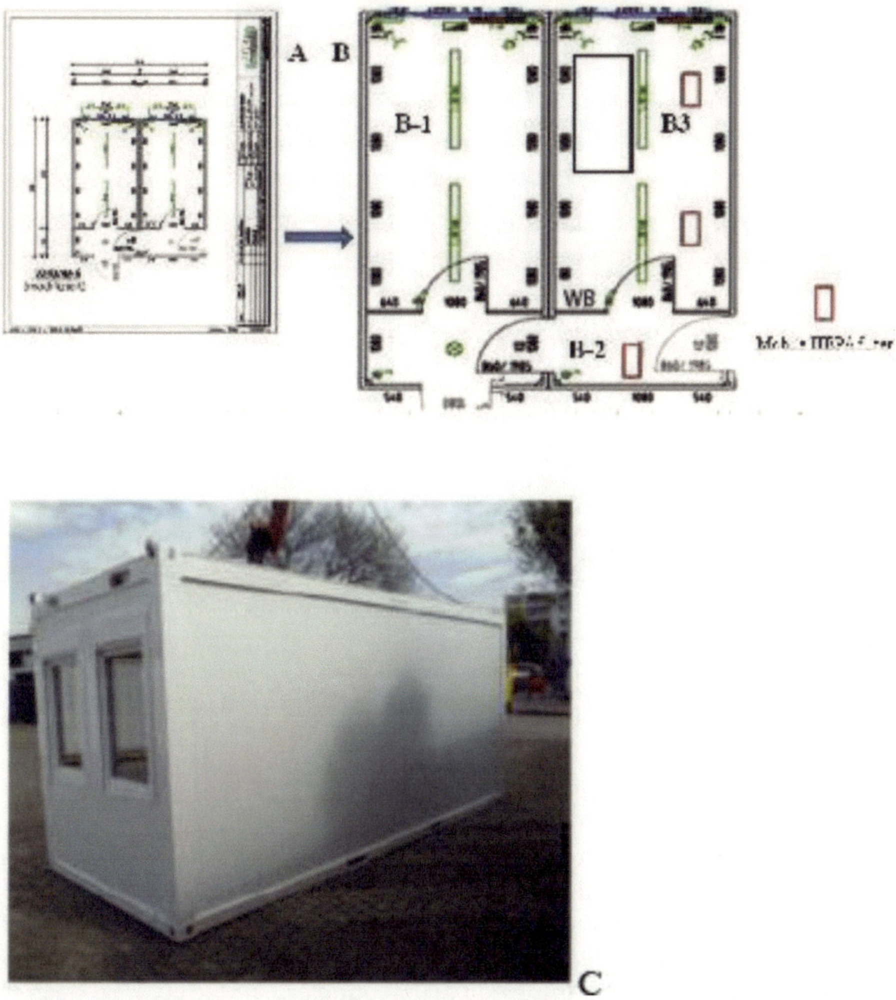

Figure 5. A hard wall clean air facility on wheels. Panel A shows the detailed plan of two containers with windows and doors from ALHO, Type Variante 6 [12]. The enlarged image (panel B) shows how the space was divided. The space was divided into: Space B-1: storage and clothes changing; Space B-2: air lock with a mobile HEPA filter (in red) for air shower; Space B-3: Main clean air facility equipped with two mobile HEPA filters (in red). A two-person-clean air bench (in black), two chairs, and a CO_2 incubator. Panel C shows one container from outside.

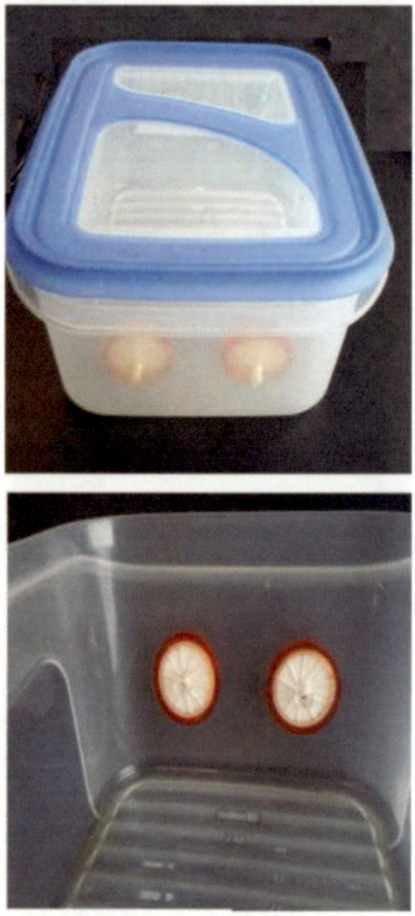

Figure 6. A clean air minibox.

3. Conclusion

It took the entire group almost two years to convert a regular cell culture laboratory in an old building located in a clinic campus into a clean air facility. It is well worth the effort to have undertaken such a project. A professional from Heraeus came to visit our clean air facility and did the particle count and particle collection himself. When he was about to leave the airlock, he shook his head and said that it is incredible that our clean air facility is so clean that even the industrial clean air facility could hardly achieve such a clean level. We consider our experiences valuable and worthwhile to share with members of the research community who wish to work in the direction of clinically-oriented projects such as ES cells and iPS. The total cost is estimated to be 10% of the commercial products, an affordable facility for many. In addition, we have moved the clean air facility into two containers and continued the work

http://dx.doi.org/10.5772/62757

there. The advantage of "clean air facility on wheels" is that it can be shared by several laboratories in a nearby region for performing similar projects such as establishment of clinical grade stem cell lines. Our research work originated in 1987 while I was a scientific member of the Basel Institute for Immunology, and continued through decades in Germany from 1996, exploring different research directions and clinical applications including this clean air facility. For an overview of our research aims and outcome, please see review articles, Discussion Forum, and a World Patent on "Methods for growing stem cells" [13–16].

Acknowledgements

The author wishes to thank the entire team members for engagement. The author is grateful for the support and understanding of Professor Ewald Beck of Biochemistry Institute and Professor K. Knorr, Dean of the Uniclinic Giessen. We thank Dr. Donald Gerson of Canada for his advice and in serving as a QA person for this project. This project was financed by WP, ISPI e.V. internal grants. Dr. Richard Pink of Switzerland helps with the reading, suggestion, and editing of this article.

Appendix 1: Examples of S.O.P.s on how to write, to operate, and to control the practice of a clean air facility. This appendix will have the following three figures:

Figure 7. S.O.P. -1. Into the clean air room

G.M.P./S.O.P.1 J.B.	J.B. / AG Chcn	written on: 11.01.01
		valid until: 11.01.03
Titel: In the clean air room		
	Filled by:________	on (date):________
	Controlled by:________	on (date):________

About one hour before the operation, one must turn on the HEPA filters to let them running (register in the log book).

In the entrance preparation room:

1. Lab coat, jacket, sweater, etc. remain in the entrance room (The clothes hangers are on the right side).

2. Step onto the sticky mattress and take off the regular shoes and change to clean air room shoes right there.

3. Disinfect hands with 70% alcohol.

4. Tight hair and put on hair net.

5. Disinfect hands, put on hand gloves, and sterilize again.

6. Wear the hair cap over the hair net. Hair must be completely covered!

7. Put on clean room gown.

8. Put on mouth mask and eye glasses (when not wearing glasses).

9. Disinfect hands.

10. The material, which should be brought into the clean room, should be disinfected and put into the basket to carry in.

***In air shower-airlock*:**

11. Enter the airlock and close the door to the preparation room.

12. Stay at least 5 min in the air shower-airlock and turn around regularly.

13. Meanwhile carry out one measurement using the particle-counter and one measurement using the air-collector.

14. Put on the clean room gown hanging on the door to the clean air room.

15. Stand on the sticky mattress, then enter the clean air room.

In clean air room (HEPA filter 1, mobile):

16. Run through once the particle counter and once the air collector.

17. Before entering the softwall clean air space (HEPA-Filter 2, hanging), step on the sticky mattress.

In softwall cleanroom (HEPA-Filter 2, hanging):

18. Before starting to work, carry out once the particle count and once the air collection.

Figure 8. S.O.P. -2. Out of clean air room

G.M.P./S.O.P.2 J.B.	J.B. / AG Chcn	written on: 16.08.01
		valid until: 16.08.03
Titel: Out of the cleanair room		
	At:________	Until:_________hr

1. The total surface area inside the soft-wall clean room should be sprayed with 70% alcohol and wait a few minutes to let it absorb.

2. Turn on UV light with the key provided.

3. Turn off clean bench.

4. Record the data in the log book (Hera safe) and put the log book back to the book shelf.

5. Bring the garbage and agar plates to the entrance-preparation room. (Let agar plates incubate and check for growth).

6. Hang the clean air coat back on the door hook.

7. Take off all other clothes in the entrance preparation room and hang on the hooks or dispose. (Dirty clothes should be autoclaved).

8. Register in the log-book.

9. Turn off HEPA filters.

10. Turn off light.

Name 1 ________

Name 2 ________

Figure 9. S.O.P.-3. Routine clean air room control

G.M.P./S.O.P.3 J.B.	J.B. / AG Chcn	written on: 15.03.01
		valid until: 15.03.03
	Performed & filled by:________	on:________
	Controlled by:________	on:________

Important: Turn on both HEPA filters in Room 130 and Room 131 at least one hour before the Monday S.O.P. routine control.

Change to clean room gown.

Room 130 **(HEPA-Filter 2,** i.e. inside the soft wall)

-Read and record the temperature and CO_2-content on a data sheet attached to the CO_2 incubator front door.

-Place one agar plate each (from the refrigerator located in the entrance preparation room) under the Flow/clean bench, on the desk, and in the CO_2 incubator.

-One hour later, bring the agar plates to the incubator in Rm 108 (research lab) and let it incubate overnight (and count the bacteria colonies).

- Afterwards store the agar plates (inside one plastic bag) for four weeks on the wood shelf located in Rm 108 at room temperature, (then count the bacteria/fungi colonies again).

-Clean the flow/clean bench and working desk

-Afterwards, turn on UV light inside the flow/clean bench for one hour

-After that, place one more agar plate under the Flow for one hour, and proceed to incubation and colony counts similar to the above procedure

-Carry out once particle count and once air collection under the Flow and in clean room (see G.M.P./S.O.P.13)

Room 130: (HEPA-Filter 1, i.e. mobile one outside the soft wall)

-Clean the working surface.

-Perform one particle count and one air collection (see G.M.P./S.O.P.13).

-Place one agar plate in the cabinet next to the door to Rm 131 and proceed as of above.

Air lock:

-Perform each time one air collection and one particle count when coming in and going out (see G.M.P./S.O.P.13).

-Place one agar plate on top of the mobile HEPA filter and proceed as of above.

Entrance preparation room

-Read and record the temperature of the refrigerator and -20°C deep freezer in the log book.

-Clean the working surface.

Author details

Una Chen

Address all correspondence to: ispiev@gmx.de

International Senior Professional Institute e.V. Grünbergerstrsse, Giessen, Germany

References

[1] http://www.fda.gov/drugs/developmentapprovalprocess/manufacturing/ucm169105.htm.

[2] http://www.nibsc.org/UK_Stem_Cell_Bank/Cell_lines/EUTCD_Cell_Lines.aspx.

[3] Cleanroom Classification/Particle Count/FS209E/ISO TC209/

[4] Cancellation of FED-STD-209E - Institute of Environmental Sciences and Technology

[5] http://www.wbdg.org/ccb/FEDMIL/notices.pdf, p. 148

[6] https://en.wikipedia.org/wiki/Cleanroom

[7] http://www.mssl.ucl.ac.uk/www_cleanroom/cleanroom/cr_standards.html#pcc

[8] http://www.mw-zander.com

[9] www.cleanairproducts.com

[10] http://www.cleanairproducts.com/product/cap-577h/

[11] https://en.wikipedia.org/wiki/Cleanroom

[12] http://www.ALHO.com

[13] Chen U, Esser R, Kotelenga K, Neis S, Anhlan D, and Szepan U (1997). Potential application of quasi-totipotent ES cells: a ten-year study of soft-tissue engineering with ES cells. Journal of Tissue Engineering 3:321–328.

[14] Chen U. (2006–2015). Methods for growing stem cells. World patent.

[15] Chen U. (2011). Some properties and applications of cell lines and clones established from tet-responsive-SV40 Tag mice and mES cell lines. Scandinavian Journal of Immunology 73, 531–535.

[16] Chen U. (2014). Do we have a workable clinical protocol for differentiating lympho-hematopoietic stem cells from the source of embryonic stem cells and induced pluripotent stem cells in culture? Scandinavian Journal of Immunology, 80, 247–249.

Chapter 9

Induced Pluripotent Stem Cells for Clinical Use

Tomohisa Seki and Keiichi Fukuda

Additional information is available at the end of the chapter

http://dx.doi.org/10.5772/62505

Abstract

Induced pluripotent stem cells (iPSCs) are expected to be a novel cell source for regenerative medicine. Although iPSCs represented a significant breakthrough, there were many initial obstacles for their clinical use such as exogenous sequence insertions, inefficient cell reprogramming, tumorigenic properties, and animal-derived culture components. However, much progress has been made in iPSC generation since their development. The first human trial of iPSC-derived cell transplantation was conducted in September 2014, in which iPSC-derived retinal pigment epithelial cells were transplanted to a patient with macular degeneration. Because multiple clinical trials using iPSCs are expected in the near future, preparation of guidelines for generating and selecting iPSC lines suitable for clinical application is a pressing issue.

For clinical use of iPSCs, many examinations for evaluating iPSC lines must be conducted before transplantation. Different combinations of reprogramming factors, gene derivation vehicles, and types of donor cells can affect the quality of iPSCs, and guidelines for selecting the most appropriate iPSC lines for clinical use are under development. Furthermore, development of time- and cost-effective selection methods is essential for expanding iPSC transplantation therapy. In this chapter, we review methods for preparing human iPSCs before clinical use and the issues that are important for defining standardization of clinical-grade iPSCs.

Keywords: induced pluripotent stem cells, regenerative medicine, transplantation therapy, standardization of stem cells, quality control of stem cells

1. Introduction

Induced pluripotent stem cells (iPSCs) are expected to be a novel cell source for regenerative therapy [1, 2]. Their capacity for self-proliferation and multilineage potential is promising for induction of regenerative cells without a natural capacity for self-renewal. In 2014, a groundbreaking advance in iPSC research occurred when iPSC-derived sheets of retinal pigment epithelium were transplanted to a patient with age-related macular degeneration, the first report of iPSC treatment in humans [3]. Since then, explosive expansion clinical iPSC treatment of patients with otherwise intractable diseases has been expected. However, expanding the clinical use of iPSCs requires a well-defined quality standard. Generating clinical-grade iPSCs for regenerative treatments presents many challenges, and concerns over safe iPSC use must be solved promptly. For instance, ensuring that culture conditions are not exposed to risk of contamination by predictable or unpredictable agents requires a great deal of investment in terms of cost and equipment. In addition, the most clinically applicable method of generating and selecting a suitable iPSC line is under debate. Furthermore, although the construction of iPSC banks for allo-transplantation has progressed [4], generating sufficient qualified cell lines for clinical use requires several years to cover a large segment of the population. In this chapter, we discuss the current issues for expanding the clinical applications of human iPSCs.

2. Existing consensus guidelines for human pluripotent stem cells

In regard to the standardization of human pluripotent stem cells, a consensus for using embryonic stem cells (ESCs) was previously announced by the International Stem Cell Banking Initiative (ISCBI) contributors and the Ethics Working Party of the International Stem Cell Forum [5]. This consensus defined general principles for human ESC banking and described quality control processes for human ESC lines. Although many of these criteria can be applied to iPSCs, standardization of iPSC lines for clinical application is not fully established because there exist various iPSC generation methods and differences between iPSC lines. The tumorigenic and differentiation properties of iPSC lines are not identical, even for those generated by the same procedure [6]. Therefore, to establish iPSC quality standards for clinical use, determining which factors can affect iPSC quality and setting up requirements for clinical application of iPSCs are critical issues. This challenge intrinsically questions whether iPSCs can be equated with ESCs. To date, the existence of epigenetic differences between iPSCs and ESCs has been shown [7], although these differences do not negate the applicability of iPSCs.

Recently, the previous consensus on human pluripotent stem cells was revisited with consideration of iPSCs [8]. However, international standardization of iPSC generation techniques and quality verification is challenging. Many problems must be solved, including the scientific validity of new insight into iPSCs, to determine their applicability to consensus guidelines. In addition, these guidelines mainly target requirements for cell banking. In the case of iPSCs, there will be clinical research using autogenic iPSCs similar to the first case in RIKEN [3]. Therefore, the number of institutions in which autogenic iPSCs are generated could increase above the number of institutions for cell banking. Whether the institutional criteria for generating autogenic iPSCs should be equal to those for cell banking remains vague. These points should respectively be verified and adjusted based on scientific acceptability.

3. Development of the definition of "clinical-grade iPSCs"

When human iPSCs are applied for clinical use as a transplantation cell source, the requirements for iPSCs to satisfy clinical conditions must be validated in advance. Although most provisions for "clinical-grade" iPSCs are associated with "safety," this term encompasses many elements at each stage of iPSC application to regenerative medicine. However, the most suitable and the safest method for generating iPSCs for clinical use has not been defined because experience in treating patients with iPSC-derived regenerative cells remains limited. Therefore, to establish novel iPSC-based regenerative therapy, all factors that might affect safety must be presented and discussed in each case.

For clinical application of iPSCs, safety is mainly divided into two considerations. The first is that iPSCs must meet standards for general cell products. To establish iPSCs of clinical grade, cell culture protocols must avoid any risk of contamination with unpredictable pathogens and meet the standards for general cell products, such as good manufacturing practice (GMP) [5, 8]. Removing animal-derived products from the culture system is important for this purpose. Furthermore, iPSCs and iPSC-derived cells must not include pathogens such as harmful viruses or bacteria before clinical use. With respect to establishing cell culture protocols, accurate sample identification must be provided throughout iPSC generation and differentiation. Therefore, to meet the standards for general cell products, the establishment of extensive systems and equipment for culturing iPSCs is required.

The second aspect of safety is meeting the high quality standards for clinical applicability. However, whereas standards for general cell products have been defined, standards for high-quality iPSCs that meet applicability to clinical use remain vague. For example, methods for denying the possible tumorigenicity of iPSCs and their derivatives remain undefined. In addition, in the course of generating iPSCs, many steps affect iPSC quality. Worldwide standardization of each stage of iPSC generation is desirable, but there are numerous problems to be solved before achieving this goal.

4. Management of safety in each stage of iPSC generation

The first step for applying iPSCs application to clinical use is sampling somatic cells from donors. Although this step appears simple, it already includes safety considerations. When treatment with iPSCs is proposed, whether iPSCs are generated from the patient's own somatic cells or brought from a pool of allogeneic iPSCs such as the iPSC bank project [4] must be decided. Using allogeneic iPSCs requires co-treatment with an immune suppressor and can lead to a risk of malignant tumor and adverse effects. Generating autogenic iPSCs for each patient is ideal, but it requires a tremendous cost and time investment. Particularly, if the patient's condition demands expediency, autogenic iPSCs might not be suitable. Even if autogenic iPSCs are available, the appropriateness of applying quality standards for the allogenic iPSC bank to autogenic iPSCs has to be considered. In addition, the choice of somatic cell sources for iPSC generation is important. Naturally, invasive cell sampling is not preferable, but the type of original cells can affect iPSC features through the residual epigenetic status of the original cells [9–12]. Therefore, this first step already requires evaluation of the appropriate choice for each case.

The step following somatic cell sampling is somatic cell reprogramming. In this stage, the suitability of the combination of reprogramming factors and gene introduction vehicles must be verified. Although achievement of residual transgene-free iPSC lines was already established [13–19], the safest and most preferred type of gene vehicle and combination of reprogramming factors for clinical use are now in discussion. In addition, there are many reagent options for culturing human iPSCs, and reagent selection must be verified in advance.

The next stage of clinical therapy using iPSCs is the induction of targeted cells from iPSCs. iPSC lines do not exhibit identical points of differentiation [6, 12]. Although efficient induction of intended differentiated cells and selection of suitable cell lines are important, ensuring the safety of iPSC derivatives is a more important consideration. In particular, avoiding contamination of undifferentiated cells is essential for achieving clinical application. However, methods for detecting residual undifferentiated cells also remain undefined. Selection of appropriate iPSC lines is also important for avoiding tumorigenesis, which is known to differ among cell lines [20].

After obtaining targeted cells for treatment, the method by which these cells are transplanted is also associated with safety concerns. An appropriate transplantation protocol must be examined and established in advance. In addition, establishment of safety nets for post-treatment patients is also important.

Therefore, although the requirements for establishing a definitive standard for clinical-grade iPSCs are unresolved, many points that affect the safety of treatments must be recognized and appropriately verified before initiating treatment.

5. Advancement of reagents for iPSC generation and culture

As described above, meeting standards for general cell products is required for clinical use of iPSCs. A culture condition for human pluripotent stem cells was first established for maintaining human ESCs [21] and contained several animal-derived products such as mouse embryonic fibroblasts for feeder layers and fatal bovine serum in culture medium. This culture system was applied to human iPSCs, and it successfully maintained their pluripotency and self-proliferation [2]. Since then, toward realizing the clinical use of human iPSCs, the need for an established, chemically defined condition for human iPSCs has attracted attention.

To this end, many researchers have challenged the removal of animal-derived feeder cells from culture conditions. The initial condition for culturing human iPSCs contained mouse embryonic fibroblasts or immortalized mouse fibroblasts such as SNL cells [2]. Although auto-fibroblasts were applied and successfully served as alternative to animal-derived feeder cells for human iPSC generation [22], the availability of human auto-fibroblasts is quantitatively limited. Therefore, to apply human iPSCs to clinical use, replacing feeder layers with a chemically defined substitute is required.

Previously, gelatinous protein mixtures were applied to culturing human pluripotent stem cells. For example, human ESCs were successfully maintained with Matrigel and chemically defined medium [23]. Although Matrigel was applicable for maintenance of human pluripotent stem cells [24–26] and achieved feeder-free human iPSC generation [27, 28], this condition was not animal product-free because the matrix is derived from Engelbreth-Holm-Swarm mouse tumor [29] and contains many types of collagens, laminin, and proteoglycans. Therefore, the essential components for human iPSC culture have been investigated.

Other types of matrices, such as CellStart [30, 31] and synthetic polymers [32, 33], were tested and successfully used as feeder cell substitutes for the maintenance and generation of human pluripotent stem cells. In addition, recombinant cell adhesion proteins have received attention as a defined alternative for feeder cells. For example, vitronectin is a glycoprotein present in the extracellular matrix that mediates cell adhesion and was shown to be an alternative for feeder cells in human pluripotent stem cell culture [34]. Laminin, a component of the basal lamin, is another possible alternative to feeder cells in maintenance and generation of human iPSCs [35, 36]. These products allow removal of animal-derived feeder layers from human iPSC cultures and, therefore, are useful for establishing xeno-free culture conditions for human iPSCs.

The initial culture medium for iPSCs also contained animal products such as fatal bovine serum. There have been many subsequent reports of xeno-free media such as TeSR2 [37], NutriStem [38], Essential E8 [34], and StemFit [39] for human iPSC generation and maintaining. The combination of these matrices and media can achieve generation of human iPSCs under completely defined conditions, thus making iPSC generation in xeno-free conditions achievable.

6. Choice of iPSC donor cell sources

There are two considerations when choosing the types of donor cells. The first is whether allogenic iPSCs or autogenic iPSCs will be used. Applying autogenic iPSCs to each patient is ideal because this method is not expected to require co-treatment with an immune suppressor [40, 41]. The first case of therapy using iPSCs was performed using autogenic iPSCs [3] and was important for reaffirming the usefulness of iPSCs as a source of autogenic regenerative cells. Nevertheless, because of the immense amount of time and effort required to make autogenic iPSCs from each patient, allogenic iPSCs that are matched in human leukocyte antigen (HLA) type are an important option for establishing treatment with iPSC-derived cells [4], especially in cases that demand expedient treatment. However, covering an entire population with HLA-matched allogenic iPSCs is nearly impossible due to the high diversity of HLA genes [42]. Therefore, to achieve complete coverage of the population with iPSC banks is an important issue. In previous reports, hypoimmunogenic human pluripotent stem cells were successfully generated through genome editing [43–45]. Although these methods could complete the missing part of the iPSC bank, whether these genome-edited pluripotent cells are safe needs further validation. At this time, the imperfect coverage of iPSC banks has to be recognized. When the time limit for generating iPSCs is not severe, autogenic iPSCs might become an important option. Therefore, whether standard guidelines for clinical-grade iPSCs can be defined equally for allogenic and autogenic iPSCs is in question. In a recent report, contaminated undifferentiated iPSC-derived cells readily grew teratomas in syngeneic conditions but not in allogenic conditions supported with an immune suppressor [46]. This difference of tumorigenesis between iPSC derivatives in autogenic and allogenic conditions could complicate definition of standards for clinical-grade iPSCs.

The other consideration about choosing types of donor cells is selection of the type of somatic cells for reprogramming. To date, there have been many efforts to minimize the invasiveness of sample acquisition. Previously, generating iPSCs from keratinocytes derived from plucked hair [47], fibroblasts derived from oral mucosa [48], and peripheral blood cells obtained by venipuncture [49–51] was established as less-invasive methods. However, the characteristics of iPSCs can be affected by the type of somatic cells used for their generation [10–12]. Whether residual epigenetic memory derived from original somatic cells is permissible in clinical-grade iPSCs must be considered. In addition, although blood cells are becoming the preferred material for iPSC generation, the best choice for generating iPSCs of high quality avoiding capture of somatic mutations and aberrant epigenetic memory remains undefined. These questions remain important issues toward standardization of iPSC quality.

7. Vehicles for gene delivery in iPSC generation

In generating iPSCs for clinical use, achieving residual transgene-free products is essential because of the possibly harmful effect of residual transgenes. Since the first report of human iPSC generation with retroviral gene introduction [2], there has been much technical progress in methods of gene introduction for iPSC generation. Currently, methods of generating transgene-free human iPSCs are established using adenovirus vectors [13], sendai virus vectors [14], transposons [16], RNA [18], recombinant protein [15, 17], or episomal vectors [19]. Even when non-viral methods such as episomal vectors are used, there is low incidence of genomic insertion of exogenous sequence. In the case of viral vectors, transposons, and episomal vectors, verification of vehicle elimination in iPSCs is necessary. Because methods for verifying the removal of these vehicles are not identical in each case, unionization and standardization of verification methods are difficult. An appropriate method must be established for each type of vehicle, or whole-genome sequencing to detect aberrant vehicle-derived sequence insertions might be essential for identifying residual transgene sequences in iPSCs.

8. Reprogramming factors and alternative molecules

In the first report of successful mouse somatic cell reprogramming with exogenous gene introduction, forced expression of *OCT3/4, SOX2, KLF4,* and *C-MYC* was introduced into mouse somatic cells [1]. Thereafter, although the resulting reprogramming efficiency was low,

http://dx.doi.org/10.5772/62505

C-MYC was shown to be dispensable for somatic cell reprogramming [52]. Another study showed that forced expression of *OCT3/4*, *SOX2*, *LIN28*, and *NANOG* also induced human somatic cell reprogramming [53]. To date, there have been many efforts to generate high-quality iPSCs, although the best combination of factors has not been established. To obtain safe iPSCs, alternative genes for *C-MYC* are needed because *C-MYC* is an oncogenic gene [54], and residual expression of *C-MYC* in iPSCs led to tumorigenesis [52]. TBX3 [55], L-MYC [56], and GLIS1 [57] were reported as alternatives to *C-MYC* with improved chimerism, germline contribution, or prognosis of iPSC-derived clone mice. Although replicating the chimerism experiments using human iPSCs is difficult because of ethical problems, these alternatives to *C-MYC* are expected to provide human iPSCs of high quality.

At the other extreme, there have been many efforts to generate iPSCs using chemical compounds. Although progress in developing gene vehicles enabled generation of residual transgene-free iPSCs, the ultimate goal is to generate iPSCs without gene introduction. Many previous reports demonstrated improved reprogramming efficiency with small molecules, with some specific small molecules serving as a substitute for reprogramming factors [58]. Finally, in mice, a combination of small molecules completely reprogrammed somatic cells into pluripotent states without forced expression of exogenous genes [59, 60]. Although these chemically generated iPSCs require further verification in terms of quality such as residual epigenetic modification of somatic cells, they have the potential to become mainstream for iPSC-associated researches.

9. Quality control of iPSC lines

The quality of mouse iPSCs has been mainly evaluated through chimerism experiments [61]. Germline contribution of iPSCs and induction of iPSC-derived mice have been the ultimate verifications of pluripotency. Previous reports of iPSC generation with TBX3 [55], L-MYC [56], or GLIS1 [57] also evaluated the quality of iPSCs through mouse chimera formation and germline contribution. However, these experiments are not applicable to human iPSCs because of ethical concerns. In addition, although an in vivo teratoma formation assay has also been used to establish the differentiation capacity of iPSCs, quantifying teratoma formation is rather difficult in contrast to in vitro differentiation because the amount of time to teratoma formation and pathological interpretation are needed. Therefore, the quality of human iPSCs has been evaluated with an in vitro differentiation assay.

For example, the in vitro differentiation assay of human iPSCs revealed that differentiation was affected by donor cell types [10–12]. This phenomenon is termed "epigenetic memory" and, interestingly, does not arise from somatic cell reprogramming with nuclear transfer. Although epigenetic memory decreases with increasing culture time [12], residual epigenetic modification of iPSC origin cells must be considered when iPSCs are applied to clinical use. In addition, the gene expression of iPSCs showing a tumorigenic tendency after neural differentiation was analyzed [20], revealing that activated expression of genes containing specific LTR7 sequences in iPSCs was statistically associated with tumorigenesis. Such predictive markers for the quality of human iPSCs are important for rapid selection of cell lines suitable for clinical use.

Recent progress in next-generation sequencing (NGS) techniques has provided platforms for exhaustive analysis of iPSC RNA, genome, and epigenome. This type of analysis can detect chromosomal aberrations in human iPSCs as sequence abnormalities [62]. Although whether somatic cell reprogramming itself can lead to genomic abnormalities in iPSCs is in discussion, long-term culture of pluripotent stem cells is known to lead to genomic abnormalities [63]. Because mutations in protein-coding regions of the genome could trigger tumorigenesis of iPSCs, analysis of iPSCs with NGS can assume a large role in evaluating iPSC quality and selecting iPSCs suitable for clinical use.

As presented above, evaluation of human iPSC quality has been performed without chimera assays. Whether all of perceptions in mouse iPSC experiments are directly applicable to human iPSCs remains a matter of debate. Therefore, investigating the chimeric contribution capacity of human iPSCs has an ultimate importance in quantifying the pluripotency of human iPSCs. In this regard, recent analysis indicated the possibility of transcending boundaries between human iPSCs and chimera formation assays. Human iPSCs and ESCs have features similar to mouse epiblast stem cells, which are in an advanced differentiation state compared to mouse ESCs [64]. Common human iPSCs and mouse epiblast stem cells are thought to be in a "primed state" distinct from the "naïve state" of mouse ESCs. Pluripotent stem cells in a primed state have difficulty in contributing chimeras in preimplantation embryos [65]. In addition, generating chimeras of human and mouse cells is ethically problematic. Therefore, methods for evaluating the quality of human iPSCs have not been standardized. However, a recent report showed that human iPSCs could contribute to chimeras in stage-matched post-implantation mouse embryos [66]. Although the observation period of chimeric embryos was limited, chimera formation experiments with human iPSCs and stage-matching post-implantation mouse embryos might represent a novel assay for evaluating the quality of human iPSCs.

10. Quality control of iPSC-derived products

iPSCs are used as a cell source for inducing intended types of differentiated cells and are not transplanted to patients directly. Therefore, as with quality control of iPSCs, quality control of iPSC-derived products is important for ensuring the safety of clinical application of iPSCs and contains two important considerations for achieving safety.

The first is purification of intended cells from a mixture of differentiated cells. Even a small contamination of undifferentiated cells in the final product could lead to teratoma development after transplantation [67]. Therefore, appropriate methods for purification are required to ensure safety. Purification of products derived from pluripotent stem cells has been previously attempted using a surface marker of pluripotent stem cells to remove undifferentiated cells [68, 69] or cell sorting targeting surface markers specific to intended cells [70, 71]. However, fluorescence-activated cell sorting with cell surface markers is difficult to apply to mass culture systems because of the large time investment required. To achieve applicability to mass culture systems, purification with medium conditions [72, 73] or with reagents with specific cytotoxic effects against undifferentiated cells [74, 75] was developed. These techniques are expected to be useful for achieving safe iPSC-derived products.

To validate the quality of products derived from iPSCs, detection of residual undifferentiated cells in the product is another essential technique for avoiding tumorigenesis after transplantation. If the methods described above for purification achieve high accuracy, methods for evaluating the elimination of undifferentiated cells and assuring safety are required. However, current validation methods for detecting residual undifferentiated cells in final products from iPSCs are limited. Examining expression of pluripotent markers in products from iPSCs with qRT-PCR [76, 77] and detecting specific glycoproteins in the cell culture supernatants [78] were reported as useful methods for evaluating elimination of undifferentiated cells. To ultimately demonstrate safety, the absence of tumorigenesis in in vivo transplantation assays is required, but the appropriate observation period and numerous transplanted cells remain evasive. In addition, whether xeno-transplantation experiments truly replicate transplantation of human cases needs further validation.

11. Building safety nets for post-treatment patients

One of the most important issues for clinical iPSC application is the establishment of safety nets for post-treatment patients. If tumorigenic cell contaminates the final iPSC-derived product and might be transplanted into patients, measures to avoid health hazard to patients must be established. Although surgical resection of iPSC-derived tumors is one conceivable method, there will be cases that cannot be managed through surgery due to the invasiveness of the operation.

Introducing a suicide system into human iPSCs before transplantation is another useful approach for ensuring safe clinical application of iPSCs. When iPSC-derived tumors occur in patients, ablation of iPSC-derived cells by switching the suicide system "on" can prevent invasive surgery in high-risk patients. For example, herpes simplex virus thymidine kinase (HSV-TK) phosphorylates ganciclovir (GCV) and leads to cytotoxicity in the presence of GCV. HSV-TK has been widely used as "suicide gene" in human ESC experiments [79]. The combination of HSV-TK and GCV was also tested in an in vivo mouse model with mouse iPSCs [80, 81]. Another possibility is the combination of inducible caspase-9 and a chemical inducer of dimerization, which was shown to work as suicide system in human iPSC derivatives [82]. Whereas the HSV-TK suicide system is cell cycle dependent, inducible caspase-9 achieves cell cycle-independent ablation of target cells. Although these systems can become an important option for treatment of iPSC-derived tumors, modified genomic introduction methods of suicide genes are required for clinical use. Because random exogenous introduction using lentiviral and retroviral vectors could break functional gene sequences in iPSCs, validation of target sites for suicide gene insertion and targeted genome editing in iPSCs is required for clinical application.

This type of strategy has a disadvantage in that all iPSC-derived cells are diminished with the suicide system. When iPSC-derived tumors occur, diminishing only tumor cells while retaining the useful cells in the engrafted treatment is ideal. With this in mind, some reports showed selective suicide systems in which the suicide gene is inserted under control of a promoter of pluripotent markers [80]. However, in this area of research, how selective removal of tumor cells should be ensured remains to be solved.

Currently, in contrast to research ensuring the safety of iPSCs and iPSC-derivatives, research establishing methods to manage cases in which iPSC-derived tumors occur in post-treatment patients is less common. To ensure safe clinical application of iPSCs, countermeasures for every possible contingency after treatment using iPSCs must be prepared in advance. Thus, this type of research is of considerable importance in the area of regenerative medicine.

12. Conclusion

iPSCs are expected to serve as a novel cell source for regenerative medicine, although there are many points that require verification before expanding their application to broad clinical uses. Standardization of iPSC quality is required, but current verification and validation procedures are not perfect. This incompleteness must be widely recognized. To establish safe iPSC use in regenerative therapy, appropriate improvements of these issues and defined guidelines for iPSCs are expected.

Author details

Tomohisa Seki[1,2*] and Keiichi Fukuda[2]

*Address all correspondence to: seki@a6.keio.jp

1 Department of Emergency and Critical Care Medicine, Keio University School of Medicine, Shinjuku-ku Tokyo, Japan

2 Department of Cardiology, Keio University School of Medicine, Shinjuku-ku Tokyo, Japan

References

[1] Takahashi K, Yamanaka S. Induction of pluripotent stem cells from mouse embryonic and adult fibroblast cultures by defined factors. Cell 2006;126(4):663–76.

[2] Takahashi K, Tanabe K, Ohnuki M, Narita M, Ichisaka T, Tomoda K, et al. Induction of pluripotent stem cells from adult human fibroblasts by defined factors. Cell 2007;131(5):861–72.

[3] 365 days: Nature's 10. Nature 2014;516(7531):311–9.

[4] Cyranoski D. Stem-cell pioneer banks on future therapies. Nature 2012;488(7410):139.

[5] The International Stem Cell Banking Initiative. Consensus guidance for banking and supply of human embryonic stem cell lines for research purposes. Stem Cell Review 2009;5(4):301–14.

[6] Kajiwara M, Aoi T, Okita K, Takahashi R, Inoue H, Takayama N, et al. Donor-dependent variations in hepatic differentiation from human-induced pluripotent stem cells. Proceedings of the National Academy of Sciences of the United States of America 2012;109(31):12538–43.

[7] Ma H, Morey R, O'Neil RC, He Y, Daughtry B, Schultz MD, et al. Abnormalities in human pluripotent cells due to reprogramming mechanisms. Nature 2014;511(7508): 177–83.

[8] Andrews PW, Baker D, Benvinisty N, Miranda B, Bruce K, Brustle O, et al. Points to consider in the development of seed stocks of pluripotent stem cells for clinical applications: International Stem Cell Banking Initiative (ISCBI). Regenerative Medicine 2015;10(2 Suppl):1–44.

[9] Miura K, Okada Y, Aoi T, Okada A, Takahashi K, Okita K, et al. Variation in the safety of induced pluripotent stem cell lines. Nature Biotechnology 2009;27(8):743–5.

[10] Kim K, Doi A, Wen B, Ng K, Zhao R, Cahan P, et al. Epigenetic memory in induced pluripotent stem cells. Nature. 2010;467(7313):285–90.

[11] Polo JM, Liu S, Figueroa ME, Kulalert W, Eminli S, Tan KY, et al. Cell type of origin influences the molecular and functional properties of mouse induced pluripotent stem cells. Nature Biotechnology 2010;28(8):848–55.

[12] Bar-Nur O, Russ HA, Efrat S, Benvenisty N. Epigenetic memory and preferential lineage-specific differentiation in induced pluripotent stem cells derived from human pancreatic islet beta cells. Cell Stem Cell 2011;9(1):17–23.

[13] Stadtfeld M, Nagaya M, Utikal J, Weir G, Hochedlinger K. Induced pluripotent stem cells generated without viral integration. Science (New York, NY) 2008;322(5903):945–9.

[14] Fusaki N, Ban H, Nishiyama A, Saeki K, Hasegawa M. Efficient induction of transgene-free human pluripotent stem cells using a vector based on Sendai virus, an RNA virus that does not integrate into the host genome. Proceedings of the Japan Academy. Series B, Physical and Biological Sciences 2009;85(8):348–62.

[15] Kim D, Kim CH, Moon JI, Chung YG, Chang MY, Han BS, et al. Generation of human induced pluripotent stem cells by direct delivery of reprogramming proteins. Cell Stem Cell 2009;4(6):472–6.

[16] Woltjen K, Michael IP, Mohseni P, Desai R, Mileikovsky M, Hamalainen R, et al. PiggyBac transposition reprograms fibroblasts to induced pluripotent stem cells. Nature 2009;458(7239):766–70.

[17] Zhou H, Wu S, Joo JY, Zhu S, Han DW, Lin T, et al. Generation of induced pluripotent stem cells using recombinant proteins. Cell Stem Cell 2009;4(5):381–4.

[18] Warren L, Manos PD, Ahfeldt T, Loh YH, Li H, Lau F, et al. Highly efficient reprogramming to pluripotency and directed differentiation of human cells with synthetic modified mRNA. Cell Stem Cell 2010;7(5):618–30.

[19] Okita K, Matsumura Y, Sato Y, Okada A, Morizane A, Okamoto S, et al. A more efficient method to generate integration-free human iPS cells. Nature Methods 2011;8(5):409–12.

[20] Koyanagi-Aoi M, Ohnuki M, Takahashi K, Okita K, Noma H, Sawamura Y, et al. Differentiation-defective phenotypes revealed by large-scale analyses of human pluripotent stem cells. Proceedings of the National Academy of Sciences of the United States of America 2013;110(51):20569–74.

[21] Thomson JA, Itskovitz-Eldor J, Shapiro SS, Waknitz MA, Swiergiel JJ, Marshall VS, et al. Embryonic stem cell lines derived from human blastocysts. Science (New York, NY) 1998;282(5391):1145–7.

[22] Takahashi K, Narita M, Yokura M, Ichisaka T, Yamanaka S. Human induced pluripotent stem cells on autologous feeders. PLoS One 2009;4(12):e8067.

[23] Ludwig TE, Levenstein ME, Jones JM, Berggren WT, Mitchen ER, Frane JL, et al. Derivation of human embryonic stem cells in defined conditions. Nature Biotechnology 2006;24(2):185–7.

[24] Xu C, Inokuma MS, Denham J, Golds K, Kundu P, Gold JD, et al. Feeder-free growth of undifferentiated human embryonic stem cells. Nature Biotechnology 2001;19(10):971–4.

[25] Totonchi M, Taei A, Seifinejad A, Tabebordbar M, Rassouli H, Farrokhi A, et al. Feeder- and serum-free establishment and expansion of human induced pluripotent stem cells. International Journal of Developmental Biology 2010;54(5):877–86.

[26] Stover AE, Schwartz PH. Adaptation of human pluripotent stem cells to feeder-free conditions in chemically defined medium with enzymatic single-cell passaging. Methods in Molecular Biology (Clifton, NJ) 2011;767:137–46.

[27] Lai WH, Ho JC, Lee YK, Ng KM, Au KW, Chan YC, et al. ROCK inhibition facilitates the generation of human-induced pluripotent stem cells in a defined, feeder-, and serum-free system. Cell Reprogram 2010;12(6):641–53.

[28] Kishino Y, Seki T, Fujita J, Yuasa S, Tohyama S, Kunitomi A, et al. Derivation of transgene-free human induced pluripotent stem cells from human peripheral T cells in defined culture conditions. PLoS One 2014;9(5):e97397.

[29] Emonard H, Grimaud JA, Nusgens B, Lapiere CM, Foidart JM. Reconstituted basement-membrane matrix modulates fibroblast activities in vitro. Journal of Cellular Physiology 1987;133(1):95–102.

[30] Ausubel LJ, Lopez PM, Couture LA. GMP scale-up and banking of pluripotent stem cells for cellular therapy applications. Methods in Molecular Biology (Clifton, NJ) 2011;767:147–59.

[31] Bergstrom R, Strom S, Holm F, Feki A, Hovatta O. Xeno-free culture of human pluripotent stem cells. Methods in Molecular Biology (Clifton, NJ) 2011;767:125–36.

[32] Mei Y, Saha K, Bogatyrev SR, Yang J, Hook AL, Kalcioglu ZI, et al. Combinatorial development of biomaterials for clonal growth of human pluripotent stem cells. Nature Materials 2010;9(9):768–78.

[33] Lu HF, Narayanan K, Lim SX, Gao S, Leong MF, Wan AC. A 3D microfibrous scaffold for long-term human pluripotent stem cell self-renewal under chemically defined conditions. Biomaterials 2012;33(8):2419–30.

[34] Chen G, Gulbranson DR, Hou Z, Bolin JM, Ruotti V, Probasco MD, et al. Chemically defined conditions for human iPSC derivation and culture. Nature Methods 2011;8(5): 424–9.

[35] Rodin S, Domogatskaya A, Strom S, Hansson EM, Chien KR, Inzunza J, et al. Long-term self-renewal of human pluripotent stem cells on human recombinant laminin-511. Nature Biotechnology 2010;28(6):611–5.

[36] Miyazaki T, Futaki S, Suemori H, Taniguchi Y, Yamada M, Kawasaki M, et al. Laminin E8 fragments support efficient adhesion and expansion of dissociated human pluripotent stem cells. Nature Communications 2012;3:1236.

[37] Meng G, Liu S, Rancourt DE. Synergistic effect of medium, matrix, and exogenous factors on the adhesion and growth of human pluripotent stem cells under defined, xeno-free conditions. Stem Cells and Development 2012;21(11):2036–48.

[38] Sugii S, Kida Y, Kawamura T, Suzuki J, Vassena R, Yin YQ, et al. Human and mouse adipose-derived cells support feeder-independent induction of pluripotent stem cells. Proceedings of the National Academy of Sciences of the United States of America 2010;107(8):3558–63.

[39] Nakagawa M, Taniguchi Y, Senda S, Takizawa N, Ichisaka T, Asano K, et al. A novel efficient feeder-free culture system for the derivation of human induced pluripotent stem cells. Scientific Reports 2014;4:3594.

[40] Araki R, Uda M, Hoki Y, Sunayama M, Nakamura M, Ando S, et al. Negligible immunogenicity of terminally differentiated cells derived from induced pluripotent or embryonic stem cells. Nature 2013;494(7435):100–4.

[41] Guha P, Morgan JW, Mostoslavsky G, Rodrigues NP, Boyd AS. Lack of immune response to differentiated cells derived from syngeneic induced pluripotent stem cells. Cell Stem Cell 2013;12(4):407–12.

[42] Zimmermann A, Preynat-Seauve O, Tiercy JM, Krause KH, Villard J. Haplotype-based banking of human pluripotent stem cells for transplantation: Potential and limitations. Stem Cells and Development 2012;21(13):2364–73.

[43] Riolobos L, Hirata RK, Turtle CJ, Wang PR, Gornalusse GG, Zavajlevski M, et al. HLA engineering of human pluripotent stem cells. Molecular Therapy: The Journal of the American Society of Gene Therapy 2013;21(6):1232–41.

[44] Rong Z, Wang M, Hu Z, Stradner M, Zhu S, Kong H, et al. An effective approach to prevent immune rejection of human ESC-derived allografts. Cell Stem Cell 2014;14(1): 121–30.

[45] Wang D, Quan Y, Yan Q, Morales JE, Wetsel RA. Targeted disruption of the beta2-microglobulin gene minimizes the immunogenicity of human embryonic stem cells. Stem Cells Translational Medicine 2015;4(10):1234–45.

[46] Kawamura A, Miyagawa S, Fukushima S, Kawamura T, Kashiyama N, Ito E, et al. Teratocarcinomas arising from allogeneic induced pluripotent stem cell-derived cardiac tissue constructs provoked host immune rejection in mice. Scientific Reports 2016;6:19464.

[47] Aasen T, Raya A, Barrero MJ, Garreta E, Consiglio A, Gonzalez F, et al. Efficient and rapid generation of induced pluripotent stem cells from human keratinocytes. Nature Biotechnology 2008;26(11):1276–84.

[48] Miyoshi K, Tsuji D, Kudoh K, Satomura K, Muto T, Itoh K, et al. Generation of human induced pluripotent stem cells from oral mucosa. Journal of Bioscience and Bioengineering 2010;110(3):345–50.

[49] Loh YH, Hartung O, Li H, Guo C, Sahalie JM, Manos PD, et al. Reprogramming of T cells from human peripheral blood. Cell Stem Cell 2010;7(1):15–9.

[50] Seki T, Yuasa S, Oda M, Egashira T, Yae K, Kusumoto D, et al. Generation of induced pluripotent stem cells from human terminally differentiated circulating T cells. Cell Stem Cell 2010;7(1):11–4.

[51] Staerk J, Dawlaty MM, Gao Q, Maetzel D, Hanna J, Sommer CA, et al. Reprogramming of human peripheral blood cells to induced pluripotent stem cells. Cell Stem Cell 2010;7(1):20–4.

[52] Nakagawa M, Koyanagi M, Tanabe K, Takahashi K, Ichisaka T, Aoi T, et al. Generation of induced pluripotent stem cells without Myc from mouse and human fibroblasts. Nature Biotechnology 2008;26(1):101–6.

[53] Yu J, Vodyanik MA, Smuga-Otto K, Antosiewicz-Bourget J, Frane JL, Tian S, et al. Induced pluripotent stem cell lines derived from human somatic cells. Science (New York, NY). 2007;318(5858):1917–20.

[54] Dang CV. MYC on the path to cancer. Cell 2012;149(1):22–35.

[55] Han J, Yuan P, Yang H, Zhang J, Soh BS, Li P, et al. Tbx3 improves the germ-line competency of induced pluripotent stem cells. Nature 2010;463(7284):1096–100.

[56] Nakagawa M, Takizawa N, Narita M, Ichisaka T, Yamanaka S. Promotion of direct reprogramming by transformation-deficient Myc. Proceedings of the National Academy of Sciences of the United States of America 2010;107(32):14152–7.

[57] Maekawa M, Yamaguchi K, Nakamura T, Shibukawa R, Kodanaka I, Ichisaka T, et al. Direct reprogramming of somatic cells is promoted by maternal transcription factor Glis1. Nature 2011;474(7350):225–9.

[58] Lin T, Wu S. Reprogramming with small molecules instead of exogenous transcription factors. Stem Cells International 2015;2015:794632.

[59] Hou P, Li Y, Zhang X, Liu C, Guan J, Li H, et al. Pluripotent stem cells induced from mouse somatic cells by small-molecule compounds. Science (New York, NY) 2013;341(6146):651–4.

[60] Zhao Y, Zhao T, Guan J, Zhang X, Fuv Y, Ye J, et al. A XEN-like state bridges somatic cells to pluripotency during chemical reprogramming. Cell 2015;163(7):1678–91.

[61] Okita K, Ichisaka T, Yamanaka S. Generation of germline-competent induced pluripotent stem cells. Nature 2007;448(7151):313–7.

[62] Gore A, Li Z, Fung HL, Young JE, Agarwal S, Antosiewicz-Bourget J, et al. Somatic coding mutations in human induced pluripotent stem cells. Nature 2011;471(7336):63–7.

[63] Baker DE, Harrison NJ, Maltby E, Smith K, Moore HD, Shaw PJ, et al. Adaptation to culture of human embryonic stem cells and oncogenesis in vivo. Nature Biotechnology 2007;25(2):207–15.

[64] Mascetti VL, Pedersen RA. Naivete of the human pluripotent stem cell. Nature Biotechnology 2014;32(1):68–70.

[65] Masaki H, Kato-Itoh M, Umino A, Sato H, Hamanaka S, Kobayashi T, et al. Interspecific in vitro assay for the chimera-forming ability of human pluripotent stem cells. Development (Cambridge, England) 2015;142(18):3222–30.

[66] Mascetti VL, Pedersen RA. Human-mouse chimerism validates human stem cell pluripotency. Cell Stem Cell 2016;18(1):67–72.

[67] Blum B, Benvenisty N. The tumorigenicity of human embryonic stem cells. Advances in Cancer Research 2008;100:133–58.

[68] Tang C, Lee AS, Volkmer JP, Sahoo D, Nag D, Mosley AR, et al. An antibody against SSEA-5 glycan on human pluripotent stem cells enables removal of teratoma-forming cells. Nature Biotechnology 2011;29(9):829–34.

[69] Ben-David U, Nudel N, Benvenisty N. Immunologic and chemical targeting of the tight-junction protein Claudin-6 eliminates tumorigenic human pluripotent stem cells. Nature Communications 2013;4:1992.

[70] Dubois NC, Craft AM, Sharma P, Elliott DA, Stanley EG, Elefanty AG, et al. SIRPA is a specific cell-surface marker for isolating cardiomyocytes derived from human pluripotent stem cells. Nature Biotechnology 2011;29(11):1011–8.

[71] Uosaki H, Fukushima H, Takeuchi A, Matsuoka S, Nakatsuji N, Yamanaka S, et al. Efficient and scalable purification of cardiomyocytes from human embryonic and induced pluripotent stem cells by VCAM1 surface expression. PLoS One 2011;6(8):e23657.

[72] Tohyama S, Hattori F, Sano M, Hishiki T, Nagahata Y, Matsuura T, et al. Distinct metabolic flow enables large-scale purification of mouse and human pluripotent stem cell-derived cardiomyocytes. Cell Stem Cell 2013;12(1):127–37.

[73] Shiraki N, Shiraki Y, Tsuyama T, Obata F, Miura M, Nagae G, et al. Methionine metabolism regulates maintenance and differentiation of human pluripotent stem cells. Cell Metabolism 2014;19(5):780–94.

[74] Matsumoto S, Nakao H, Kawabe K, Nonaka M, Toyoda H, Takishima Y, et al. A cytotoxic antibody recognizing lacto-N-fucopentaose I (LNFP I) on human induced pluripotent stem (hiPS) cells. Journal of Biological Chemistry 2015;290(33):20071–85.

[75] Tateno H, Onuma Y, Ito Y, Minoshima F, Saito S, Shimizu M, et al. Elimination of tumorigenic human pluripotent stem cells by a recombinant lectin-toxin fusion protein. Stem Cell Reports 2015;4(5):811–20.

[76] Kuroda T, Yasuda S, Kusakawa S, Hirata N, Kanda Y, Suzuki K, et al. Highly sensitive in vitro methods for detection of residual undifferentiated cells in retinal pigment epithelial cells derived from human iPS cells. PLoS One 2012;7(5):e37342.

[77] Kuroda T, Yasuda S, Sato Y. In vitro detection of residual undifferentiated cells in retinal pigment epithelial cells derived from human induced pluripotent stem cells. Methods in Molecular Biology (Clifton, NJ) 2014;1210:183–92.

[78] Tateno H, Onuma Y, Ito Y, Hiemori K, Aiki Y, Shimizu M, et al. A medium hyperglycosylated podocalyxin enables noninvasive and quantitative detection of tumorigenic human pluripotent stem cells. Scientific Reports 2014;4:4069.

[79] Schuldiner M, Itskovitz-Eldor J, Benvenisty N. Selective ablation of human embryonic stem cells expressing a "suicide" gene. Stem Cells (Dayton, Ohio) 2003;21(3):257–65.

[80] Chen F, Cai B, Gao Y, Yuan X, Cheng F, Wang T, et al. Suicide gene-mediated ablation of tumor-initiating mouse pluripotent stem cells. Biomaterials 2013;34(6):1701–11.

[81] Lim TT, Geisen C, Hesse M, Fleischmann BK, Zimmermann K, Pfeifer A. Lentiviral vector mediated thymidine kinase expression in pluripotent stem cells enables removal of tumorigenic cells. PLoS One 2013;8(7):e70543.

[82] Ando M, Nishimura T, Yamazaki S, Yamaguchi T, Kawana-Tachikawa A, Hayama T, et al. A safeguard system for induced pluripotent stem cell-derived rejuvenated T cell therapy. Stem Cell Reports 2015;5(4):597–608.

Section 4

Toxicology

Chapter 10

Pluripotent Stem Cells in Toxicity Testing: An Omics Approach

Smita Jagtap, Kesavan Meganathan and Vilas Wagh

Additional information is available at the end of the chapter

http://dx.doi.org/10.5772/63929

Abstract

Traditional approaches to toxicological testing are expensive and time consuming usually involving exposure of chemicals to large numbers of animals during the crucial period of organ development. In order to provide cost-efficient and high-throughput methods, various in vitro test systems have been proposed to access toxicity for environmental toxicants and many drugs. Although effective, these platforms are based on in vitro cell cultures and ex vivo models using embryo cultures and often do not accurately interpret results for human safety because of interspecies difference and/or the inability to reproduce human physiology. To address this problem, a humanized system, pluripotent stem cells were introduced to study toxicity of drugs.

Keywords: pluripotent stem cells, toxicity platform, toxicogenomics, differentiation

1. Introduction

Pluripotent stem cells are endowed with the capacity to self-replicate, to keep their pluripotent state, and ability to differentiate into specialized somatic cells [1–5]. The ability of pluripotent stem cells to differentiate into various types of somatic cells opened up a new era of in vitro toxicity testing. It has been recognized that pluripotent stem cells are not only promising for a potential cell therapy of degenerative diseases but also can be applied for toxicological in vitro assays during drug development. Omics technologies have been used to decipher the networks of signaling events and perturb molecular signaling to identify the new therapeutic targets [6]. In particular, the combination of pluripotent stem cells and "-omics" technologies is extremely promising as a model system for toxicology since at every time point of differentiation, thousands of genes are differentially expressed, thereby significantly increasing the probability of identifying sensitive mRNA or protein markers of toxicity [7, 8].

2. Pluripotent stem cells and its somatic cells differentiation during toxicity testing

Embryonic stem cells (ESCs) are pluripotent cells; when yamanaka factors [9] that maintain them as pluripotent and provide suitable environment to differentiate are removed, these ESCs can differentiate into endodermal, mesodermal, and ectodermal lineages. During embryogenesis the developmental processes occurring in vivo can be partly recapitulated by the use of cultured ESCs in vitro. Progressive-directed differentiation leading to the formation of tissue-specific cells, for example, cardiac, neural, and hepatic, is controlled by differential gene expression, apoptosis, intercellular communication, and cell-matrix interactions [1, 2]. The temporal gene expression and morphological changes of the cultured ESCs occur hierarchically, first developing epiblast cells, then germ layers, and finally somatic cells. Progressive differentiation of human ESCs is coordinated by multiple genes and pathways involved in biological processes, such as cell proliferation and cell death. A balance between these mechanisms is essential for normal embryonic development in vivo. In vitro models combined with appropriate experimental protocols allow the identification of genes participating in developmental processes in vivo. Differentiation of ESCs can partially reproduce early human embryonic development [1, 2]. Therefore, stem cells are a suitable tool to assess toxicant profiles to understand and predict the damage caused by potential therapeutic agents (**Figure 1**).

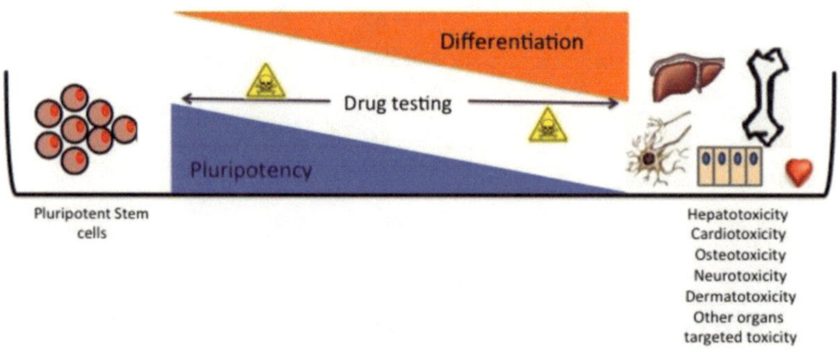

Figure 1. Graphical representation of in vitro toxicogenomics approach using pluripotent stem cells.

Traditional approaches to toxicological testing typically involve exposure of large number of animals to chemicals during the crucial period of organ development and further investigations of fetuses for visceral and skeletal developmental defects. These approaches are expensive and time consuming [3–4, 10]. A magnitude of in vitro test systems have been proposed in order to provide cost-efficient and high-throughput methods to assess the developmental toxicity of candidate drugs and environmental toxicants in the past 20 years. These platforms include primary in vitro cell cultures and ex vivo models using embryo cultures [11]. ESCs-based differentiation systems toward neuronal, cardiac, hepatic and, in general, multiple lineage differentiation have been utilized to monitor the toxic nature of known developmental toxicants on either a mechanistic or functional level [1, 2, 12–16].

3. Directed differentiation of pluripotent stem cell (embryoid-body-mediated differentiation)

In order to uncover multiple embryonic development perturbations in presence of potential toxicants, it is critical to determine optimal differentiation of ESC. Protocols for different cell lineages have been described that exhibit variable success. In most cases, the in vitro differentiation recapitulates the stepwise stages of embryological development for the cell type of interest. In vitro differentiation of stem cells to three germ layers mimics the sequential stages of embryonic development, a brief description of the key steps is provided, and factors in multilineage development are highlighted below in **Figure 2** [1, 2, 12].

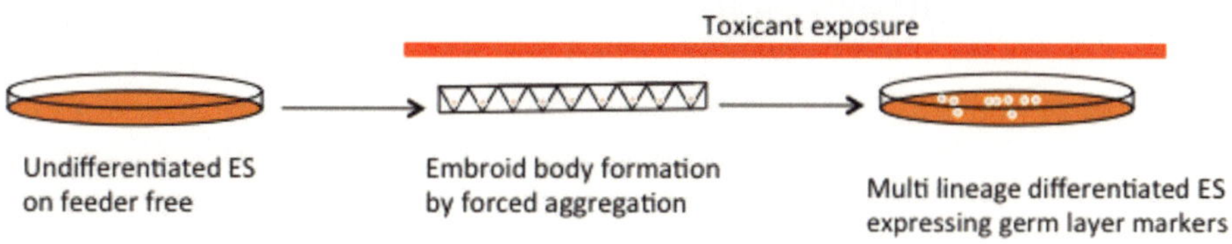

Figure 2. A schematic representation of multilineage differentiation.

hESC were cultured and passaged on irradiated mouse embryonic fibroblasts in knockout-DMEM-F12, KO serum replacement, nonessential amino acids, penicillin/streptomycin, b-mercaptoethanol supplemented with basic fibroblast growth factor. Prior to differentiation, the cells were maintained for 5 days on hESC-qualified matrix-coated tissue culture plates. Cultures were passaged and maintained in feeder-free conditions with conditioned medium supplemented with 8 ng bFGF.

For multilineage differentiation, embryoid bodies were formed by forced aggregation and hESC colony aggregates typically formed rounded discrete structures and acquired an embryoid-body-like appearance within 2–3 days [1, 2]. For differentiation, embryoid bodies were cultured in DMEM-F12, KO serum replacement, nonessential amino acids, penicillin/, streptomycin, and β-mercaptoethanol. These multilineage-differentiated embryoid bodies were analyzed for the presence of markers of the three germ layers. Multilineage differentiation of hESCs in embryoid bodies resulted in downregulation of pluripotency markers, such as POU5F1 and NANOG, and subsequent increase in the germ layer markers. The germ layer markers expressed on day 6 gradually developed and became stronger upon reaching day 12 and day 14. Markers for endoderm (FGA, AFP and DCN), ectoderm (SOX3 and MAP2), and mesoderm (HAND1 POSTN, PITX2) were expressed in these embryoid bodies.

4. Neuronal differentiation

Developmental neurotoxicity (DNT) and many forms of reproductive toxicity (RT) often manifest themselves in functional deficits that are not necessarily based on cell death but rather on minor changes relating to cell differentiation or communication. The fields of DNT/RT would greatly benefit from in vitro tests that allow the identification of toxicant-induced changes of the cellular proteostasis or of its underlying transcriptome network. Despite its high relevance, DNT is one of the least studied forms of toxicity [13, 17]. It is also particularly difficult to study because DNT is not necessarily caused by cell death. In fact, chemically induced changes in the proportions of neural cells, positioning, or connectivity may be sufficient to cause DNT [13, 18]. ESC-based systems recapitulate early neuronal development in vitro, including neural patterning, neurogenesis, and gliogenesis.

For neural differentiation, the cells were thawed and cultured in suspension in T75 flasks with N2B27 medium. From day 2 to day 7, the cells were incubated in N2B27 medium supplemented with 10 μM anti-TGF-beta and 2 μM dorsomorphin. From day 8 to day 32, medium replacement was performed with N2B27 medium only. On day 33, the generated spheres were dissociated as single cells and cultured in N2B27 medium in poly-ornithine and laminin-coated 6-well plates. On day 36, the cells were detached and frozen in N2B27 medium in different aliquots. To test neurotoxicity of chemical compounds, an aliquot was thawed in PLO and laminin-coated 6-well plates. The cells were cultured in a neuronal differentiation medium made of neurobasal medium, B-27 supplement as well as BDNF, recombinant human glial cell-derived neurotrophic factor, and 10 μM ROCK inhibitor. After 1 day of recovery, the cells were incubated with the neurotoxicant in neurobasal medium without ROCK inhibitor for 2 days and then the material was collected for analysis (**Figure 3**).

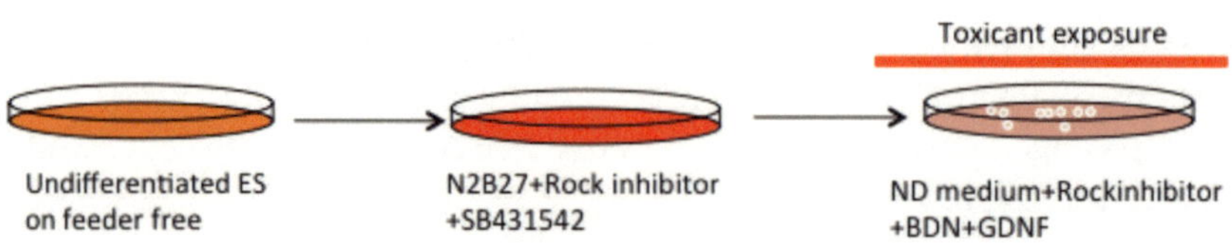

Figure 3. A schematic representation of the early neurogenesis differentiation.

The neural induction differentiation is represented in **Figure 4**. ESCs were differentiated by dual SMAD inhibition. ESCs were plated as single cells with mitomycin C-inactivated mouse embryonic fibroblasts, containing 10 μM ROCK inhibitor Y-27632 and 10 ng/ml bFGF. Differentiation was initiated 3 days after replating on the day of differentiation by changing the medium to knockout serum replacement medium supplemented with 35 ng/ml noggin, 600 nM dorsomorphin, and 10 μM SB-431642. From day 4 onward, KSR was replaced stepwise with N2 medium starting with 25% N2 medium on day 4. To assess the chemical effects on RNA expression, the cells were differentiated in the presence or absence of the chemicals.

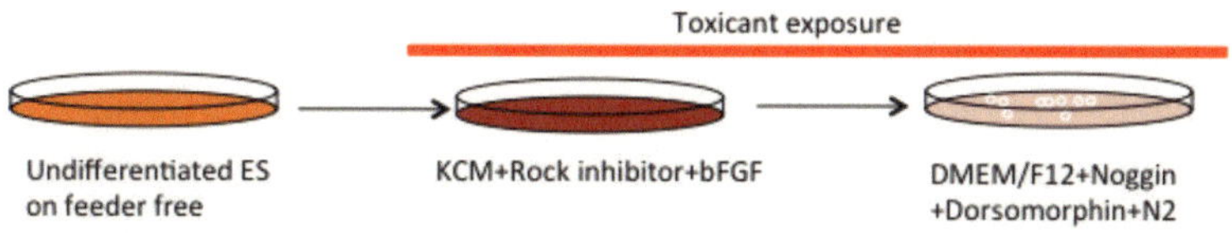

Figure 4. A schematic representation of the neural induction.

5. Cardiac and hepatic differentiation

Liver and cardiovascular disease modeling, mechanisms and therapies can be enhanced using ESC technology. Once pluripotency is established, ESCs can be differentiated into cardiomyocytes [15, 19] and hepatocyte-like cells [20, 21] by using various methods. Further, an embryoid-body-based technique that uses a defined proprietary medium, in addition to ascorbic acid, hypoxic conditions, and a rigorously defined growth factor regiment can be a way to differentiate cardiac cells [19]. Another method claims ESCs are cultured directly on stromal cells, such as murine OP9 or visceral endoderm-like cells [22, 23], and a third approach is a monolayer-based method that uses a medium consisting of RPMI 1640, the supplement B-27, and growth factors activin A and BMP4 with high-density hESC cultured on Matrigel [24, 25]. In all the above techniques, contracting cells are witnessed after 8–12 days.

The only cure for a critically failing organ is liver transplant, despite major breakthrough in liver diseases. The use of ESCs can be explored to developing new treatments leading to a better understanding of the disease process, offering new treatment or preventative strategies.

The protocol [20, 21] describes efficient differentiation of pluripotent stem cells to hepatocytes different tissue culture. Upon hepatic specification, cells express albumin, display cytochrome P450 activity and upregulation of "liver transcription factors," such as CAR, FXR, and PXR.

In summary, the in vitro testing systems demonstrate the ability of ESCs to be used as an alternative to traditional toxicity testing on animals. ESCs not only better reflect the human physiology but also help avoid interspecies differences making it an ideal tool to explore disease mechanism, cures, and to predict toxicity of the drugs.

6. Omics approaches to screen toxic drugs and their advantages to decipher molecular mechanism and signaling pathways perturbed during exposure of drug

A single in vitro assay system may not be able to mimic the complex interactions during human early development. For which, high-density technologies scrutinizing the genome, transcriptome and proteome such as microarray-based technologies and next-generation sequencing technologies come into play, having impressively enriched our knowledge on dynamics related to effects of toxicants during early development [26].

6.1. Detection of different transcriptional responses to toxicity modeling using ESCs

In vivo, the susceptibility of the developing embryo and fetus to chemical exposures during prenatal and early postnatal life may result in important effects on gene expression, thus ensuing functional defects and increased risks of disease later in life [27].

These chemicals are often environmental factors or therapeutic drugs. Since exposure to chemicals during development may cause irreversible effects, it is important to understand their specific mechanisms of toxicity [28].

Toxicogenomics and toxicoproteomics focus on modulation of gene expression in response to exposure and explore the basic mechanisms of toxicity and assessment of the proteome in organs and biofluids such as liver and blood, respectively. DNA microarrays can be used to measure the differential expression of thousands of genes at the same time. Cellular response and organ structure induced by the toxicant well correlates with changes in gene expression levels. A variety of array platforms such as cDNA arrays, high-density oligonucleotide and oligonucleotide bead arrays have been used [29, 30].

The discovery of key modified proteins mainly utilizes toxicoproteomics as early biomarkers for the prediction of an adverse effect. Although time consuming, using high-density LC-MS/MS platforms and its complementary methods can accelerate the toxicoproteomics analysis.

Multilineage differentiation expresses all the three germ layer transcripts. Expression of germ line markers during multilineage differentiation is influenced by the toxicant, resulting in potential prediction of developmental toxicity. The time course multilineage differentiation expression patterns reveal temporal changes in gene expression. Further, upregulation of the markers involving germ layer lineage, such as FGA, FGB, AFP, SLIT2, COL5A2, MYL4, COL6A3, MSX1, MYL7, BMP4, COL1A1, and COL5A1, is seen [1, 2]. Cytosine arabinoside, a strong teratogen, induces developmental toxicity in murine and rat embryonic development. A high concentration of cytosine arabinoside treatment produced cleft palate and lip abnormalities in vivo. Toxicogenomics study revealed cytosine arabinoside repressed gene expression of MSX1, COL3A1, COL1A1, COL1A2, AXUD1, MGP, CDH11, SPARC, POSTN, and BMP4 during bone morphogenesis and osteogenesis in multilineage differentiation of ESCs. Further, cytosine arabinoside could stimulate the expression of neuronal genes and pathways that consequently result in increased neurogenesis and in parallel suppress expression of mesodermal markers [1].

In another study, thalidomide, a teratogen and developmental toxicant, was found to perturb heart and limb development after analyzing microarray data. Thalidomide toxicity affected multilineage differentiation to reveal an upregulation of skeletal, neuronal, and respiratory development and 33 proteins along the reelin pathway. The ESCs differentiation and -omics assisted to uncover inhibition of RANBP1, which participated in the nucleocytoplasmic trafficking of proteins and inhibition of glutathione transferases (GSTA1, GSTA2) as a novel mechanism for thalidomide toxicity [2].

A relationship between cytotoxic response and developmental neurotoxicity-specific transcriptome is analyzed during neuronal differentiation [12]. Toxicity of methyl mercury and valproic acid was accessed using different neuronal differentiation protocols that employed transcriptomics to further investigate the technical feasibility of using transcriptomics as a major endpoint to characterize responses of ESCs-based test systems. Statistically overrepresented GO terms were identified to obtain an overview of the biological processes and displayed for each test system and condition. The genes downregulated by valproic acid in each of the neuronal differentiation system pointed to effects of the toxicant on RNA processing, and on chromatin modification/histone acetylation, consistent with the known activity of valproic acid as a histone deacetylase inhibitor. Transcriptome changes during differentiation were also able to identify upregulation of axonogenesis and ventral forebrain-associated genes, such as SLIT1, SEMA3A, DLX2/4, and GAD2. Further, valproic acid induced expression of miR-378, which was identified to target forebrain markers [12, 16].

Toxicity testing in vitro increasingly utilizes high-throughput screening (HTS) assays. Although HTS assays can test many chemicals, they have a limited use in the regulatory arena. This is because of the need to undergo rigorous, time-consuming formal validation. These HTS assays are additionally used to identify high-concern chemicals. The high-concern chemicals could then be tested sooner, and validation process would ensure the dependability and significance of assays (**Table 1**).

Toxicity	Differentiation	Chemicals	Readout	Markers	Ref.
Developmental toxicity	Multilineage	Cytarabine arabinoside	Microarray, RT-PCR	Induction of neurons and mesoderm inhibition.	[1]
		Thalidomide	Microarray, proteomics, RT-PCR	perturbed heart and limb development	[2]
	Undifferentiated human fibroblast	Arsenic + inhibitor, Busulfan, hydroxyurea	RT-PCR, cyQuant proliferation assay	POU5F1, NANOG, TDGF1, NES, NEFH, TUBB3, CCND1, CCRK, BAX, CASP3, DNMT3B, BCL2	[31, 32]
Neurotoxicity	Early neurogenesis	Methyl mercury, valproic acid	Microarray, RT-PCR		[12]
	Neural induction	Methyl mercury, valproic acid	Microarray, RT-PCR		[12]
	Dopaminergic neurons	pirenzepine, amiodarone, selamectin, clofoctol, perhexilline, griseofulvin, chloroactoxy quinoline, menadione, hexetidine	ATP measurement	OCT4, NESTIN, TH, beta-III tubulin	[33]
Cardiotoxicity	Cardiomyocytes, Na and Ca channel functions	TTX, diltiazem, nifedipine	MEA, Patch clamp	Expression of Na, Ca and HCN channel RNA	[34]

Toxicity	Differentiation	Chemicals	Readout	Markers	Ref.
	Cardiomyocytes, long QT model	cisapride, nifedipine, pinacidil, ranolazine	MEA, Patch clamp	Troponin I, alpha actinin and connexion 43 staining, expression of NKX2-5, MYL2, MYH6, MYH7, KCNH2	[35]
Hepatotoxicity	Hepatocytes	Not tested	accumulation of glycogen, metabolism of indocyanine green, accumulation of lipid	80% of cells expressed albumin, expression of hepatocyte-specific genes	[36]

Table 1. Differentiation protocols and test systems for toxico-omics relevant models.

7. Concerns: challenges and unresolved issues in stem cells toxicity testing

A species difference during animal and human study is a key important factor endpoints in toxicity study suffer (e.g., cardiotoxicity, hepatoxicity, developmental toxicity, neurotoxicity). To avoid such interspecies, variance toxicity tests that are based on human stem cells have been developed. A high-throughput toxicity screening demands unlimited supply of homogeneous population of cells, which is yet another challenge. Stem cells and its differentiation, being of nonmalignant origin and human origin, have the advantage to generate several types of cells of toxicological relevance. Although pluripotent stem cells have enormous potential for toxicity testing, many unresolved issues need to be addressed. The toxicity dose prediction for humans using in vitro system and challenges to yield consistent somatic cell types from pluripotent stem cells are in their preliminary stages. However, integration of the stem cells testing in combination with omics technology and in vivo animal testing could save millions of dollars and years of time in drug development and could demonstrate the mechanism of the toxicity.

Author details

Smita Jagtap[1], Kesavan Meganathan[2] and Vilas Wagh[1*]

*Address all correspondence to: vwagh@mgh.harvard.edu

1 Massachusetts General Hospital, Harvard Medical School, Boston, MA, USA

2 Department of Developmental Biology, Washington University School of Medicine, St. Louis, MO, USA

References

[1] Jagtap S, Meganathan K, Gaspar J, Wagh V, Winkler J et al. Cytosine arabinoside induces ectoderm and inhibits mesoderm expression in human embryonic stem cells during multilineage differentiation. Br J Pharmacol 162: 1743–1756, 2011.

[2] Meganathan K, Jagtap S, Wagh V, Winkler J, Gaspar JA, Hildebrand D, Trusch M, Lehmann K, Hescheler J, Schlüter H, Sachinidis A. Identification of thalidomide-specific transcriptomics and proteomics signatures during differentiation of human embryonic stem cells. PloS One 7(8): e44228, 2012.

[3] Kraljevic S, Stambrook PJ, Pavelic K. Accelerating drug discovery. EMBO Rep 5: 837–842, 2004.

[4] Seidenberg JM, Anderson DG, Becker RA. Validation of an in vivo developmental toxicity screen in the mouse. Teratog Carcinog Mutag 6: 361–374, 1986.

[5] Adler S, Pellizzer C, Hareng L, Hartung T, Bremer S. First steps in establishing a developmental toxicity test method based on human embryonic stem cells. Toxicol In Vitro 22: 200–211, 2008.

[6] Yao Z, Petschnigg J, Ketteler R, Stagljar I. Application guide for omics approaches to cell signaling. Nat Chem Biol 11: 387-389, 2015.

[7] Desbaillets I, Ziegler U, Groscurth P, Gassmann M. Embryoid bodies: an in vitro model of mouse embryogenesis. Exp Physiol 85: 645–651, 2000.

[8] Gissel C, Voolstra C, Doss MX et al. An optimized embryonic stem cell model for consistent gene expression and developmental studies: a fundamental study. Thromb Haemost 94: 719–727, 2005.

[9] Liu X, Huang J, Chen T, Wang Y, Xin S, Li J, Pei G, Kang J. Yamanaka factors critically regulate the developmental signaling network in mouse embryonic stem cells. Cell Res, 18: 1177–1189, 2008.

[10] Hardin BD, Schuler RL, Burg JR, Booth GM, Hazelden KP et al. Evaluation of 60 chemicals in a preliminary developmental toxicity test. Teratog Carcinog Mutagen 7: 29–48, 1987.

[11] Augustine-Rauch K, Zhang CX, Panzica-Kelly JM. In vitro developmental toxicology assays: a review of the state of the science of rodent and zebrafish whole embryo culture and embryonic stem cell assays. Birth Defects Res Part C-Embryo Today-Rev 90: 87–98, 2010.

[12] Krug AK, Kolde R, Gaspar JA, Rempel E, Balmer MV, Meganathan K, Vojnits K, Baquie M, Waldmann T, Ensenat-Waser R, Jagtap S, Evans RM, Julien S, Peterson H, Zagoura D, Kadereit S, Gerhard D, Sotiriadou I, Heke M, Natarajan K, Henry M, Winkler J, Marchan R, Stoppini L, Bosgra S, Westerhout J, Verwei M, Vilo J, Kortenkamp A, Hescheler J, Hothorn L, Bremer S, van Thriel C, Kraus K-H, Hengstler JG, Rahnenfuhrer

J, Leist M, Sachinidis A. Human embryonic stem cell-derived test systems for developmental neurotoxicity: a transcriptomics approach. Arch Toxicol 87: 123–143, 2013. DOI: 10.1007/s00204-012-0967-3 TOXICOGENOMICS

[13] Kadereit S, Zimmer B, van Thriel C, Hengstler JG, Leist M. Compound selection for in vitro modeling of developmental neurotoxicity. Front Biosci 17: 2442–2460, 2012.

[14] Liang P, Lan F, Lee A, Gong T, Sanchez-Freire V, Wang Y, Diecke S, Sallam K, Knowles J, Wang P, Nguyen P, Bers D, Robbins R, Wu J. Drug screening using a library of human induced pluripotent stem cell-derived cardiomyocytes reveals disease-specific patterns of cardiotoxicity. Circulation 127: 1677–1691, 2013. DOI 10.1161/CIRCULATIONAHA.113.001883

[15] Mordwinkin N, Burridge P, Wu J. A review of human pluripotent stem cell-derived cardiomyocytes for high-throughput drug discovery, cardiotoxicity screening, and publication standards. J Cardiovasc Trans Res 6: 22–30, 2013. DOI 10.1007/s12265-012-9423-2

[16] Meganathan K, Jagtap S, Srinivasan SP, Wagh V, Hescheler J, Hengstler J, Leist M, Sachinidis A. Neuronal developmental gene and miRNA signatures induced by histone deacetylase inhibitors in human embryonic stem cells. Cell Death Dis 6: e1756, 2015 May 7.

[17] Makris SL, Raffaele K, Allen S et al. A retrospective performance assessment of the developmental neurotoxicity study in support of OECD test guideline 426. Environ Health Perspect 117(1): 17–25, 2009.

[18] Kuegler PB, Zimmer B, Waldmann T et al. Markers of murine embryonic and neural stem cells, neurons and astrocytes: reference points for developmental neurotoxicity testing. AL-TEX 27(1): 17–42, 2010.

[19] Keller, GM. In vitro differentiation of embryonic stem cells. Curr Opin Cell Biol 7(6): 862–869, 1995.

[20] Szkolnicka D, Farnworth S, Lucendo-Villarin B, Hay DC. Deriving functional hepatocytes from pluripotent stem cells. MRC Centre for Regenerative Medicine, University of Edinburgh, Edinburgh, United Kingdom. Curr Protoc Stem Cell Biol 1: 30, 2014.

[21] Godoy P, Schmidt-Heck W, Natarajan K, Lucendo-Villarin B, Szkolnicka D, Asplund A, Björquist P, Widera A, Stöber R, Campos G, Hammad S, Sachinidis A, Chaudhari U, Damm G, Weiss T, Nüssler A, Synnergren J, Edlund K, Küppers-Munther B, Hay D, Hengstler J. Gene networks and transcription factor motifs defining the differentiation of stem cells into hepatocyte-like cells. J Hepatol 63(4): 934–942, 2015.

[22] Mummery C et al. Differentiation of human embryonic stem cells to cardiomyocytes: role of coculture with visceral endoderm-like cells. Circulation 107(21): 2733–2740, 2003.

[23] Nakano T, Kodama H, Honjo T. Generation of lympho hematopoietic cells from embryonic stem cells in culture. Science 265(5175): 1098–1101, 1994.

[24] Murry CE, Keller G. Differentiation of embryonic stem cells to clinically relevant populations: lessons from embryonic development. Cell 132(4): 661–680, 2008.

[25] Nishikawa SI et al. Progressive lineage analysis by cell sorting and culture identifies FLK1 + VE-cadherin + cells at a diverging point of endothelial and hemopoietic lineages. Development 125(9): 1747–1757, 1998.

[26] Derks KWJ, Hoeijmakers JHJ, Pothof J. The DNA damage response: the omics era and its impact. DNA Repair (Amst) 19: 214–220, 2014.

[27] Grandjean P, Bellinger D, Bergman A, Cordier S, Davey-Smith G, Eskenazi B et al. The faroes statement: human health effects of developmental exposure to chemicals in our environment. Basic Clin Pharmacol Toxicol 102: 73–75, 2008.

[28] Sioka C, Kyritsis AP. Central and peripheral nervous system toxicity of common chemotherapeutic agents. Cancer Chemother Pharmacol 63: 761–767, 2009.

[29] Boorman GA, Anderson SP, Casey WM, Brown RH, Crosby LM, Gottschalk K, Easton M, Ni H, Morgan KT. Toxicogenomics, drug discovery, and the pathologist. Toxicol Pathol 30: 15–27, 2002.

[30] Bulera SJ, Eddy SM, Ferguson E, Jatkoe TA, Reindel JF, Bleavins MR, De La Iglesia FA. RNA expression in the early characterization of hepatotoxicants in Wistar rats by high-density DNA microarrays. Hepatology 33: 1239–1258, 2001.

[31] Flora SJ, Mehta A. Monoisoamyl dimercaptosuccinic acid abrogates arsenic-induced developmental toxicity in human embryonic stem cell derived embryoid bodies: comparison with in vivo studies. Biochem Pharmacol 78: 1340–1349, 2009.

[32] Mehta A, Konala VB, Khanna A, Majumdar AS. Assessment of drug induced developmental toxicity using human embryonic stem cells. Cell Biol Int 32: 1412–1424, 2008.

[33] Han Y, Miller A, Mangada J, Liu Y, Swistowski A, Zhan M, Rao MS, Zeng X. Identification by automated screening of a small molecule that selectively eliminates neural stem cells derived from hESCs but not dopamine neurons. PLoS One 4: e7155, 2009.

[34] Satin J, Kehat I, Caspi O, Huber I, Arbel G, Itzhaki I, Magyar J, Schroder EA, Perlman I, Gepstein L. Mechanism of spontaneous excitability in human embryonic stem cell derived cardiomyocytes. J Physiol 559: 479–496, 2004.

[35] Itzhaki I, Maizels L, Huber I, Zwi-Dantsis L, Caspi O, Winterstern A, Feldman O, Gepstein A, Arbel G, Hammerman H, Boulos M, Gepstein L. Modelling the long QT syndrome with induced pluripotent stem cells. Nature 471: 225–229, 2011.

[36] Si-Tayeb K, Noto FK, Nagaoka M, Li J, Battle MA, Duris C, North PE, Dalton S, Duncan SA. Highly efficient generation of human hepatocyte-like cells from induced pluripotent stem cells. Hepatology 51: 297–305, 2010.

Section 5

Disease model

Chapter 11

Nontargeted Metabolite Profiling of Induced Pluripotent Stem Cells (iPSCs) Derived Neural Cells: Insights Into Mechanisms of Brain Diseases

Šárka Lehtonen, Marko Lehtonen and
Jari Koistinaho

Additional information is available at the end of the chapter

http://dx.doi.org/10.5772/62851

Abstract

Since the discovery that introduction of four specific genes encoding transcription factors convert mature human somatic cells into induced pluripotent stem cells (iPSCs), there has been an enormous increase in the applications of iPSC technologies to medical sciences, especially in modeling human diseases. In this review, we summarize recent advances in applying human iPSC reprogramming to generate patient-specific neural subtypes in order to reveal molecular pathways affected in various neurodegenerative diseases. Metabolites provide a functional readout of various cellular states enabling identification of biomarker candidates for early diagnosis, segregation of patient cohorts, and to follow-up disease progression or disease responses to novel therapies. With emerging technologies, that is, mass spectrometry and nuclear magnetic resonance applications to metabolomics and various software solutions of bioinformatics, it has become possible to measure thousands of metabolites simultaneously. These fascinating new techniques provide a powerful tool for setting scientific hypotheses and linking cellular pathways to biological mechanism. This review focuses on mass spectrometry-based metabolomics as a tool for iPSC research.

Keywords: iPS cell-derived neural cells, metabolomics, mass spectrometry, Alzheimer's disease, Parkinson's disease

1. Introduction

Stem cell reprogramming has provided a completely new way to explore human development and diseases. This novel technology is especially useful when investigating diseases in which affected tissue are not easily accessible prior to autopsy. The central nervous system is a good example of a tissue that falls into this category. Using induced pluripotent stem cells (iPSCs) to model human brain diseases has remarkable potential to generate insights into understanding disease mechanisms and opening new avenues for the development of effective clinical intervention. Importantly, iPSCs may allow for searching potential biomarkers of brain diseases, especially neurodegenerative diseases which are well known to be diagnosed far too late to be efficiently treated. Disease-specific iPSCs are now available from patients of several major neurodegenerative diseases.

Metabolomics combines strategies to identify and quantify endogenous small molecules that are products of biochemical reactions, and thereby to reveal connections between different pathways that operate within a living cell. Nontargeted metabolite profiling has become a powerful analytical tool to reveal molecular mechanism in various physiological and pathophysiological stages. Even though there are a lot of publications related to these techniques, there are only a few reports combining iPSCs technologies and nontargeted metabolite profiling.

In this chapter, we first describe the use of nervous system disease-specific iPSCs to model human neurodegenerative diseases. Next, we summarize key metabolites from human cohort studies of Alzheimer's disease (AD) and Parkinson's disease (PD). Finally, the recent findings in mass spectrometry-based nontargeted metabolite profiling are discussed to reveal the details of cellular mechanism in iPS cell-derived neurons and astrocytes.

2. Human-induced pluripotent stem cells (iPSCs) in studies of nervous system diseases

2.1. Application of iPSCs in neurodegenerative diseases

Chronic neurodegenerative diseases are characterized by the slow and progressive loss of neuronal functions, which in turn results in memory loss, cognitive deficits, and/or motor coordination impairment or even loss of motor functions. Neurodegenerative diseases, such as AD, amyotrophic lateral sclerosis (ALS), Huntington's disease (HD), PD, and spinal muscular atrophy (SMA), are typically of sporadic origin or are caused by very rare gene mutations (**Table 1**). Although intensive efforts have been made, effective treatments for neurodegenerative diseases have not been yet discovered.

Disease	Gene	Cell type	Phenotype	References
Alzheimer's disease	APP, PS1, PS2, sporadic	Cortical neurons	Increased deposition of amyloid β; activated GSK-3β; endosomal abnormalities; APP mutations increased total and pTau protein, whereas PS1 mutations did not	[1–6]
Amyotropic lateral sclerosis	SOD1, VAPB, TDP43, FUS, C9ORF72, sporadic	Motor neurons, astroglia	Reduced survival, cytoplasmic and nuclear protein aggregation, altered expression in genes encoding cytoskeletal proteins, neurofilament aggregation, mitochondrial defects, increased oxidative stress, activation of ER and UPR	[7–16]
Huntington's disease	HTT	Glutamatergic neurons, GABAergic neurons, astrocytes	Reduced survival, neurite outgrowth and firing capacity, enhanced lysosomal activity, changes in actin cytoskeleton, decreased cell-cell adhesion properties, and decreased intracellular ATP, decreased ATP/ADP ratios; increased susceptibility to stressors, vacuolation phenotype of astrocytes	[17–23]
Parkinson's disease	LRRK2, GBA, PINK1, SNCA, PARKIN, sporadic	Dopaminergic neurons	Reduced number of neurons and their branches, mitochondrial dysfunction, elevated α-synuclein, reduced synthesis and release of dopamine, increased MAOB expression, impaired autophagy, increased susceptibility to stressors, activated ERK, impaired intrinsic network activity, reduced GBA activity	[24–32]
Spinal muscular atrophy	SMN1, CAG repeats	Motor neurons	Reduced number of neurons, abnormality in neurite outgrowth, impaired clustering of AchR	[33–35]

AchR, acetylcholine receptor; APP, amyloid precursor protein; C9ORF72, chromosome 9 open reading frame 72; ER, endoplasmic reticulum; ERK, extracellular signal-regulated kinases; FUS, fused in sarcoma; GBA, glucosidase beta acid; GSK-3β, glycogen synthase kinase-3 beta; HTT, hungtintin; LRRK2, leucine-rich repeat kinase 2; MAOB, monoamine oxidase B; PINK1, PTEN-induced putative kinase 1; PS1, presenilin 1; PS2, presenilin 2; SMN1, survival of motor neuron 1; SNCA, alpha-synuclein; SOD1, superoxide dismutase; TDP43, transactive response DNA-binding protein 43; UPR, unfolded protein response; VAPB, vesicle-associated protein B.

Table 1. Neurodegenerative diseases modeled with iPSCs.

Possible reasons for failing to develop efficient therapies include the lack of appropriate disease models of human neurons and a limited understanding of the etiological and neurobiological mechanisms of neurodegenerative diseases. Recent advances in PSC research have now opened the path to the generation of iPSCs starting from somatic cells, thus offering an unlimited source of patient-specific disease-relevant neuronal cells. By applying sophisticated differentiation protocols, iPSCs can be directed to differentiate to functional neuronal subtypes, such as glutamatergic cortical neurons, striatal GABAergic neurons, midbrain dopaminergic (DA) neurons, or motor neurons of the spinal ventral horn. Since patient-specific cells are used to generate neuronal population, the resulting phenotype of these cells can be used to model the pathology of neurons or other neural cells within the same individual.

2.1.1. Modeling AD with disease-specific iPSCs

The central nervous system has a limited capacity to regenerate after acute injuries or during chronic degenerative diseases. Neurodegeneration is probably most studied and characterized in AD and PD.

AD is the most prevalent age-related disorder characterized by dysfunction and deterioration of neurons within the neocortex and limbic system, resulting in gradual progressive memory loss and cognitive decline. More than 48 million patients are afflicted with AD worldwide. AD is the most common cause of dementia and may contribute to 60–70% of the cases. There are currently no medications for preventing the disease progression. Even though the pathogenesis of AD has been extensively studied during the last decades, the mechanisms underlying the neuronal defects and synaptic damage in this disease are still unclear. Mutations in the presenilin 1 (PS1), presenilin 2 (PS2), and the amyloid precursor protein (APP) genes account for most of the familial early onset cases of AD by enhancing the production of pathological amyloid beta (Aβ), especially Aβ42, which has a greater tendency to form fibrillary amyloid deposits. The large amounts of Aβ accumulate and form senile plaques in the brains of AD patients.

The "amyloid hypothesis" states that extracellular Aβ deposits are the cause of the disease [36]. This theory has been difficult to verify in living neurons of patients, using ordinary models and technologies. Fortunately, iPSCs technology and the recent method's development toward AD iPS cell-derived cortical neurons provides access to cell types that were previously unobtainable [37]. Presenilin-1 (PS1) A246E and presenilin-2 (PS2) N141I iPSC-derived neurons in conventional 2D cultures produced twice the normal level of highly toxic Aβ42 [1]. In one report, the Aβ accumulation was also seen in neuronal cells derived from sporadic AD patient-derived iPSCs [2]. In addition to Aβ, the accumulation of phosphorylated Tau was observed in neurons of sporadic AD patients, but not in iPS cell-derived neurons that had a PS1 or PS2 gene mutation [2]. Moreover, cerebral neurons derived from iPSCs of APP-mutant patients (E693D, V717L) and sporadic AD patients showed an accumulation of Aβ oligomers associated with endoplasmic reticulum and increased oxidative stress [3]. Intracellular stress and neuronal death were inhibited with the unsaturated fatty acid docosahexaenoic acid (C22:6n3) in neurons derived from one patient carrying mutant APP and one patient with sporadic AD. These findings point out a diversity of pathologies in AD and indicate the need for individual treatment strategies. In line with this conclusion go also the results from a very recent study where neurons derived from patients with various forms of AD differed in APP processing [6]. While in iPSC models, APP mutations appeared to increase the levels of total and phosphorylated Tau, PS1 mutations failed to do so. Finally, AD pathology was recapitulated in a single 3D human neural cell culture system where human neural progenitor cells overexpressed mutant APP or both mutant APP and mutant PS1. These mutations induced robust deposition of Aβ and exhibited high levels of pTau [4]. ApoE4 is the major known genetic risk factor for sporadic AD. Defining the precise functions of ApoE that lead to AD pathology is likely to allow for more specific therapeutic treatment. Combination of iPSC technology and genome editing may shed light on the role of ApoE in AD.

In summary, disease-specific iPSCs have provided the field of neurodegeneration with an exciting tool for research. Revealing the mechanisms underlying AD and other brain diseases is essential for advancing our understanding of those disorders with the goal of finding effective therapies that would restrain or stop disease progression.

2.1.2. Modeling PD with disease-specific iPS cells

PD is the third most common neurodegenerative disorder after AD and dementia with Lewy bodies and affects approximately 1–2% of the population over 60. The pathology is characterized by the progressive degeneration of dopaminergic (DA) neurons in the *substantia nigra pars compacta* and by intracellular inclusions known as Lewy bodies [38]. The loss of DA neurons

http://dx.doi.org/10.5772/62851

results in motor symptoms, including resting tremor, cogwheel rigidity, bradykinesia, and later in the course of the disease, loss of postural reflexes. Despite extensive research, there is no cure for this devastating disorder. The majority of PD cases are sporadic and of unknown origin, with up to 10% of patients presenting with familial monogenetic forms of the disease. Combination of genetic and environmental factors is likely to play an important role in the pathogenesis of PD. Although monogenic forms of PD account for a small percentage of PD cases, the most successful reports involving PD modeling to date utilize DA neurons from patients with monogenetic mutations in which the gene mutation has been characterized (**Table 1**). Understanding how mutations of these genes cause the degeneration of DA neurons is critically important for the mechanistic studies as well as for the identification of disease-modifying drugs. In human molecular genetic studies, at least 18 loci and 11 genes leading to the development of PD have been identified. Mutations in SNCA, UCHL1, LRRK2, PINK1, DJ-1, and ATP13A2 can lead to monogenic forms of PD [39]. Furthermore, mutations in genes including those that code for SNCA, LRRK2, and GBA have been found to be risk factors for sporadic PD [40], and in many cases people carrying these mutations will develop PD. Overall, the most extensively studied PD-related genes are SNCA and LRRK2. DA neurons derived from iPSCs of PD patients with triplication of the SNCA gene produced double the amount of α-synuclein protein when compared with the normal controls [24]. Also, DA neurons bearing the LRRK2 G2019S mutation showed an increased level of α-synuclein and greater oxidative stress [25]. These diseased DA neurons also exhibit an increased number of autophagy vacuoles and impaired synaptic and neuritic morphology [35]. In the latest study, gene correction of LRRK2 G2019S mutation in iPSCs resulted in phenotypic rescue of differentiated DA neurons [30].

In summary, studies of PD using iPSCs technology have shown the presence of PD-associated abnormalities in mitochondrial function, autophagy/lysosomal pathway, axonal transport, and neurite extension. These findings open up an opportunity to use iPSC-derived human midbrain DA neurons in studies of disease mechanisms, biomarker search, and drug development, even in much larger scale.

3. Metabolomics

The metabolome of an organism refers to the complete set of endogenous and exogenous molecules present in biological sample. Metabolite concentrations reflect the phenotype of tissue or cells and provide insights into the biochemistries of the disease [41, 42]. There are two general approaches used in metabolomics study; targeted and nontargeted metabolomics. The targeted approach measures only a defined set of metabolites, whereas the nontargeted profiling, or global approach, measures as many metabolites as possible [41, 43]. The nontargeted metabolomics cover very wide concentration range of the metabolites present in the cells (i.e., from millimolar (mM, 10^{-3} mol/L) to attomolar (aM, 10^{-18} mol/L)) with the different chemical and structural diversities of compounds. Recent developments in mass spectrometry (MS) and bioinformatics-related software solutions have enabled to measure thousands of metabolites simultaneously from biological sample.

MS is an analytical technique that sorts molecules as ions, based on their mass-to-charge (*m/z*) ratio. As a detection method, MS provides the quantitation with extremely high selectivity and sensitivity, reproducibility, and a wide dynamic range from highly complex biological samples such as iPSC-derived neurons and astrocytes. In addition, MS has the ability to give a specific information about molecular structure, which together with the above-mentioned qualities makes MS the technique of choice for metabolomics [43, 44]. Biological samples with extreme molecular complexity can be detected by MS combined with gas chromatography (GC) or liquid chromatography (LC) [45–49]. In addition to chromatographic techniques, MS has been combined to capillary electrophoresis (CE) and flow injection analysis (FIA) used in shotgun lipidomics [45, 46].

Targeted approach measurements are hypothesis-driven and focus on one or more metabolic pathways of interest [41]. A specified list of metabolites is measured by highly optimized protocols in regard to sampling, sample preparation, and instrumental analysis of these metabolic classes [52–54]. The most widely used MS technique for these types of analyses is the triple quadrupole mass spectrometer in combination with atmospheric pressure ionization techniques (i.e., electrospray ionization (ESI) and atmospheric pressure chemical ionization (APCI) [43–55].

Nontargeted metabolite profiling aims to simultaneously measure as many metabolites as possible from a biological specimen, in this case from iPSC-derived neurons or glia. These studies are often hypothesis-generating, and therefore it is important to carefully construct an experimental design that maximizes the number of metabolites detected and their quantitative reproducibility [41]. The workflow of nontargeted metabolite profiling used in our laboratory to study iPSC-derived neurons and astrocytes is presented in **Figure 1**. We adapted previously published protocols to generate iPSC-derived mesencephalic dopaminergic neurons [56, 57] and astrocytes [58]. To maximize the number of molecular features obtained from cell culture samples, it is possible to optimize several steps in the nontargeted metabolite profiling workflow (i.e., sample extraction, chromatographic separation, ionization, mass filtering, and detection of metabolites) [42, 43, 55, 59, 60]. Samples are measured with ultra-high performance liquid chromatography (UHPLC), combined with high accuracy and resolution mass spectrometers with fast scanning capabilities, like the quadrupole time-of-flight (QTOF) mass spectrometer [59, 61]. In our laboratory, two chromatographic separation techniques (i.e., reversed phase (RP) and hydrophilic interaction chromatography (HILIC)) are used to cover both hydrophilic and hydrophobic metabolites. In addition, both ionization polarities (i.e., positive and negative) in ESI are employed to widen the coverage of metabolites present in the sample. After data acquisition, the molecular features are extracted and the results delivered for downstream univariate and multivariate statistical analyses. The identification of metabolites is based on the comparison of the accurate mass and isotope information (i.e., ratios, abundances, and spacing), as well as product ion spectra (MS/MS) to metabolite databases (e.g., the Human Metabolome Database (HMDB) and METLIN). This tentative identification is subsequently confirmed by comparing tandem mass spectrometry (MS/MS) data, together with chromatographic retention time with reference standards.

http://dx.doi.org/10.5772/62851

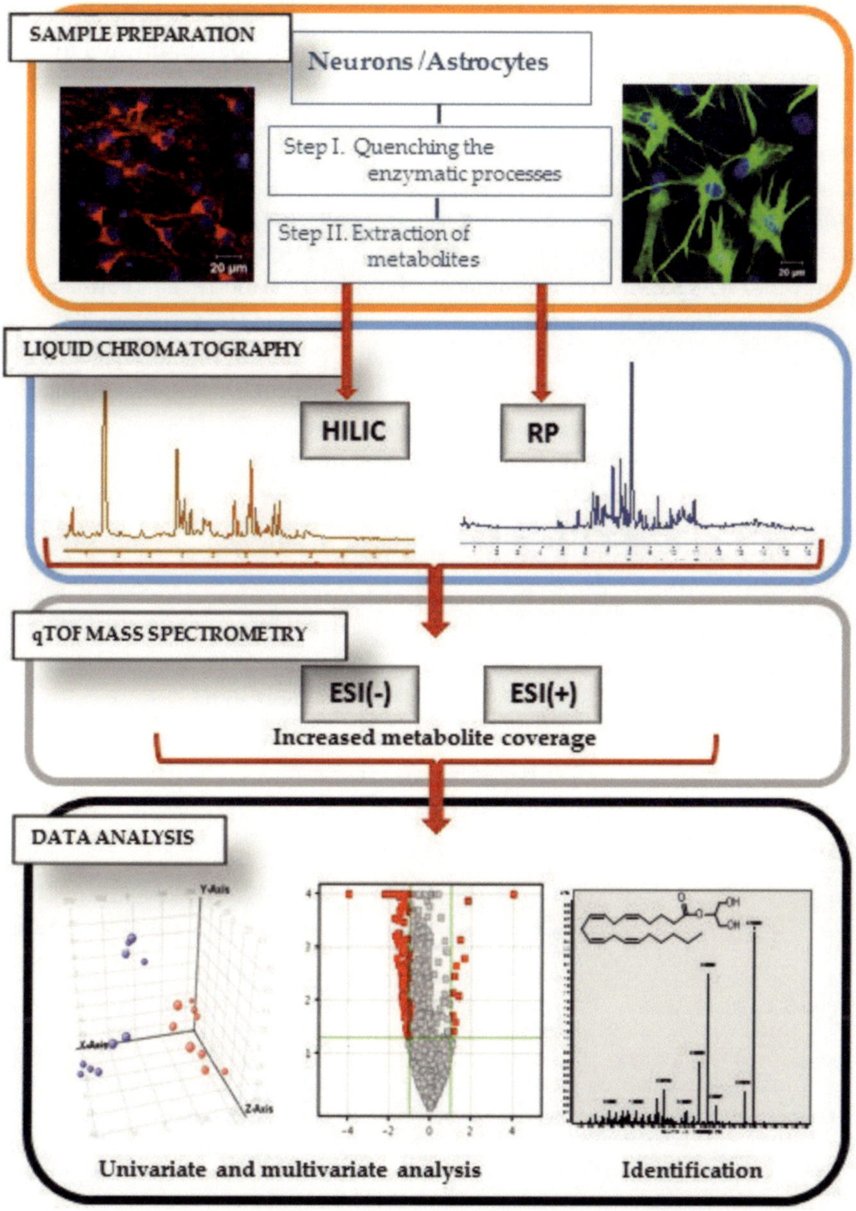

Figure 1. The proposed workflow. The proposed workflow of the nontargeted metabolite profiling for iPS cell-derived neurons and astrocytes. After data acquisition, the results are processed by bioinformatics software for peak-picking, nonlinear retention time alignment, visualization, relative quantification, and statistical analysis. ESI(−), negative mode electrospray ionization; ESI(+), positive mode electrospray ionization; HILIC, hydrophilic interaction chromatography; RP, reversed phase.

Using nontargeted metabolite profiling, in concert with the iPSC technology, presents the new opportunity to better understand the biological processes inside of a cell and extend our view of AD and PD pathophysiology. In our hands, this technique provides thousands of molecular features from the neurons and astrocytes (unpublished data). In conclusion, MS-based metabolite profiling has a central role in revealing the metabolic pathways in healthy and disease-specific iPSCs.

3.1. Metabolomic profiling of AD and PD

Metabolomics is a powerful tool measuring the downstream products of the –omics cascade that reflects genomic, transcriptomic, and proteomic changes in a given biological sample. Metabolomics characterizes the endogenous small molecules and reveals connection between different pathways within a living cell in health and disease [41]. More than 3000 metabolites have been reported, but this number is not ultimate, as many metabolites need to be still identified and characterized. Out of 3000 metabolites, approximately 550 metabolites are of particular interest in studies of brain diseases [62]. Among the metabolites of interest are obviously neurotransmitters, but it is especially phosphates and lipids that are likely to reveal novel features of neurodegenerative diseases with the potential to help drug discovery or identification of novel biomarkers for brain diseases.

Disease	Sample	Instrumentation	Key metabolites	Ref.
Alzheimer's disease	Brain (post mortem)	UPLC-QTOF-MS	Unidentified metabolites	[49]
	CSF (post mortem)	HPLC-EC	Up: methoxy-hydroxymandelate, 5-hydroxytryptophan, methoxy-hydroxyphenylglycol, Down: alpha tocopherol, norepinephrine, ascorbate, 3-methoxytyramine	[64]
	CSF	GC–MS LC–MS/MS	Up: pyruvate, creatinine, cysteine, tyrosine, serine, phenylalanine, methionine, cortisol, dopamine Down: uridine	[45]
	CSF	CE–MS	Up: choline, valine, serine Down: carnitine	[51]
	CSF	HPLC-EC	Up: 5-hydroxyindoleacetic acid, methionine, xanthosine, vanillylmandelic acid, gluthatione Down: gluthatione/ methionine, 5-hydroxyindoleacetic acid/5-hydroxytryptophan	[65]
	Plasma	MDMS-SL	Up: ceramide down: sphingomyelin	[50]
	Plasma	UPLC-MS/ GCxGC-MS	Up: 2,4-dihydroxybutanoic acidDown: ether phospholipids, phosphatidylcholines, sphingomyelins, sterols	[46]
	Plasma	LC–MS/MS UPLC-QTOF-MS	Down: serotonin, phenylalanine, proline, lysine, taurine, phosphatidylcholine, acylcarnitine	[47]
Parkinson's disease	CSF (post mortem)	UHPLC-MS/MS GS–MS	Up: 3-hydroxykynurenine down: oxidized glutathione, N-acetylated amino acids, trimethylglycine, corticosteron	[48]
	Plasma	HPLC-EC	Up: glutathione, 2-hydroxy-2-deoxyguanosine Down: uric acid	[67]
	Plasma	HPLC-EC	Down: hypoxanthine, uric acid, hypoxanthine/xanthosine, xanthosine/xanthine, hypoxanthine/uric acid	[68]

CE–MS, capillary electrophoresis–mass spectrometry; CSF, cerebrospinal fluid; GC–MS, gas chromatography–mass spectrometry; HPLC-EC, high-performance liquid chromatography electrochemical detector; LC–MS/MS, liquid chromatography-tandem mass spectrometry; MDMS-SL, multidimensional mass spectrometry-based shotgun lipidomics; UPLC-QTOF-MS, ultra-performance liquid chromatography quadrupole time-of-flight mass spectrometry.

Table 2. Summary of key metabolites from human cohort studies of AD and PD patients.

Studies utilizing different analytical platforms have already identified several AD pathology-related metabolites in cerebrospinal fluid (CSF) and plasma from human cohorts (**Table 2**). Importantly, by using metabolomics set enrichment analyses, it is possible to identify and quantify metabolite changes [63]. The changes of neurotransmitters (e.g., dopamine, adrenaline, and noradrenaline) are tightly linked to phenylalanine and tyrosine metabolism pathways and have been found to be altered in CSF of AD patients [45]. While the decrease of noradrenaline may correlate with the severity of the disease [64], the increased levels of vanillylmandelic acid seem to correspond to a degradation of xanthine and dopamine [65]. The studies using serum of AD patients have revealed the decrease in several lipid classes, especially sphingomyelins [46, 50]. The most recent study followed elderly adults for 4 years, at a time by which some from them had developed AD [47]. The authors identified 10 metabolite profile features, phospholipids that play a role in integrity and functionality of cell membranes and were predictive of the occurrence of mild cognitive impairment or AD within 2–3 year time frame. The finding supported the link between dyslipidemia in AD and APOE4 variant as a major risk allele for AD that leads to disruption in sphingolipid metabolism [66]. It is reasonable to assume that global metabolite profiling directly from the brain cells (i.e., iPSC-derived material) of corresponding patients with AD or carrying APOE4 gene variant would tremendously increase the sensitivity of the method.

It is known that different areas of brain are affected during progression of neurodegenerative diseases. The composition of metabolites in CSF reflects the neuronal integrity as well as the underlying pathology of brain that leads to changes in brain metabolism. Studies from CSF of postmortem PD patients have shown the activation of kynurenine pathway by increase of 3-hydroxykynurenine concentration [48].

While 3-hydroxykynurenine contributes to oxidative stress, glutathione is known for its antioxidant proprieties. The levels of glutathione were found to be increased in PD plasma samples reinforcing the contribution of oxidative stress [67]. In the same study, the levels of uric acid were negatively associated with PD progression. The aberrations in the purine pathway that might appear upstream from uric acid have been studied in LRRK2 PD and sporadic PD [68]. Based on the plasmatic changes of nine metabolites involved in purine metabolism, the authors were able to predict whether the PD was secondary to LRRK2 mutation or of sporadic origin. A key perspective in metabolic phenotyping of brain diseases like AD and PD depends on the metabolic features interpretation: a single metabolite is unlikely to be a biomarker of a disease, while combinations of altered metabolites could form a diagnostic metabolic signature of the disease.

3.2. Cell-type specific metabolomics

Nontargeted metabolite profiling so far has focused to predict diseases by measuring the changes in specific metabolites in brain and CSF or systemically from plasma or urine. Essentially, the next step would be to move to the study of cell-type specific metabolic variations. iPS cell-derived neurons and astrocytes of AD and PD patients can be used to study cell-type specific metabolites which could determine whether the brain pathology changes originate from one cell type or another. This approach will also provide insights into neuron/ glia cell co-metabolome and will help to refine more targeted treatments for those diseases.

The causes of AD and PD diseases suggest a combination of genetic and environmental factors, with possible mechanisms ranging from mitochondrial dysfunction, oxidative stress, protein aggregation, impaired protein degradation, dysregulated autophagy, and inflammation. So far, there are only unpublished data available from one metabolome analysis of PARK2 iPS cell-derived neurons briefly discussed in a review by Okano and Yamanaka [69]. The changes seen in glycolytic pathways, the tricarboxylic acid cycle (TCA), and pentose phosphate pathways suggest that mitochondria are the sites of dysfunctions in neurons harboring PARK2 mutation (H.O. unpublished data). Abnormalities in mitochondria were further confirmed by morphology and impaired mitochondrial turnover (unpublished data). As mitochondria are required for the energy demands of the brain cells including neurons, mitochondrial alterations can promote neuron degeneration. Mitochondrial complex I activity and its regulation by transcriptional factors are both seen to be altered also in PD patients [70]. Reduced level of mitochondrial α-ketoglutarate dehydrogenase, a rate-limiting enzyme in TCA cycle, was reported in the brain of PD patients [71–73], suggesting the involvement of bioenergetics defects as well reduced mitochondrial complex activity in PD pathology. The results from postmitotic human dopaminergic neurons (DA) exposed to mitochondrial respiratory chain inhibitor MPP^+ (1-methyl-4-phenylpyridinium) are in line with those findings. Altogether, 190 metabolites have been found altered in MPP^+-treated DA neurons [74]. In energy metabolism, decreased levels of intracellular glucose leads to increase of pyruvate and lactate, while consumption of phosphorylated creatine and accumulation of creatine suggest depletion of the cellular energy accompanied by gradual increase of ADP, AMP, and adenine [74]. Decreased total intracellular ATP levels have also been found from fibroblast cells of PD patients with mutations in LRRK2 accompanied with decreased mitochondrial membrane potential [75]. Moreover, increase in methionine sulfoxide formation in MPP^+-treated DA neurons suggested an increase in oxidative stress and ROS production [74]. To better understand the effects of oxidative stress pathways on cellular metabolism, we propose to apply a combined strategy, consisting of nontargeted metabolite profiling together with gene expression analyses to characterize DA neurons and astrocytes derived from iPSCs of PD patients, both sporadic and familiar form. This will allow us to explore how the cells of interest generate reducing potential to compensate oxidative stress and how this could be impaired in genetically perturbed cells.

Several studies have suggested an important role of mitochondria also in the onset and progression of AD [76]. The fact that mitochondria are prime targets for APP as well for Aβ [77] has been underlined in studies of neuronal cells overexpressing APP and Aβ that leads to abnormal mitochondrial dynamics via modulation of mitochondrial fission/fusion proteins [78]. PET imaging has shown decreased levels of brain glucose metabolism in AD brain [79], while another study has demonstrated decreased expression of genes involved in glucose delivery, oxidative phosphorylation, and energy consumption in the brains of AD patients [80]. Postmortem studies on AD brain and fibroblasts from AD patients have revealed changes in TCA, pyruvate dehydrogenase, isocitrate dehydrogenase, and α-ketoglutarate dehydrogenase [81–83]. In addition, the levels of ATP and activities of cytochrome oxidase were reported to be decreased. Apart from direct mitochondrial respiratory chain defects, increased autophagic degradation of mitochondria has also been reported [84].

While AD and PD pathology from diseased patients show late stages of these neurodegenerative diseases, patient-specific derived iPSCs have a potential to model early presymptomatic stage of AD and PD. In 2D cultures where the cells are grown in a monolayer, cell phenotype might be amplified and seen much more earlier compared to the moment when patients are diagnosed. On the other hand, 2D system lacks interaction between the other cells present normally in human brain. In 3D culture models, cells interact with each other, but culture is rather heterogeneous and compensation capability is higher than that in 2D culture. Therefore, it would be necessary to study the metabolic changes in neurons and astrocytes first separately as well as in co-culture, in order to understand the mechanisms behind the pathology. Genetic background needs to be taken into account as well. Genetic correction of patient mutations or introduction of mutation to healthy line allows validation of the role of mutations in the observed phenotype.

4. Conclusions

In this review, we have described the current status of using disease-specific iPSCs to model the onset and initial stages of neurodegenerative diseases such as AD and PD. Human iPSCs improve disease modeling in previously inaccessible cells, such as neurons and astrocytes that are the key cell types of the brain. Emerging evidence suggests that brain energy metabolism goes beyond synthesis of neurotransmitters and that different pathways within neuronal cells contribute to brain disorders. Combining the technologies of iPCS and untargeted metabolomics may provide a powerful tool to better understand the biological processes inside of cell. In the future, brain metabolite profiles could be correlated to DNA polymorphisms, and together with transcriptomics/proteomics data could provide a platform for new neuroprotective strategies.

Author details

Šárka Lehtonen[1], Marko Lehtonen[2] and Jari Koistinaho[1*]

*Address all correspondence to: jari.koistinaho@uef.fi

1 Department of Neurobiology, A. I. Virtanen Institute for Molecular Sciences, University of Eastern Finland, Kuopio, Finland

2 School of Pharmacy, University of Eastern Finland, Kuopio, Finland

References

[1] Yagi T, Ito D, Okada Y, Akamatsu W, Nihei Y, Yoshizaki T, et al. Modeling familial Alzheimer's disease with induced pluripotent stem cells. Hum Mol Genet. 2011;20(23): 4530–9.

[2] Israel MA, Yuan SH, Bardy C, Reyna SM, Mu Y, Herrera C, et al. Probing sporadic and familial Alzheimer's disease using induced pluripotent stem cells. Nature. 2012;482(7384):216–20.

[3] Kondo T, Asai M, Tsukita K, Kutoku Y, Ohsawa Y, Sunada Y, et al. Modeling Alzheimer's disease with iPSCs reveals stress phenotypes associated with intracellular Aβ and differential drug responsiveness. Cell Stem Cell. 2013;12(4):487–96.

[4] Choi SH, Kim YH, Hebisch M, Sliwinski C, Lee S, D'Avanzo C, et al. A three-dimensional human neural cell culture model of Alzheimer's disease. Nature. 2014;515(7526): 274–8.

[5] Muratore CR, Rice HC, Srikanth P, Callahan DG, Shin T, Benjamin LNP, et al. The familial Alzheimer's disease APPV717I mutation alters APP processing and Tau expression in iPSC-derived neurons. Hum Mol Genet. 2014;23(13):3523–36.

[6] Moore S, Evans LDB, Andersson T, Portelius E, Smith J, Dias TB, et al. APP metabolism regulates tau proteostasis in human cerebral cortex neurons. Cell Rep. 2015;11(5):689–96.

[7] Dimos JT, Rodolfa KT, Niakan KK, Weisenthal LM, Mitsumoto H, Chung W, et al. Induced pluripotent stem cells generated from patients with ALS can be differentiated into motor neurons. Science. 2008;321(5893):1218–21.

[8] Mitne-Neto M, Machado-Costa M, Marchetto MCN, Bengtson MH, Joazeiro CA, Tsuda H, et al. Downregulation of VAPB expression in motor neurons derived from induced pluripotent stem cells of ALS8 patients. Hum Mol Genet. 2011;20(18):3642–52.

[9] Bilican B, Serio A, Barmada SJ, Nishimura AL, Sullivan GJ, Carrasco M, et al. Mutant induced pluripotent stem cell lines recapitulate aspects of TDP-43 proteinopathies and reveal cell-specific vulnerability. Proc Natl Acad Sci U S A. 2012;109(15):5803–8.

[10] Egawa N, Yamamoto K, Inoue H, Hikawa R, Nishi K, Mori K, et al. The endoplasmic reticulum stress sensor, ATF6α, protects against neurotoxin-induced dopaminergic neuronal death. J Biol Chem. 2011;286(10):7947–57.

[11] Burkhardt MF, Martinez FJ, Wright S, Ramos C, Volfson D, Mason M, et al. A cellular model for sporadic ALS using patient-derived induced pluripotent stem cells. Mol Cell Neurosci. 2013;56:355–64.

[12] Serio A, Bilican B, Barmada SJ, Ando DM, Zhao C, Siller R, et al. Astrocyte pathology and the absence of non-cell autonomy in an induced pluripotent stem cell model of TDP-43 proteinopathy. Proc Natl Acad Sci USA. 2013;110(12):4697–702.

[13] Kiskinis E, Sandoe J, Williams LA, Boulting GL, Moccia R, Wainger BJ, et al. Pathways disrupted in human ALS motor neurons identified through genetic correction of mutant SOD1. Cell Stem Cell. 2014;14(6):781–95.

[14] Meyer K, Ferraiuolo L, Miranda CJ, Likhite S, McElroy S, Renusch S, et al. Direct conversion of patient fibroblasts demonstrates non-cell autonomous toxicity of astrocytes to motor neurons in familial and sporadic ALS. Proc Natl Acad Sci U S A. 2014;111(2):829–32.

[15] Chen H, Qian K, Du Z, Cao J, Petersen A, Liu H, et al. Modeling ALS with iPSCs reveals that mutant SOD1 misregulates neurofilament balance in motor neurons. Cell Stem Cell. 2014;14(6):796–809.

[16] Liu X, Chen J, Liu W, Li X, Chen Q, Liu T, et al. The fused in sarcoma protein forms cytoplasmic aggregates in motor neurons derived from integration-free induced pluripotent stem cells generated from a patient with familial amyotrophic lateral sclerosis carrying the FUS-P525L mutation. Neurogenetics. 2015;16(3):223–31.

[17] Park I-H, Arora N, Huo H, Maherali N, Ahfeldt T, Shimamura A, et al. Disease-specific induced pluripotent stem cells. Cell. 2008;134(5):877–86.

[18] An MC, Zhang N, Scott G, Montoro D, Wittkop T, Mooney S, et al. Genetic correction of Huntington's disease phenotypes in induced pluripotent stem cells. Cell Stem Cell. 2012;11(2):253–63.

[19] Camnasio S, Delli Carri A, Lombardo A, Grad I, Mariotti C, Castucci A, et al. The first reported generation of several induced pluripotent stem cell lines from homozygous and heterozygous Huntington's disease patients demonstrates mutation related enhanced lysosomal activity. Neurobiol Dis. 2012;46(1):41–51.

[20] HD iPSC Consortium. Induced pluripotent stem cells from patients with Huntington's disease show CAG-repeat-expansion-associated phenotypes. Cell Stem Cell. 2012;11(2):264–78.

[21] Jeon I, Lee N, Li J-Y, Park I-H, Park KS, Moon J, et al. Neuronal properties, in vivo effects, and pathology of a Huntington's disease patient-derived induced pluripotent stem cells. Stem Cells Dayt Ohio. 2012;30(9):2054–62.

[22] Juopperi TA, Kim WR, Chiang C-H, Yu H, Margolis RL, Ross CA, et al. Astrocytes generated from patient induced pluripotent stem cells recapitulate features of Huntington's disease patient cells. Mol Brain. 2012;5:17.

[23] Zhang QC, Yeh T, Leyva A, Frank LG, Miller J, Kim YE, et al. A compact beta model of huntingtin toxicity. J Biol Chem. 2011;286(10):8188–96.

[24] Devine MJ, Ryten M, Vodicka P, Thomson AJ, Burdon T, Houlden H, et al. Parkinson's disease induced pluripotent stem cells with triplication of the α-synuclein locus. Nat Commun. 2011;2:440.

[25] Nguyen HN, Byers B, Cord B, Shcheglovitov A, Byrne J, Gujar P, et al. LRRK2 mutant iPSC-derived DA neurons demonstrate increased susceptibility to oxidative stress. Cell Stem Cell. 2011;8(3):267–80.

[26] Seibler P, Graziotto J, Jeong H, Simunovic F, Klein C, Krainc D. Mitochondrial Parkin recruitment is impaired in neurons derived from mutant PINK1 induced pluripotent stem cells. J Neurosci Off J Soc Neurosci. 2011;31(16):5970–6.

[27] Cooper O, Seo H, Andrabi S, Guardia-Laguarta C, Graziotto J, Sundberg M, et al. Pharmacological rescue of mitochondrial deficits in iPSC-derived neural cells from patients with familial Parkinson's disease. Sci Transl Med. 2012;4(141):141ra90.

[28] Imaizumi Y, Okada Y, Akamatsu W, Koike M, Kuzumaki N, Hayakawa H, et al. Mitochondrial dysfunction associated with increased oxidative stress and α-synuclein accumulation in PARK2 iPSC-derived neurons and postmortem brain tissue. Mol Brain. 2012;5:35.

[29] Jiang H, Ren Y, Yuen EY, Zhong P, Ghaedi M, Hu Z, et al. Parkin controls dopamine utilization in human midbrain dopaminergic neurons derived from induced pluripotent stem cells. Nat Commun. 2012;3:668.

[30] Reinhardt P, Schmid B, Burbulla LF, Schöndorf DC, Wagner L, Glatza M, et al. Genetic correction of a LRRK2 mutation in human iPSCs links parkinsonian neurodegeneration to ERK-dependent changes in gene expression. Cell Stem Cell. 2013;12(3):354–67.

[31] Woodard CM, Campos BA, Kuo S-H, Nirenberg MJ, Nestor MW, Zimmer M, et al. iPSC-derived dopamine neurons reveal differences between monozygotic twins discordant for Parkinson's disease. Cell Rep. 2014;9(4):1173–82.

[32] Sánchez-Danés A, Richaud-Patin Y, Carballo-Carbajal I, Jiménez-Delgado S, Caig C, Mora S, et al. Disease-specific phenotypes in dopamine neurons from human iPS-based models of genetic and sporadic Parkinson's disease. EMBO Mol Med. 2012;4(5):380–95.

[33] Ebert AD, Yu J, Rose FF, Mattis VB, Lorson CL, Thomson JA, et al. Induced pluripotent stem cells from a spinal muscular atrophy patient. Nature. 2009;457(7227):277–80.

[34] Chang T, Zheng W, Tsark W, Bates S, Huang H, Lin R-J, et al. Brief report: phenotypic rescue of induced pluripotent stem cell-derived motoneurons of a spinal muscular atrophy patient. Stem Cells. 2011;29(12):2090–3.

[35] Yoshida M, Kitaoka S, Egawa N, Yamane M, Ikeda R, Tsukita K, et al. Modeling the early phenotype at the neuromuscular junction of spinal muscular atrophy using patient-derived iPSCs. Stem Cell Rep. 2015;4(4):561–8.

[36] Hardy J, Allsop D. Amyloid deposition as the central event in the aetiology of Alzheimer's disease. Trends Pharmacol Sci. 1991;12:383–8.

[37] Byrne JA. Developing neural stem cell-based treatments for neurodegenerative diseases. Stem Cell Res Ther. 2014;5(3):72.

[38] Fahn S. Description of Parkinson's disease as a clinical syndrome. Ann N Y Acad Sci. 2003;991:1–14.

[39] Lesage S, Brice A. Parkinson's disease: from monogenic forms to genetic susceptibility factors. Hum Mol Genet. 2009;18(R1):R48–59.

[40] Neumann J, Bras J, Deas E, O'Sullivan SS, Parkkinen L, Lachmann RH, et al. Glucocerebrosidase mutations in clinical and pathologically proven Parkinson's disease. Brain. 2009;132(7):1783–94.

[41] Patti GJ, Yanes O, Siuzdak G. Innovation: metabolomics: the apogee of the omics trilogy. Nat Rev Mol Cell Biol. 2012;13(4):263–9.

[42] Want EJ, Masson P, Michopoulos F, Wilson ID, Theodoridis G, Plumb RS, et al. Global metabolic profiling of animal and human tissues via UPLC-MS. Nat Protoc. 2013;8(1): 17–32.

[43] Want EJ, Nordström A, Morita H, Siuzdak G. From exogenous to endogenous: the inevitable imprint of mass spectrometry in metabolomics. J Proteome Res. 2007;6(2): 459–68.

[44] Nordström A, Want E, Northen T, Lehtiö J, Siuzdak G. Multiple ionization mass spectrometry strategy used to reveal the complexity of metabolomics. Anal Chem. 2008;80(2):421–9.

[45] Czech C, Berndt P, Busch K, Schmitz O, Wiemer J, Most V, et al. Metabolite profiling of Alzheimer's disease cerebrospinal fluid. PLoS One. 2012;7(2):e31501.

[46] Orešič M, Hyötyläinen T, Herukka S-K, Sysi-Aho M, Mattila I, Seppänan-Laakso T, et al. Metabolome in progression to Alzheimer's disease. Transl Psychiat. 2011;1:e57.

[47] Mapstone M, Cheema AK, Fiandaca MS, Zhong X, Mhyre TR, MacArthur LH, et al. Plasma phospholipids identify antecedent memory impairment in older adults. Nat Med. 2014;20(4):415–8.

[48] Lewitt PA, Li J, Lu M, Beach TG, Adler CH, Guo L, et al. 3-Hydroxykynurenine and other Parkinson's disease biomarkers discovered by metabolomic analysis. Mov Disord. 2013;28(12):1653–60.

[49] Graham SF, Chevallier OP, Roberts D, Hölscher C, Elliott CT, Green BD. Investigation of the human brain metabolome to identify potential markers for early diagnosis and therapeutic targets of Alzheimer's disease. Anal Chem. 2013;85(3):1803–11.

[50] Han X, Rozen S, Boyle SH, Hellegers C, Cheng H, Burke JR, et al. Metabolomics in early Alzheimer's disease: identification of altered plasma sphingolipidome using shotgun lipidomics. PLoS One. 2011;6(7):e21643.

[51] Ibáñez C, Simó C, Martín-Álvarez PJ, Kivipelto M, Winblad B, Cedazo-Mínguez A, et al. Toward a predictive model of Alzheimer's disease progression using capillary electrophoresis-mass spectrometry metabolomics. Anal Chem. 2012;84(20):8532–40.

[52] Astarita G, Piomelli D. Lipidomic analysis of endocannabinoid metabolism in biological samples. J Chromatogr B Analyt Technol Biomed Life Sci. 2009;877(26):2755–67.

[53] Bajad S, Shulaev V. LC–MS-based metabolomics. Methods Mol Biol. 2011;708:213–28.

[54] Dudley E, Yousef M, Wang Y, Griffiths WJ. Targeted metabolomics and mass spectrometry. Adv Protein Chem Struct Biol. 2010;80:45–83.

[55] Patti GJ. Separation strategies for untargeted metabolomics. J Sep Sci. 2011;34(24):3460–9.

[56] Kriks S, Shim J-W, Piao J, Ganat YM, Wakeman DR, Xie Z, et al. Dopamine neurons derived from human ES cells efficiently engraft in animal models of Parkinson's disease. Nature. 2011;480(7378):547–51.

[57] Xi J, Liu Y, Liu H, Chen H, Emborg ME, Zhang S-C. Specification of midbrain dopamine neurons from primate pluripotent stem cells. Stem Cells. 2012;30(8):1655–63.

[58] Krencik R, Zhang S-C. Directed differentiation of functional astroglial subtypes from human pluripotent stem cells. Nat Protoc. 2011;6(11):1710–7.

[59] Yanes O, Tautenhahn R, Patti GJ, Siuzdak G. Expanding coverage of the metabolome for global metabolite profiling. Anal Chem. 2011;83(6):2152–61.

[60] Zhang X, Clausen MR, Zhao X, Zheng H, Bertram HC. Enhancing the power of liquid chromatography–mass spectrometry-based urine metabolomics in negative ion mode by optimization of the additive. Anal Chem. 2012;84(18):7785–92.

[61] Wikoff WR, Anfora AT, Liu J, Schultz PG, Lesley SA, Peters EC, et al. Metabolomics analysis reveals large effects of gut microflora on mammalian blood metabolites. Proc Natl Acad Sci U S A. 2009;106(10):3698–703.

[62] Pawson AJ, Sharman JL, Benson HE, Faccenda E, Alexander SPH, Buneman OP, et al. The IUPHAR/BPS Guide to PHARMACOLOGY: an expert-driven knowledge base of drug targets and their ligands. Nucleic Acids Res. 2014;42(Database issue):D1098–106.

[63] Xia J, Wishart DS. MSEA: a web-based tool to identify biologically meaningful patterns in quantitative metabolomic data. Nucleic Acids Res. 2010;38(Web Server issue):W71–7.

[64] Kaddurah-Daouk R, Rozen S, Matson W, Han X, Hulette CM, Burke JR, et al. Metabolomic changes in autopsy-confirmed Alzheimer's disease. Alzheimers Dement. 201;7(3):309–17.

[65] Kaddurah-Daouk R, Zhu H, Sharma S, Bogdanov M, Rozen SG, Matson W, et al. Alterations in metabolic pathways and networks in Alzheimer's disease. Transl Psychiat. 2013;3:e244.

[66] Kanekiyo T, Xu H, Bu G. ApoE and Aβ in Alzheimer's disease: accidental encounters or partners? Neuron. 2014;81(4):740–54.

[67] Bogdanov M, Matson WR, Wang L, Matson T, Saunders-Pullman R, Bressman SS, et al. Metabolomic profiling to develop blood biomarkers for Parkinson's disease. Brain J Neurol. 2008;131(Pt 2):389–96.

[68] Johansen KK, Wang L, Aasly JO, White LR, Matson WR, Henchcliffe C, et al. Metabolomic profiling in LRRK2-related Parkinson's disease. PLoS One. 2009;4(10):e7551.

[69] Okano H, Yamanaka S. iPS cell technologies: significance and applications to CNS regeneration and disease. Mol Brain. 2014;7:22.

[70] Devi L, Raghavendran V, Prabhu BM, Avadhani NG, Anandatheerthavarada HK. Mitochondrial import and accumulation of alpha-synuclein impair complex I in human dopaminergic neuronal cultures and Parkinson disease brain. J Biol Chem. 2008;283(14):9089–100.

[71] Mizuno Y, Matuda S, Yoshino H, Mori H, Hattori N, Ikebe S. An immunohistochemical study on alpha-ketoglutarate dehydrogenase complex in Parkinson's disease. Ann Neurol. 1994;35(2):204–10.

[72] Gibson GE, Kingsbury AE, Xu H, Lindsay JG, Daniel S, Foster OJF, et al. Deficits in a tricarboxylic acid cycle enzyme in brains from patients with Parkinson's disease. Neurochem Int. 2003;43(2):129–35.

[73] Salminen A, Jouhten P, Sarajärvi T, Haapasalo A, Hiltunen M. Hypoxia and GABA shunt activation in the pathogenesis of Alzheimer's disease. Neurochem Int. 2016;92:13–24.

[74] Krug AK, Gutbier S, Zhao L, Pöltl D, Kullmann C, Ivanova V, et al. Transcriptional and metabolic adaptation of human neurons to the mitochondrial toxicant MPP+. Cell Death Dis. 2014;5(5):e1222.

[75] Mortiboys H, Johansen KK, Aasly JO, Bandmann O. Mitochondrial impairment in patients with Parkinson disease with the G2019S mutation in LRRK2. Neurology. 2010;75(22):2017–20.

[76] Swerdlow RH, Burns JM, Khan SM. The Alzheimer's disease mitochondrial cascade hypothesis. J Alzheimers Dis JAD. 2010;20(Suppl 2):S265–79.

[77] Pavlov PF, Hansson Petersen C, Glaser E, Ankarcrona M. Mitochondrial accumulation of APP and Abeta: significance for Alzheimer disease pathogenesis. J Cell Mol Med. 2009;13(10):4137–45.

[78] Wang X, Su B, Siedlak SL, Moreira PI, Fujioka H, Wang Y, et al. Amyloid-beta overproduction causes abnormal mitochondrial dynamics via differential modulation of mitochondrial fission/fusion proteins. Proc Natl Acad Sci USA. 2008;105(49):19318–23.

[79] Rapoport SI. In vivo PET imaging and postmortem studies suggest potentially reversible and irreversible stages of brain metabolic failure in Alzheimer's disease. Eur Arch Psychiatry Clin Neurosci. 1999;249(Suppl 3):46–55.

[80] Chandrasekaran K, Hatanpää K, Brady DR, Rapoport SI. Evidence for physiological down-regulation of brain oxidative phosphorylation in Alzheimer's disease. Exp Neurol. 1996;142(1):80–8.

[81] Huang H-M, Zhang H, Xu H, Gibson GE. Inhibition of the alpha-ketoglutarate dehydrogenase complex alters mitochondrial function and cellular calcium regulation. Biochim Biophys Acta. 2003;1637(1):119–26.

[82] Bubber P, Haroutunian V, Fisch G, Blass JP, Gibson GE. Mitochondrial abnormalities in Alzheimer brain: mechanistic implications. Ann Neurol. 2005;57(5):695–703.

[83] Shi Q, Xu H, Yu H, Zhang N, Ye Y, Estevez AG, et al. Inactivation and reactivation of the mitochondrial α-ketoglutarate dehydrogenase complex. J Biol Chem. 2011;286(20):17640–8.

[84] Moreira PI, Siedlak SL, Wang X, Santos MS, Oliveira CR, Tabaton M, et al. Increased autophagic degradation of mitochondria in Alzheimer disease. Autophagy. 2007;3(6):614–5.

Chapter 12

Modeling Hypertrophic Cardiomyopathy with Human Induced Pluripotent Stem Cells

Marisa Ojala and Katriina Aalto-Setälä

Additional information is available at the end of the chapter

http://dx.doi.org/10.5772/62649

Abstract

Research of genetic cardiovascular diseases has lacked of good disease models because rodents, which are primarily used, differ greatly from humans. The ability to derive human induced pluripotent stem cells (hiPSCs) from patients carrying inherited cardiac diseases has revolutionized research in the cardiovascular field. The aim for this chapter is to review the current hiPSC reprogramming methods and methods for differentiating human pluripotent stem cells (hPSCs) into cardiomyocytes. The chapter focuses on the published hiPSC models for hypertrophic cardiomyopathy (HCM) and discusses the challenges related to modeling this interesting disease using hiPSC technology.

Keywords: cardiac differentiation, cardiomyocyte, disease modeling, human induced pluripotent stem cell, hypertrophic cardiomyopathy

1. Introduction

The derivation of human embryonic stem cells (hESCs) [1] and, more recently, the invention of human induced pluripotent stem cells (hiPSCs) [2] have opened new opportunities for research and cellular therapies in regenerative medicine. These cells, collectively called human pluripotent stem cells (hPSCs), have the ability to self-renew indefinitely and to differentiate into derivatives of all three germ layers [1, 2]. Thus, hPSCs provide a potential source of cells for regenerative medicine applications as well as in vitro modeling of genetic diseases and drug screening.

Traditionally, hiPSCs have been reprogrammed from skin fibroblasts by virally transferring four pluripotency factors, specifically octamer-binding transcription factor 3/4 (*OCT3/4*), sex-determining region Y-box 2 (*SOX2*), Kruppel-like factor 4 (*KLF4*), and myelocytomatosis viral oncogene homolog (*c-MYC*), which integrate into the genome of the target cell [2]. More recent methods have aimed to produce hiPSCs using non-integrative viral transfection [3], integrative vectors that can be excised after reprogramming [4], or non-viral delivery methods, such as the introduction of episomal vectors into target cells using electroporation [5].

hPSCs can be differentiated into cardiomyocytes under laboratory conditions. During the recent years, the cardiac differentiation methods have developed from embryoid body (EB) formation [6] and co-culturing hPSCs with mouse endodermal-like cells (END-2) [7] to more direct differentiation on a monolayer using different growth factors and small molecules [8–10].

Hypertrophic cardiomyopathy (HCM) is one of the most common genetic cardiac diseases, with a worldwide prevalence of 1:500. In HCM, the cardiac muscle tissue mainly in the interventricular septum is thickened. The most severe symptoms of HCM are progressive heart failure and sudden cardiac death [11]. HCM is caused by more than 1400 mutations, which reside primarily in genes coding for sarcomeric proteins [12]. The clinical phenotype of the disease is variable, and most of the patients carrying the mutations live their lives without any symptoms [11]. Currently, there have been five reports published using HCM patient-specific hiPSCs for modeling the disease [13–17]. Results from these reports will be reviewed thoroughly in this chapter.

2. Human pluripotent stem cells

hPSCs are defined as undifferentiated cells which have the ability to self-renew indefinitely and to differentiate into derivatives of all three germ layers: endoderm, mesoderm, and ectoderm [1, 2]. The first hESC line was derived by Thomson and co-workers in 1998 [1]. Traditionally, hESCs are derived from the inner cell mass (ICM) of the blastocyst but early blastomeres or morula stage embryos have also been used [18, 19].

For a long time, the scientific community believed that cell differentiation was a one-way route and that there was no turning back when a cell had passed specific differentiation stages or reached the fully differentiated state. However, in 2006, Yamanaka and co-workers were able to reprogram already fully differentiated mouse cells back into the pluripotent stage via retroviral induction with specific pluripotency factors: *Oct3/4*, *Sox2*, *Klf4*, and *c-Myc*. These four transcription factors, so called "Yamanaka factors," were able to force endogenous pluripotency genes to be turned on in the transfected cell, changing the cell back to a pluripotent state [20]. In 2007, the team repeated the reprogramming using human fibroblasts, creating hiPSCs [2]. Similar to hESCs, hiPSCs also have the ability to self-renew and give rise to all somatic cell types. Although the discovery of hiPSCs has been a revolutionary invention in the stem cell field, it would not have been possible without earlier research, including somatic cell nuclear transfer (SCNT) [21–23], which suggested that the nuclear status of a differentiated cell could be reverted back to totipotency by factors in the cytoplasm of an oocyte [24].

Because of their unique properties, hPSCs can be maintained in the laboratory for extended periods of time in their undifferentiated state and differentiated into various cell types, including cardiomyocytes [6], retinal pigmented epithelial (RPE) cells [25], neural cells [26], and hepatocytes [27] (**Figure 1**). Thus, hPSCs represent a limitless cell source for regenerative medicine applications as well as for studying developmental processes and genetic diseases.

The groundbreaking discovery of hiPSCs has opened completely new opportunities for disease modeling and drug screening also in the cardiac field. Conventionally, cardiac diseases have been modeled using animal models or genetically engineered cell lines. These models often correlate poorly with the results from human studies. In addition, obtaining cardiac tissue

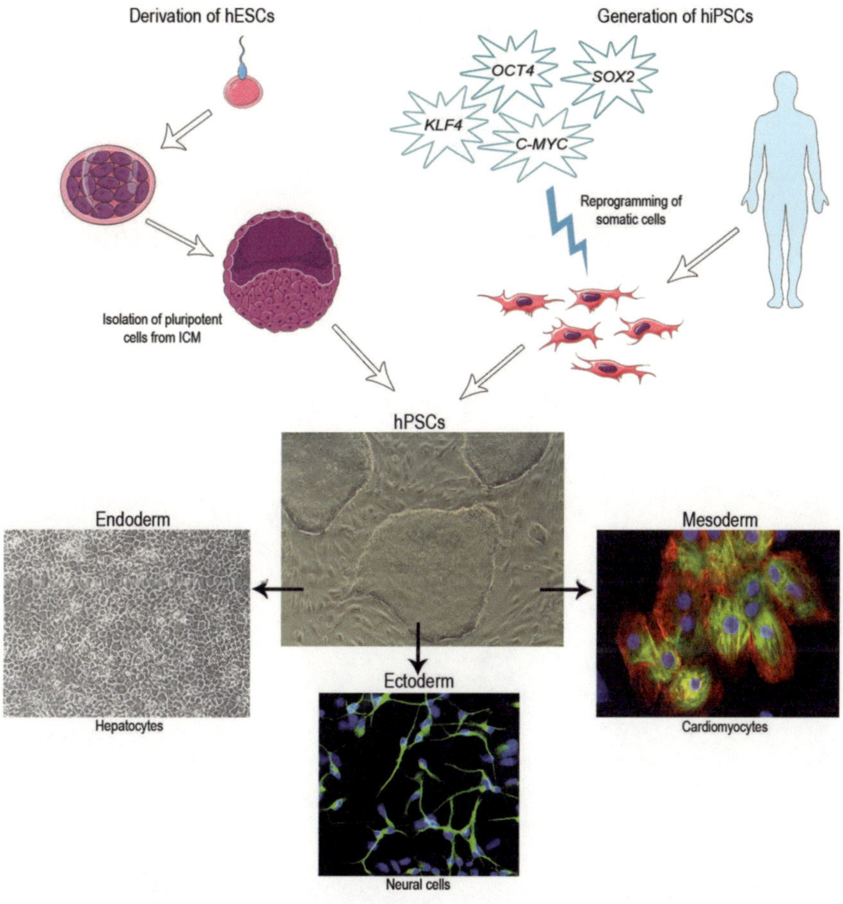

Figure 1. Human pluripotent stem cells (hPSCs). Human embryonic stem cells (hESCs) are derived from the inner cell mass (ICM) of a blastocyst. Somatic cells from adult individual are reprogrammed into human induced pluripotent stem cells (hiPSCs) by transferring exogenous pluripotency factors into the cells. hESCs and hiPSCs, collectively called hPSCs, have the ability to form all three germ layers. In the laboratory, hPSCs can be cultured for extended periods of time and differentiated into derivatives of different germ layers, such as hepatocytes, neural cells and cardiomyocytes. The figure was composed of images from the Servier Medical Art image bank (www.servier.com/Powerpont-image-bank) and cell pictures from BioMediTech.

directly from patients for research purposes is difficult, and adult human cardiomyocytes dedifferentiate rapidly under cell culture conditions and lose their characteristic properties. With the hiPSC technique, we are able to derive cells directly from a patient and transfer the same genetic information and mutations into hiPSC-derived cardiomyocytes. Therefore, hiPSCs have great potential to revolutionize the research of cardiovascular diseases.

2.1. Generation of human induced pluripotent stem cells

In the first conversion of mouse and human fibroblasts into iPSCs, Yamanaka's group used the viral transduction of four transcription factors (*OCT3/4, SOX2, KLF4,* and *c-MYC*) [2, 20]. This method involved the integration of viral genes into the host cell genome, which involves a risk of tumorigenicity due to the insertional mutagenesis and uncontrolled gene expression as well as potential reactivation of the virus [28, 29]. To circumvent these problems, a variety

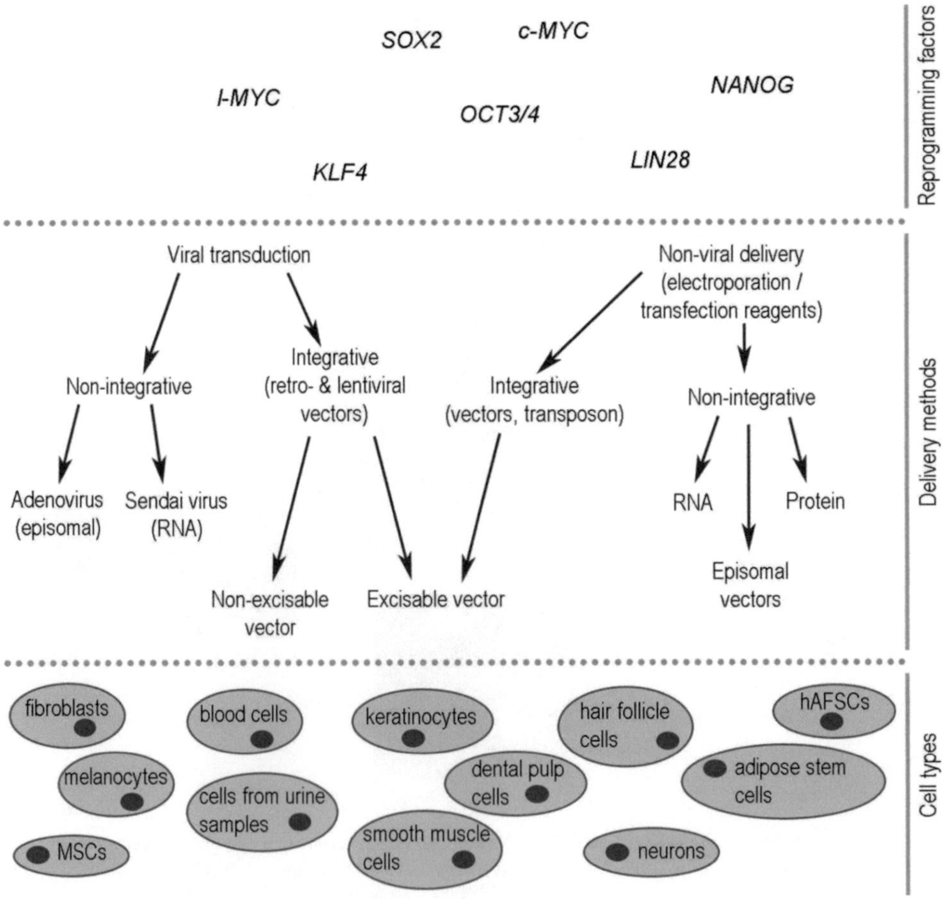

Figure 2. Reprogramming factors, different delivery methods, and donor cell types used in hiPSC generation. The data presented in the figure were collected from review articles [28, 30]. MSCs, mesenchymal stem cells; hAFSCs, human amniotic fluid stem cells.

http://dx.doi.org/10.5772/62649

of new methods using different non-viral and non-integrative methods have been developed. Technological options for hiPSC transduction are presented in **Figure 2**.

Viral transduction can be achieved using either integrating or non-integrating viral vectors. In 2007, two distinct research groups published the first generation of hiPSCs. In the first paper, the delivery was accomplished using retroviral pMXs vectors [2], while in the second paper, the transduction was performed using lentiviruses [31]. The proteins, which are needed for the additional rounds of virus replication and packaging, are deleted from the pMXs vectors. Retroviral vectors are able to target cells according to their envelope pseudotype and they only transduce actively dividing cells. Lentiviruses, in contrast, can also transduce non-dividing cells [28]. Both retro- and lentivirally transferred genes are expected to be silenced during the reprogramming process through methylation and epigenetic modification [32]. Sometimes, however, the process is incomplete, which results in partially reprogrammed hiPSC lines [20, 33, 34].

To overcome the problems related to transgene integration into the host genome, excisable vectors have been engineered based on, for example, Cre-recombinase-mediated excision [4]. In this method, the sequence of the gene to be integrated into the genome is inserted between two loxP sites in the LTR (long terminal repeat) region of the vector. The integrated transgenes can be afterwards excised from the genome by transfecting the hiPSCs with Cre-recombinase [4]. Reprogramming factors in distinct vectors are integrated at independent sites in the genome, which can lead to genomic instability and genome reorganization when Cre-recombinase is introduced into the cells. Therefore, polycistronic vectors, which express all reprogramming factors in one vector separated by 2A sequences, are favored when using this method [35, 36].

Non-integrative viral methods include the use of Sendai viral and adenoviral vectors. With these two non-integrative viral methods, the Sendai virus has turned out to be more efficient for the generation of hiPSCs. Sendai virus vectors replicate their single-stranded RNA in the cytoplasm without entering the nucleus of the infected cell [3]. In addition, they are able to infect a wide variety of cell species and tissues by attaching to sialic acid receptors, which are present on the surface of various cell types [3]. Adenoviral vectors, in contrast, contain DNA, which is transported to the nucleus of the target cell. However, this adenoviral DNA is not integrated into the genome, and the expression of the adenoviral genes is thus transient [37, 38].

Non-viral methods are based on delivering genes (DNA), RNA copies of the genes or proteins to target cells. Different delivery carriers, such as transposons, and methods, including electroporation and transfection reagents, have been reported. The previously mentioned polycistronic vectors can also be transferred into target cells without viral delivery through electroporation [4, 36]. The PiggyBac transposon and transposase system is another integrative non-viral method used in hiPSC generation [39]. In this method, the transposase enzyme cleaves the delivered genes from specific cleavage sites in the PiggyBac vector and transfers them into the target genome. The same enzyme can be used to excise exogenous genes from the genome after reprogramming [39].

Methods based on episomal vectors, messenger RNA (mRNA) molecules, or purified protein have been developed for non-integrative non-viral hiPSC production. Episomal vectors derived from Epstein–Barr virus (oriP/EBNA1) can be used to introduce reprogramming factors into cells without a need for viral packaging [5, 40]. The expression of the exogenes is transient and they disappear from the transfected cells during culture [5]. Lastly, nanoparticles have been used to improve the efficiency of reprogramming, particularly with mRNA molecules and proteins [41, 42].

3. Differentiation of human pluripotent stem cells into cardiomyocytes

The heart is the first functional organ that develops during embryogenesis. It develops from the mesoderm, although signals from adjacent cell populations, especially from the endoderm, have a significant role in cardiogenesis. The entirety of the development of the human heart is not yet fully understood; however, many molecular events and factors taking part in the early stages of cardiomyogenesis have been identified. The three main growth factor families thought to participate in early mesodermal induction and cardiomyogenesis are the wingless/INT proteins (WNTs), the fibroblast growth factors (FGFs), and members of transforming growth factor-beta (TGFβ) superfamily, which include bone morphogenetic protein 4 (BMP4), Nodal and Activin A [43]. The expression of these factors or their inhibitors in the adjacent

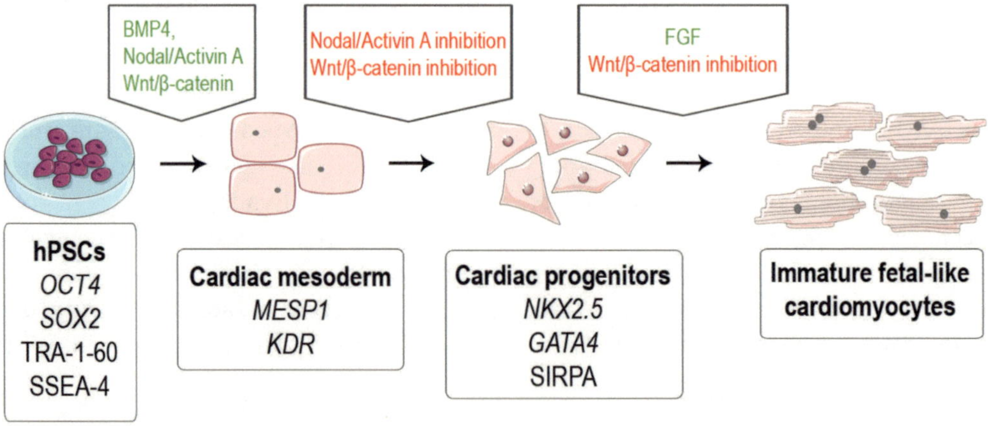

Figure 3. Model of different stages in cardiac differentiation. Transcription factors are presented in cursive and cell surface markers in normal text. The first precardiac mesoderm cells express the T-box factor Brachyury T and the homedomain protein Mixl1 (*MIXL1*). These cells subsequently activate the mesoderm posterior 1 protein (*MESP-1*), and the cardiac mesoderm is formed. NK2 transcription factor-related gene, locus 5 (*NKX2.5*) and GATA-bonding protein 4 (*GATA4*) direct the development of embryonic cardiomyocytes. Finally, *NKX2.5*, together with certain other growth factors, activates cardiac structural genes, including actin, myosins, and troponins. By activating (green) and inhibiting (red) signaling pathways, hPSCs can be directed toward cardiac differentiation in vitro [43, 44]. The figure was composed of images from the Servier Medical Art image bank (www.servier.com/Powerpont-image-bank).

http://dx.doi.org/10.5772/62649

endoderm occurs at different times, and their combination eventually leads to the induction of the cardiac mesoderm. After receiving these initial signals, development is directed to more specific and highly conserved cardiogenesis [43–46].

In vitro differentiation methods mimic the phases of heart development, from the mesoderm to the cardiac mesoderm, cardiac progenitors, and finally, cardiomyocytes [44]. This process is regulated by alternating activation and inhibition of the signaling pathways participating in the cardiomyogenesis (**Figure 3**). Current cardiac differentiation methods are based on EB differentiation in suspension, co-culturing the cells with END-2 cells or inducing cardiac differentiation with different growth factors on a monolayer. The details of the most important differentiation methods are collected in **Table 1**.

Platform	Basal medium	GFs/SMs	Description/breakthrough	Efficiency	Reference
EB-based differentiation methods					
Low-attachment	DMEM + ko-SR	–	First generation of EBs from hESCs	–	Itskovitz-Eldor et al. [6]
Low-attachment	DMEM + FBS	–	First functional cardiomyocytes from EBs	30% (cTnI, FC)	Kehat et al. [47]
V-96 well	RPMI + FBS/HSA	PVA, BMP4, bFGF, insulin, AA	Forced aggregation strategy	60–90% (cTnI, FC)	Burridge et al. [48]
3D microwell	DMEM/F-12 + FBS	–	Engineered 3D microwell coated with Matrigel	18–20% (beating EBs)	Mohr et al. [49]
Low-attachment	StemPro34	AA, BMP4, bFGF, Activin A, VEGF, DKK1	Serum-free, GF-based EB method	40–50% (cTnT, FC)	Yang et al. [50]
Low-attachment	StemPro34	BMP4, AA, IWR-1, blebbistatin, Activin A	Serum-free, GF-based EB method	90% (cTnT, FC)	Karakikes et al. [51]
Methods based on END-2 cells					
END-2	DMEM + FBS	–	First co-culture with END-2 cells	35% (beating wells)	Mummery et al. [7]
END-2	DMEM + ko-SR	AA	Addition of AA and serum-free medium in END-2 co-culture	5–20% (α-actinin, ICC)	Passier et al. [52]
Low-attachment	END-2 CM	SB203580	Differentiation with medium conditioned on END-2 cells	22% (α-MHC, cytospin)	Graichen et al. [53]
Low-attachment	END-2 CM	Prostaglandin I2, AA, SB203580	Differentiation of EBs in END-2 CM, inhibitory	6–9%	Xu et al. [54]

Platform	Basal medium	GFs/SMs	Description/breakthrough	Efficiency	Reference
			effect of insulin	(α-MHC, cytospin)	
Monolayer methods					
Matrigel	RPMI + B27	Activin A, BMP4	First monolayer differentiation method	30% (β-MHC, ICC)	Laflamme et al. [9]
Matrigel	RPMI + B27	Activin A, BMP4, bFGF	Sandwhich method, Matrigel is applied below and on top of cells	40–98% (cTnT, FC)	Zhang et al. [55]
Matrigel	RPMI + B27	CHIR99021, IWP2/ IWP4	Protocol exploiting small molecules modulating Wnt signaling pathway	80–95% (cTnT, FC)	Lian et al. [10]
Vitronectin	RPMI + albumin	AA, CHIR99021, WNT-C59	Serum- and animal-component free, defined differentiation protocol	80–90% (cTnT, FC)	Burridge et al. [8]

GFs, growth factors; SMs, small molecules; ko-SR, knockout serum replacement; FC, flow cytometry; HSA, human serum albumin; PVA, polyvinyl alcohol; AA, ascorbic acid; ICC, immunocytochemistry; CM, conditioned medium.

Table 1. Overview of the development of cardiac differentiation methods.

The first hESC-derived cardiomyocytes were isolated from spontaneously formed EBs [6, 47]. EBs are spherical, multicellular, three-dimensional aggregates, which are formed when hPSCs are detached from the feeder cell layer or the culture matrix and cultured in suspension. In these EB structures, the hPSCs spontaneously differentiate into all three germ layers. In more recent versions of EB differentiation protocols, the aim has been to generate more uniformly-sized EBs, for example, via forced-aggregation using centrifugation [48, 56] or by culturing hPSCs in microwells coated with Matrigel prior to EB formation [49]. Due to the increased knowledge of the heart development in recent years, more guided EB differentiation methods using different growth factors and serum-free basal medium have been developed. Yang and co-workers were able to generate a population of cardiac progenitor cells by inducing EB differentiation with Activin A, BMP4, bFGF, vascular endothelial growth factor (VEGF), and Dickkopf homolog 1 (Dkk-1) [50]. With the protocol of Karakikes et al. [51], almost 100% of the EBs were beating and 90% of the cells were cTnT positive. In this protocol, the EB formation is induced by BMP4 and blebbistatin, which inhibits the actin-myosin contraction and suppress the dissociation-induced apoptosis. After this initial step, the EBs are induced to cardiac lineages by ascorbic acid, BMP4, Activin A, and finally with IWR-1 [51].

As discussed above, anterior endoderm is located directly posterior to mesoderm in the early development and the signals from adjacent endoderm initiate the early cardiogenesis. Based on this fact, Mummery et al. developed a differentiation method in 2003, in which the hPSCs are plated on top of mitotically inactivated END-2 cells [7, 52]. END-2 cells are derived from mouse P19 embryonal carcinoma cells, and they provide cell-to-cell contacts and produce

http://dx.doi.org/10.5772/62649

factors that induce cardiac differentiation. However, medium conditioned on END-2 cells alone has been used in cardiac differentiation, which suggest that cell-to-cell contacts might not be needed in the cardiac induction [53, 54]. The exact mechanism how END-2 cells induce the cardiac differentiation is still unclear. However, a valuable details of cardiac differentiation has been discovered using these cells. For example, END-2 cells have been shown to remove insulin from the medium and insulin has been shown to inhibit the differentiation of hPSCs into cardiac lineages [54, 57]. Although, the efficiency of END-2 co-culture method is quite low, the method has turned out to be reliable in generating cardiomyocytes from various hPSC lines.

The increase in understanding the regulatory signaling pathways related to cardiogenesis in conjunction with advances in hPSC culture methods has enabled the development of more defined and guided monolayer-based differentiation methods. Monolayer methods begin with the feeder cell-free culture of hPSCs, in which the feeders are substituted with Matrigel or Geltrex, or matrix composed of different extracellular matrix (ECM) proteins. The advantage of monolayer differentiation methods is that the cells are in uniform monolayers and there are no diffusional barriers, which would prevent the function of growth factors. Thus, differentiation should be easier to control and reproduce, than in EB or in co-culture methods [44]. In the first monolayer method published in 2007 by Laflamme and co-workers [9], cells were directed toward cardiac differentiation by a combination of Activin A and BMP4. One of the more recent methods, the so-called sandwich method, is based on the combination of ECM with growth factor signaling [55]. Cells were seeded on Matrigel matrix, and after reaching 90% confluence, Matrigel was added on top of the cells. The sequential application of Activin A, BMP4, and bFGF on the matrix sandwich resulted in 40–90% pure cardiomyocyte populations [55].

As in the case of hPSC culture development, cardiac differentiation is also moving toward easily scalable, chemically defined, and xeno-free conditions. A group of small molecules has been identified and applied to replace the recombinant cytokines and unknown factors in the serum. These molecules either activate or inhibit the WNT and TGFβ-signaling pathways. Two recent publications are based on the sequential activation and inhibition of the WNT signaling pathway. At first, the formation of the mesoderm is induced using small molecules, such as CHIR99021 and BIO for WNT signaling activation [10, 58]. After that, more specific cardiac differentiation is induced by inhibiting the WNT signaling pathway using small molecules, such as KY02111 and IWP2 [10, 58].

In 2014, Burridge and co-workers [8] published a chemically defined cardiac differentiation method. The medium consisted of Roswell Park Memorial Institute (RPMI) basal medium supplemented with HSA and L-ascorbic acid. Cardiac differentiation was further induced by the sequential activation and inhibition of WNT signaling by CHIR99021 and WNT-C59, respectively. They also tested various defined matrices (E-cadherin, vitronectin, vitronectin peptide, laminin-521, laminin-511, fibronectin, and fibronectin peptide) in combination with differentiation medium. Laminins were the most promising, but because they are extremely expensive for large-scale applications, vitronectin was selected for further studies. The final protocol resulted in 80–90% pure cardiomyocyte populations for the multiple hPSC lines tested [8].

Although the differentiation methods for hPSC-derived cardiomyocytes have developed greatly, the efficiency of differentiation varies to a large degree when using different methods and cell lines and none of the methods result in a homogenous population of cardiomyocytes [8, 44, 59]. Homogenous cell populations would be needed, for example, to obtain reliable results from drug-screening assays [44]. Thus, there is a need for efficient purification methods.

Manual dissection [7] and Percoll gradient separation [60] were the first published methods for cardiomyocyte purification. However, the yield of pure cardiomyocytes with these methods was quite low. There are a few cell surfer markers that can be used for fluorescence activated cell sorting (FACS)-based purification of cardiomyocytes. These markers include signal regulatory protein α (SIRPA), which is expressed both in cardiac progenitor cells and in hPSC-derived cardiomyocytes [61], and vascular cell adhesion molecule 1 (VCAM1), which functions in leukocyte-endothelial cell adhesion but is also expressed in hPSC-derived cardiomyocytes [62]. One of the most recent method takes advantage of cardiomyocytes' ability to use lactate as an energy source. When culturing cells in glucose-depleted and lactate-abundant conditions other cells than cardiomyocytes will not survive [63]. Another interesting option is the microRNA (miRNA) switch technology, in which the cells are purified based on their miRNA activity. Heterogenous cell population is transfected with synthetic mRNA, which comprise cell type specific miRNA target site and fluorescent protein. Cells in which the target miRNA is absent will translate the fluorescent protein and can be sorted out from the cardiomyocyte population. Alternatively, miR-Bim-switch can be used, which selectively induces the apoptosis of cells in which the miRNA is not present [64].

3.1. Characterization of hPSC-derived cardiomyocytes

hPSC-derived cardiomyocytes can be characterized by their structural and biochemical, as well as by their functional features. The most apparent characteristic of hPSC-derived cardiomyocytes is their ability to contract spontaneously in culture [7, 47]. The organization of the internal structures can be studied and analyzed after immunolabeling the sarcomeric proteins [65]. The ultrastructural features of hPSC-derived cardiomyocytes can be studied in more detail using electron microscopy (EM). Laurila et al. has recently published a thorough review on different methods used in functional analysis of hiPSC-derived cardiomyocytes [66]. Shortly, the electrical properties of hPSC-derived cardiomyocytes can be studied on a cell cluster level using the microelectrode array (MEA) platform approach [47, 67] or on the single-cell level using the patch clamp-method [68–70]. Calcium plays a major role in excitation-contraction coupling process by which an electrical signal is transformed into a mechanical contraction [71]. The intracellular Ca^{2+} signaling in hPSC-derived cardiomyocytes on a single-cell level is studied in vitro using specific fluorescent probes for Ca^{2+} ions, such as fura-2 [72]. The mechanical beating behavior of hPSC-derived cardiomyocytes can be analyzed using methods based on video imaging [73].

http://dx.doi.org/10.5772/62649

4. Hypertrophic cardiomyopathy

An adult cardiomyocyte is composed of evenly distributed and organized myofibrils, which are divided into approximately 2.2-μm-long contractile units called sarcomeres. Sarcomeres are composed of thin actin and thick myosin filaments and Z-discs. The thin filaments are attached to Z-discs, which separate the sarcomeres from each other. The thin filament is composed of repeating actin molecules, troponin (Tn) complexes, and α-tropomyosin (TPM1) molecules. Tn complexes consist of TnT, TnI, and TnC. The Tn complex works together with TPM1 during cardiac contraction. The thick filament consists of myosin molecules, which are built upon two units of α- and β-myosin heavy chain (α-MHC, β-MHC) and four myosin light chain (MLC) molecules. Among the other proteins of the thick filament, myosin-binding protein C (MYBPC) plays the most important role in the contraction. It contributes to actin-myosin interactions and cross-bridge formation [74].

When the Ca^{2+} concentration in the cytosol increases, the Ca^{2+} binds to TnC leading to a conformational change in the Tn complex. This leads to TPM1 moving from its inhibitory position, allowing the head region of the MHC to bind to actin, forming a cross-bridge. Then, myosin hydrolyzes adenosine triphosphate (ATP), causing the sliding of actin and myosin filaments and muscle contraction. In addition to the structural proteins mentioned above, the sarcomere consists of many other important proteins, which together form a stabilized and organized structure [71].

Cardiomyopathies are diseases that affect the heart muscle and can lead to progressive heart failure and cardiac death. Cardiomyopathies can either be genetic or acquired, and they can be divided into groups based on their morphological and functional characteristics. Cardiomyopathies include, among others, HCM, dilated cardiomyopathy (DCM), and arrhythmogenic right ventricular cardiomyopathy (ARVC) [75]. HCM is one of the most common genetic cardiac diseases, with a worldwide prevalence of 1:500, and is the most common cause of sudden cardiac death among young competing athletes. HCM is inherited in an autosomal dominant pattern, and the mutations are mainly located in the sarcomeric proteins, which are responsible for the contraction and relaxation of the cardiomyocyte. The clinical manifestation of the disease is extremely variable: it has age-related penetrance, and the clinical symptoms can vary within the same family having the same gene mutation. Together, these facts indicate that there might be other factors in addition to the actual gene mutation, for example, epigenetic and environmental factors, that determine the clinical outcomes of the disease. Although a large number of mutations have now been identified and related to HCM, the pathophysiological mechanisms of the disease are still largely unknown [11].

In HCM, the cardiac muscle is thickened (≥15 mm) most commonly in the interventricular septum separating the right and left ventricles. This thickening, i.e., hypertrophy, can lead to outflow tract obstruction, which indicates that the passage to the aorta from the left ventricle becomes narrow and obstructive, disturbing the blood flow. This narrowed outflow induces a massive overload to the heart and can lead to progressive heart failure. Other severe complications related to HCM include arrhythmias and sudden cardiac death. However, most individuals remain asymptomatic for their whole lives, and it has been estimated that the actual prevalence of the disease might even be 1:200 in the general population [11]. Penetrance indicates the percentage of mutation carriers who experience the phenotype of the disease. In HCM, the penetrance is highly variable: it can be age-related or even incomplete and related to gender [76].

There is no specific cure for this disease and all the complications related to HCM should be treated individually. Patients are typically asymptomatic for a long time. Often, the first sign is diastolic heart failure, while systolic heart failure can develop later. Treatments include beta-blockers or Ca^{2+}-channel blockers for relieving symptoms such as chest pain and shortness of breath and implantable cardioverter defibrillators (ICDs) for those patients who have survived cardiac arrest. The actual hypertrophy can be treated by surgical myectomy, which indicates removing a small portion of the thickened cardiac tissue; via ethanol ablation, in which a myocardial infarction is induced in the septal area; or at the end-stage via heart transplantation [11]. Histologically, HCM is characterized by myocyte hypertrophy (diameter >40 μm); disorganization of myocyte bundles, individual myocytes, and sarcomeres; and fibrosis of heart tissue. Nuclei are hyperchromatic (contain an abundance of chromatin), pleomorphic (vary in size and shape), and often enlarged [77].

HCM is inherited in an autosomal dominant pattern and is caused by over 1400 mutations found in eleven or more genes coding primarily for sarcomeric proteins. Approximately 70% of patients have a gene mutation either in the *MYH7* or *MYBPC3* genes while mutations in other genes are far less common [11, 12]. Rarer mutations are located in Z-disc genes or calcium handling and regulation genes. One interesting feature of HCM mutations is that they are almost all identified only in one or a few families [76]. Typically, patients carry only one heterozygous mutation in a single allele. However, lately, double or even greater numbers of mutations have been reported to be found in one patient, which might affect the clinical variance related to the disease [76, 78]. Most pathogenic mutations are missense mutations in which a single nucleotide is changed, resulting in an amino acid substitution. Missense mutations are thought to act in a dominant negative manner. Thus, the mutated protein is produced and interferes with the normal function of the sarcomere. Nonsense mutations, in contrast, lead to a premature stop codon and truncated proteins. These mutations are thought to result in haploinsufficiency, in which the mutated protein is either degraded or not produced at all. The majority of the gene mutations in *MYBPC3* are thought to act through this mechanism [74, 76].

In Finland, two founder mutations in genes coding for MYBPC and TPM1 proteins account for approximately 18% of Finnish HCM cases, and these *MYBPC3-Gln1061X* and *TPM1-Asp175Asn* mutations are relatively uncommon in other countries besides Finland [79]. *MYBPC3-Gln1061X* is a nonsense mutation leading to a premature stop-codon and a truncated MYBPC protein lacking the binding sites for both myosin and titin [80]. This mutation, similar to other mutations in *MYBPC3*, is characterized by age-related penetrance and late onset of the disease [80–82]. *TPM1-Asp175Asn* is a missense mutation that leads to the substitution of aspartic acid with asparagine in codon 175 [83]. Originally, *MYBPC3-Gln1061X* was associated with a clinically mild phenotype, and *TPM1-Asp175Asn* was associated with a clinically intermediate phenotype with a substantial risk for sudden cardiac death. Additionally, patients with *MYBPC3-Gln1061X* were suggested to be more prone to cardiac dilation and heart failure [84]. However, in the most recent studies, both mutations were associated with variable left ventricular hypertrophy, and no clear genotype-phenotype correlations could be verified in clinical studies [79, 85].

Different animal models have been used to study HCM in vitro. These animals include cats, which have certain naturally occurring HCM mutations, and genetically engineered mice, rats, rabbits, and Drosophila [86]. Additionally, HCM patient tissues obtained from myectomy samples have been studied [87]. One of the major difficulties in studying the pathophysiological mechanisms of HCM has been the lack of tissue samples at early stages of disease development. While animal models have provided valuable insight into disease mechanisms, they contain only the mutated gene and not the rest of the genome, which might have effects on disease phenotype and progression. Thus, given that they contain the whole genomes of HCM patients, in addition to the fact that they are cell of human origin, hiPSC-derived cardiomyocytes represent a valuable new tool for modeling HCM in vitro.

4.1. Human induced pluripotent stem cell models for studying hypertrophic cardiomyopathy

To date, morphological and functional characteristics of cardiomyocytes, derived from HCM patient-specific hiPSCs, have been studied in five different reports [13–17]. The results presented in these publications are collected in **Table 2**. Cardiomyocytes have been obtained with different cardiac differentiation methods in each publication. In our report, we have differentiated hiPSCs into cardiomyocytes using END-2 co-culture method [15]. In addition, HCM hiPSC-derived cardiomyocytes carrying *MYBPC3-c.2373dupG* mutation were used in one report, in which the effects of serum on hypertrophic phenotypes were studied [88]. In this particular study, serum was found to mask the hypertrophic phenotype of hiPSC-derived cardiomyocytes with the *MYBPC3-c.2373dupG* mutation [88]. Serum increased the cell size of neonatal rat cardiomyocytes as well as cardiomyocytes derived from hESCs and control hiPSCs. However, the size of the hiPSC-derived cardiomyocytes with the *MYBPC3-c.2373dupG* mutation was smaller than that of hiPSC-derived control cardiomyocytes in the presence of serum. Under serum-free conditions, hiPSC-derived cardiomyocytes with the *MYBPC3-c.2373dupG* mutation were significantly larger than hiPSC-derived control cardiomyocytes [88]. Conversely, in our study, cardiomyocytes carrying the *MYBPC3-Gln1061X* mutation were enlarged, despite the 20% FBS used in the culture medium, challenging the role of serum in masking the hypertrophic phenotype of hiPSC-derived cardiomyocytes [15].

Patients, mutations and hiPSCs	HCM phenotype in hiPSC-CMs	Drug treatments/other findings
Lan et al. [16]	**Morphological properties**	**Calcineurin-NFAT signaling**
Family with HCM	Cellular enlargement, multinucleation,	Blockade by cyclosporin A and FK506
–5 pers. with MYH7-Arg663His	increased myofibril content, disorganized sarcomeres	reduced hypertrophy
		β-adrenergic stimulation
–5 pers. without mutation	**Biochemical properties**	Isoproterenol increased cell size and

Patients, mutations and hiPSCs	HCM phenotype in hiPSC-CMs	Drug treatments/other findings
(used as control) Timepoints: 20, 30, 40 days	Upregulation of *ANF*, *TNNT2*, *MYL2*, *MYH7*, *GATA4* and *MEF2c* Elevation of *MYH7*/*MYH6* ratio Nuclear translocation of NFAT **Ca^{2+} handling properties** Ca^{2+} transient irregularities Elevation of intracellular [Ca^{2+}] Smaller SR Ca^{2+} release **Electrophysiological and mechanical properties** Arrhythmic waveforms including frequent DADs Irregular beating observed in video recordings	Amount of irregular Ca^{2+} transients and arrhythmia Isoproterenol together with propanolol abolished Ca^{2+} abnormalities, arrhytmia and hypertrophy **Blockade of L-type Ca^{2+} channel** Treatment with verapamil for 5 days ameliorated HCM phenotype Diltiazem abolished Ca^{2+} abnormalities and arrhythmia
Han et al. [14] One patient: - *MYH7-Arg442Gly* Two control hiPSC lines from unrelated donors	**Morphological changes** Cellular enlargement, disorganized sarcomeres Disorganized Z lines in TEM **Biochemical changes** Changes in (whole transcriptome sequencing): wnt/β-catenin pathway Notch signaling pathway FGF pathway Nuclear translocation of NFAT Decreased level of *RYR2*, *SERCA2* **Ca^{2+} handling properties** Ca^{2+} transient irregularities elevation of intracellular [Ca^{2+}] Smaller SR Ca^{2+} release Delayed Ca^{2+} transient decay time **Electrophysiological properties** Prolonged and dispersed interspike intervals and increase of arrhythmogenic events in MEA Irregular contractility in real-time cell analyzer APD prolongation Changes in the shape of AP Increased Ca^{2+}, Na^{+} and outward K^{+} currents	**β-adrenergic stimulation** Isoproterenol elevated premature beats and irregular beating rates Isopoterenol together with metoprolol decreased beating irregularity and arrhythmia **Blockade of L-type Ca^{2+}channel** Treatment with verapamil for 4 days reduced arrhythmia and Ca^{2+} handling abnormalities **K_{ATP} channel opener** Antihypertensive drug pinacidil induced irregular interspike intervals **Inhibition of histone deacetylase activity** Treatment with trichostatin A for 3 days decreased cell size, nuclear translocation of NFAT, suppressed Ca^{2+} abnormalities and decreased resting [Ca^{2+}]
Tanaka et al. [17] Three patients:	**Morphology without stimulation** Mildly but significantly larger cell size	**Stimulation with hypertrophic factors** Angiotensin II, IGF-1, phenylephrine

Patients, mutations and hiPSCs	HCM phenotype in hiPSC-CMs	Drug treatments/other findings
–*MYBPC3-GLY999-Gln1004del* –*TMP1-Arg91Cys* (identified later) –Mutation unknown Two control hiPSC lines from Unrelated donors Timepoints: 30, 60, 90 days culture as EBs	No time-dependent changes in cell size Myofibrillar disarray in EM and cTnT staining (>60d, >90d) cTnT and ANP protein levels higher MYBPC level lower in hiPSC-CMs with *MYBPC3* mutation Mildly disorganized contractile form in video analysis	No difference to non-stimulated cells Endothelin 1 (ET-1) Increased cell size and disarray Nuclear translocation of NFAT Disorganized contraction Similar response in mouse MYBPC$^{+/-}$ **Blocking of ET-1 signaling** ETA-b was able to block ET-1 induced hypertrophic phenotype ETB-b had no effect
Birket et al. [13] (hiPSC lines published in Dambrot et al. [88] Three patients: –*MYBPC3-c.2373.dupG* –One hiPSC line from each One control hiPSC line from unrelated donor	**Biochemical properties** Decreased level of MYBPC relative to α-actinin	**Stimulation with (T3, IGF-1, Dex) TID** Decreased traction stress Decreased traction force No difference in the cell size or contraction frequencies when compared to controls Similar findings in MYBPC knockdown hESC-derived cardiomyocytes
Ojala et al. [15] Four patients: –2 pers. with *MYBPC3-Gln1061X* –2 pers. with *TPM1-Asp175Asn* Two control hiPSC lines from unrelated donors Timepoints: 1w, 3w, 6w (as single cells)	**Morphological properties** Cellular enlargement, multinucleation **Biochemical properties** Upregulation of *MYBPC3, TNNT2, ACTN2, TTN* *MYL7* and *MYL9* MYBPC level slightly reduced in *MYBPC3-Gln1061X* cells **Ca^{2+} handling properties** Ca^{2+} transient irregularities **Electrophysiological and mechanical properties** Arrhythmic waveforms including frequent EADs and DADs	**Differences between mutations** *MYBPC3-Gln1061X* cells significantly larger Differences in expressions of various cardiac genes More abnormal Ca^{2+} signals in *TPM1-Asp175Asn* mutation More prolonged APD in *TPM1-Asp175Asn* mutation *MYBPC3-Gln1061X* mutation was not detected in mRNA or protein level

hiPSC-CM, human induced pluripotent stem cell-derived cardiomyocyte; SR, sarcomeric reticulum; DAD, delayed after depolarization; EM, electron microscopy; MEA, microelectrode array; APD, action potential duration; AP, action potential; T3 triiodothyronine hormone; IGF, insulin-like growth factor 1; Dex, dexamethasone; EAD, early after depolarization.

Table 2. Summary of results from published reports using HCM patient-specific hiPSCs.

Lan et al. published the first report of HCM hiPSC-derived cardiomyocytes in 2013. In their publication, hiPSCs were established via the lentiviral infection of fibroblasts derived from five patients carrying the *MYH7-Arg663His* mutation and from five related healthy individuals. Although the two youngest patients had not developed the clinical phenotype of HCM, the hiPSC-derived cardiomyocytes from all patients were significantly larger than the control cardiomyocytes [16]. In 2014, Han et al. published a report in which hiPSCs were derived from a single patient carrying a *MYH7-Arg442Gly* mutation. Control hiPSCs were derived from two unrelated healthy individuals. Fibroblasts were used as a cell source for all established hiPSC lines, and the infection was performed using retroviruses [14]. Tanaka et al. [17] derived hiPSCs from three unrelated HCM patients and three healthy volunteers in 2014. One of the HCM patients carried the *MYBPC3-Gly999-Gln1004del* mutation, while in the two other patients, the mutations were unknown. Two control hiPSC lines were generated from dermal fibroblasts using retroviruses, while all patient hiPSC lines and one control hiPSC line were derived from T lymphocytes or peripheral blood with Sendai viruses. They used a mixture of EBs derived from all three patients in their experiments [17]. In our publication, we derived hiPSC lines from two patients carrying the *MYBPC3-Gln1061X* mutation and from two patients carrying the *TPM1-Asp175Asn* mutation. The ages and clinical symptoms of our patients varied from asymptomatic to a patient suffering from atrial fibrillation accompanied with substantial thickening of interventricular septum. Two unrelated hiPSC lines derived from healthy volunteers were used as control cells. hiPSCs were derived from fibroblasts using either retro- or Sendai viruses. All hiPSC-derived HCM cardiomyocytes were significantly larger than hiPSC-derived control cardiomyocytes [15].

In addition to cell size, myofibrillar disarray has been studied in three of the publications. However, in all publications, different methods and criteria have been used to quantify the disarray, and the results can be subjective. Han et al. showed that HCM hiPSC-derived cardiomyocytes have more disrupted sarcomeres than cardiomyocytes derived from control hiPSCs. However, the authors did not present any criteria to determine how the disruption was qualified [14]. In Tanaka et al. [17], cardiomyocytes were qualified with myofibrillar disarray if over 50% of the myofibrils intersected with each other. In Lan et al. [16], cardiomyocytes that had more than 25% of their cell area exhibiting punctate TnT distribution were considered disorganized. The overall morphology of hiPSC-derived cardiomyocytes is not mature, and their structure is often unorganized; this is also the case in cardiomyocytes derived from control hiPSCs. We did not report myofibrillar disarray due to the lack of proper quantitation criteria for this phenomenon.

Electrophysiological properties of hiPSC-derived cardiomyocytes have been studied in three of the publications. In our study, the APD90 was significantly increased in hiPSC-derived cardiomyocytes carrying the *TPM1-Asp175Asn* mutation or the *MYBPC3-Gln1061X* mutation. Additionally, the action potential duration 90 (APD90) was significantly longer in cardiomyocytes carrying the *TPM1-Asp175Asn* mutation compared to cardiomyocytes carrying the *MYBPC3-Gln1061X* mutation. Although the numbers of cardiomyocytes exhibiting arrhythmias (delayed after depolarizations (DADs) and early after depolarizations (EADs) were similar for both mutations, the DAD rate was higher in cardiomyocytes carrying the *MYBPC3-*

Gln1061X mutation than in control cardiomyocytes [15]. Lan et al. observed more cardiomyocytes exhibiting DADs in hiPSC-derived HCM cell populations carrying the *MYH7-Arg663His* mutation than in control cardiomyocyte populations. In addition, the DAD rate was significantly higher in these cells. Significant differences were observed only 30 days after the initiation of cardiac differentiation [16]. Han and co-workers [14] demonstrated a marked prolongation of APD in HCM hiPSC-derived cardiomyocytes carrying the *MYH7-Arg442Gly* mutation but no increased DAD ratio.

Changes in Ca^{2+} handling properties are considered to be one of the earliest pathophysiological mechanisms in HCM. In our study, the proportion of cardiomyocytes with abnormal Ca^{2+} transients was approximately 20% for hiPSC-derived control cardiomyocytes, 42% for cardiomyocytes carrying the *TPM1-Asp175Asn* mutation and 21% for cardiomyocytes carrying the *MYBPC3-Gln1061X* mutation at basic conditions. Lan and co-workers [16] reported that approximately 20% of hiPSC-derived HCM cardiomyocytes carrying the *MYH7-Arg663His* mutation had irregularities in their Ca^{2+} handling properties on day 30 and approximately 30% on day 40 after initiating cardiac differentiation, while in control cardiomyocytes, the proportion was approximately 5% at all timepoints. Han et al. [14] reported that approximately 20% of hiPSC-derived cardiomyocytes (*MYH7-Arg442Gly*) had abnormal Ca^{2+} handling properties, while no control hiPSC-derived cardiomyocytes had abnormalities. Similar to the characterization of disrupted sarcomeres, these data are also subjective, and the qualification or quantification criteria are rarely stated in publications. Thus, the results obtained in different publications are not directly comparable. For example, the portion of control hiPSC-derived cardiomyocytes, which have abnormalities in their Ca^{2+} handling properties varies from 0% to 20% in different publications. However, it is apparent that cardiomyocytes with HCM mutation have more abnormalities in their Ca^{2+} transients than control cardiomyocytes.

The publication from Birket et al. is to date the only one focused on studying the contraction forces of HCM hiPSC-derived cardiomyocytes. They studied cardiomyocytes derived from three individual patient hiPSC lines carrying *MYBPC3-c.2373dupG* mutation and two unrelated control hiPSC lines [13]. These hiPSC lines, derived from fibroblasts by lentiviruses, have been published earlier by Dambrot et al [88]. To be able to detect the beating forces from hiPSC-derived cardiomyocytes, Birket et al. [13] used triiodothyronine hormone (T3), insulin-like growth factor 1 (IGF-1), and dexamethasone to enhance their bioenergetics and contractile force generation. In these optimized conditions, decreased force generation was observed in hiPSC-derived cardiomyocytes carrying the *MYBPC3-c.2373dupG* mutation. The impact of mutation was confirmed by studying the generation of force in cardiomyocytes derived from *MYBPC3* knockdown hESCs [13].

4.2. Variable phenotype of hypertrophic cardiomyopathy

The primary cause of HCM is a mutation in a sarcomeric gene, while changes in Ca^{2+} handling properties, energy deficiency, ion channel remodeling, and microvascular dysfunction are thought to be the earliest pathophysiological mechanisms that play a role in disease progression [89]. However, the reasons why progressive changes initiated by these primary mutations occur in one individual and not in others are still largely unknown. Many mechanisms related to the variable phenotype of HCM have been proposed, including additional mutations, genetic modifiers, epigenetic factors, environment and function of protein quality systems. These additional disease modifiers are thought to impact on the development of mild HCM phenotype to end-stage, in which the functionality of the heart is disrupted [90]. Conventionally, HCM has been studied with animal models or patient samples obtained from surgical myectomy. However, animal models carry only the mutated gene, and although they contain the entire genome of an HCM patient, myectomy samples are obtained from patients in the late stage of HCM development. While HCM leads to progressive heart failure, the major difficulty in studying the pathophysiological mechanisms leading to HCM has been the lack of tissue at the early stages of disease development. Thus, hiPSC models represent a valuable tool to study HCM in vitro.

In our study, we could not detect the mutated MYBPC protein in hiPSC-derived cardiomyocytes carrying the *MYBPC3-Gln1061X* mutation, which is in line with previous studies using human myectomy samples and hiPSC-derived cardiomyocytes [17, 87, 91, 92]. The total amount of the MYBPC protein was slightly reduced in our study in cardiomyocytes carrying the *MYBPC3-Gln1061X* mutation compared to hiPSC-derived control cardiomyocytes [15]. Similarly, Birket et al. [13] observed a decrease in MYBPC protein relative to α-actinin in their hiPSC-derived cardiomyocytes carrying *MYBPC3-c.2373dupG* mutation, which they suggested to be the reason for the decreased traction force. The dosage of the mutated gene is one of the factors that has been proposed to affect the severity of the observed clinical phenotype. HCM is inherited in an autosomal dominant pattern, and thus, both mutated and wildtype proteins are expected to be incorporated into the sarcomere [90]. However, the expression of the mutated protein can be regulated on many levels. In particular, the *MYBPC3* mutations leading to truncated proteins can be directed toward degradation by nonsense-mediated mRNA decay, ubiquitin proteasome system (UPS), and the autophagy/lysosomal pathway, which leads to haploinsufficiency [93]. The age-related decline of these protein quality systems has been suggested to effect the progression of HCM [90]. The amount of the wildtype MYBPC protein varies in patient tissue samples, which might correlate with the clinical phenotypes observed in patients [92, 92]. Another mechanism affecting the dosage of the mutated gene and thus disease severity could be allelic imbalance, which indicates differences in the expression levels of mutated and wildtype alleles [93, 94].

Protein phosphorylation is one of the most important post-translational modifications, which has also been suggested to affect disease development. For example, the reduced phosphorylation of TnI and MYBPC has been related to increased myofilament Ca^{2+} sensitivity, which is a common feature of HCM [95]. However, Ca^{2+} sensitivity has also been suggested to be a primary consequence of HCM. Additionally, Ca^{2+} sensitivity has been proposed to increase arrhythmia sensitivity either by increasing the Ca^{2+} binding affinity in the cytosol, which could lead to remodeling of action potentials and thus trigger arrhythmias, or by affecting energy consumption and increasing arrhythmia susceptibility via stress [90, 96, 97].

Some HCM patients have been shown to carry more than just one mutation in their genotype, and patients carrying multiple mutations have been associated with more severe symptoms or earlier onset of disease [76, 78]. Other genetic mechanisms include genetic modifiers, which can be either near or distantly located DNA variants that influence the expression of the mutated gene [76]. Additionally, epigenetic changes, which cannot be explained by the DNA sequence itself, have been suggested to contribute to the progression of HCM. These mechanisms include the methylation of GpC islands by DNA methyltransferases, histone modification, miRNAs and long non-coding RNAs (lncRNAs), which can lead to the altered regulation of genes [76].

4.3. Challenges in modeling hypertrophic cardiomyopathy with human induced pluripotent stem cells

The major limitation when using hiPSC-derived cardiomyocytes in disease modeling is the immature nature of these cells. The characteristics of hPSC-derived cardiomyocytes and the differences between these and adult human cardiomyocytes have recently been reviewed [98]. Compared to adult cardiomyocytes, the sarcomeric structures of hPSC-derived cardiomyocytes are unorganized, they lack clear T tubules, they have different ion channel profiles, and their Ca^{2+} handling is immature. In addition, the shapes of hPSC-derived cardiomyocytes vary from circular to star-shaped, while adult human cardiomyocytes are rod-shaped in vivo. Overall, the phenotype of hPSC-derived cardiomyocytes is thought to more closely resemble that of fetal cardiomyocytes than adult cardiomyocytes [98]. At the moment, much effort is being expended to obtain more mature cardiomyocytes. In addition to culture conditions and the long-term culture of cardiomyocytes, other techniques such as electrical and mechanical stimulation, as well as engineered heart tissue (EHT) structures, have been developed [99–102].

Another issue related to the study of HCM with hiPSC-derived cardiomyocytes is related to the assumption that cell types other than cardiomyocytes might be directly involved in the progression of HCM. For example, microvascular dysfunction is thought to be the primary reason for replacement-type fibrosis observed in HCM patients [90]. Thus, it is important to consider whether studying HCM at the single cell level is sufficient to obtain an exact picture of the disease mechanisms. EHT structures and cardiovascular constructs, consisting of various cell types could be useful in this context [102, 103].

As discussed above, in addition to primary HCM gene mutations, gene modifiers and epigenetic changes might also have an effect on disease development and progression. In all published reports, the phenotypes of HCM hiPSC-derived cardiomyocytes have been compared to control hiPSC-derived cardiomyocytes established from related or non-related healthy individuals [13–17]. Currently, useful gene-editing approaches, including clustered regularly interspaced short palindromic repeats (CRISPR) and transcription activator-like effector nuclease (TALEN) techniques, are available, and they can be used either to create mutations or correct existing mutations in an hiPSC line, thus generating genotype-matched isogenic control lines [104]. When considering the variable phenotypes of HCM patients, these isogenic control hiPSCs would be useful when further studying HCM disease mechanisms.

5. Conclusions

Although, the current methods for studying cardiomyopathies in vitro with patient-specific cardiomyocytes are far from optimal, we and others have been able to create hiPSC models with HCM phenotype. hiPSC-derived cardiomyocytes can also be utilized to study the effects of genetic modifiers and epigenetic factors on disease progression between different individuals, which has been difficult when using animal models or samples from surgical myectomy.

However, particularly in the case of HCM associated with highly variable phenotypes, it would be important to optimize cardiac differentiation and cell culture conditions. The study design becomes highly valuable, and the culture conditions should be similar for both control and disease-specific cardiomyocytes. When experiments are thoroughly designed, the results obtained from these studies would be more robust and reliable. Nonetheless, hiPSC-derived HCM in vitro models represent a valuable tool to study the pathophysiological mechanisms of HCM as well as to test novel drug therapies developed to prevent disease progression and potentially optimize treatments in a mutation-specific manner.

Acknowledgements

We thank Anu Hyysalo for image of neural cells and Mostafa Kiamehr for image of hepatocytes. Parts of the text and some of the images have been published earlier in Acta Universitatis Tamperensis series as a part of a PhD thesis.

Author details

Marisa Ojala[1*] and Katriina Aalto-Setälä[1,2,3]

*Address all correspondence to: marisa.ojala@uta.fi

1 BioMediTech, University of Tampere, Tampere, Finland

2 School of Medicine, University of Tampere, Tampere, Finland

3 Heart Hospital, Tampere University Hospital, Tampere, Finland

References

[1] Thomson J.A., Itskovitz-Eldor J., Shapiro S.S., Waknitz M.A., Swiergiel J.J., Marshall V.S., et al. Embryonic stem cell lines derived from human blastocysts. Science. 1998;282(5391):1145–1147.

[2] Takahashi K., Tanabe K., Ohnuki M., Narita M., Ichisaka T., Tomoda K., et al. Induction of pluripotent stem cells from adult human fibroblasts by defined factors. Cell. 2007;131(5):861–872. DOI: 10.1016/j.cell.2007.11.019

[3] Fusaki N., Ban H., Nishiyama A., Saeki K., Hasegawa M. Efficient induction of transgene-free human pluripotent stem cells using a vector based on Sendai virus, an RNA virus that does not integrate into the host genome. Proceedings of the Japan Academy Series B, Physical and Biological Sciences. 2009;85(8):348–362.

[4] Soldner F., Hockemeyer D., Beard C., Gao Q., Bell G.W., Cook E.G., et al. Parkinson's disease patient-derived induced pluripotent stem cells free of viral reprogramming factors. Cell. 2009;136(5):964–977. DOI: 10.1016/j.cell.2009.02.013

[5] Yu J., Hu K., Smuga-Otto K., Tian S., Stewart R., Slukvin I.I., et al. Human induced pluripotent stem cells free of vector and transgene sequences. Science. 2009;324(5928): 797–801. DOI: 10.1126/science.1172482

[6] Itskovitz-Eldor J., Schuldiner M., Karsenti D., Eden A., Yanuka O., Amit M., et al. Differentiation of human embryonic stem cells into embryoid bodies compromising the three embryonic germ layers. Molecular Medicine. 2000;6(2):88–95.

[7] Mummery C., Ward-vanOostwaard D., Doevendans P., Spijker R., van den Brink S., Hassink R., et al. Differentiation of human embryonic stem cells to cardiomyocytes: role of coculture with visceral endoderm-like cells. Circulation. 2003;107(21):2733–2740. DOI: 10.1161/01.CIR.0000068356.38592.68

[8] Burridge P.W., Matsa E., Shukla P., Lin Z.C., Churko J.M., Ebert A.D., et al. Chemically defined generation of human cardiomyocytes. Nature Methods. 2014;11(8):855–860. DOI: 10.1038/nmeth.2999

[9] Laflamme M.A., Chen K.Y., Naumova A.V., Muskheli V., Fugate J.A., Dupras S.K., et al. Cardiomyocytes derived from human embryonic stem cells in pro-survival factors enhance function of infarcted rat hearts. Nature Biotechnology. 2007;25(9):1015–1024. DOI: 10.1038/nbt1327

[10] Lian X., Hsiao C., Wilson G., Zhu K., Hazeltine L.B., Azarin S.M., et al. Robust cardiomyocyte differentiation from human pluripotent stem cells via temporal modulation of canonical Wnt signaling. Proceedings of the National Academy of Sciences of the United States of America. 2012;109(27):E1848–E1857. DOI: 10.1073/pnas.1200250109

[11] Maron B.J., Ommen S.R., Semsarian C., Spirito P., Olivotto I., Maron M.S. Hypertrophic cardiomyopathy: present and future, with translation into contemporary cardiovascular medicine. Journal of the American College of Cardiology. 2014;64(1):83–99. DOI: 10.1016/j.jacc.2014.05.003

[12] Roma-Rodrigues C., Fernandes A.R. Genetics of hypertrophic cardiomyopathy: advances and pitfalls in molecular diagnosis and therapy. The Application of Clinical Genetics. 2014;7:195–208. DOI: 10.2147/TACG.S49126

[13] Birket M.J., Ribeiro M.C., Kosmidis G., Ward D., Leitoguinho A.R., van de Pol V. Contractile defect caused by mutation in MYBPC3 revealed under conditions optimized for human PSC-cardiomyocyte function. Cell Reports. 2015;13(4):733–745. DOI: 10.1016/j.celrep.2015.09.025

[14] Han L., Li Y., Tchao J., Kaplan A.D., Lin B., Li Y. Study familial hypertrophic cardiomyopathy using patient-specific induced pluripotent stem cells. Cardiovascular Research. 2014;104(2):258–269. DOI: 10.1093/cvr/cvu205

[15] Ojala M., Prajapati C., Pölönen R-P., Rajala K., Pekkanen-Mattila M., Rasku J., et al. Mutation-specific phenotypes in hiPSC-derived cardiomyocytes carrying either myosin-binding protein C or α-tropomyosin mutation for hypertrophic cardiomyopathy. Stem Cells International. 2016; Article ID: 1684792. DOI: 10.1155/2016/1684792

[16] Lan F., Lee A.S., Liang P., Sanchez-Freire V., Nguyen P.K., Wang L., et al. Abnormal calcium handling properties underlie familial hypertrophic cardiomyopathy pathology in patient-specific induced pluripotent stem cells. Cell Stem Cell. 2013;12(1):101–113. DOI: 10.1016/j.stem.2012.10.010

[17] Tanaka A., Yuasa S., Mearini G., Egashira T., Seki T., Kodaira M., et al. Endothelin-1 induces myofibrillar disarray and contractile vector variability in hypertrophic cardiomyopathy-induced pluripotent stem cell-derived cardiomyocytes. Journal of the American Heart Association. 2014;3(6):e001263. DOI: 10.1161/JAHA.114.001263

[18] Klimanskaya I., Chung Y., Becker S., Lu S-J., Lanza R. Human embryonic stem cell lines derived from single blastomeres. Nature. 2006;444(7008):481–485. DOI: 10.1038/nature05142

[19] Strelchenko N., Verlinsky O., Kukharenko V., Verlinsky Y. Morula-derived human embryonic stem cells. Reproductive Biomedicine Online. 2004;9(6):623–629.

[20] Takahashi K., Yamanaka S. Induction of pluripotent stem cells from mouse embryonic and adult fibroblast cultures by defined factors. Cell. 2006;126(4):663–676. DOI: 10.1016/j.cell.2006.07.024

[21] Briggs R., King T.J. Transplantation of living nuclei from blastula cells into enucleated frogs' eggs. PNAS. 1952;38(5):455–463.

[22] Gurdon J.B. The developmental capacity of nuclei taken from intestinal epithelium cells of feeding tadpoles. Journal of Embryology and Experimental Morphology. 1962;10:622–640.

[23] Wilmut I., Schnieke A.E., McWhir J., Kind A.J., Campbell K.H. Viable offspring derived from fetal and adult mammalian cells. Nature. 1997;385(6619):810–813. DOI: 10.1038/385810a0

[24] Yamanaka S. Pluripotency and nuclear reprogramming. Philosophical Transactions of the Royal Society of London. Series B, Biological Sciences. 2008;363(1500):2079–2087. DOI: 10.1098/rstb.2008.2261

[25] Vaajasaari H., Ilmarinen T., Juuti-Uusitalo K., Rajala K., Onnela N., Narkilahti S., et al. Toward the defined and xeno-free differentiation of functional human pluripotent stem cell-derived retinal pigment epithelial cells. Molecular Vision. 2011;17:558–575.

[26] Nat R., Nilbratt M., Narkilahti S., Winblad B., Hovatta O., Nordberg A. Neurogenic neuroepithelial and radial glial cells generated from six human embryonic stem cell lines in serum-free suspension and adherent cultures. Glia. 2007;55(4):385–399. DOI: 10.1002/glia.20463

[27] Hay D.C., Fletcher J., Payne C., Terrace J.D., Gallagher R.C.J., Snoeys J., et al. Highly efficient differentiation of hESCs to functional hepatic endoderm requires ActivinA and Wnt3a signaling. PNAS. 2008;105(34):12301–12306. DOI: 10.1073/pnas.0806522105

[28] Bayart E., Cohen-Haguenauer O. Technological overview of iPS induction from human adult somatic cells. Current Gene Therapy. 2013;13(2):73–92.

[29] Park H-J., Shin J., Kim J., Cho S-W. Nonviral delivery for reprogramming to pluripotency and differentiation. Archives of Pharmacal Research. 2014;37(1):107–119. DOI: 10.1007/s12272-013-0287-z

[30] Hu K. All roads lead to induced pluripotent stem cells: the technologies of iPSC generation. Stem Cells and Development. 2014;23(12):1285–1300. DOI: 10.1089/scd.2013.0620

[31] Yu J., Vodyanik M.A., Smuga-Otto K., Antosiewicz-Bourget J., Frane J.L., Tian S., et al. Induced pluripotent stem cell lines derived from human somatic cells. Science. 2007;318(5858):1917–1920. DOI: 10.1126/science.1151526

[32] Matsui T., Leung D., Miyashita H., Maksakova I.A., Miyachi H., Kimura H., et al. Proviral silencing in embryonic stem cells requires the histone methyltransferase ESET. Nature. 2010;464(7290):927–931. DOI: 10.1038/nature08858

[33] Mikkelsen T.S., Hanna J., Zhang X., Ku M., Wernig M., Schorderet P., et al. Dissecting direct reprogramming through integrative genomic analysis. Nature. 2008;454(7200): 49–55. DOI: 10.1038/nature07056

[34] Sridharan R., Tchieu J., Mason M.J., Yachechko R., Kuoy E., Horvath S., et al. Role of the murine reprogramming factors in the induction of pluripotency. Cell. 2009;136(2): 364–377. DOI: 10.1016/j.cell.2009.01.001

[35] Chang C-W., Lai Y-S., Pawlik K.M., Liu K., Sun C-W., Li C., et al. Polycistronic lentiviral vector for "hit and run" reprogramming of adult skin fibroblasts to induced pluripotent stem cells. Stem Cells. 2009;27(5):1042–1049. DOI: 10.1002/stem.39

[36] Kaji K., Norrby K., Paca A., Mileikovsky M., Mohseni P., Woltjen K. Virus-free induction of pluripotency and subsequent excision of reprogramming factors. Nature. 2009;458(7239):771–775. DOI: 10.1038/nature07864

[37] Stadtfeld M., Nagaya M., Utikal J., Weir G., Hochedlinger K. Induced pluripotent stem cells generated without viral integration. Science. 2008;322(5903):945–949. DOI: 10.1126/science.1162494

[38] Zhou W., Freed C. Adenoviral gene delivery can reprogram human fibroblasts to induced pluripotent stem cells. Stem Cells. 2009;27(11):2667–2674. DOI: 10.1002/stem.201

[39] Woltjen K., Michael I.P., Mohseni P., Desai R., Mileikovsky M., Hämäläinen R., et al. piggyBac transposition reprograms fibroblasts to induced pluripotent stem cells. Nature. 2009;458(7239):766–770. DOI: 10.1038/nature07863

[40] Okita K., Matsumura Y., Sato Y., Okada A., Morizane A., Okamoto S., et al. A more efficient method to generate integration-free human iPS cells. Nature Methods. 2011;8(5):409–412. DOI: 10.1038/nmeth.1591

[41] Khan M., Narayanan K., Lu H., Choo Y., Du C., Wiradharma N., et al. Delivery of reprogramming factors into fibroblasts for generation of non-genetic induced pluripotent stem cells using a cationic bolaamphiphile as a non-viral vector. Biomaterials. 2013;34(21):5336–5343. DOI: 10.1016/j.biomaterials.2013.03.072

[42] Lee C.H., Kim J-H., Lee H.J., Jeon K., Lim H., Choi H., et al. The generation of iPS cells using non-viral magnetic nanoparticle based transfection. Biomaterials. 2011;32(28): 6683–6691. DOI: 10.1016/j.biomaterials.2011.05.070

[43] Später D., Hansson L., Zangi L., Chien K.R. How to make a cardiomyocyte. Development. 2014;141(23):4418–4431. DOI: 10.1242/dev.091538

[44] Mummery C., Zhang J., Ng E.S., Elliott D.A., Elefanty A.G., Kamp T.J. Differentiation of human embryonic stem cells and induced pluripotent stem cells to cardiomyocytes: a methods overview. Circulation Research. 2012;111(3):344–358. DOI: 10.1161/CIRCRESAHA.110.227512

[45] Rajala K., Pekkanen-Mattila M., Aalto-Setälä K. Cardiac differentiation of pluripotent stem cells. Stem Cells International. 2011; Article ID: 383709. DOI: 10.4061/2011/383709

[46] Verma V., Purnamawati K., Manasi S.W. Steering signal transduction pathway towards cardiac lineage from human pluripotent stem cells: a review. Cellular Signalling. 2013;25(5):1096–1107. DOI: 10.1016/j.cellsig.2013.01.027

[47] Kehat I., Kenyagin-Karsenti D., Snir M., Segev H., Amit M., Gepstein A., et al. Human embryonic stem cells can differentiate into myocytes with structural and functional properties of cardiomyocytes. The Journal of Clinical Investigation. 2001;108(3):407–414. DOI: 10.1172/JCI12131

[48] Burridge P.W., Thompson S., Millrod M., Weinberg S., Yuan X., Peters A., et al. A universal system for highly efficient cardiac differentiation of human induced pluripotent stem cells that eliminates interline variability. PloS One. 2011;6(4):e18293. DOI: 10.1371/journal.pone.0018293

[49] Mohr J.C., Zhang J., Azarin S.M., Soerens A.G., de Pablo J.J., Thomson J.A., et al. The microwell control of embryoid body size in order to regulate cardiac differentiation of human embryonic stem cells. Biomaterials. 2010;31(7):1885–1893. DOI: 10.1016/j.biomaterials.2009.11.033

[50] Yang L., Soonpaa M.H., Adler E.D., Roepke T.K., Kattman S.J., Kennedy M., et al. Human cardiovascular progenitor cells develop from a KDR+ embryonic-stem-cell-derived population. Nature. 2008;453(7194):524–528. DOI: 10.1038/nature06894

[51] Karakikes I., Senyei G.D., Hansen J., Kong C-W., Azeloglu E.U., Stillitano F., et al. Small molecule-mediated directed differentiation of human embryonic stem cells toward ventricular cardiomyocytes. Stem Cells Translational Medicine. 2014;3(1):18–31. DOI: 10.5966/sctm.2013-0110

[52] Passier R., Oostwaard D.W., Snapper J., Kloots J., Hassink R.J., Kuijk E., et al. Increased cardiomyocyte differentiation from human embryonic stem cells in serum-free cultures. Stem Cells. 2005;23(6):772–780. DOI: 10.1634/stemcells.2004-0184

[53] Graichen R., Xu X., Braam S.R., Balakrishnan T., Norfiza S., Sieh S., et al. Enhanced cardiomyogenesis of human embryonic stem cells by a small molecular inhibitor of p38 MAPK. Differentiation. 2008;76(4):357–370. DOI: 10.1111/j.1432-0436.2007.00236.x

[54] Xu X.Q., Graichen R., Soo S.Y., Balakrishnan T., Rahmat S.N.B., Sieh S., et al. Chemically defined medium supporting cardiomyocyte differentiation of human embryonic stem cells. Differentiation. 2008;76(9):958–970. DOI: 10.1111/j.1432-0436.2008.00284.x

[55] Zhang J., Klos M., Wilson G.F., Herman A.M., Lian X., Raval K.K., et al. Extracellular matrix promotes highly efficient cardiac differentiation of human pluripotent stem cells: the matrix sandwich method. Circulation Research. 2012;111(9):1125–1136. DOI: 10.1161/CIRCRESAHA.112.273144

[56] Burridge P.W., Anderson D., Priddle H., Barbadillo Muñoz M.D., Chamberlain S., Allegrucci C., et al. Improved human embryonic stem cell embryoid body homogeneity and cardiomyocyte differentiation from a novel V-96 plate aggregation system highlights interline variability. Stem Cells. 2007;25(4):929–938. DOI: 10.1634/stemcells.2006-0598

[57] Freund C., Ward-vanOostwaard D., Monshouwer-Kloots J., van den Brink S., van Rooijen M., Xu X., et al. Insulin redirects differentiation from cardiogenic mesoderm and endoderm to neuroectoderm in differentiating human embryonic stem cells. Stem Cells. 2008;26(3):724–733. DOI: 10.1634/stemcells.2007-0617

[58] Minami I., Yamada K., Otsuji T.G., Yamamoto T., Shen Y., Otsuka S., et al. A small molecule that promotes cardiac differentiation of human pluripotent stem cells under defined, cytokine- and xeno-free conditions. Cell Reports. 2012;2(5):1448–1460. DOI: 10.1016/j.celrep.2012.09.015

[59] Osafune K., Caron L., Borowiak M., Martinez R.J., Fitz-Gerald C.S., Sato Y., et al. Marked differences in differentiation propensity among human embryonic stem cell lines. Nature Biotechnology. 2008;26(3):313–315. DOI: 10.1038/nbt1383

[60] Xu C., Police S., Rao N., Carpenter M. K. Characterization and enrichment of cardiomyocytes derived from human embryonic stem cells. Circulation Research. 2002;91(6): 501–508.

[61] Dubois N.C., Craft A.M., Sharma P., Elliott D.A., Stanley E.G., Elefanty A.G., et al. SIRPA is a specific cell-surface marker for isolating cardiomyocytes derived from human pluripotent stem cells. Nature Biotechnology. 2011;29(11):1011–1018. DOI: 10.1038/nbt.2005

[62] Uosaki H., Fukushima H., Takeuchi A., Matsuoka S., Nakatsuji N., Yamanaka S., et al. Efficient and scalable purification of cardiomyocytes from human embryonic and induced pluripotent stem cells by VCAM1 surface expression. PLoS One. 2011;6(8):e23657. DOI: 10.1371/journal.pone.0023657

[63] Tohyama S., Hattori F., Sano M., Hishiki T., Nagahata Y., Matsuura T., et al. Distinct metabolic flow enables large-scale purification of mouse and human pluripotent stem cell-derived cardiomyocytes. Cell Stem Cell. 2013;12(1):127–137. DOI: 10.1016/j.stem.2012.09.013

[64] Miki K., Endo K., Takahashi S., Funakoshi S., Takei I., Katayama S., et al. Efficient detection and purification of cell populations using synthetic microRNA switches. Cell Stem Cell. 2015;16(6):699–711. DOI: 10.1016/j.stem.2015.04.005

[65] Kartasalo K., Pölönen R-P., Ojala M., Rasku J., Lekkala J., Aalto-Setälä K., et al. CytoSpectre: a tool for spectral analysis of oriented structures on cellular and subcellular levels. BMC Bioinformatics. 2015;16:344. DOI: 10.1186/s12859-015-0782-y

[66] Laurila E., Ahola A., Hyttinen J., Aalto-Setälä K. Methods for in vitro functional analysis of iPSC derived cardiomyocytes—special focus on analyzing the mechanical beating behavior. Biochimica et Biophysica Acta. 2015. DOI: 10.1016/j.bbamcr.2015.12.013

[67] Reppel M., Pillekamp F., Lu Z.J., Halbach M., Brockmeier K., Fleischmann B.K., et al. Microelectrode arrays: a new tool to measure embryonic heart activity. Journal of Electrocardiology. 2004;37:104–109.

[68] Hamill O., Marty A., Neher E., Sakmann B., Sigworth F. Improved patch-clamp techniques for high-resolution current recording from cells and cell-free membrane patches. Pflügers Archiv: European Journal of Physiology. 1981;2(391):85–100.

[69] Sakmann B., Neher E. Patch clamp techniques for studying ionic channels in excitable membranes. Annual Review of Physiology. 1984;46:455–472. DOI: 10.1146/annurev.ph.46.030184.002323

[70] Zilberter Y.I., Timin E.N., Bendukidze Z.A., Burnashev N.A. Patch-voltage-clamp method for measuring fast inward current in single rat heart muscle cells. Pflügers Archiv: European Journal of Physiology. 1982;394(2):150–155.

[71] Bers D.M. Calcium cycling and signaling in cardiac myocytes. Annual Review of Physiology. 2008;70:23–49. DOI: 10.1146/annurev.physiol.70.113006.100455

[72] Grynkiewicz G., Poenie M., Tsien R.Y. A new generation of Ca2+ indicators with greatly improved fluorescence properties. The Journal of Biological Chemistry. 1985;260(6): 3440–3450.

[73] Ahola A., Kiviaho A.L., Larsson K., Honkanen M., Aalto-Setälä K., Hyttinen J. Video image-based analysis of single human induced pluripotent stem cell derived cardiomyocyte beating dynamics using digital image correlation. Biomedical Engineering Online. 2014;7:13–39. DOI: 10.1186/1475-925X-13-39

[74] Lopes L.R., Elliott P.M. A straightforward guide to the sarcomeric basis of cardiomyopathies. Heart. 2014;100(24):1916–1923. DOI: 10.1136/heartjnl-2014-305645

[75] Elliott P., Andersson B., Arbustini E., Bilinska Z., Cecchi F., Charron P., et al. Classification of the cardiomyopathies: a position statement from the European Society of Cardiology Working Group on Myocardial and Pericardial Diseases. European Heart Journal. 2008;29(2):270–276. DOI: 10.1093/eurheartj/ehm342

[76] Ho C.Y., Charron P., Richard P., Girolami F., van Spaendonck-Zwarts K.Y., Pinto Y. Genetic advances in sarcomeric cardiomyopathies: state of the art. Cardiovascular Research. 2015;105(4):397–408. DOI: 10.1093/cvr/cvv025

[77] Kocovski L., Fernandes J. Sudden cardiac death: a modern pathology approach to hypertrophic cardiomyopathy. Archives of Pathology & Laboratory Medicine. 2015;136(3):413–416. DOI: 10.5858/arpa.2013-0489-RS

[78] Maron B.J., Maron M.S., Semsarian C. Double or compound sarcomere mutations in hypertrophic cardiomyopathy: a potential link to sudden death in the absence of conventional risk factors. Heart Rhythm. 2012;9(1):57–63. DOI: 10.1016/j.hrthm. 2011.08.009

[79] Jääskeläinen P., HeliöT., Aalto-Setälä K., Kaartinen M., Ilveskoski E., Hämäläinen L., et al. Two founder mutations in the alpha-tropomyosin and the cardiac myosin-binding protein C genes are common causes of hypertrophic cardiomyopathy in the Finnish population. Annals of Medicine. 2013;45(1):85–90. DOI: 10.3109/07853890.2012.671534

[80] Jääskeläinen P., Kuusisto J., Miettinen R., Kärkkäinen P., Kärkkäinen S., Heikkinen S., et al. Mutations in the cardiac myosin-binding protein C gene are the predominant cause of familial hypertrophic cardiomyopathy in eastern Finland. Journal of Molecular Medicine. 2002;80(7):412–422. DOI: 10.1007/s00109-002-0323-9

[81] Maron B.J., Niimura H., Casey S.A., Soper M.K., Wright G.B., Seidman J.G., et al. Development of left ventricular hypertrophy in adults in hypertrophic cardiomyop-

athy caused by cardiac myosin-binding protein C gene mutations. Journal of the American College of Cardiology. 2001;38(2):315–321.

[82] Niimura H., Bachinski L.L., Sangwatanaroj S., Watkins H., Chudley A.E., McKenna W., et al. Mutations in the gene for cardiac myosin-binding protein C and late-onset familial hypertrophic cardiomyopathy. The New England Journal of Medicine. 1998;338(18): 1248–1257. DOI: 10.1056/NEJM199804303381802

[83] Thierfelder L., Watkins H., MacRae C., Lamas R., McKenna W., Vosberg H.P., et al. Alpha-tropomyosin and cardiac troponin T mutations cause familial hypertrophic cardiomyopathy: a disease of the sarcomere. Cell. 1994;77(5):701–712.

[84] Jääskeläinen P., Miettinen R., Kärkkäinen P., Toivonen L., Laakso M., Kuusisto J. Genetics of hypertrophic cardiomyopathy in eastern Finland: few founder mutations with benign or intermediary phenotypes. Annals of Medicine. 2004;36(1):23–32.

[85] Jääskeläinen P., Heliö T., Aalto-Setälä K., Kaartinen M., Ilveskoski E., Hämäläinen L., et al. A new common mutation in the cardiac beta-myosin heavy chain gene in Finnish patients with hypertrophic cardiomyopathy. Annals of Medicine. 2014;46(6):424–429. DOI: 10.3109/07853890.2014.912834

[86] Duncker D.J., Bakkers J., Brundel B.J., Robbins J., Tardiff J.C., Carrier L. Animal- and in silico models for the study of sarcomeric cardiomyopathies. Cardiovascular Research. 2015;105(4):439–448. DOI: 10.1093/cvr/cvv006

[87] Marston S., Copeland O., Jacques A., Livesey K., Tsang V., McKenna W.J., et al. Evidence from human myectomy samples that MYBPC3 mutations cause hypertrophic cardiomyopathy through haploinsufficiency. Circulation Research. 2009;105(3):219–222. DOI: 10.1161/CIRCRESAHA.109.202440

[88] Dambrot C., Braam S.R., Tertoolen L.G.J., Birket M., Atsma D.E., Mummery C.L. Serum supplemented culture medium masks hypertrophic phenotypes in human pluripotent stem cell derived cardiomyocytes. Journal of Cellular and Molecular Medicine. 2014;18(8):1509–1518. DOI: 10.1111/jcmm.12356

[89] Tardiff J.C., Carrier L., Bers D.M., Poggesi C., Ferrantini C., Coppini R., et al. Targets for therapy in sarcomeric cardiomyopathies. Cardiovascular Research. 2015;105(4): 457–470. DOI: 10.1093/cvr/cvv023

[90] Van der Velden J., Ho C.Y., Tardiff J.C., Olivotto I., Knollmann B.C., Carrier L. Research priorities in sarcomeric cardiomyopathies. Cardiovascular Research. 2015;105(4):449–456. DOI: 10.1093/cvr/cvv019

[91] Rottbauer W., Gautel M., Zehelein J., Labeit S., Franz W.M., Fischer C., et al. Novel splice donor site mutation in the cardiac myosin-binding protein-C gene in familial hypertrophic cardiomyopathy: Characterization of cardiac transcript and protein. The Journal of Clinical Investigation. 1997;100(2):475–482. DOI: 10.1172/JCI119555

[92] Van Dijk S.J., Dooijes D., Remedios C.D., Michels M., Lamers J.M.J., Winegrad S., et al. Cardiac myosin-binding protein C mutations and hypertrophic ardiomyopathy

haploinsufficiency, deranged phosphorylation, and cardiomyocyte dysfunction. Circulation. 2009;119(11):1473–1483. DOI: 10.1161/CIRCULATIONAHA.108.838672

[93] Helms A.S., Davis F.M., Coleman D., Bartolone S.N., Glazier A.A., Pagani F., et al. Sarcomere mutation-specific expression patterns in human hypertrophic cardiomyopathy. Circulation. Cardiovascular Genetics. 2014;7(4):434–443. DOI: 10.1161/CIRCGENETICS.113.000448

[94] Tripathi S., Schultz I., Becker E., Montag J., Borchert B., Francino A., et al. Unequal allelic expression of wild-type and mutated β-myosin in familial hypertrophic cardiomyopathy. Basic Research in Cardiology. 2011;106(6):1041–1055. DOI: 10.1007/s00395-011-0205-9

[95] Sequeira V., Wijnker P.J.M., Nijenkamp L.L.M., Kuster D.W.D., Najafi A., Witjas-Paalberends E.R., et al. Perturbed length-dependent activation in human hypertrophic cardiomyopathy with missense sarcomeric gene mutations. Circulation Research. 2013;112(11):1491–1505. DOI: 10.1161/CIRCRESAHA.111.300436

[96] Huke S., Venkataraman R., Faggioni M., Bennuri S., Hwang H.S., Baudenbacher F., et al. Focal energy deprivation underlies arrhythmia susceptibility in mice with calcium-sensitized myofilaments. Circulation Research. 2013;112(10):1334–1344. DOI: 10.1161/CIRCRESAHA.113.301055

[97] Schober T., Huke S., Venkataraman R., Gryshchenko O., Kryshtal D., Hwang H.S., et al. Myofilament Ca sensitization increases cytosolic Ca binding affinity, alters intracellular Ca homeostasis, and causes pause-dependent Ca-triggered arrhythmia. Circulation Research. 2012;111(2):170–179. DOI: 10.1161/CIRCRESAHA.112.270041

[98] Van den Heuvel N.H.L., van Veen T.A.B., Lim B., Jonsson M.K.B. Lessons from the heart: mirroring electrophysiological characteristics during cardiac development to in vitro differentiation of stem cell derived cardiomyocytes. Journal of Molecular and Cellular Cardiology. 2014;67:12–25. DOI: 10.1016/j.yjmcc.2013.12.011

[99] Kreutzer J., Ikonen L., Hirvonen J., Pekkanen-Mattila M., Aalto-Setälä K., Kallio P. Pneumatic cell stretching system for cardiac differentiation and culture. Medical Engineering & Physics. 2014;36(4):496–501. DOI: 10.1016/j.medengphy.2013.09.008

[100] Kujala K., Ahola A., Pekkanen-Mattila M., Ikonen L., Kerkelä E., Hyttinen J., et al. Electrical field stimulation with a novel platform: effect on cardiomyocyte gene expression but not on orientation. International Journal of Biomedical Science: IJBS. 2012;8(2):109–120.

[101] Nunes S.S., Miklas J.W., Liu J., Aschar-Sobbi R., Xiao Y., Zhang B., et al. Biowire: a platform for maturation of human pluripotent stem cell-derived cardiomyocytes. Nature Methods. 2013;10(8):781–787. DOI: 10.1038/nmeth.2524

[102] Stoehr A., Neuber C., Baldauf C., Vollert I., Friedrich F.W., Flenner F., et al. Automated analysis of contractile force and Ca2+ transients in engineered heart tissue. AJP: Heart

and Circulatory Physiology. 2014;306(9):H1353-H1363. DOI: 10.1152/ajpheart.00705.2013

[103] Vuorenpää H., Ikonen L., Kujala K., Huttala O., Sarkanen J.R., Ylikomi T., et al. Novel in vitro cardiovascular constructs composed of vascular-like networks and cardiomyocytes. In vitro Cellular & Developmental Biology. Animal. 2014;50(4):275–286. DOI: 10.1007/s11626-013-9703-4

[104] Li M., Suzuki K., Kim N.Y., Liu G-H., Izpisua Belmonte J.C. A cut above the rest: targeted genome editing technologies in human pluripotent stem cells. The Journal of Biological Chemistry. 2014;289(8):4594–4599. DOI: 10.1074/jbc.R113.488247

Chapter 13

Stem Cells for Modeling Human Disease

Martín F, Sánchez-Gilabert A,
Tristán-Manzano M and Benabdellah K

Additional information is available at the end of the chapter

http://dx.doi.org/10.5772/63231

Abstract

Human pluripotent stem cells (PSCs) in the form of human embryonic stem cells (hESCs) or induced pluripotent stem cells (iPSCs) are capable of growing indefinitely *in vitro*, maintaining their capacity to differentiate into the three primary germ layers: mesoderm, endoderm and ectoderm. Different protocols have been developed to differentiate PSCs into almost any cellular type with different degree of success. This technology has allowed scientists to use patient-derived iPSCs to study the physiopathology of the disease by analyzing the phenotype of the cells derived from these iPSCs. However, control iPSCs obtained from healthy individuals will always have different genomic environment than patient's iPSCs, making it difficult the interpretation of the cells phenotype. The recent appearance of specific nucleases [zinc-finger nucleases (ZFNs), the transcription activator-like effector nucleases (TALENs) and the clustered regularly interspaced short palindromic repeats (CRISPR)] has made it possible to edit the genome of PSCs. We can now generate syngeneic hESCs or iPSCs harboring the desired mutation and comparing the emerging cells with those derived from genetically identical PSCs that will differ only in the mutated gene. In this chapter, we summarize the progress made in this field and discuss the different approaches that have been used recently for the generation of syngeneic human pluripotent cellular models for different pathologies.

Keywords: human pluripotent stem cells, disease models, ZFNs, TALEN, CRISPR

1. Introduction

Disease models are an essential tool for elucidating the molecular basis of several pathologies, allowing the development of novel therapies. Historically, and taking into account the scarcity of the patient cells, the use of model organism made possible the clarification of cellular mechanisms underlying human disease. *Drosophila melanogaster, Caenorhabditis elegans* and zebrafish have been very helpful in dissecting basic disease mechanisms [1–3]. However, the simplicity of these organisms from the physiological point of view and their phylogenic distance from human limit their use as models for human disease. The most popular animal models for studying human diseases are based on genetically engineered mice. For an updated list of available models, see http://www.informatics.jax.org/humanDisease.shtml. A quantum leap was made by the introduction of "humanized" mouse models in which various types of human cells are engrafted and they function as they would in human's organs [4–6]. This humanized mice models can harbor human hematopoietic stem cells (HSCs) facilitating the analysis of human hematology and immunology disease *in vivo*. In spite of the potential of humanized animal models, several issues remain to be overcome, such as the insufficient intercellular relationships or the physiological differences between humans and the animal models. These differences could explain in part why some drugs tested in animal models fail in the corresponding clinical trials. Recent progress in the field of regenerative medicine allows the generation of patient-specific stem cells that are suitable for generation of human disease models. This can be done by two different approaches: the first approach is to drive stem cells from patient with the target disease, and the second is to genetically alter human stem cells with gene editing tools.

2. Disease modeling with human stem cells derived from patients

The advantage of using human stem cells derived from patients is that they can be isolated at different stages of disease severity. For example, by isolating stem cells from patients with end-stage disease, we could study those phenotypes resulting from a combination of insults in the patient that take the patient to that particular stage. These particular phenotypes are practically impossible to mimic by other means such as genetic or epigenetic manipulations.

2.1. Human adult stem cells

Adult stem cells are multipotent cells found in all adult tissues, and they participate in the physiological regeneration of the tissues where they belong. The adult stem cell better characterized and with better perspectives for use as human models of disease are the neural progenitor cells (NPCs), mesenchymal stromal cells (MSCs) and the HSCs:

2.1.1. Neural progenitor cells

NPCs comprise relatively undifferentiated cell population in the central nervous systems (CNSs) that give rise to a broad array of specialized cells, including neurons and glial cells. These cells can be isolated and cultured *in vitro*. NPCs driven from patients with known mutations associated with a specific disease allow an excellent "in dish disease modeling" for neurogenetic disease, allowing a direct study of the cellular pathogenic mechanisms. Amyotrophic lateral sclerosis (ALS), commonly referred as Lou Gehrig's disease, is a fatal neurodegenerative disease characterized by loss of motor neurons (MNs) in the motor cortex, brain stem and spinal cord, resulting in muscle paralysis and ultimately death due to respiratory failure. A human cellular model for ALS has been developed from NPCs derived from post-mortem spinal cord NPCs from ALS patients. The derived astrocytes provide the first *in vitro* model system to investigate common disease mechanisms and evaluate potential therapies for ALS [7]. A second model of a neurodegenerative disease was established by the use of immature neural cells derived from adult tissues from post-mortem brain tissue of a 25-year-old man with fragile X syndrome (FXS) [8]. The cultured fragile X cells displayed many of the characteristics of NPCs, as well as the biochemical hallmarks of FXS, including CGG repeat expansion. These two models allowed the study of the efficacy of new therapeutic agents.

2.1.2. Mesenchymal stromal cells

MSCs represent another interesting alternative for disease modeling. The International Society for Cellular Therapy (ISCT) established that MSCs must be purified from stromal populations based on plastic adherence and must be positive for CD105, CD90 and CD73, negative for MHC-II, CD11b, CD14, CD34, CD45 and CD31 and express low levels of MHC-I. In addition, MSCs must differentiate *in vitro* into osteocytes, chondrocytes and adipocytes. MSCs can be obtained from bone marrow, adipose tissue, synovial membranes, dental pulp, Wharton's jelly, umbilical cord blood, liver tissue, etc.

MSCs are a very attractive source for disease modeling because they are able to give rise to different tissue types. Therefore, MSCs derived from patients could be used to model diseases affecting tissues to which MSCs can be derived to. Contrary to induced pluripotent stem cells (iPSCs), MSCs are not induced by the expression of gene involved in oncogenic cells transformation (discussed later). Due to their relative abundance within the body, the MSCs could be used as novel human *in vitro* models for several diseases. Dossena *et al* [9] isolated MSCs from the adipose tissue of spinal and bulbar muscular atrophy (SBMA), a late-onset progressive neurodegenerative disorder caused by a trinucleotide (CAG) repeat expansion within the coding region. CAG expansions encode for longer polyQ chains in the produced protein [10]. Dossena *et al* showed that MSCs isolated from the adipose tissue of SBMA patients form nuclear polyQ inclusions producing a robust pathogenic polyQ phenotype *in vitro* [9]. Unfortunately, there is still some controversy regarding the differentiation of neuron from MSCs, limiting the relevance of this model to study neurodegeneration. The same limitation stand for other diseases affecting other tissues/organs to which MSCs cannot differentiate to, such as liver, blood or muscle.

2.1.3. Hematopoietic stem cells

HSCs have the potential to give rise to all hematopoietic cells *in vitro* and are therefore a potential source to mimic diseases affecting the hematopoietic system including primary immunodeficiencies (PIDs) and autoimmune diseases. Most applications of human HSCs in disease modeling have focused on the development of humanized mice models. In these systems, human HSCs are inoculated into immunodeficient mice engrafting and regenerating most of the hematopoietic system of the mice with human cells. The engrafted cells can be used to study behaviors of the different populations or to study therapeutic interventions. These models have been used for modeling infectious diseases such as HIV-1 [11], Ebola virus [12], PID [13] and other disorders of the hematopoietic systems [14].

2.2. Human embryonic stem cells from embryos with genetic diseases

Human embryonic stem cells (hESCs), derived from pre-implantation embryos, were the first human pluripotent stem cells (PSCs) to be isolated. Thanks to the pre-implantation genetic diagnosis (PGD), it is now possible to generate hESCs from monogenetic diseases [15–17]. Using these approaches, hESCs derived from embryos with FXS, Huntington's disease (HD) and familial dysautonomia (FD) have been generated [18–21]. The authors observed that *in vitro* differentiation of FXS-hESCs into neurons resulted in abnormal neurogenesis and poor neuronal maturation mimicking the developmental events taking place during neurogenesis in FXS patients. Similarly, Feyeux *et al* [22] demonstrated a down-regulation of the Huntingtin (HTT) gene in HD hESCs-derived neurons and identified early-stage mitochondrial dysfunction during development. In the case of the FD or Riley-Day syndrome, by using FD-diagnosed embryos, Lefler and colleagues [21] have found that IKAP is likely a vesicular-like protein involved in neuronal transport and synaptic integrity in FD hESCs, probably reflecting some peripheral nervous system (PNS) neuronal dysfunction observed in FD. However, the scarcity of PGD, legal concerns in relation to the parental consent for embryo donation and some ethical consideration have made this approach very difficult.

2.3. Induced pluripotent stem cells

The development of the iPSCs technologies [23] bypassed these limitations since we can now generate hESC-like cells from patients. The ability of iPSCs to differentiate into virtually any tissue or cell type makes these cells an excellent tool for modeling human disease. These cells allow the generation of patient-specific disease-relevant cells in virtually a continuous manner. This is imperative in the case where the disease is a result of an interplay genetics risk, instead of a punctual mutation. This is the case of the vast majority of the neurodegenerative disorders where only the 5–10% are Mendelian disorders caused by a punctual mutation, the rest are a result of multifactorial genetic association. Several iPSCs models have been developed for neurodegenerative diseases including HD [24–29], Alzheimer's disease [30] and spinal muscular atrophy [31–36]. Several inherited bone marrow failure (BMF) syndromes have also been modeled *in vitro* by the derivation of the corresponding iPSCs, and the corresponding differentiation protocols. This is the case of defective telomere elongation [37], Fanconi anemia [38–40] or congenital megakaryocytic thrombocytopenia [41]. Following the same procedure, Wang and colleagues were able to found new insights in the Barth syndrome (BTHS), an X-linked cardiac and skeletal mitochondrial myopathy caused by mutation of the TAZ gene, obtaining BTHS iPSC-derived cardiomyocytes [42].

http://dx.doi.org/10.5772/63231

3. Disease modeling with gene editing tools and stem cells

iPSCs and hESCs from embryos with genetic diseases (hESCs-GD) provide powerful tools for modeling human diseases, although the different genetic background of the control cells versus the patient-specific cells makes the interpretation of the results difficult [43]. A solution to this problem is the use of genome editing (GE) technologies to generate the desired mutations into iPSCs or hESCs [44]. This approach allows the direct comparison of PSCs harboring the desired mutation with isogenic control cells lines. However, until recently, this approach presents low efficiency and specificity. This situation changed considerably with the appearance of specific nucleases (SNs). The introduction of double-strand breaks (DSBs) in the targeted DNA sequence has dramatically improved the homologous recombination up to 10,000. Several SNs have been described such as the meganuclease I-SceI, the zinc-finger nucleases (ZFNs), the transcription activator-like effector nucleases (TALENs) and the clustered regularly interspaced short palindromic repeats (CRISPR)-associated system 9 (CRISPR/Cas9). Meganucleases, also called homing endonucleases, are a class of highly sequence-specific enzymes that recognize a relative large long DNA sequence ranging from 12 to 30 bp. The recognition site of MNs can be engineered in order to target specific sites within the genome. One of the major advantages of the meganucleases is their small size making them amenable to be packaged into single viral vector and allowing efficient delivery. However, this technology requires good knowledge of protein engineering and has been limited to some DNA targets [45]. A major breakthrough arises with the discovery of ZFNs [46–48], the first SNs capable to target almost any DNA sequence in the human genome. ZFNs are chimeric proteins that combine a nuclease domain (FokI) and a zinc-finger domain (ZFD) that recognizes the targeted DNA sequence. The specificity is therefore determined by the ZFD that harbors four-six zinc fingers of 30 amino acids. ZFNs are easier to construct in comparison with MNs, but still required intensive labor to obtain site-specific ZFNs. Soon after the appearance of ZFNs, a new protein-based SNs, TALENs, were described [49 51]. TALENs are a combination of the catalytic domain of an endonuclease (FokI) fused with a DNA-binding domain derived from transcription activator-like effectors from plant pathogen *Xanthomonas* species. TALENs recognize specific DNA sequences via DNA-binding domains composed of nearly identical 34-amino acid repeated unit. As ZFNs, site-specific TALENs can be derived in almost any laboratory with a good molecular biology expertise. However, the wide distribution of ZFNs and TALENs was hindered by the complexity of their designs. The latest described SN system is CRISPR/Cas9. This system, contrary to the previously described nucleases, uses a short molecule of RNA, called guide RNA (gRNA) to direct the Cas9 (nuclease) to the target sequence. This simplicity has opened the use of SNs to almost any laboratory in cell and molecular biology. All the endonucleases can cause DSBs and, subsequently, insertion or deletion at the site of the genomic DSB can be induced by imprecise non-homologous end-joining (NHEJ)-mediated repair or by precise editing using HDR.

All these GE technologies have opened up the possibility to obtain almost any mutation in any cell type. Therefore, different groups have already applied these technologies to model human disease by generating disease-causing mutations in primary stem cells:

3.1. Editing hHSCs for disease modeling

Gene therapy targeting HSCs is a real option now for several genetic diseases, including severe combined immunodeficiency (SCID) and other severe non-SCID PID. The PID patient has been treated classically by allogeneic HSCs transplantation or by the correction of the patient's own HSCs by the insertion of a functional copy of the affected gene by a viral vector [52, 53]. However, to further improve HSCs-based gene therapy, HSCs cellular models capable to recapitulate the disease phenotype are needed. The low frequency of these disorders and the difficulties to obtain HSCs from the patients preclude the use of patient HSCs as standard tools for disease modeling and preclinical testing. As an alternative, GE tools can be used to generate the disease-causing mutation in HSCs obtained from umbilical cord. For example, a X-linked severe combined immunodeficiency (SCID-X1) induced by a mutation in the IL2RG gene using specific ZFNs has been recently modeled *in vitro* [54]. Sickle-cell disease (SCD), another rare in the β-globin gene, has also been modeled using this strategy. SCD can be cured by allogenic HSC transplant. However, this is only possible when a matched donor is available, making the development of gene therapy using autologous HSCs a highly desirable alternative [55]. In order to improve the gene therapy treatment for this rare disease, a model was required. In this sense, the electroporation of healthy HSCs cells with mRNA-ZFN that specifically targeted the α-globin gene-generated SCD-like HSCs. The author shows that the SCD-like HSCs generated similar numbers and patterns of erythroid and myeloid clones in comparison to HSCs from SCD patients. The SCD-like HSCs cells maintained the repopulation capacity into immune-deficient NSG mice, representing an excellent disease model for SCD [56]. Finally, it is noteworthy to mention that the gene editing approach using stem cells was used not only for disease modeling but also for disease treatment. This has been done by knocking down a clinically relevant gene CCR5 principal co-receptor involved in human immunodeficiency virus (HIV) infection [57]. A recent clinical trial shows for the first time that CCR5 ablation in immune cells of 12 patients with HIV enhanced their resistance to the HIV infection [58].

3.2. Editing hESCs for disease modeling

As we mentioned before, hESCs-GD can be generated and could in theory be an excellent tool for any disease. We also mentioned the problems this strategy face if we want to extend its use to model any disease [43]. GE technologies have allowed the development of a very potent alternative: the generation of gene-edited hESCs harboring the disease-causing mutation. This strategy has the additional advantage of count with isogenic hESCs that only differed in the mutation and are otherwise genetically identical.

Then, we will describe some examples of how different groups have used GE to model different diseases:

Cancer. Chromosomal translocations are signatures of numerous cancers and lead to expression of fusion genes that act as oncogenes. This aberrant translocation can be induced using gene editing tools *in vitro* in hESCs. For instance, the translocation related with Ewing sarcoma and anaplastic large cell lymphoma (ALCL) was recently reproduced *in vitro* by the way of

ZFN and TALEN [59]. Breakpoint junctions recovered MSCs derived from GE hESCs fully recapitulating the genomic landscape found in tumor cells from Ewing sarcoma patients.

Monogenetic diseases. Disrupting a single gene by GE is easier than promoting translocations or multiple gene mutations. Human cellular models of monogenetic diseases were the first to be generated by GE as an alternative to iPSCs from patients. An elegant demonstration of the importance of using isogenic cell lines to model disease was provided by Reinhardt *et al* [60]. These authors generated iPSCs lines from patients with the *LRRK2* mutation (Parkinson's disease-related mutation) and healthy individuals without the mutation. In addition, they used GE to correct the mutation generating isogenic cell lines that only differed in the LRRK2 mutation. Gene expression profile clustering of hESCS-derived neurons showed that the healthy and mutant iPSC lines did not cluster in separate groups (as would be expected). Rather, one of the healthy lines clustered closely to a mutant line and one healthy line was very different to all of the other lines. Therefore, no conclusion could be taken with these disease-specific iPSCs lines [60]. Interestingly, the only cell lines that clustered close together were pairs of mutant lines with and without correction of the mutation by GE. Taking these data into consideration, the authors managed to demonstrate that mutation in LRRK2 induced dysregulation of CPNE8, MAP7, UHRF2, ANXA1 and CADPS2 genes. Other good examples of the relevance of using isogenic cell lines were showed by Li *et al* [61] and by our group [62]. Li *et al* generated a model for the Rett Syndrome by mutating the MECP2 gene. MECP2 mutant neurons mimic the defects observed in this disorder and unbiased global gene expression analyses (thanks to the use of isogenic cell lines) showed that MECP2 protein functions as a global gene activator in neurons but not in neural precursors. Similarly, our group has generated human cellular models for Wiskott-Aldrich syndrome (WAS) by mutating the WAS gene in hESCs using ZFNs [62]. Using these models, we uncovered that the absence of WAS protein also affected early hematopoiesis and megakaryocyte development, a phenotype that could not be observed when using iPSCs from WAS patients [63].

Several other monogenetic diseases have been modeled by GE of hESCs (see **Table 1**). Martinez *et al* used TALENs to knock out the KCTD13 in hESCs to model the Timothy syndrome, a variant of an autism disorder, allowing a rapid drug screening [80]. Vanuytsel *et al* [79] generated FANCA-deficient hESCs, a human model for FANCA, a recessive disorder characterized by progressive BMF, congenital abnormalities and development of malignancies. The authors used ZFNs to introduce a selection cassette into exon 2 of FANCA, disrupting its expression in hESCs. Interestingly, they couldn't obtain homozygous FANCA mutants due to growth arrest of these cells. Only heterozygous population could be obtained. Zou *et al* have generated hESCs models of Paroxysmal nocturnal hemoglobinuria (PNH) [77] by mutating the PIG-A gene. Recently, a hepatocyte-like cell line has been derived from hESCs treated with TALEN targeting the SORT1 gene, allowing to model coronary artery disease *in vitro* [81]. The same group has also generated derived cell lines with specific mutations in AKT2 gene, which performs an important role in the development of metabolic syndrome, and also in PLIN1, which encodes the perilipin protein and when altered, is responsible for the autosomal-dominant subtype of partial lipodystrophy [81].

Disease	Gene	Gene editing tools	Cell type	References
Congenital cataracts	PITX	ZFNs TALEN	hESCs iPSCs	[64]
Cancer	Chromosomal translocation	ZFNs TALEN	hESCs MSCs	[59, 65]
Cancer	Oct4	ZFNs	hESCs iPSCs	[64]
Parkinson's disease	*LRRK2*	ZFN	iPSC	[60]
Hemophilia A	Coagulation factor VIII (F8) gene	ZFNs TALEN	iPSC	[66, 67]
AIDS	Chemokine (C-C motif) receptor 5 (CCR5)	ZFN	iPSC; hESCs	[68–71]
Imaging purpose	AAVS1	ZFN	iPSC, hESCs	[72]
Neonatal alloimmune thrombocytopenia	Human platelet alloantigens	CRISPR/Cas9	iPSC	[73]
Autism spectrum disorders (ASDs)	CHD8	CRISPR/Cas9	iPSCs	[74]
Fragile X syndrome (FXS)	FMR1	CRISPR/Cas9	iPSCs	[75]
Epilepsies	SCN1A	CRISPR/Cas9 and TALEN	iPSCs	[76]
Paroxysmal nocturnal hemoglobinuria,	PIG-A	ZFNs	hiPSCs, hESCs	[77]
Schizophrenia		CRISPR/Cas9	iPSCs	[78]
Ewing sarcoma and anaplastic large cell lymphoma	t(11;22)(q24;q12) and t(2;5)(p23;q35) translocations	TALEN and ZFNs	hESCs	[59]
FANCA	FANCA	ZFNs	hESCs	[79]
Autism	KCTD13	TALEN	hESCs	[80]
Amyotrophic lateral sclerosis		CRISPR/Cas9	iPSC, hESCs	[85]
Coronary artery disease	SORT1	TALEN	hESCs	[81]
X-linked severe combined immunodeficiency	IL2RG	ZFNs	HSCs	[54]
Wiskott-Aldrich Syndrome	WAS	ZFNs	hESC	[62]
Sickle-cell disease	β-globin	ZFNs	HSCs	[56]
Metabolic syndrome	AKT2	TALEN	hESCs	[81]
Autosomal-dominant subtype of partial lipodystrophy	PLIN1	TALEN	hESCs	[81]
Barth syndrome	TAZ	CRISPR/Cas9	iPSCs	[42]

Table 1. Summary of pluripotent stem cells and gene editing tools to generate human disease models.

http://dx.doi.org/10.5772/63231

3.3. Editing iPSCs for disease modeling

As mentioned before, iPSCs can be obtained from any patient representing very potent tools for disease modeling. However, several considerations should be taken into account when interpreting the phenotypes of iPSCs from patients and controls since they have different genetic backgrounds. Indeed, a major concern to interpret the data coming out from a disease-associated iPSCs is the choice of appropriate controls. In addition, several studies have shown that the generation and expansion of different iPSCs from some patients' lines shows some genetic aberration ranging from single nucleotide variant, chromosomal deletion and differences in the methylation pattern of their genomes. An interesting solution to this problem is the GE of iPSCs from healthy individuals, as we have described previously for hESCs, in order to generate "home-made" disease-specific iPSCs **Figure 1**. This approach allows the direct comparison of "gene-edited" iPSCs cell line with its isogenic control. Then, we will describe some examples of modeling human disease using this approach:

Parkinson's disease (PD) iPSCs were generated by mutating the LRRK2 gene in iPSCs lines from a healthy control using ZFNs. Neurons derived from gene-edited PD iPSCs exhibit

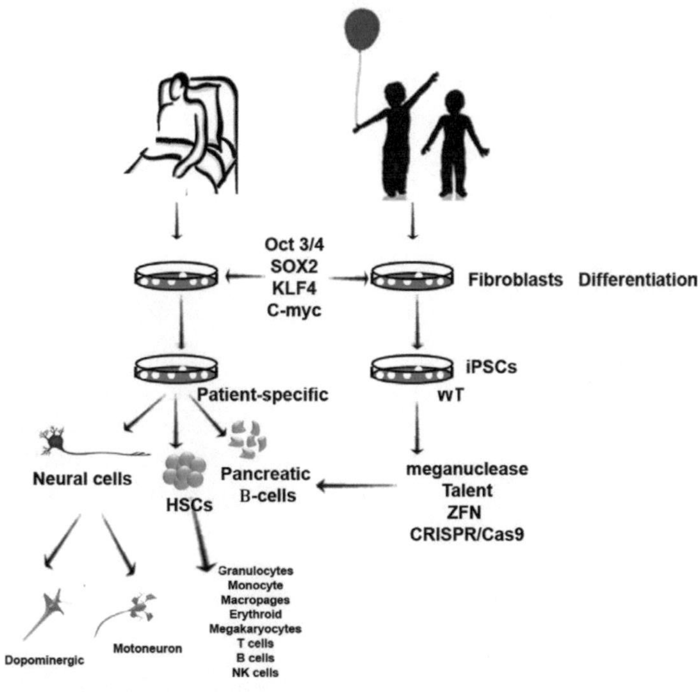

Figure 1. Schematic representation of the generation of iPSCs for drug screening, disease modeling and therapy. Somatic cells from patient (left) and healthy donor (right) are treated with four reprogramming factor (OCT3/4, SOX2, Klf4 and C-Myc). In the case of the control donor, the obtained iPSCs are gene edited using specific nuclease (Maganuclease, TALEN, ZFN or CRISPR/Cas9). Finally, the patient or control gene-edited iPSCs can be differentiated into a variety of cells type including primitive hematopoietic cells, pancreatic beta-cells and several different neuronal cell types.

reduced neurite outgrowth and increased apoptosis in response to oxidative stress when compared with neuron derived from isogenic control iPSCs [60, 82]. Park *et al* generated hemophilia A iPSCs using a TALEN pair to invert a 140Kbp DNA segment of the F8 gene (a mutation shared by half of the hemophilia A patients) in human control hiPSCs. The authors demonstrated that TALENs could be used to generate disease models associated with chromosomal rearrangements [66]. The chemokine (C-C motif) receptor 5 (CCR5) serves as an HIV-1 co-receptor and is essential for cell infection with HIV-1 virus. Loss of this receptor confers protection against HIV-1 infection. In this sense, several groups have generated CCR5-negative iPSCs using CRISPR/Cas9 [68] or ZFNs [69]. HSCs-like cells derived from the CCR5-iPSCs could be used in the future to model HIV-1 resistance to infection [69, 70, 83]. Antibodies against the human platelet alloantigens (HPAs) cause severe alloimmune bleeding disorders. HPA-1b/b platelets (variant Pro33 within the Integrin β3) are relatively rare in the population and difficult to obtain for transfusion or diagnosis. Zhang *et al* [73] used CRISPR/Cas9 GE to generate Integrin β3-Pro33 (HPA-1b) iPSCs to model this variant. Megakaryocyte progenitors derived from these iPSCs expressed the HPA-1b alloantigenic epitope. In a similar approach, Liu *et al* [76] used CRISPR/Cas9 and TALENs to generate iPSC deficient for the SCN1A protein to model epilepsies. Using SCN1A-deficients iPSCs, they were able to explore the mechanism of epilepsy caused by loss of SCN1A protein, showing reduction of the amplitudes and enhancement of the action potential thresholds. A novel model of Barth syndrome derived from the control human line PGP1 iPSC was obtained by CRISPR/Cas9 and mutated in exon 6 of TAZ gene, resulting valid to study this monogenic mitochondrial cardiomyopathy [42, 84]. Several other diseases have been modeled *in vitro* using one of the previously described GE tools; for more details, see **Table 1**.

Acknowledgements

This work has been financed by Fondo de Investigaciones Sanitarias ISCIII (Spain) and Fondo Europeo de Desarrollo Regional (FEDER) from the European Union through the research grants PI12/01097 (F.M.) and by the CICE and CS de la Junta de Andalucía FEDER/Fondo de Cohesion Europeo (FSE) de Andalucía 2007–2013 through research grants P09-CTS-04532, PI-57069, PI-0001/2009 and PAIDI-Bio-326 (F.M.); PI-0160/2012 (K.B.).

Author details

Martín F, Sánchez-Gilabert A, Tristán-Manzano M and Benabdellah K*

*Address all correspondence to: karim.benabdel@genyo.es

Genomic Medicine Department. GENYO, Centre for Genomics and Oncological Research, Pfizer-University of Granada-Andalusian Regional Government, Parque Tecnológico Ciencias de la Salud, Av. de la Ilustración Granada, Spain

http://dx.doi.org/10.5772/63231

References

[1] Casci, I. and U.B. Pandey, *A fruitful endeavor: modeling ALS in the fruit fly.* Brain Res, 2014. 1607: p. 47–74.

[2] Lepesant, J.A., *The promises of neurodegenerative disease modeling.* C R Biol, 2015. 338(8–9): p. 584–92.

[3] Jucker, M., *The benefits and limitations of animal models for translational research in neurodegenerative diseases.* Nat Med, 2010. 16(11): p. 1210–4.

[4] Legrand, N., K. Weijer, and H. Spits, *Experimental models to study development and function of the human immune system in vivo.* J Immunol, 2006. 176(4): p. 2053–8.

[5] Shultz, L.D., F. Ishikawa, and D.L. Greiner, *Humanized mice in translational biomedical research.* Nat Rev Immunol, 2007. 7(2): p. 118–30.

[6] Zhang, B., Z. Duan, and Y. Zhao, *Mouse models with human immunity and their application in biomedical research.* J Cell Mol Med, 2009. 13(6): p. 1043–58.

[7] Haidet-Phillips, A.M., et al., *Astrocytes from familial and sporadic ALS patients are toxic to motor neurons.* Nat Biotechnol, 2011. 29(9): p. 824–8.

[8] Schwartz, P.H., et al., *Neural progenitor cells from an adult patient with fragile X syndrome.* BMC Med Genet, 2005. 6: p. 2.

[9] Dossena, M., et al., *Human adipose-derived mesenchymal stem cells as a new model of spinal and bulbar muscular atrophy.* PLoS One, 2014. 9(11): p. e112746.

[10] Orr, H.T., *Cell biology of spinocerebellar ataxia.* J Cell Biol, 2012. 197(2): p. 167–77.

[11] Joseph, A., et al., *Inhibition of in vivo HIV infection in humanized mice by gene therapy of human hematopoietic stem cells with a lentiviral vector encoding a broadly neutralizing anti-HIV antibody.* J Virol, 2010. 84(13): p. 6645–53.

[12] Ludtke, A., et al., *Ebola virus disease in mice with transplanted human hematopoietic stem cells.* J Virol, 2015. 89(8): p. 4700–4.

[13] Scaramuzza, S., et al., *Preclinical safety and efficacy of human CD34(+) cells transduced with lentiviral vector for the treatment of Wiskott-Aldrich syndrome.* Mol Ther, 2013. 21(1): p. 175–84.

[14] Levasseur, D.N., et al., *Correction of a mouse model of sickle cell disease: lentiviral/antisickling beta-globin gene transduction of unmobilized, purified hematopoietic stem cells.* Blood, 2003. 102(13): p. 4312–9.

[15] Mateizel, I., et al., *Derivation of human embryonic stem cell lines from embryos obtained after IVF and after PGD for monogenic disorders.* Hum Reprod, 2006. 21(2): p. 503–11.

[16] Ben-Yosef, D., M. Malcov, and R. Eiges, *PGD-derived human embryonic stem cell lines as a powerful tool for the study of human genetic disorders.* Mol Cell Endocrinol, 2008. 282(1–2): p. 153–8.

[17] Verlinsky, Y., et al., *Human embryonic stem cell lines with genetic disorders.* Reprod Biomed Online, 2005. 10(1): p. 105–10.

[18] Niclis, J., et al., *Human embryonic stem cell models of Huntington disease.* Reprod Biomed Online, 2009. 19(1): p. 106–13.

[19] Eiges, R., et al., *Developmental study of fragile X syndrome using human embryonic stem cells derived from preimplantation genetically diagnosed embryos.* Cell Stem Cell, 2007. 1(5): p. 568–77.

[20] Niclis, J.C., et al., *Characterization of forebrain neurons derived from late-onset Huntington's disease human embryonic stem cell lines.* Front Cell Neurosci, 2013. 7: p. 37.

[21] Lefler, S., et al., *Familial dysautonomia (FD) Human embryonic stem cell derived PNS neurons reveal that synaptic vesicular and neuronal transport genes are directly or indirectly affected by IKBKAP downregulation.* PLoS One, 2015. 10(10): p. e0138807.

[22] Feyeux, M., et al., *Early transcriptional changes linked to naturally occurring Huntington's disease mutations in neural derivatives of human embryonic stem cells.* Hum Mol Genet, 2012. 21(17): p. 3883–95.

[23] Takahashi, K. and S. Yamanaka, *Induction of pluripotent stem cells from mouse embryonic and adult fibroblast cultures by defined factors.* Cell, 2006. 126(4): p. 663–76.

[24] Arber, C., et al., *Activin A directs striatal projection neuron differentiation of human pluripotent stem cells.* Development, 2015. 142(7): p. 1375–86.

[25] Juopperi, T.A., et al., *Astrocytes generated from patient induced pluripotent stem cells recapitulate features of Huntington's disease patient cells.* Mol Brain, 2012. 5: p. 17.

[26] Mattis, V.B., et al., *HD iPSC-derived neural progenitors accumulate in culture and are susceptible to BDNF withdrawal due to glutamate toxicity.* Hum Mol Genet, 2015. 24(11): p. 3257–71.

[27] Mattis, V.B. and C.N. Svendsen, *Modeling Huntington's disease with patient-derived neurons.* Brain Res, 2015. doi: 10.1016/j.brainres.2015.10.001 [Epub ahead of print].

[28] Russo, F.B., et al., *Induced pluripotent stem cells for modeling neurological disorders.* World J Transplant, 2016. 5(4): p. 209–21.

[29] Zhang, N., et al., *iPSC-based drug screening for Huntington's disease.* Brain Res, 2015. doi: 10.1016/j.brainres.2015.09.020 [Epub].

[30] Sullivan, S.E. and T.L. Young-Pearse, *Induced pluripotent stem cells as a discovery tool for Alzheimer's disease.* Brain Res, 2015. doi: 10.1016/j.brainres.2015.10.005 [Epub ahead of print].

[31] Frattini, E., et al., *Pluripotent stem cell-based models of spinal muscular atrophy.* Mol Cell Neurosci, 2014. 64: p. 44–50.

[32] Liu, H., et al., *Spinal muscular atrophy patient-derived motor neurons exhibit hyperexcitability.* Sci Rep, 2015. 5: p. 12189.

[33] Ohuchi, K., et al., *Established stem cell model of spinal muscular atrophy is applicable in the evaluation of the efficacy of thyrotropin-releasing hormone analog.* Stem Cells Transl Med, 2015. 5(2): p. 152–63.

[34] Valetdinova, K.R., S.P. Medvedev, and S.M. Zakian, *Model systems of motor neuron diseases as a platform for studying pathogenic mechanisms and searching for therapeutic agents.* Acta Naturae, 2015. 7(1): p. 19–36.

[35] Yoshida, M., et al., *Modeling the early phenotype at the neuromuscular junction of spinal muscular atrophy using patient-derived iPSCs.* Stem Cell Reports, 2015. 4(4): p. 561–8.

[36] Fuller, H.R., et al., *Spinal muscular atrophy patient iPSC-derived motor neurons have reduced expression of proteins important in neuronal development.* Front Cell Neurosci, 2016. 9: p. 506.

[37] Winkler, T., et al., *Defective telomere elongation and hematopoiesis from telomerase-mutant aplastic anemia iPSCs.* J Clin Invest, 2013. 123(5): p. 1952–63.

[38] Raya, A., et al., *A protocol describing the genetic correction of somatic human cells and subsequent generation of iPS cells.* Nat Protoc, 2010. 5(4): p. 647–60.

[39] Liu, G.H., et al., *Modelling Fanconi anemia pathogenesis and therapeutics using integration-free patient-derived iPSCs.* Nat Commun, 2014. 5: p. 4330.

[40] Suzuki, N.M., et al., *Pluripotent cell models of fanconi anemia identify the early pathological defect in human hemoangiogenic progenitors.* Stem Cells Transl Med, 2015. 4(4): p. 333–8.

[41] Hirata, S., et al., *Congenital amegakaryocytic thrombocytopenia iPS cells exhibit defective MPL-mediated signaling.* J Clin Invest, 2013. 123(9): p. 3802–14.

[42] Wang, G., et al., *Modeling the mitochondrial cardiomyopathy of Barth syndrome with induced pluripotent stem cell and heart-on-chip technologies.* Nat Med, 2014. 20(6): p. 616–23.

[43] Saha, K. and R. Jaenisch, *Technical challenges in using human induced pluripotent stem cells to model disease.* Cell Stem Cell, 2009. 5(6): p. 584–95.

[44] Musunuru, K., *Genome editing of human pluripotent stem cells to generate human cellular disease models.* Dis Model Mech, 2013. 6(4): p. 896–904.

[45] Arnould, S., et al., *The I-CreI meganuclease and its engineered derivatives: applications from cell modification to gene therapy.* Protein Eng Des Sel, 2011. 24(1–2): p. 27–31.

[46] Bibikova, M., et al., *Targeted chromosomal cleavage and mutagenesis in Drosophila using zinc-finger nucleases.* Genetics, 2002. 161(3): p. 1169–75.

[47] Bibikova, M., et al., *Enhancing gene targeting with designed zinc finger nucleases.* Science, 2003. 300(5620): p. 764.

[48] Urnov, F.D., et al., *Highly efficient endogenous human gene correction using designed zinc-finger nucleases.* Nature, 2005. 435(7042): p. 646–51.

[49] Wood, A.J., et al., *Targeted genome editing across species using ZFNs and TALENs.* Science, 2011. 333(6040): p. 307.

[50] Boch, J., et al., *Breaking the code of DNA binding specificity of TAL-type III effectors.* Science, 2009. 326(5959): p. 1509–12.

[51] Moscou, M.J. and A.J. Bogdanove, *A simple cipher governs DNA recognition by TAL effectors.* Science, 2009. 326(5959): p. 1501.

[52] Fischer, A., et al., *Gene therapy for primary immunodeficiencies.* Clin Genet, 2015. 88(6): p. 507–15.

[53] Ott de Bruin, L.M., S. Volpi, and K. Musunuru, *Novel genome-editing tools to model and correct primary immunodeficiencies.* Front Immunol, 2015. 6: p. 250.

[54] Genovese, P., et al., *Targeted genome editing in human repopulating haematopoietic stem cells.* Nature, 2014. 510(7504): p. 235–40.

[55] Urbinati, F., et al., *Potentially therapeutic levels of anti-sickling globin gene expression following lentivirus-mediated gene transfer in sickle cell disease bone marrow CD34+ cells.* Exp Hematol, 2015. 43(5): p. 346–51.

[56] Hoban, M.D., et al., *Correction of the sickle cell disease mutation in human hematopoietic stem/progenitor cells.* Blood, 2015. 125(17): p. 2597–604.

[57] Mandal, P.K., et al., *Efficient ablation of genes in human hematopoietic stem and effector cells using CRISPR/Cas9.* Cell Stem Cell, 2014. 15(5): p. 643–52.

[58] Tebas, P., et al., *Gene editing of CCR5 in autologous CD4 T cells of persons infected with HIV.* N Engl J Med, 2014. 370(10): p. 901–10.

[59] Piganeau, M., et al., *Cancer translocations in human cells induced by zinc finger and TALE nucleases.* Genome Res, 2013. 23(7): p. 1182–93.

[60] Reinhardt, P., et al., *Genetic correction of a LRRK2 mutation in human iPSCs links Parkinsonian neurodegeneration to ERK-dependent changes in gene expression.* Cell Stem Cell, 2013. 12(3): p. 354–67.

[61] Li, Y., et al., *Global transcriptional and translational repression in human-embryonic-stem-cell-derived Rett syndrome neurons.* Cell Stem Cell, 2014. 13(4): p. 446–58.

[62] Toscano, M.G., et al., *Absence of WASp enhances hematopoietic and megakaryocytic differentiation in a human embryonic stem cell model.* Mol Ther, 2015. 24(2): p. 342–53.

[63] Ingrungruanglert, P., et al., *Wiskott-Aldrich syndrome iPS cells produce megakaryocytes with defects in cytoskeletal rearrangement and proplatelet formation*. Thromb Haemost, 2015. 113(4): p. 792–805.

[64] Hockemeyer, D., et al., *Efficient targeting of expressed and silent genes in human ESCs and iPSCs using zinc-finger nucleases*. Nat Biotechnol, 2009. 27(9): p. 851–7.

[65] Brunet, E., et al., *Chromosomal translocations induced at specified loci in human stem cells*. Proc Natl Acad Sci U S A, 2009. 106(26): p. 10620–5.

[66] Park, C.Y., et al., *Targeted inversion and reversion of the blood coagulation factor 8 gene in human iPS cells using TALENs*. Proc Natl Acad Sci U S A, 2014. 111(25): p. 9253–8.

[67] Lee, H.J., et al., *Targeted chromosomal duplications and inversions in the human genome using zinc finger nucleases*. Genome Res, 2012. 22(3): p. 539–48.

[68] Kang, H., et al., *CCR5 disruption in induced pluripotent stem cells using CRISPR/Cas9 provides selective resistance of immune cells to CCR5-tropic HIV-1 virus*. Mol Ther Nucleic Acids, 2015. 4: p. e268.

[69] Ramalingam, S., et al., *Generation and genetic engineering of human induced pluripotent stem cells using designed zinc finger nucleases*. Stem Cells Dev, 2012. 22(4): p. 595–610.

[70] Ru, R., et al., *Targeted genome engineering in human induced pluripotent stem cells by penetrating TALENs*. Cell Regen (Lond), 2013. 2(1): p. 5.

[71] Lei, Y., et al., *Gene editing of human embryonic stem cells via an engineered baculoviral vector carrying zinc-finger nucleases*. Mol Ther, 2011. 19(5): p. 942–50.

[72] Wang, Y., et al., *Genome editing of human embryonic stem cells and induced pluripotent stem cells with zinc finger nucleases for cellular imaging*. Circ Res, 2012. 111(12): p. 1494–503.

[73] Zhang, N., et al., *CRISPR/Cas9-mediated conversion of human platelet alloantigen allotypes*. Blood, 2015. doi: 10.1182/blood-2015-10-675751 [Epub].

[74] Wang, P., et al., *CRISPR/Cas9-mediated heterozygous knockout of the autism gene CHD8 and characterization of its transcriptional networks in neurodevelopment*. Mol Autism, 2015. 6: p. 55.

[75] Park, C.Y., et al., *Reversion of FMR1 methylation and silencing by editing the triplet repeats in fragile X iPSC-derived neurons*. Cell Rep, 2015. 13(2): p. 234–41.

[76] Liu, J., et al., *CRISPR/Cas9 facilitates investigation of neural circuit disease using human iPSCs: mechanism of epilepsy caused by an SCN1A loss-of-function mutation*. Transl Psychiatry, 2016. 6: p. e703.

[77] Zou, J., et al., *Gene targeting of a disease-related gene in human induced pluripotent stem and embryonic stem cells*. Cell Stem Cell, 2009. 5(1): p. 97–110.

[78] Flaherty, E.K. and K.J. Brennand, *Using hiPSCs to model neuropsychiatric copy number variations (CNVs) has potential to reveal underlying disease mechanisms*. Brain Res. 2015

Nov 12. pii: S0006-8993(15)00852-5. doi: 10.1016/j.brainres.2015.11.009. [Epub ahead of print].

[79] Vanuytsel, K., et al., *FANCA knockout in human embryonic stem cells causes a severe growth disadvantage.* Stem Cell Res, 2014. 13(2): p. 240–50.

[80] Martinez, R.A., et al., *Genome engineering of isogenic human ES cells to model autism disorders.* Nucleic Acids Res, 2015. 43(10): p. e65.

[81] Ding, Q., et al., *A TALEN genome-editing system for generating human stem cell-based disease models.* Cell Stem Cell, 2012. 12(2): p. 238–51.

[82] Soldner, F., et al., *Generation of isogenic pluripotent stem cells differing exclusively at two early onset Parkinson point mutations.* Cell, 2011. 146(2): p. 318–31.

[83] Yao, Y., et al., *Generation of CD34+ cells from CCR5-disrupted human embryonic and induced pluripotent stem cells.* Hum Gene Ther, 2011. 23(2): p. 238–42.

[84] Wang, Y., et al., *Genome editing of isogenic human induced pluripotent stem cells recapitulates long QT phenotype for drug testing.* J Am Coll Cardiol, 2014. 64(5): p. 451–9.

[85] Liu, Y. and W. Deng, *Reverse engineering human neurodegenerative disease using pluripotent stem cell technology.* Brain Res. 2016 May 1; 1638(Pt A):30-41. doi: 10.1016/j.brainres. 2015.09.023. Epub 2015 Sep 28.

Chapter 14

Modelling Neurodegenerative Diseases Using Human Pluripotent Stem Cells

Vanessa J. Hall

Additional information is available at the end of the chapter

http://dx.doi.org/10.5772/63521

Abstract

Neurodegenerative diseases are being modelled in-vitro using human patient-specific, induced pluripotent stem cells and transgenic embryonic stem cells to determine more about disease mechanisms, as well as to discover new treatments for patients. Current research in modelling Alzheimer's disease, frontotemporal dementia and Parkinson's disease using pluripotent stem cells is described, along with the advent of gene-editing, which has been the complimentary tool for the field. Current methods used to model these diseases are predominantly dependent on 2D cell culture methods. Outcomes reveal that only some of the phenotype can be observed in-vitro, but these phenotypes, when compared to the patient, correlate extremely well. Many studies have found novel molecular mechanisms involved in the disease and therefore elucidate new potential targets for reversing the phenotype. Future research that includes studying more complex 3D cell cultures, as well as accelerating aging of the neurons, may help to yield stronger phenotypes in the cultured cells. Thus, the use and application of pluripotent stem cells for modelling disease have already shown to be a powerful approach for discovering more about these diseases, but will lead to even more findings in the future as gene and cell culture technology continues to develop.

Keywords: Disease modelling, Alzheimer's disease, frontotemporal dementia, Parkinson's disease, pluripotent stem cells

1. Introduction

The ability for researchers to model diseases in a dish has accelerated during the past decade, thanks to the discovery of a new stem cell type, the induced pluripotent stem cell (iPSC). This is an artificially created cell that recapitulates all the features of embryonic stem cells (ESCs)

isolated from the early pre-implantation embryo. The production of this cell type in 2006 was a remarkable finding which led its founder, Shinya Yamanaka, to receive the Nobel Prize in Physiology and Medicine, just 6 years after its discovery, in 2012. The prize at that time was also shared with Sir John Gurdon who uncovered the mechanism of reprogramming in the late 1950s. These iPSCs were first produced from mouse fibroblasts by the transduction of four transcription factors, which when overexpressed, could completely change the fibroblast's phenotype into that of an embryonic stem cell-like cell, capable of forming all cell types in the body, upon differentiation [1]. Today, iPSCs are being produced from human cells and other species in many labs across the world, and production of these has been streamlined using more refined reprogramming techniques as well as different combinations of either genes, proteins, small molecules or miRNAs that can replace the function of the transgenes [2]. What makes these cells so useful for studying disease is that they can easily be produced from patients suffering from the disease (patient-specific iPSCs) and be differentiated into the cell type/s affected by the disease, thus modulating and mimicking the disease in a dish.

The ability to produce patient-specific iPSCs has a number of advantages for both learning more about the disease itself and also in improving therapies and treatments. The ability to produce autologous stem cell populations from easy-to-access cells from the patient (e.g. blood cells and skin biopsies) overcomes the ethical conundrum of having to first produce a cloned human embryo using a donor cell from the patient and host de-nucleated oocyte and then having to destroy the cloned embryo, to harvest the pluripotent ESCs within [3]. This in itself is an enormous breakthrough. There are several benefits in being able to have autologous cells from the patient. In patients that have degenerative diseases (e.g. diabetes, heart disease, osteoporosis, atherosclerosis and varying neurodegenerative diseases), the potential opportunity to have healthy cells transplanted back into the site affected is particularly appealing. The patient's own cells can in fact also be corrected, in cases where genetic mutations induce the disease pathology. Alternately, autologous iPSCs derived from patients can also be used to improve the patient's own medical treatment. In this case, the iPSC-derived cells can be screened in-vitro to determine which drugs prove most beneficial for the patients. This is one aspect of many approaches for developing tailored-specific treatments for patients, known as 'personalized medicine'. The iPSCs, when differentiated into the target cells affected in the disease, can also be used to screen the potential new drugs being developed by Pharma, or potentially even used to discover new biomarkers of the disease. One of the latest developing fields in medical research includes the development of nanoparticles for treating disease, which are particularly attractive for use in brain diseases as they may pass easily through the blood-brain-barrier [4].

Despite the forefront in iPSC research, human ESC research is still in practice today for modelling neurodegenerative disease. Cell lines can be gene-targeted to induce familial-linked mutations, and in this way can be compared to genetically matched, unmodified control cell lines which are similar, if not more stringent controls than isogenic controls produced from iPSCs (see section on gene-editing below). Human ESCs are derived following the culture of the inner cell mass isolated from a pre-implantation embryo [5]. Hundreds of lines have been produced over the years for research purposes for the study of cell pluripotency and regen-

http://dx.doi.org/10.5772/63521

eration, and these can be easily sourced from stem cell banks, registries or commercial companies. Generally, human ESCs are non-autologous, unless derived by somatic cell nuclear transfer from the patient. Only a handful of studies have produced autologous human ESCs from patients with disease [6, 7], and none exist at present for neurodegenerative diseases. Reasons for this are likely due to the ethical dilemmas in producing cloned human embryos and the technical challenges in cloned embryo production, compared to the ease in production of iPSCs.

Neurodegenerative diseases are characterized by progressive dysfunction of the nervous system as a result of loss of neuronal function in either the brain or the spinal cord and include Alzheimer's disease (AD), frontotemporal dementia (FTD), Parkinson's disease (PD), Huntington's disease, amylotrophic lateral sclerosis and multiple sclerosis. Pluripotent stem cells (PSCs) have shown to be of particular promise for studying these diseases, since they can be expanded exponentially, hence providing much cell material for study. This is useful since it is particularly difficult to obtain tissue from the brain from patients suffering from the disease. In this review, we focus on the use of both iPSCs and PSCs in modelling AD, FTD and PD. In order for PSCs to deliver on their promises, it is important that clinical grade and safe cells can be produced for potential cell therapy. It is also important that these cells can modulate the disease accurately in the dish. That is, the cells must show the same pathology linked to the disease. In this review, we focus on how well iPSCs can model disease in a dish. We discuss how far the field has come in correcting the familial forms of AD, FTD and PD and how important the corrected mutations are for these diseases in relation to both the in-vitro studies and the potential for future cell therapy. Finally, we discuss what more is required to improve modelling in a dish, and where the current research is heading.

2. Use of gene-editing in modelling disease

Gene-editing involves insertion, deletion or replacement of DNA in the genome of an organism using engineered nucleases. This field has advanced considerably in just over a decade, thanks to the discovery and application of nucleases, combined with the latest molecular technology, which both enhance and improve the editing process. Gene-editing using designer nucleases was first applied to PSCs in 2007 [8]. Since then, its application and use on PSCs have become widespread. It is currently being used by researchers to correct disease-causing mutations (endogenous gene correction) found within patient-specific iPSCs. In the case of PD, corrected autologous iPSCs through gene-editing are particularly promising for future cell transplantation studies, where diseased cells are genetically corrected and transplanted back into the patient's brain. Another application for gene-editing in modelling disease is to produce genotypically matched control cell lines of disease iPSC lines. That is, because comparisons with age-matched healthy control lines from non-related persons are genetically different, which may impede on both phenotype and even differentiation capabilities [9]. This use of gene-editing technology thus enables eloquent comparable studies in-vitro of disease phenotypes, which is directly linked to the mutation *per se* and not to other compounding factors. One other application is to use gene-editing for gene knockout studies [10]. PSCs are especially amenable for gene-editing, since they can be cultured in-vitro for a very long duration without changing either their genotype or their phenotype. To date, three custom-engineered nuclease technologies have been developed. This technology shares a common background. That is, all introduced engineered nucleases are able to introduce double-strand breaks (DSBs) in the DNA, which trigger DNA repair either via an error-prone non-homologous end joining (NHEJ) or via the preferred route of precise homology-directed repair (HDR) [9].

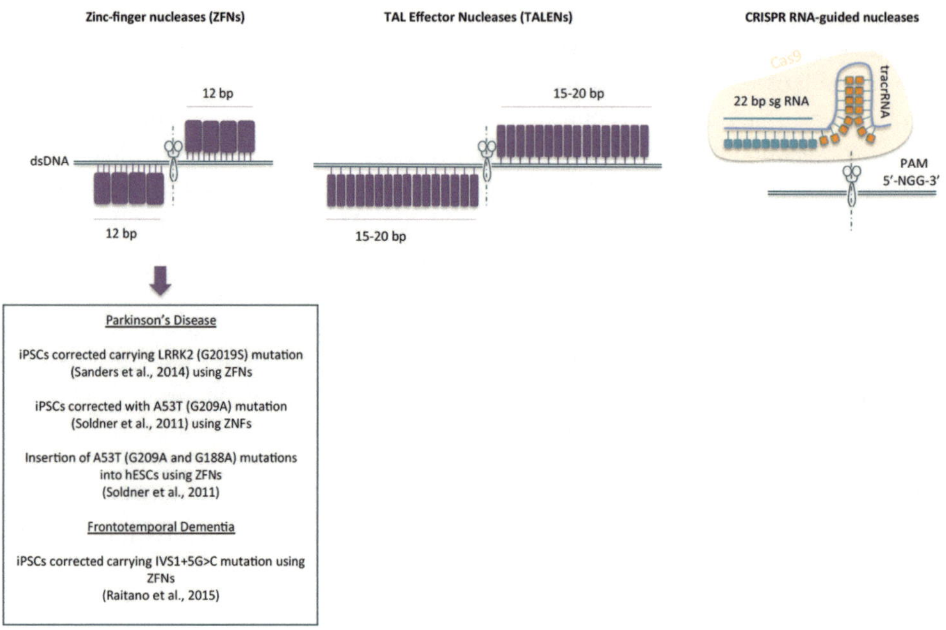

Figure 1. Three custom-engineered nuclease technologies exist, including zinc-finger nucleases (ZFNs), transcription activator-like effector nucleases (TALENS) and clustered, regularly interspaced, short palindromic repeats (CRISPR) RNA-guided repeats, although only ZFNs have so far been used to correct point mutations in two neurodegenerative diseases.

The first generation of engineered nucleases produced were the zinc-finger nucleases (ZFNs), which were initially developed as chimeric restriction enzymes [11]. These are modular proteins containing a *FokI* endonuclease catalytic domain fused to several Cys_2-His_2 zinc finger (ZF) DNA-binding motifs [12]. They function as dimers, with each monomer consisting of a non-specific cleavage domain from the *Fok1* endonuclease, fused to a ZF array that is designed to bind to the target sequence of interest [13]. Each ZF domain consists of a 3bp subsite that can be constructed into monomers that recognize up to 24bp of the target site (**Figure 1**) [13]. Several studies have shown that ZFNs can target endogenous genes in both human ESCs and iPSCs with variable efficiency (from >1% up to 94%) [14]. In addition, they have been used to target and insert gene cassettes within the *AAVS1* locus in both human ESCs and iPSCs, which is a commonly targeted locus for long-term stable transgene expression in mammalian cells [14]. ZFNs can be designed and produced using two different methods, including modular assembly (mix-and-match combination of several individual pre-characterized ZFs) and cell-based selection using a public platform called 'OPEN' provided by the ZF consortium [15,

16]. The length and design of the ZFs, as well as unwanted homodimer binding and cleavage, can result in off-target binding and cleavage of DNA, which can lead to unwanted alterations of the genome elsewhere. Thus, to overcome this, newer strategies have been developed such as 'obligate heterodimers', which include modified *Fok1* nucleases, which only cleave when heterodimers form [16] as well as another manipulated form of *Fok1*, called zinc finger nickases, which stimulate HDR and produce fewer off-target effects [17].

Another gene-editing tool is the transcription activator-like effector nucleases (TALENs). These are composed of a sequence-specific DNA-binding domain and a non-specific DNA cleavage module [18]. The DNA binding domain contains a series of tandem repeats comprising 33–35 amino acids, similar to tandem repeats first discovered in the plant pathogen *Xanthomonas* [19]. The DNA recognition is conferred by the highly variable amino acids at positions 12 and 13 [20]. Like ZFNs, TALENs form dimers on either side of the DNA strand and use the non-specific cleavage effect of the *Fok1* cleavage domain to produce a DSB (**Figure 1**) [20]. TALENS can be designed to almost any sequence due to their simple protein-DNA code. The only requirement is for the presence of thymine at each 5' end of the DNA recognition site [10]. TALENS generally also have fewer off-target cuts due to their longer recognition motifs and they are also less cytotoxic when compared to ZFNs, which is an attractive feature of this technology [10, 20]. To date, TALENs have been used for generating gene reporter lines, biallelic knock-out of genes and repair and introduction of point mutations in human PSCs [9].

The most recently developed method, which is even easier to use than TALENs and ZNFs is clustered, regularly interspaced, short palindromic repeats/Cas9-mediated genome-editing method (CRISPR/Cas9). This method consists of a specialized two-RNA structure containing CRISPR RNA (crRNA) and trans-activating RNA (tracrRNA), which are able to bind as a monomer to DNA strands next to a protospacer adjacent motif (PAM) composed of the sequence, 5'NGG'3 [18]. This chimeric RNA is known as a guide RNA (gRNA) and facilitates a DSB by guiding an endonuclease (first derived from *Streptococcus pyogenes* (Cas9)) to induce cleavage near the PAM site (**Figure 1**). The target recognition site is also typically 22 bp, which is shorter than the recognition site of TALENs. The CRISPR/Cas9 system has already been used successfully in gene-editing PSCs and has been shown to have a higher efficacy when compared to TALENs and ZFNs for gene knock-out studies [21]. It is also comparable to TALENs for HDR-mediated gene-editing of PSCs [22]. Importantly, for precise editing of point mutations (both deletions and insertions), dependence on HDR is required and use of an exogenous DNA template is needed such as single-stranded oligodeoxynucleotides or plasmid DNA templates [9]. An advantage of both TALENs and CRISPRs is that they do not leave a trace in the DNA, following the genome-editing process. Despite these advantages, there are some drawbacks in the case of CRISPRs. Some constraints exist in the target design, due to requirement for a PAM motif in the target site. Another potential disadvantage is the potential increased off-target binding and cleavage compared to TALENs, due to its monomeric action, shorter target sequences and greater chance of binding to identical target sites elsewhere in the genome [9, 23]. Like ZFNs, nickases have been effectively used to prevent off-target binding and cleavage [18].

To date, despite a surge in literature in gene-editing technologies, only three reports have been published that have led to the correction of iPSCs from patients with neurodegenerative diseases or alternately, insertion of disease-causing mutations into healthy PSCs. Two of these have been in the field of PD, and one in the field of FTD and all cases used ZFNs [24, 25] (**Figure 1**). In the case of PD, insertion of point mutations, A53T and G188A located in the gene, α-synuclein (*SNCA*), which leads to familial onset of PD, was successfully achieved in ESCs [24]. In the same study, repair of the A53T and point mutation in *SCNA* mutation was also successfully performed in patient-specific iPSCs. Genome-wide analyses did not reveal any off-target effects following the ZFN targeting in all engineered lines. A follow-up study of the repaired A53T iPSCs showed that defects in dopaminergic neurons and mitochondrial dysfunction originally observed in the patient-iPSC lines could be reversed in the corrected iPSCs [26]. This mitochondrial dysfunction, which led to apoptotic death, was induced by nitrosative stress and could be directly linked with impaired functioning of the MEF2C-PGC1α pathway. This research was therefore unable to uncover a novel pathway, which could be used for the development of new drugs. Another study was able to correct the G2019S LRRK2 mutation in PD patient-specific iPSCs and rescued phenotypic and genotypic dysfunction [27]. This study was also able to find a new molecular pathway responsible for the phenotypes found in-vitro. In this case, ERK signaling was dysfunctioning and could be repaired through correction of the mutation. In the field of FTD, one recent study revealed that iPSCs derived from patients carrying a familial inherited mutation in the progranulin gene (*GRN*) were unable to form cortical neurons in-vitro. However, this phenotype could be restored when the mutation was corrected using ZFNs [28]. This helped to verify that the phenotype, which had not been described before, was directly related to the mutation and disease and not due to other unrelated, technical or cell culture factors.

Together, these studies show that gene-editing is a particularly helpful tool for modelling diseases with iPSCs, as well as for helping to determine new molecular pathways involved in the disease by repair of the mutation and assessment for amelioration of the phenotype. It is likely that an increase in similar studies will soon emerge for PD and FTD as well as other neurodegenerative diseases, such as AD in the near future.

3. Current approaches to developing neural cells in a dish

Investigations on iPSC-derived neural cells can be performed either in two-dimensional (2D) models or three-dimensional (3D) models. Traditional stem cell research has been performed in 2D, predominantly by culturing cells that adhere to a plastic surface, and which form a flattened monolayer across the plastic surface. The advantage of this technique is that it is low cost and an easy system to use. Stem cells, however, can alternately be cultured in 3D. One popular 3D method is to culture cells in small spheres in suspension within the media, termed either 'embryoid bodies', or to differentiate them into neural progenitor cells that also cluster together in small spheres, termed 'neurospheres'. These can be further differentiated into more mature neural cell types, which are often termed 'engineered neural tissue' or 'organoid' cultures. The cells can be cultured as described in aggregates, but can also be cultured in either the presence of microcarriers, on alginate microencapsulates, in thermoreversible hydrogel or in scaffolds [29]. It is accepted today that neural stem cells (NSCs) isolated from primary tissue from foetal tissue or brain differ to neurospheres differentiated from EBs, as the former spheres tend to contain radial glial-like stem cells that are unable to form complex neural tissues such as the layered cortical neuroepithelium and complex pattern formations [30]. In contrast, the stem cell-derived neurospheres can be instructed to form specific neural regions of the developing brain when exposed to potent mitogens/morphogens [30]. Alternately, neural cells can also be cultured within artificially produced 3D scaffolds or in microwells formed within the plastic substrate that help to re-create a microenvironmental cue for the cells to form in 3D clusters. In fact, 3D cultures originated in the NSC field in the early 1990s when the first suspension cultures of rodent brain NSCs was performed [31], and have become a standardized way of culturing NSCs in-vitro in labs across the world. Today, there are several types of 3D scaffolds available, including metal, synthetic organic types made from polymers, synthetic inorganic materials, natural organic materials, natural inorganic material types and even nanostructure scaffolds [29]. All of these have their advantages. For example, microcarrier systems allow for good diffusion properties and induce cells of high quality. Also, encapsulated cells in gels allow them to be protected from shear force-induced cell death, and thermoreversible hydrogel allows for rapid expansion of cells [29].

Comparative studies of 2D versus 3D cultures suggest that 3D culturing may improve the quality of the cell expression profile of the cultured cells as cells are influenced by the biochemical, mechanical and physical surface properties of the surrounding matrix in which they normally reside [32]. One such comparative study showed neural-derived ESCs expressed more neural markers and greater neurite outgrowth when cultured in a 3D scaffold than the equivalent neural cells cultured in 2D [33]. Furthermore, timing in differentiation appears to differ between 2D versus 3D cultures. In fact, stem cells appear to differentiate earlier in 2D culture when compared to culture in extracellular matrix gel or as spheres, shown by the earlier upregulation of differentiation markers [34]. Whether this is abnormal or not has not yet been determined. Cell size and proliferation can also be altered by culture in 3D. One study has illustrated human ESCs cultured in 3D within microwells were smaller in size and divided more slowly compared to equivalent cells grown in 2D [35].

There has also been a recent surge in developing 3D models that better recapitulate the 3D complexity of the tissue in the body and which contain several cell types. The recent discovery that a human foetal-like brain could be recapitulated in the dish after culture of neural stem cells for several weeks was a remarkable discovery. This tissue was termed as a 'cerebral organoid' and was formed by embedding neural aggregates into Matrigel® droplets and culturing these in a bioreactor for 75 days [36]. The tissue contained both early-born and late-born cortical neurons, suggesting more complex cortical neural development could be recapitulated in-vitro. It also contained interneurons, suggesting a mix of different progenitor origins were present in the tissue. Other researchers have also produced complex neural tissue with cortical layer patterning in 3D neural cultures, which depict both proliferative cell populations and post-mitotic cortical cell types; however, the complex stratification of the cortical layers has not yet been replicable [30]. The addition of extracellular matrix molecules to both the substrate of 2D and within 3D culture systems may also be particularly advantageous for the growth and cellular expression of neural cell types, as shown by Lancaster and colleagues where neural aggregates were cultured in Matrigel® droplets [36].

There appears to be improvement in the cellular expression and cell function when cultured in 3D, as well as other physical and changes in size and growth. However, some drawbacks in using 3D scaffolds are the difficulties of performing molecular analyses on the tissue, which are related to problems in extracting the cells from the scaffolds or light refraction that emanate from the scaffold structures and which interfere with fluorescence microscopy. In addition, there are also seeding issues related to the complexity of some scaffold structures. Furthermore, 3D culturing is more labour-intensive and can also be difficult to scale up. Bioreactors help, in part, to solve this issue when cells are grown in spheres or small scaffolds and they also help through their spinning properties to distribute medium evenly throughout the culture. Use of bioreactors, however, requires extensive volumes of media, which can be costly when large volumes of cytokines or growth factors are required in the culture medium.

Despite the given advantages in use of 3D culture over 2D, disease modelling studies for AD, PD and FTD using PSCs have been performed using 2D cultures. One interesting research article relevant for AD demonstrates the advantages of producing 3D cultures. Choi and colleagues showed that transgenic human NSCs overexpressing either *APP* or *PSEN1* could be differentiated into 3D tissue, which contained the classical hallmarks of the disease, including amyloid plaques and aggregates of phosphorylated tau (p-tau) [37]. These hallmark pathologies have not yet been demonstrated in iPSC models of AD. What seems apparent is that 3D modelling may recapitulate the in-vivo environment better. Use of more complex models might be better for modulating and studying the brain and the disadvantages of 3D modelling described are outweighed by the advantages.

4. Modelling Alzheimer's disease using pluripotent stem cells

Alzheimer's disease is the most prevalent type of dementia, which in most cases (approximately 90%) arises in patients with no known genetic link. However, some risk factor genes (e.g. presence of the allele ε4 of apolipoprotein E4 (*APOE* ε4) appear to play a role in more than half of these cases [38]. See section under "Sporadic cases with mutations in APOE", for more details. The disease induces loss of memory and impairs cognitive function, but is also known to induce loss in olfaction, hearing and even some motor function [39]. Depression, agitation, apathy, social withdrawal, insomnia, delusion and emotional/physical outbursts are all typical symptoms associated with AD. Ultimately, the disease culminates in loss of respiration function, leading to death. The disease is induced by protein aggregation and has two distinct histopathological signatures, which lead to neural degeneration of the tissue. The first includes the development of extracellular amyloid plaques, which are clusters of overproduced toxic forms of amyloid-beta (Aβ) peptides, deposited outside of the cell. The common theory accepted today is that these plaques develop first, prior to the development of the second signature, the intracellular neurofibrillary tangles (NFTs), which are composed of a predominant protein, tau in a hyperphosphorylated state. The disease spreads throughout the patient's brain and is present pathologically years before the first symptoms appear. It arises first in the periallococortical transentorhinal region of the temporal mesocortex before spreading to the entorhinal cortex and then the hippocampus before later spreading to many regions of the brain, including the temporal neocortex, the insular cortex, the medial temporal gyrus and superior temporal gyru and then the occipital lobe [40]. Since many parts of the brain are affected, modulating the disease in a dish is a difficult task. In addition, the likelihood of the success of stem cell therapy is very low considering the widespread nature of the pathology. Instead, there is more interest in understanding AD pathogenesis and for developing new and more effective therapies by using PSCs [41]. Most researchers model the cortical tissue, which is affected later on in the disease; however, NSCs derived from PSCs have also been studied. These progenitor cells are attractive as they might reveal early mechanisms of the disease.

4.1. Sporadic AD with no known genetic mutations

The main pathology of sporadic AD is typical for the disease with the accumulation of toxic forms of Aβ peptides and tau protein. However, other proteins are also known to accumulate, including *SNCA, TDP-43* and *ACTIN* which form Lewy-like bodies and Hirano bodies [42]. In sporadic AD, the symptoms occur later than in the familial forms [42]. However, the complexity and heterogeneity of both symptoms and pathology are widespread in sporadic AD, which may be due to multiple unknown risk factors, mutations in varying genes which have not yet described, but potentially also the diverse effects of environmental factors and potential interaction with genes [42]. To date, three studies have been performed to modulate sporadic AD using patient-specific iPSCs. From these studies, no clear pathology for altered processing APP was observed. The patient iPSC-derived neurons did not reveal a change in the levels of either Aβ40 or Aβ42 [43, 44]. Tau pathology was reported only from one of the studies, with increased p-tau and enlarged endosomes observed in neurons, but only in one of the analysed patient's cell lines [43]. Other pathology, however, was reported. One study found altered WNT signalling and glutamate metabolism in mixed cell cultures from one patient, as well as, gene expression changes related to proteosome function, reactive oxygen species (ROS) and cell death [45]. Another study found elevated oxidative stress-related gene expression and also elevated ROS in the iPSC-derived cortical neurons [44]. Interestingly, this study that consequently found no altered expression in Aβ expression in the neurons found elevated Aβ in iPSC-derived astrocytes.

4.2. Sporadic cases with mutations in APOE

The polymorphism of the *APOE* gene is a risk factor for the disease and presence of the ε4 allelle (*APOE4*) has been linked with AD [42]. One copy of this allele increases the risk of AD twofold, whereas two copies increases the risk of AD by 12-fold [46]. The ε4 allele has been shown to be less efficient in transporting cholesterol from neurons [42]. One iPSC study performed on patients with a APOε3/ε4 genotype found elevated Aβ42 in neurons derived from two of the three patients studied. Two of the patients also showed increased cell stress following glutamate-induced excitation [46]. No investigation of cholesterol transport was performed, which could have been useful to confirm this specific phenotype related to *APOE* dysfunction. There are also no iPSC studies to date performed on patients with the Apoε4/ε4 genotype.

5. Familial AD

Pathology common to all familial mutations includes an earlier onset of the disease and increased amyloid plaque formation when compared to sporadic AD in many, but not all cases [47]. Plaques tend to predominantly contain Aβ42 with often no increase in Aβ40 observed, contrasting that seen in sporadic AD [47]. A summary of the PSC-derived neural cell pathology and comparative pathology known in familial AD patients is summarized in **Table 1**.

Gene	Mutation	Pathology in Patients	Study	Pathology in PSCs	Study
PSEN1	A246E (ex7)	Unknown		Increased Aβ42:Aβ40 No increase in Aβ42	Yagi et al., 2011; Liu et al., 2014; Mahairaki et al., 2014 Duan et al., 2014
	D385N	Unknown		Decreased Aβ40	Koch et al., 2012
	L166P (ex6)	Amyloid plaques Cotton wool plaques	Moehlmann et al., 2002	Decreased Aβ40	Koch et al., 2012
	H163R (ex6)	Increased Aβ42 Lewy bodies Atrophy of substantia nigra and cerebral cortex Cotton wool plaques Corticospinal degeneration SNCA deposits Neurofibrillary tangles Amyloid plaques	Ishikawa et al., 2005,84	Increased Aβ42:Aβ40	Liu et al., 2014
	M146L (ex5)	Amyloid plaques Pick bodies	Halliday et al., 2005	Increased Aβ42:Aβ40	Liu et al., 2014
PSEN2	N141I	Increased Aβ42:Aβ40 *SNCA*deposits Amyloid Plaques Neurofibrillary tangles Some cases with Lewy body pathology in frontal cortex and amydala	Levy-Lahad et al., 1995; Rogaev et al., 1995 Jayadev et al., 2010	Increased Aβ42:Aβ40	Yagi et al., 2011

Gene	Mutation	Pathology in Patients	Study	Pathology in PSCs	Study
APP	duplication	Intercerebral haemorrhage, Diffuse brain atrophy Cerebral ventricular dilation, Intraneural Aβ40 Cerebellar purkinje cell atrophy Amyloid plaques	Cabrejo et al., 2006	Increased Aβ40 Increased p-tau (Thr231)	Israel et al., 2012 Israel et al., 2012
	E693Δ	Early stages—limited brain atrophy and little accumulation of Aβ. Cerebellar ataxia Aβ oligomers No plaques	Shimada et al., 2011	Decreased Aβ40 and Aβ42 Elevated Aβ oligomers Oxidative stress	Kondo et al., 2013
	V717L	Unknown		Increased Aβ42:Aβ40 Oxidative stress	Kondo et al., 2013
	V717I	Tau pathology Neurofibrillary tangles Plaque neurites Neuropil threads Cortical and subcortical Lewy bodies	Lantos et al., 1992	Increased Aβ42:Aβ40 Increased Aβ38 Increased total tau Increased p-tau (S262)	Muratore et al., 2014

Abbreviations: Aβ - amyloid beta; ex - exon; p-tau - phosphorylated tau; *SNCA* – alpha synuclein gene

Table 1. Pathology in patients with familial Alzheimer's disease patients and respective pluripotent stem cell (PSC) studies.

5.1. PSEN1 mutations

Over 170 mutations in *PSEN1* have been described, making this the most common cause of autosomal dominant early onset AD [48, 49]. Patients with mutations in the *PSEN1* locus have the earliest age of onset (AOO). These patients often have seizures, myoclonus, paraparesis and cerebeller signs [50]. For all *PSEN1* cases, classic hallmark pathologies are observed, including amyloid plaques, NFTs, tissue atrophy, neuronal loss and inflammation [47]. However, pathological differences do exist depending on the gene affected, and even depending on the location of the mutation within the gene. For example, Lewy bodies and other parkinsonian pathologies have been described in *PSEN1* H163R carriers, and pick bodies have been described in M146L carriers (see **Table 1**). Pathological mechanisms common to both *PSEN1* and *PSEN2* are altered Aβ peptide metabolism induced by disrupted γ-secretase cleavage, which results in increased Aβ42 production. In some *PSEN1* mutations, severe neurodegeneration has been described without any Aβ pathology, which contradicts the hypothesis that amyloid pathology arises prior to tau pathology. Other pathological differences described include some plaques containing predominantly Aβ40. These Aβ40 plaques have been observed in the cortex of some *PSEN1* mutation patients [47]. Studies have also highlighted that soluble and insoluble levels of Aβ42 is higher in familial AD brain tissue compared to sporadic AD [47]. It is thus apparent that given the diversities in the patient's pathology, a similar diversity in iPSC-derived neurons might also be evident.

The majority of iPSC lines modelling AD have in fact been produced from patients carrying mutations in *PSEN1*. These studies have revealed that most of the neurons generated from *PSEN1* AD patients also have increased Aβ42 [51–54], although decreases in Aβ40 have also been described in two particular mutations [55]. However, little other pathology has been described. In one study, no tau accumulation or tangle formation was observed [51]. One other study has reported small changes in gene function in *PSEN1* mutant iPSC-derived neurons, including increased *NLRP2, ASB9* and *NDP* [53]. In addition, overexpression of *PSEN1* in human ESCs led to increased Aβ42/Aβ40 and Aβ43/Aβ40 ratios as well as synaptic dysfunction in neurons expressing NeuN and BIII-tubulin [56].

5.2. PSEN2 mutations

There are 23 known DNA variants reported in the *PSEN2* gene. Patients with mutations in *PSEN2* have a delayed AOO, suffer from disorientation and endure a long duration of the disease [50]. Pathologically, similar to *PSEN1*, mutations affect Aβ peptide metabolism by γ-secretase cleavage which results in Aβ42 production.

To date, only one study has investigated the pathology from neurons derived from a patient carrying the N141I mutation in *PSEN2*. This study reported increased Abβ2:Aβ40, but no tau accumulation or NFT formation [51]. Interestingly, some patients with this mutation do also have Lewy body pathology and SNCA deposition within neurons [49]. Clearly, more research is required to investigate Aβ metabolism, γ-secretase function and whether neural atrophy, inflammation, tau, NFT and even parkinsonism pathology can be observed in-vitro from patient cells carrying this mutation.

5.3. APP mutations

Patients with *APP* mutations frequently are more aggressive and those with an *APP* duplication frequently have apraxia [50], cerebellar dysfunction and some cases have cerebral haemorrhaging [57]. Different *APP* mutations induce neural death by different mechanisms. For example, some mutations induce an increased production of AICD and other C-terminal APP fragments directly regulate apoptosis [47]. Other mutations affect intracellular mechanisms which increase oxidative stress and death [47]. Studies have also highlighted that soluble and insoluble levels of Aβ42 are reportedly also higher in *APP* mutation cases compared to sporadic AD cases.

Three studies have investigated the pathology in neurons derived from patients carrying mutations in *APP*. These studies have conflicting results in relation to *APP* processing, likely related to diverse pathology dependent on the mutation. One study found elevated Aβ40 [43], another study found one patient had increased Aβ42 whilst a second patient had decreased Aβ40 [44], and a third study revealed a patient with both increased Aβ42 and Aβ38 [58]. Two

http://dx.doi.org/10.5772/63521

of these studies showed good correlation with known patient pathology, whilst the study reporting increased Aβ42:Aβ40 levels cannot be validated to the patient pathology since this mutation has not been well characterized. One study also reported an increase in Aβ oligomers in the analysed neural cells and astrocytes [44], which again correlates well with the patient's phenotype [59]. Increased p-tau has been observed in patients from two different studies [43, 58] and total tau has also been reported [58]. Increased αGSK-3β has also been reported [43]. No studies have looked at AICD function; however, oxidative stress has been partially investigated in one study, which revealed elevated ROS and oxidative stress-related genes in the cortical neurons and also elevated ROS in astrocytes [44]. No pathology related to cerebellar dysfunction or parkinsonism-related pathology described in some of these mutations has been reported.

6. Other models of AD

6.1. Trisomy 21

Trisomy 21 (also known as Down's syndrome) results in the duplication of the APP gene, which is located on chromosome 21. People with this syndrome develop symptoms and pathology early in life, which are strikingly similar to AD [60]. The extra copy of APP is

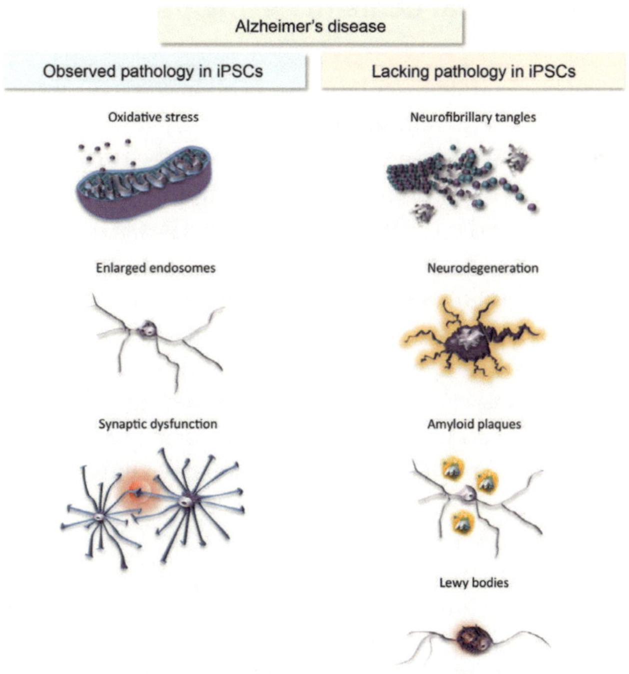

Figure 2. A summary of pathology observed in published induced pluripotent stem cell (iPSC) models of Alzheimer's disease (AD) and pathology found in AD patients but currently lacking in the iPSC models.

considered the major factor in the AD-like symptoms and, in addition, duplication of Dyrk1A kinase (which is also located on chromosome 21), which phosphorylates tau, may also contribute to the pathology and symptoms [60]. Increased Aβ peptides can be observed in early childhood which are the main candidate thought to induce the early onset of dementia [60]. Since duplications of APP are observed in AD patients, trisomy 21 has also been used to model AD in many studies. One study has produced iPSCs from patients with trisomy 21 and neurons derived from the iPSCs showed perturbed Aβ processing, including increased Aβ40 and Aβ42 in long-term cultured neurons, as well as Aβ42 intracellular and extracellular aggregates [60]. Furthermore, this study also reported increased p-tau and total tau, as well as increased cell death [60]. This is the only study to date which reports cell death in an iPSC model of AD, which suggests this may be a relevant and worthy model of APP duplication and study of AD-like dementia.

To conclude, iPSCs from AD models tend to show early features of the disease in the dish, rather than distinct histopathological hallmarks (**Figure 2**). The most common observations include altered expression levels of Aβ and increased levels of tau. It might be that the main pathological hallmarks only develop after many years of protein aggregation and build-up in the cell.

7. Modelling frontotemporal dementia using pluripotent stem cells

Frontotemporal dementia accounts for a large proportion (50% of dementia cases that arise before the age of 60, and is the second most common early-onset dementia). This disease is characterized by the progressive loss and degeneration of the cortical neuron population, in the frontal and temporal lobes of the brain. Common symptoms include altered behaviour and deterioration in both speech and cognition [61, 62]. This disease has a much stronger genetic link than AD and PD. Approximately 40% of the cases are attributed to mutations in one of three genes, including microtubule-associated protein tau (*MAPT*), progranulin (*GRN*) and *C9ORF72* [28]. However, many other genes carrying mutations have also been linked to the disease, including, valosin-containing protein (*VCP*), charged multivesicular body protein 2B (*CHMP2B*), ubiquilin 2, transactive response DNA-binding protein (*TDP-43*), dynactin (*DCTN1*) and fused in sarcoma (*FUS*) [63]. The large genetic diversity is reflected by diverse symptoms and pathology amongst the patients that differ from the common symptoms. Hence, the disease is stratified into a behavioural variant of FTD (bvFTD), two language variants (semantic dementia and progressive nonfluent aphasia (PNFA)) and an overlap of these with atypical parkinsonian disorders corticobasal syndrome (CBS) and progressive supranuclear palsy (PSP) [63]. In addition, in some cases of FTD, shared pathological features with motor neuron disease (MND)/amyotrophic lateral sclerosis (ALS) are observed, including the accumulation of proteins TDP-43 and FUS [63]. Thus, another variant (FTD-MND/ALS) is described and is due to mutations in *VCP* and *CHMP2B, C9ORF72* and *UBQLN2*, which can lead to the development of either FTD or ALS [64]. A summary of the familial FTD studies modelled using PSCs and respective known pathology from the patients is shown in **Table 2**.

http://dx.doi.org/10.5772/63521

Gene	Mutation	Pathology in Patients	Study	Pathology in PSCs	Study
C9ORF72RF72	expansion repeat	Brain atrophy Hippocampal sclerosis TDP-43 accumulation Argyrophilic grain disease in limbic areas and orbital frontal cortex Tau pathology Neurofibrillary tangles Atrophy also in substantia nigra, brain stem and spinal cord Lewy bodies RNA foci	Shinagawa et al., 2014 DeJesus-Hernandez et al., 2011	Cellular stress RNA foci Glutamate excitotoxicity	Almeida et al., 2013 Almeida et al., 2013; Donnelly et al., 2013 Donnelly et al., 2013
MAPTT	ex10 N279K FTDP-17-1	Hyperphosphorylated tau in DA neurons and glia and in brain stem and temporal cortex Neurofibrillary tangles Increased 4R tau isoform	Ehrlich et al., 2015; Wren et al., 2015	Increased expression of 4R tau isoform, increased tau fragmentation Neurite shortening Oxidative stress Cellular stress Enlarged vesicles Early maturation Altered axonal mitochondrial transport	Ehrlich et al., 2015: Iovino et al., 2015 Ehrlich et al., 2015 Wren et al., 2015 Iovino et al., 2015
	Ex12 V337M FTDP-17-2	Frontotemporal atrophy Moderate parietal cortical atrophy Hippocampal atrophy Astrogliosis Atrophy of substantia nigra Tau pathology	Domoto-Reilly et al., 2016	Increased tau fragmentation Neurite shortening Oxidative stress	Ehrlich et al., 2015
	P301L	Frontotemporal atrophy Astrogliosis Atrophy of substantia nigra Tau pathology Pick bodies	Spillantini et al., 1998	Early maturation Altered axonal mitochondrial transport *SNCA* deposition 4R tau	Iovino et al., 2015
GRN	IVS1+5G>C	Frontotemporal atrophy Caudate nucleus atrophy Substantia nigra atrophy Gliosis	Brouwers et al., 2007	Impaired corticogenesis Impaired WNT signalling	Raitano et al., 2015

Gene	Mutation	Pathology in Patients	Study	Pathology in PSCs	Study
		Ubiquitin inclusions containing TDP-43 Lewy bodies Amyloid plaques			
	S116X	Unknown		Cellular stress	Almeida et al., 2012
TARDBP	A90V	Unknown		Staurosporine-induced cellular stress	Zhang et al., 2013
	M337V	TDP-43 accumulation	Tamaoka et al., 2010	Decreased survival	Bilican et al., 2012

Abbreviations: *SNCA* – alpha synuclein gene; TDP-43 – TAR DNA-binding protein 43 gene

Table 2. Pathology in familial frontotemporal dementia patients and respective pluripotent stem cell (PSC) studies.

7.1. Sporadic FTD

Sporadic FTD has been modulated in-vitro by two independent studies to date [65, 66]. Brain atrophy is greater in the anterior cingulate compared to familial FTD cases [67]. In one study, no characterization of the phenotype was performed. Another study also characterized patient iPSCs from a sporadic case of FTD and the cultured neurons showed greater cellular stress and oxidative stress compared to control cells [66]. Whether all cases of sporadic FTD are in fact undiscovered genetic mutations is a distinct possibility, since there have been many recent discoveries of gene mutations lying behind the disease [68].

7.2. C90RF72 mutations

A GGGGCC repeat expansion in the noncoding region of *C9ORF72* is a common mutation for both FTD and ALS, representing approximately 40% of FTD cases and familial ALS [69]. This mutation leads to specific pathology including non-ATG-initiated translation of RAN peptides [69] and formation of nuclear RNA foci which lead to bvFTD with diverse phenotypes, even within the same family [68]. The AOO appears to be earlier in carriers of this mutation and have higher prevalence of delusional psychotic symptoms and hallucinations [70].

Two studies have produced patient-specific iPSCs from patients carrying this mutation. Characterization of the neurons derived from the iPSCs revealed they were more susceptible to cellular stress compared to control neurons [64]. In addition, RNA foci were observed in the in-vitro produced neurons [69], but there is controversial evidence that suggests patient neurons have these [64]. Both studies also showed cytoplasmic expression of RAN. Of interest was one other study that used patient-specific iPSCs to discover potential binding partners, e.g. ADARB2 to the expanded repeat region in an attempt to discover more about the mechanisms that lead to disease onset with this mutation [69] and to find out more about the gene's actual function.

7.3. MAPT mutations

Mutations in *MAPT* account for approximately half of the familial cases of FTD and tend to present symptoms typical of FTD, but may also include Parkinson's disease-like symptoms; therefore, these patients are termed frontotemporal dementia with parkinsonism related to chromosome 17 tau (FTDP-17T). There are at least 50 family kindreds carrying mutations in this gene, which includes nine missense mutations, one deletion mutation, two transition mutations within exons 9, 10, 12 and 13 and five intronic mutations leading to alternate splicing of exon 10 [71]. The missense mutations give rise to pathology very similar to AD and include the formation of NFTs, whereas those that lead to alternatively spliced exon 10 show progressive PSP, corticobasal degeneration and Pick's disease [71]. The N279K substitution is an intronic mutation and one of the most common types of FTDP-17T mutations, along with the mutation *P301L* [72]. The N279K mutation leads to overproduction of tau protein isoforms containing four tandem microtubule-binding domain repeats (4R-tau) which induces disease pathogenesis, including accumulation of neurotoxic tau aggregates and NFTs in both neurons and glia [73]. In addition, MAPT mutations (in particular the N279K mutation) are also associated with degeneration of the basal ganglia and depigmentation of the substantia nigra [71].

Two eloquent studies have generated iPSCs from patients carrying the N279K mutation and compared the in-vitro pathology directly to pathology in the deceased patient brain. Here, the cultured neurons had increased expression of the 4R tau isoform and fragmentation of tau [74], which helped to confirm the phenotype; however, neurite shortening, oxidative stress [74], cellular stress and enlarged vesicles [72] were also observed in the cultured neurons. The NFTs, however, were not able to be recapitulated in-vitro. Another study reported similar pathology in iPSC-derived neurons also carrying the N279K mutation [75]. In addition, this same study also compared the N279K iPSCs to iPSCs carrying the *MAPT* mutation P301L and found that they shared some cellular phenotypes and differed in others [75]. Specifically, neurons from both mutation backgrounds matured earlier compared to controls and had altered axonal mitochondrial transport, whereas the P301L iPSCs showed SNCA and 4R tau deposition in varicosity-like structures in the neurons.

7.4. GRN mutations

Patients carrying mutations in *GRN* tend to display Parkinson-like symptoms. Haploinsufficiency of GRN induces the disease and typically does not present with tau pathology, but instead, patients have cytoplasmic ubiquitin inclusions and intranuclear inclusions, comprising of TDP-43 both in neurons and in microglia [76]. They are often characterized as frontotemporal lobar degeneration (FTLD-TDP) and generally increased expression of *GRN* is associated with pathogenesis. However, patients tend to have varying expression of *GRN* in the brain at late stages of disease [77]. In another study, neurons were characterized from a patient carrying a novel nonsense mutation in *GRN* (S116X) [66]. The patient had Parkinson-like symptoms. The neurons were found to be more susceptible to cellular stress and oxidative stress compared to control neurons, which could be reversed upon induced expression of *GRN*

[66]. A follow-up study on the same neurons used these neurons to test different approaches of rescuing *GRN* expression as a way of finding new tools to treat the disease [78].

7.5. TARDBP mutations

Mutations in the *TARDBP* gene tend to result in aggregation and mislocalization of its protein TDP-43 to the cytoplasm [79] and many cases can be characterized as semantic dementia [80]. One study has produced patient-specific iPSCs carrying a M337V *TARDBP* mutation, and found during long-term culture that these cells had decreased cell survival [81]. A later study on these same cell lines found that application of two different autophagy-inducing molecules (which help to clear the accumulating TDP-43) could help to increase cell survival in these patient iPSCs [82]. Another study produced iPSCs from patients carrying the A90V mutation, and here cellular stress could be detected in the cells following exposure to staurosporine [83]. No other pathology was reported.

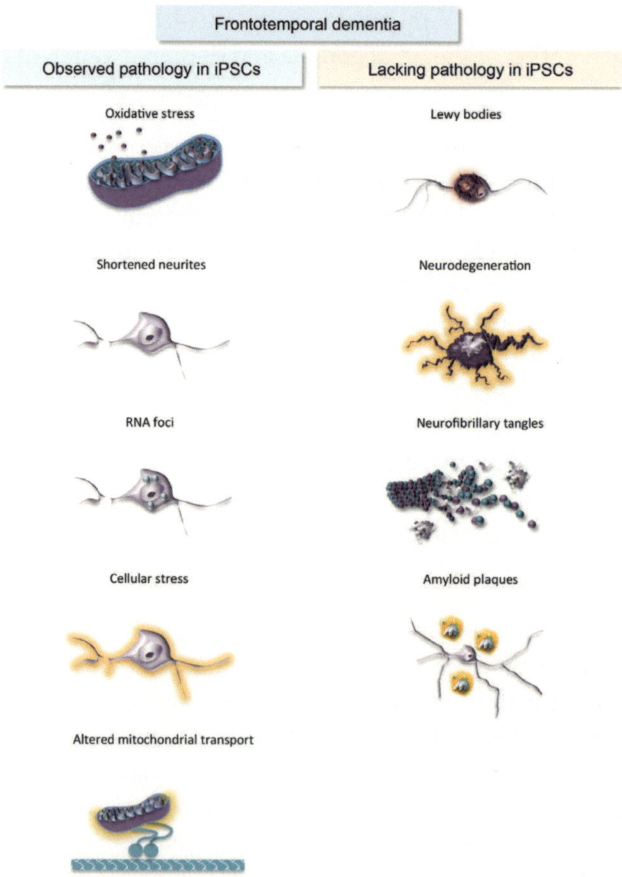

Figure 3. A summary of pathology observed in published induced pluripotent stem cell (iPSC) models of frontotemporal dementia (FTD) and pathology found in FTD patients but currently lacking in the iPSC models.

http://dx.doi.org/10.5772/63521

To date, only a fraction of the familial FTD mutations have been modelled using PSCs. Given the expanse of different mutations that exist as well as the broad pathology from each of the FTD variants, as well as within the variants themselves, it is important that the in-vitro studies can be correlated to the known pathology in the patients. Many different phenotypes can be observed in the dish, but the classical hallmarks appear to be missing in modelling the disease using iPSC-derived cells (**Figure 3**). How this might be improved upon is discussed more at the end of the chapter in the section under "Current limitations in modelling neurodegenerative disease using pluripotent stem cells".

8. Modelling Parkinson's disease using pluripotent stem cells

Parkinson's disease is the second most common neurodegenerative disease, which is both sporadic and monogenic in form. The inherited monogenic form accounts for the minority of cases with approximately 5–10% of presented cases. The genetic contribution to Parkinson's disease has been firmly characterized in the past few years to be directly induced by over 15 mutations in PARK loci, which are located within six genes: *SNCA*, *LRRK2*, *PARK2*, *PINK1*, *PARK7* and *ATP13A2*. Furthermore, several genetic risk factors are linked to the onset of the disease [84]. Dependent on the gene affected, the disease may initiate in juveniles, early in adult life or in late adult life. Some of these genes are autosomal dominant (*SNCA* and *LRRK2*), whilst the others are autosomal recessive (*PARK2*, *PINK1*, *PARK7* and *ATP13A2*).

The common idiopathic features of the disease are motor disturbances including resting tremor, rigidity and bradykinesia, as well as non-motor symptoms such as cognitive impairment, autonomic dysregulation, sleep deterioration and neuropyschiatric symptoms [84]. These symptoms and disturbances arise due to the loss of nigrostriatal dopaminergic neurons in the substantia nigra pars compacta and the development of Lewy bodies in surviving neurons. This makes PD a particularly easy disease to modulate in a dish, as one predominant neuron type is affected. Despite this, there still lacks perfect differentiation protocols that result in the A9 type dopaminergic neuron in high proportion. Given the simplicity in the tissue affected by the disease, it has been considered that PSC-derived nigrostriatal dopaminergic neurons from healthy donors or genetically corrected iPSCs could be used for transplantation either into the striatum where they migrate to or in the substantia nigra where the cell bodies lie. In this case, many studies have attempted to improve the production and numbers of nigrostriatal dopaminergic neurons from PSCs [85]. A summary of the PSC studies that model familial PD in a dish are shown in **Table 3** along with known pathology in the patients.

Gene	Mutation	Pathology in Patients	Study	Pathology in PSCs	Study
SNCAA	A53T (G209A)	Lewy body pathology	Golbe et al., 1990	Oxidative stress Mitochondrial dysfunction	Ryan et al., 2013

Gene	Mutation	Pathology in Patients	Study	Pathology in PSCs	Study
				Cell death	
	A53T (G188A)				Soldner et al., 2011
	Triplication	Lewy body pathology, hippocampal neuronal loss, temporal lobe vacuolation	Farrer et al., 2004; Muenter et al., 1998	Increased α-syn Oxidative stress	Devine et al., 2011 Byers et al., 2011 Byers et al., 2011
LRRK2	G2019S	Lewy bodies in brain stem Lewy body + AD pathology Transitional Lewy body disease SN neuronal loss, tau pathology, + AD pathology SN neuronal loss Lewy body pathology	Giasson et al., 2006; Ross et al., 2006 Giasson et al ., 2006 Ross et al., 2006 Rajput et al., 2006 Gaig et al., 2007 Gilks et al., 2005	Oxidative stress Altered ERK signalling Neurite shortening Increased sensitivity to toxins Increased expression of MAPT and p-tau Mitochondrial DNA damage Increased autophagy Dysregulated epigenome	Nguyen et al., 2011 Reinhardt et al., 2013 Reinhardt et al., 2013; Sanchez-Danes et al., 2012 Reinhardt et al., 2013; Cooper et al., 2012 Reinhardt et al., 2013 Sanders et al., 2014 Sanches-Danes et al., 2012 Fernandez-Santiago et al., 2015
	R1441C substitution	Not known		Increased sensitivity to toxins	Cooper et al., 2012
PARK2K2	del_40bp ex3/del ex4+5	Not known		Oxidative stress	Aboud et al., 2012
	del_40bp ex3	Lewy body pathology	Farrer et al., 2001		
	del_ex7	Lewy body pathology	Pramstaller et al., 2005		
	del ex4	SN neuronal loss, no Lewy bodies	Mori et al., 1998 Hayashi et al., 2000		

Gene	Mutation	Pathology in Patients	Study	Pathology in PSCs	Study
	del ex5	Only clinical features known		Proteosome dysfunction Oxidative stress Increased *SNCA*	Chang et al., 2016
PINK1	del_ex7/c.1488+1 G>A	Lewy body pathology	Samaranch et al. 2010		
	c.1366C>T;p.Q456X	Only clinical features known	Hedrich, et al., 2006	Oxidative stress	Seibler et al.,2011
	c.509T>G;pV170G				
	Q456X homozygote			Increased sensitivity to toxins Oxidative stress	Cooper et al., 2012
	D525N/W577R compound				Cooper et al., 2012

Abbreviations: AD – Alzheimer's disease; del – deletion; ex – exon; p-tau – phosphorylated tau; *SNCA* – alpha synuclein gene; SN – substantia nigra

Table 3. Pathology in familial Parkinson's disease patients and respective pluripotent stem cell (PSC) studies.

8.1. Sporadic PD

One study that produced iPSC lines from seven patients with idiopathic PD revealed that dopaminergic neurons produced in-vitro had reduced numbers of neurites, neurite arbourisation and increased autophagy [86]. Another study observed methylation alterations in sporadic PD iPSC-derived dopaminergic neurons [87], which was the first to describe epigenetic dysregulation in the disease.

8.2. SNCA mutations

There are at least five mutations characterized within *SNCA,* which induce PD which also include either duplication or triplication of the gene. Mutations in *SNCA* lead to an early-onset parkinsonism with or without the development of dementia. The pathology generally contains the presence of Lewy bodies, and tau pathology has also been observed some of the mutations.

Induced PSC lines have been derived from PD patients carrying either a triplication of *SNCA* [88–90] or a mutation in A53T [24]. In two articles, triplication of *SCNA* led to increased production of *SNCA* [88, 89]. Oxidative stress has also been reported in iPSC-derived neurons containing triplication of *SCNA* [89] and A53T mutations [26]. This has been shown to contribute to mitochondrial dysfunction and apoptotic cell death [26].

8.3. LRRK2 mutations

There have been five identified mutations in the *LRRK2* gene and many of these have either brainstem-predominant Lewy body pathology, diffuse Lewy body disease or no Lewy body pathology. Three research groups have produced iPSCs from patients carrying two of the known mutations. One study found increased oxidative stress gene expression and increased production of SNCA protein [91]. One research group was able to uncover altered cell signalling of several genes, some of which were involved in ERK signalling; and repression of ERK signalling could reverse certain pathology in neurons including prevention of neurodegeneration, more cell robustness when treated with oxidative stress and a reversal in the shortening of neurites. Genetically corrected iPSCs also revealed a reversal in neurite shortening and sensitivity to neurotoxins that were observed in the patient iPSC-derived neurons [27]. Furthermore, this same study also found increased expression of *MAPT* expression and p-tau^{181}. The *LRRK2* lines have also been used to more closely evaluate mitochondrial function. One study found increased mitochondrial DNA damage in neural cells from patients, which was reversed when the mutation was genetically corrected [25]. Similar to sporadic PD iPSC-derived dopaminergic neurons, it has been shown that epigenetic dysregulation also occurs in *LRRK2*-associated PD patients [87]. Interestingly, the commonly described Lewy body pathology has yet to be observed in the iPSC-derived neurons.

8.4. PARK2 mutations

Mutations in *PARK2* lead to an early-onset parkinsonism. Seven different mutations have been identified, and general pathology in most of the mutations does not include Lewy bodies. One study has evaluated the effects of manganese exposure on neural progenitor cells derived from *PARK2* mutation PD iPSCs and discovered increased ROS generation upon exposure in the patient cells compared to healthy controls [92]. Another study which evaluated iPSCs produced from patients with a deleted exon 5 in *PARK2* showed the iPSC-derived neurons had proteasome dysfunction, oxidative stress and increased expression of *SNCA* [93].

8.5. PINK1 mutations

A mutation in *PINK1* leads to early-onset parkinsonism. Here, Lewy body pathology has been identified. Two studies have produced patient-specific iPSCs harbouring mutations in *PINK1*. In one study, mitochondrial function was analysed and revealed increased mitochondrial copy numbers and increased expression of PGC-1α in the patients' in-vitro-produced dopaminergic neurons [94]. Another research article found *PINK1* mutation iPSCs were more vulnerable to chemical toxins. Mitochondrial perturbation was also discovered, with a lower basal oxygen consumption rate and an increased bidirectional motility of neurons in the proximal axon of neurons [95].

Together, the literature reveals a vast array of disease modelling studies using PSCs to model PD and a number of phenotypes have emerged in the dish although the classic hallmark, Lewy bodies and neurodegeneration, remain absent in the in-vitro cultures (**Figure 4**). Oxidative stress is a clear phenotype observed in many of the different studies and appears common for

http://dx.doi.org/10.5772/63521

many of the mutations, but it is clear that the disease pathology is diverse and therefore the pathology in the iPSC-derived cells would also expect to be diverse.

9. Current limitations in modelling neurodegenerative disease using pluripotent stem cells

Modelling AD and FTD is definitely more challenging than modelling PD, since many different cell types and regions within the brain are affected by these diseases compared to PD.

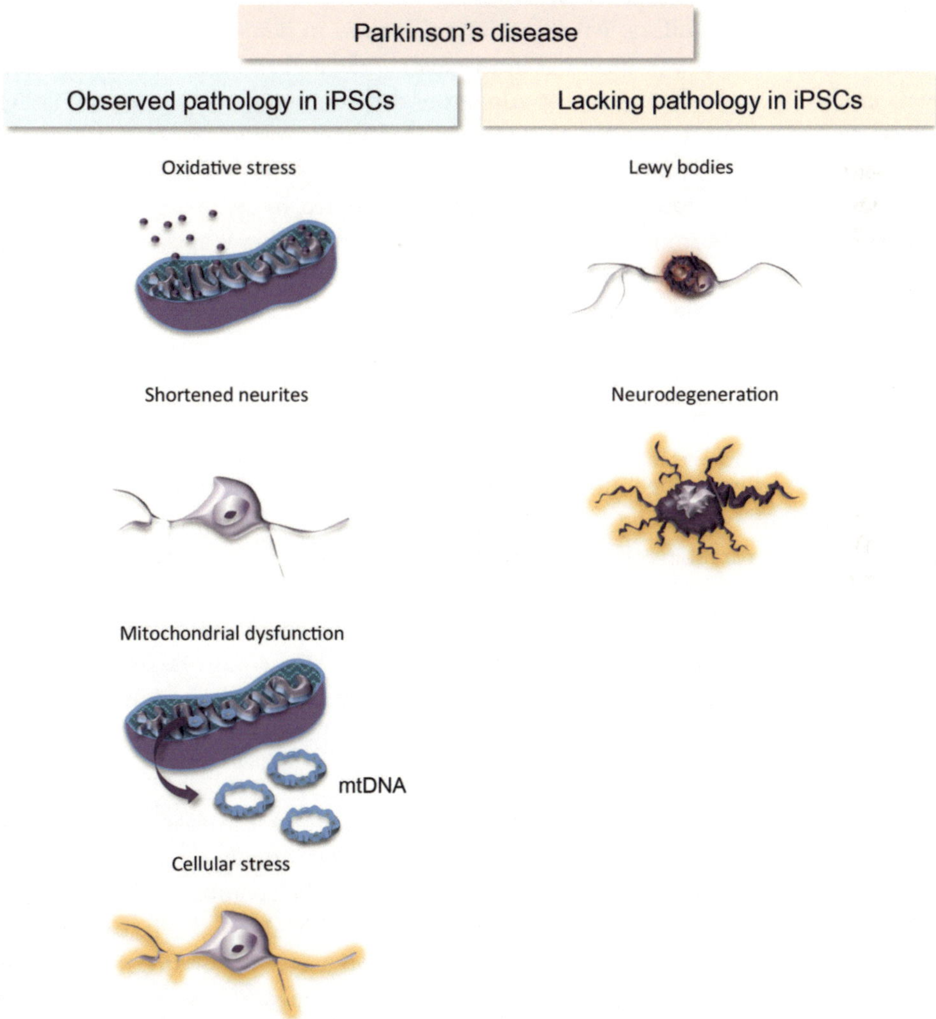

Figure 4. A summary of pathology observed in published induced pluripotent stem cell (iPSC) models of Parkinson's disease (PD) and pathology found in PD patients but currently lacking in the iPSC models.

Therefore, these former diseases are considered less attractive for cell therapy compared to PD. In fact, there is much discussion in the scientific community on the embarkment of clinical trials for PD using human PSCs and it is therefore only a matter of time before these trials begin to emerge. However, the majority of the studies to date have been performed to understand more about the mechanisms of the disease and to find new targets that could be used for discovering new and more effective medicines for braking the disease and for improving symptoms and quality of the patient's life.

One of the limitations observed to date in modelling the diseases is producing the right type of cells that are affected by the disease following differentiation of the iPSCs. In the case of AD and FTD, most studies have focused on evaluation of MAP2/TUJ1-positive neurons, cortical neurons and neural progenitors, which are often derived in heterogenic cultures. Whether these are the best cell types to examine in the dish is debatable. MAP2/TUJ1 expression is relatively unspecific and can label a vast number of different neuron subtypes, so it might be more important to use more specific antibodies to identify the specific neuron subtype that is being analysed. Cortical neurons are also numerous and some research, but not all, has identified the cortical subtype that has been produced in-vitro. In the case of AD, for example, the superficial cortical neurons are affected earlier on in the disease and the deeper cortical neurons are affected later [96]. Therefore, knowing which cortical neuron subtype is produced in the dish will help to understand better the disease mechanisms and whether the pathology being observed is related to early or late stages of the disease. Some researchers are pursuing the development of more complex laminated cortical layer cultures in-vitro [97], which could be a great source of tissue for performing further studies on. The study of neural progenitors might be more relevant for looking at potential early mechanisms of the disease and to evaluate regions of the brain where progenitors exist, as well as the effect of the disease on the cell cycle. In the case of PD, protocols for the generation of dopaminergic neurons have been refined over recent years allowing for relatively high proportions of the correct ventral-mesencephalon type following differentiation [85]. However, it is easy to forget the other cell types that can be affected in PD and the role they may also have in the disease. Therefore, the identity of the cell types produced needs to be more carefully defined in studies to help reveal more details about how the disease affects that particular cell type. More complex in-vitro models could also help to mimic the in-vivo environment better, which might help to reveal more phenotypes associated with the disease.

It is clear from the studies that the phenotypes, and in particular, the classic hallmark pathologies, are not represented in-vitro. The reasons for this are not really clear, but may relate to the fact that neural subtypes may be relatively young in the dish compared to the neural cells found in the patients. The evidence so far reveals that the iPSC-derived cells can model early changes related to the disease and therefore might prove useful for finding ways to reverse the disease or slow it down in the early or pre-symptomatic stages. If we are to evaluate some of the classic hallmarks, which are generally missing in culture (i.e. neurodegeneration, amyloid plaques, NFTs and Lewy bodies), then it might be needed to accelerate the aging of neurons in-vitro using artificial methods. One approach successfully used to age neurons in a dish was demonstrated using iPSCs from PD patients [98]. Overexpression of progerin (a gene linked to a disease of accelerated aging, progeria) was performed, which resulted in a pronounced PD phenotype of the dopaminergic neurons. The neurons were able to show much more pathology compared to non-aged neurons, including dendrite degeneration, loss of tyrosine hydroxylase expression (a typical marker of dopaminergic neurons) and Lewy-body-precursor inclusions within the neurons. This has not yet been applied to AD or FTD, but is an intriguing tool that could help to elaborate more of the classic pathologies/phenotypes in the dish.

10. Conclusions

To conclude, disease modelling of neurodegenerative diseases using PSCs has developed dramatically over a short period of time (a space of about 5 years). Already new mechanisms related to AD, FTD and PD have been discovered and these will likely lead to the development and trial of new medicines for the disease. There are many reported phenotypes that have been linked to the disease that can be reversed when familial mutations are genetically corrected using gene-editing technology. However, not all phenotypes have been reported so far, which may be linked to the cell types evaluated, the relatively simple systems used and also the relatively young neural cell types analysed. New studies developing 3D cultures and more complex tissue types may help move the field forward. In addition, new technologies that accelerate aging in the dish are also likely to help overcome these limitations. Thus, iPSCs have already been useful for uncovering some of the mysteries surrounding neurodegenerative disease and the future will likely lead to uncovering more about the disease mechanisms and how we can repair and treat the dysfunctioning cells before they are lost in the patient.

Acknowledgements

The author's current research and activities are currently funded from the People Programme (Marie Curie Actions) of the European Union Seventh Framework Program FP7/2007-2013/ under REA grant agreement n°PIAPP-GA-2012-324451 (STEMMAD), as well as the Innovation Fund Denmark supported project, BrainStem (a Stem Cell Center of Excellence in Neurology).

Author details

Vanessa J. Hall

Address all correspondence to: vh@sund.ku.dk

University of Copenhagen, Institute of Veterinary Clinical and Animal Sciences, Frederiksberg C, Denmark

References

[1] Takahashi, K. and S. Yamanaka, *Induction of pluripotent stem cells from mouse embryonic and adult fibroblast cultures by defined factors.* Cell, 2006. 126(4): pp. 663–76.

[2] Singh, V.K., et al., *Mechanism of induction: induced pluripotent stem cells (iPSCs).* J Stem Cells, 2015. 10(1): pp. 43–62.

[3] Hall, V.J., P. Stojkovic, and M. Stojkovic, *Using therapeutic cloning to fight human disease: a conundrum or reality?* Stem Cells, 2006. 24(7): pp. 1628–37.

[4] Jang, S.F., et al., *Nanomedicine-based neuroprotective strategies in patient specific-iPSC and personalized medicine.* Int J Mol Sci, 2014. 15(3): pp. 3904–25.

[5] Thomson, J.A., et al., *Embryonic stem cell lines derived from human blastocysts.* Science, 1998. 282(5391): pp. 1145–7.

[6] Hwang, W.S., et al., *Patient-specific embryonic stem cells derived from human SCNT blastocysts.* Science, 2005. 308(5729): pp. 1777–83.

[7] Fan, Y., et al., *Derivation of cloned human blastocysts by histone deacetylase inhibitor treatment after somatic cell nuclear transfer with beta-thalassemia fibroblasts.* Stem Cells Dev, 2011. 20(11): pp. 1951–9.

[8] Lombardo, A., et al., *Gene editing in human stem cells using zinc finger nucleases and integrase-defective lentiviral vector delivery.* Nat Biotechnol, 2007. 25(11): pp. 1298–306.

[9] Hendriks, W.T., C.R. Warren, and C.A. Cowan, *Genome Editing in Human Pluripotent Stem Cells: Approaches, Pitfalls, and Solutions.* Cell Stem Cell, 2016. 18(1): pp. 53–65.

[10] Merkert, S. and U. Martin, *Targeted genome engineering using designer nucleases: State of the art and practical guidance for application in human pluripotent stem cells.* Stem Cell Res, 2016. 16(2): pp. 377–386.

[11] Kim, Y.G., J. Cha, and S. Chandrasegaran, *Hybrid restriction enzymes: zinc finger fusions to Fok I cleavage domain.* Proc Natl Acad Sci U S A, 1996. 93(3): pp. 1156–60.

[12] Carroll, D., *Genome engineering with zinc-finger nucleases.* Genetics, 2011. 188(4): pp. 773–82.

[13] Maeder, M.L., et al., *Rapid "open-source" engineering of customized zinc-finger nucleases for highly efficient gene modification.* Mol Cell, 2008. 31(2): pp. 294–301.

[14] Collin, J. and M. Lako, *Concise review: putting a finger on stem cell biology: zinc finger nuclease-driven targeted genetic editing in human pluripotent stem cells.* Stem Cells, 2011. 29(7): pp. 1021–33.

[15] Jo, Y.I., H. Kim, and S. Ramakrishna, *Recent developments and clinical studies utilizing engineered zinc finger nuclease technology.* Cell Mol Life Sci, 2015. 72(20): pp. 3819–30.

[16] Hauschild-Quintern, J., et al., *Gene knockout and knockin by zinc-finger nucleases: current status and perspectives.* Cell Mol Life Sci, 2013. 70(16): pp. 2969–83.

[17] Ramirez, C.L., et al., *Engineered zinc finger nickases induce homology-directed repair with reduced mutagenic effects.* Nucleic Acids Res, 2012. 40(12): pp. 5560–8.

[18] Niu, J., B. Zhang, and H. Chen, *Applications of TALENs and CRISPR//Cas9 in human cells and their potentials for gene therapy.* Mol Biotechnol, 2014. 56(8): pp. 681–8.

[19] Boch, J., et al., *Breaking the code of DNA binding specificity of TAL-type III effectors.* Science, 2009. 326(5959): pp. 1509–12.

[20] Sun, N. and H. Zhao, *Transcription activator-like effector nucleases (TALENs): a highly efficient and versatile tool for genome editing.* Biotechnol Bioeng, 2013. 110(7): pp. 1811–21.

[21] Ding, Q., et al., *Enhanced efficiency of human pluripotent stem cell genome editing through replacing TALENs with CRISPRs.* Cell Stem Cell, 2013. 12(4): pp. 393–4.

[22] Yang, L., et al., *Optimization of scarless human stem cell genome editing.* Nucleic Acids Res, 2013. 41(19): pp. 9049–61.

[23] Wu, Y., et al., *Correction of a genetic disease in mouse via use of CRISPR-Cas9.* Cell Stem Cell, 2013. 13(6): pp. 659–62.

[24] Soldner, F., et al., *Generation of isogenic pluripotent stem cells differing exclusively at two early onset Parkinson point mutations.* Cell, 2011. 146(2): pp. 318–31.

[25] Sanders, L.H., et al., *LRRK2 mutations cause mitochondrial DNA damage in iPSC-derived neural cells from Parkinson's disease patients: reversal by gene correction.* Neurobiol Dis, 2014. 62: pp. 381–6.

[26] Ryan, S.D., et al., *Isogenic human iPSC Parkinson's model shows nitrosative stress-induced dysfunction in MEF2-PGC1alpha transcription.* Cell, 2013. 155(6): pp. 1351–64.

[27] Reinhardt, P., et al., *Genetic correction of a LRRK2 mutation in human iPSCs links parkinsonian neurodegeneration to ERK-dependent changes in gene expression.* Cell Stem Cell, 2013. 12(3): pp. 354–67.

[28] Raitano, S., et al., *Restoration of progranulin expression rescues cortical neuron generation in an induced pluripotent stem cell model of frontotemporal dementia.* Stem Cell Reports, 2015. 4(1): pp. 16–24.

[29] Meng, X., et al., *Stem cells in a three-dimensional scaffold environment.* Springerplus, 2014. 3: p. 80.

[30] Karus, M., S. Blaess, and O. Brustle, *Self-organization of neural tissue architectures from pluripotent stem cells.* J Comp Neurol, 2014. 522(12): pp. 2831–44.

[31] Reynolds, B.A. and S. Weiss, *Generation of neurons and astrocytes from isolated cells of the adult mammalian central nervous system.* Science, 1992. 255(5052): pp. 1707–10.

[32] Chen, G., et al., *Matrix mechanics and fluid shear stress control stem cells fate in three dimensional microenvironment.* Curr Stem Cell Res Ther, 2013. 8(4): pp. 313–23.

[33] Zare-Mehrjardi, N., et al., *Differentiation of embryonic stem cells into neural cells on 3D poly (D, L-lactic acid) scaffolds versus 2D cultures.* Int J Artif Organs, 2011. 34(10): pp. 1012–23.

[34] Pineda, E.T., R.M. Nerem, and T. Ahsan, *Differentiation patterns of embryonic stem cells in two- versus three-dimensional culture.* Cells Tissues Organs, 2013. 197(5): pp. 399–410.

[35] Azarin, S.M., et al., *Effects of 3D microwell culture on growth kinetics and metabolism of human embryonic stem cells.* Biotechnol Appl Biochem, 2012. 59(2): pp. 88–96.

[36] Lancaster, M.A., et al., *Cerebral organoids model human brain development and microcephaly.* Nature, 2013. 501(7467): pp. 373–9.

[37] Choi, S.H., et al., *A three-dimensional human neural cell culture model of Alzheimer's disease.* Nature, 2014. 515(7526): pp. 274–8.

[38] Michaelson, D.M., *APOE epsilon4: the most prevalent yet understudied risk factor for Alzheimer's disease.* Alzheimers Dement, 2014. 10(6): pp. 861–8.

[39] Albers, M.W., et al., *At the interface of sensory and motor dysfunctions and Alzheimer's disease.* Alzheimers Dement, 2015. 11(1): pp. 70–98.

[40] Braak, H., et al., *Vulnerability of cortical neurons to Alzheimer's and Parkinson's diseases.* J Alzheimers Dis, 2006. 9(3 Suppl): pp. 35–44.

[41] Chen, W.W. and M. Blurton-Jones, *Concise review: can stem cells be used to treat or model Alzheimer's disease?* Stem Cells, 2012. 30(12): pp. 2612–8.

[42] Chakrabarti, S., et al., *Metabolic risk factors of Sporadic Alzheimer's Disease: implications in the pathology, pathogenesis and treatment.* Aging Dis, 2015. 6(4): pp. 282–99.

[43] Israel, M.A., et al., *Probing sporadic and familial Alzheimer's disease using induced pluripotent stem cells.* Nature, 2012. 482(7384): pp. 216–20.

[44] Kondo, T., et al., *Modeling Alzheimer's disease with iPSCs reveals stress phenotypes associated with intracellular Abeta and differential drug responsiveness.* Cell Stem Cell, 2013. 12(4): pp. 487–96.

[45] Hossini, A.M., et al., *Induced pluripotent stem cell-derived neuronal cells from a sporadic Alzheimer's disease donor as a model for investigating AD-associated gene regulatory networks.* BMC Genomics, 2015. 16: p. 84.

[46] Duan, L., et al., *Stem cell derived basal forebrain cholinergic neurons from Alzheimer's disease patients are more susceptible to cell death.* Mol Neurodegener, 2014. 9: p. 3.

[47] Shepherd, C., H. McCann, and G.M. Halliday, *Variations in the neuropathology of familial Alzheimer's disease.* Acta Neuropathol, 2009. 118(1): pp. 37–52.

[48] Larner, A.J. and M. Doran, *Clinical phenotypic heterogeneity of Alzheimer's disease associated with mutations of the presenilin-1 gene.* J Neurol, 2006. 253(2): pp. 139–58.

[49] Jayadev, S., et al., *Alzheimer's disease phenotypes and genotypes associated with mutations in presenilin 2.* Brain, 2010. 133(Pt 4): pp. 1143–54.

[50] Shea, Y.F., et al., *A systematic review of familial Alzheimer's disease: Differences in presentation of clinical features among three mutated genes and potential ethnic differences.* J Formos Med Assoc, 2015.

[51] Yagi, T., et al., *Modeling familial Alzheimer's disease with induced pluripotent stem cells.* Hum Mol Genet, 2011. 20(23): pp. 4530–9.

[52] Liu, Q., et al., *Effect of potent gamma-secretase modulator in human neurons derived from multiple presenilin 1-induced pluripotent stem cell mutant carriers.* JAMA Neurol, 2014. 71(12): pp. 1481–9.

[53] Sproul, A.A., et al., *Characterization and molecular profiling of PSEN1 familial Alzheimer's disease iPSC-derived neural progenitors.* PLoS One, 2014. 9(1): pp. e84547.

[54] Mahairaki, V., et al., *Induced pluripotent stem cells from familial Alzheimer's disease patients differentiate into mature neurons with amyloidogenic properties.* Stem Cells Dev, 2014. 23(24): pp. 2996–3010.

[55] Koch, P., et al., *Presenilin-1 L166P mutant human pluripotent stem cell-derived neurons exhibit partial loss of gamma-secretase activity in endogenous amyloid-beta generation.* Am J Pathol, 2012. 180(6): pp. 2404–16.

[56] Honda, M., et al., *The modeling of Alzheimer's disease by the overexpression of mutant Presenilin 1 in human embryonic stem cells.* Biochem Biophys Res Commun, 2016. 469(3): pp. 587–92.

[57] Cabrejo, L., et al., *Phenotype associated with APP duplication in five families.* Brain, 2006. 129(Pt 11): pp. 2966–76.

[58] Muratore, C.R., et al., *The familial Alzheimer's disease APPV717I mutation alters APP processing and Tau expression in iPSC-derived neurons.* Hum Mol Genet, 2014. 23(13): pp. 3523–36.

[59] Shimada, H., et al., *Clinical course of patients with familial early-onset Alzheimer's disease potentially lacking senile plaques bearing the E693Delta mutation in amyloid precursor protein.* Dement Geriatr Cogn Disord, 2011. 32(1): pp. 45–54.

[60] Shi, Y., et al., *A human stem cell model of early Alzheimer's disease pathology in Down syndrome.* Sci Transl Med, 2012. 4(124): p. 124ra29.

[61] Boxer, A.L. and B.L. Miller, *Clinical features of frontotemporal dementia.* Alzheimer Dis Assoc Disord, 2005. 19 Suppl 1: pp. S3–6.

[62] Goedert, M., B. Ghetti, and M.G. Spillantini, *Frontotemporal dementia: implications for understanding Alzheimer disease.* Cold Spring Harb Perspect Med, 2012. 2(2): p. a006254.

[63] Rohrer, J.D. and J.D. Warren, *Phenotypic signatures of genetic frontotemporal dementia.* Curr Opin Neurol, 2011. 24(6): pp. 542–9.

[64] Almeida, S., et al., *Modeling key pathological features of frontotemporal dementia with C9ORF72 repeat expansion in iPSC-derived human neurons.* Acta Neuropathol, 2013. 126(3): pp. 385–99.

[65] Lee, H.K., et al., *Induced pluripotent stem cells (iPSCs) derived from frontotemporal dementia patient's peripheral blood mononuclear cells.* Stem Cell Res, 2015. 15(2): pp. 325–7.

[66] Almeida, S., et al., *Induced pluripotent stem cell models of progranulin-deficient frontotemporal dementia uncover specific reversible neuronal defects.* Cell Rep, 2012. 2(4): pp. 789–98.

[67] Whitwell, J.L., et al., *Brain atrophy over time in genetic and sporadic frontotemporal dementia: a study of 198 serial magnetic resonance images.* Eur J Neurol, 2015. 22(5): pp. 745–52.

[68] Cerami, C., et al., *Frontotemporal lobar degeneration: current knowledge and future challenges.* J Neurol, 2012. 259(11): pp. 2278–86.

[69] Donnelly, C.J., et al., *RNA toxicity from the ALS//FTD C9ORF72 expansion is mitigated by antisense intervention.* Neuron, 2013. 80(2): pp. 415–28.

[70] Floris, G., et al., *Constructional apraxia in frontotemporal dementia associated with the C9orf72 mutation: broadening the clinical and neuropsychological phenotype.* Amyotroph Lateral Scler Frontotemporal Degener, 2015. 16(1-2): pp. 8–15.

[71] Reed, L.A., Z.K. Wszolek, and M. Hutton, *Phenotypic correlations in FTDP-17.* Neurobiol Aging, 2001. 22(1): pp. 89–107.

[72] Wren, M.C., et al., *Frontotemporal dementia-associated N279K tau mutant disrupts subcellular vesicle trafficking and induces cellular stress in iPSC-derived neural stem cells.* Mol Neurodegener, 2015. 10: p. 46.

[73] van Swieten, J. and M.G. Spillantini, *Hereditary frontotemporal dementia caused by Tau gene mutations.* Brain Pathol, 2007. 17(1): pp. 63–73.

[74] Ehrlich, M., et al., *Distinct neurodegenerative changes in an induced pluripotent stem cell model of frontotemporal dementia linked to mutant tau protein.* Stem Cell Reports, 2015. 5(1): pp. 83–96.

[75] Iovino, M., et al., *Early maturation and distinct tau pathology in induced pluripotent stem cell-derived neurons from patients with MAPT mutations.* Brain, 2015. 138(Pt 11): pp. 3345–59.

[76] Baker, M., et al., *Mutations in progranulin cause tau-negative frontotemporal dementia linked to chromosome 17.* Nature, 2006. 442(7105): pp. 916–9.

[77] Chen-Plotkin, A.S., et al., *Brain progranulin expression in GRN-associated frontotemporal lobar degeneration.* Acta Neuropathol, 2010. 119(1): pp. 111–22.

[78] Lee, W.C., et al., *Targeted manipulation of the sortilin-progranulin axis rescues progranulin haploinsufficiency.* Hum Mol Genet, 2014. 23(6): pp. 1467–78.

[79] Neumann, M., et al., *Ubiquitinated TDP-43 in frontotemporal lobar degeneration and amyotrophic lateral sclerosis.* Science, 2006. 314(5796): pp. 130–3.

[80] Floris, G., et al., *Clinical phenotypes and radiological findings in frontotemporal dementia related to TARDBP mutations.* J Neurol, 2015. 262(2): pp. 375–84.

[81] Bilican, B., et al., *Mutant induced pluripotent stem cell lines recapitulate aspects of TDP-43 proteinopathies and reveal cell-specific vulnerability.* Proc Natl Acad Sci U S A, 2012. 109(15): pp. 5803–8.

[82] Barmada, S.J., et al., *Autophagy induction enhances TDP43 turnover and survival in neuronal ALS models.* Nat Chem Biol, 2014. 10(8): pp. 677–85.

[83] Zhang, Z., et al., *Downregulation of microRNA-9 in iPSC-derived neurons of FTD//ALS patients with TDP-43 mutations.* PLoS One, 2013. 8(10): p. e76055.

[84] Crosiers, D., et al., *Parkinson disease: insights in clinical, genetic and pathological features of monogenic disease subtypes.* J Chem Neuroanat, 2011. 42(2): pp. 131–41.

[85] Kirkeby, A., et al., *Generation of regionally specified neural progenitors and functional neurons from human embryonic stem cells under defined conditions.* Cell Rep, 2012. 1(6): pp. 703–14.

[86] Sanchez-Danes, A., et al., *Disease-specific phenotypes in dopamine neurons from human iPS-based models of genetic and sporadic Parkinson's disease.* EMBO Mol Med, 2012. 4(5): pp. 380–95.

[87] Fernandez-Santiago, R., et al., *Aberrant epigenome in iPSC-derived dopaminergic neurons from Parkinson's disease patients.* EMBO Mol Med, 2015. 7(12): pp. 1529–46.

[88] Devine, M.J., et al., *Parkinson's disease induced pluripotent stem cells with triplication of the alpha-synuclein locus.* Nat Commun, 2011. 2: p. 440.

[89] Byers, B., et al., *SNCA triplication Parkinson's patient's iPSC-derived DA neurons accumulate alpha-synuclein and are susceptible to oxidative stress.* PLoS One, 2011. 6(11): p. e26159.

[90] Yagi, T., et al., *Establishment of induced pluripotent stem cells from centenarians for neurodegenerative disease research.* PLoS One, 2012. 7(7): p. e41572.

[91] Nguyen, H.N., et al., *LRRK2 mutant iPSC-derived DA neurons demonstrate increased susceptibility to oxidative stress.* Cell Stem Cell, 2011. 8(3): pp. 267–80.

[92] Aboud, A.A., et al., *Genetic risk for Parkinson's disease correlates with alterations in neuronal manganese sensitivity between two human subjects.* Neurotoxicology, 2012. 33(6): pp. 1443–9.

[93] Chang, K.H., et al., *Impairment of proteasome and anti-oxidative pathways in the induced pluripotent stem cell model for sporadic Parkinson's disease.* Parkinsonism Relat Disord, 2016. 24: pp. 81–8.

[94] Seibler, P., et al., *Mitochondrial Parkin recruitment is impaired in neurons derived from mutant PINK1 induced pluripotent stem cells.* J Neurosci, 2011. 31(16): pp. 5970–6.

[95] Cooper, O., et al., *Pharmacological rescue of mitochondrial deficits in iPSC-derived neural cells from patients with familial Parkinson's disease.* Sci Transl Med, 2012. 4(141): p. 141ra90.

[96] Romito-DiGiacomo, R.R., et al., *Effects of Alzheimer's disease on different cortical layers: the role of intrinsic differences in Abeta susceptibility.* J Neurosci, 2007. 27(32): pp. 8496–504.

[97] Pasca, A.M., et al., *Functional cortical neurons and astrocytes from human pluripotent stem cells in 3D culture.* Nat Methods, 2015. 12(7): pp. 671–8.

[98] Miller, J.D., et al., *Human iPSC-based modeling of late-onset disease via progerin-induced aging.* Cell Stem Cell, 2013. 13(6): pp. 691–705.

[99] Moehlmann, T., et al., *Presenilin-1 mutations of leucine 166 equally affect the generation of the Notch and APP intracellular domains independent of their effect on Abeta 42 production.* Proc Natl Acad Sci U S A, 2002. 99(12): pp. 8025–30.

[100] Ishikawa, A., et al., *A mutant PSEN1 causes dementia with Lewy bodies and variant Alzheimer's disease.* Ann Neurol, 2005. 57(3): pp. 429–34.

[101] Halliday, G.M., et al., *Pick bodies in a family with presenilin-1 Alzheimer's disease.* Ann Neurol, 2005. 57(1): pp. 139–43.

[102] Levy-Lahad, E., et al., *Candidate gene for the chromosome 1 familial Alzheimer's disease locus.* Science, 1995. 269(5226): pp. 973–7.

[103] Rogaev, E.I., et al., *Familial Alzheimer's disease in kindreds with missense mutations in a gene on chromosome 1 related to the Alzheimer's disease type 3 gene.* Nature, 1995. 376(6543): pp. 775–8.

[104] Lantos, P.L., et al., *Familial Alzheimer's disease with the amyloid precursor protein position 717 mutation and sporadic Alzheimer's disease have the same cytoskeletal pathology.* Neurosci Lett, 1992. 137(2): pp. 221–4.

[105] Shinagawa, S., et al., *Clinicopathological study of patients with C9ORF72-associated frontotemporal dementia presenting with delusions.* J Geriatr Psychiatry Neurol, 2015. 28(2): pp. 99–107.

[106] DeJesus-Hernandez, M., et al., *Expanded GGGGCC hexanucleotide repeat in noncoding region of C9ORF72 causes chromosome 9p-linked FTD and ALS.* Neuron, 2011. 72(2): pp. 245–56.

[107] Domoto-Reilly, K., et al., *Unusually long duration and delayed penetrance in a family with FTD and mutation in MAPT (V337M).* Am J Med Genet B Neuropsychiatr Genet, 2016.

[108] Spillantini, M.G., et al., *Tau pathology in two Dutch families with mutations in the microtubule-binding region of tau.* Am J Pathol, 1998. 153(5): pp. 1359–63.

[109] Brouwers, N., et al., *Alzheimer and Parkinson diagnoses in progranulin null mutation carriers in an extended founder family.* Arch Neurol, 2007. 64(10): pp. 1436–46.

[110] Tamaoka, A., et al., *TDP-43 M337V mutation in familial amyotrophic lateral sclerosis in Japan.* Intern Med, 2010. 49(4): pp. 331–4.

[111] Golbe, L.I., et al., *A large kindred with autosomal dominant Parkinson's disease.* Ann Neurol, 1990. 27(3): pp. 276–82.

[112] Farrer, M., et al., *Comparison of kindreds with parkinsonism and alpha-synuclein genomic multiplications.* Ann Neurol, 2004. 55(2): pp. 174–9.

[113] Muenter, M.D., et al., *Hereditary form of parkinsonism--dementia.* Ann Neurol, 1998. 43(6): pp. 768–81.

[114] Giasson, B.I., et al., *Biochemical and pathological characterization of Lrrk2.* Ann Neurol, 2006. 59(2): pp. 315–22.

[115] Ross, O.A., et al., *Lrrk2 and Lewy body disease.* Ann Neurol, 2006. 59(2): pp. 388–93.

[116] Rajput, A., et al., *Parkinsonism, Lrrk2 G2019S, and tau neuropathology.* Neurology, 2006. 67(8): pp. 1506–8.

[117] Gaig, C., et al., *G2019S LRRK2 mutation causing Parkinson's disease without Lewy bodies.* J Neurol Neurosurg Psychiatry, 2007. 78(6): pp. 626–8.

[118] Farrer, M., et al., *Lewy bodies and parkinsonism in families with parkin mutations.* Ann Neurol, 2001. 50(3): pp. 293–300.

[119] Pramstaller, P.P., et al., *Lewy body Parkinson's disease in a large pedigree with 77 Parkin mutation carriers.* Ann Neurol, 2005. 58(3): pp. 411–22.

[120] Mori, H., et al., *Pathologic and biochemical studies of juvenile parkinsonism linked to chromosome 6q.* Neurology, 1998. 51(3): pp. 890–2.

[121] Hayashi, S., et al., *An autopsy case of autosomal-recessive juvenile parkinsonism with a homozygous exon 4 deletion in the parkin gene.* Mov Disord, 2000. 15(5): pp. 884–8.

[122] Samaranch, L., et al., *PINK1-linked parkinsonism is associated with Lewy body pathology.* Brain, 2010. 133(Pt 4): pp. 1128–42.

[123] Hedrich, K., et al., *Clinical spectrum of homozygous and heterozygous PINK1 mutations in a large German family with Parkinson disease: role of a single hit?* Arch Neurol, 2006. 63(6): pp. 833–8.

Chapter 15

From Human Pluripotent Stem Cells to Peripheral Neurons

Roxana Nat

Additional information is available at the end of the chapter

http://dx.doi.org/10.5772/63674

Abstract

Intense research using vertebrate model organisms has gained considerable knowledge regarding the origin of peripheral neurons, such as neural crest and cranial placodes induction and diversification. However, early development in human embryos has remained largely uncharacterized, despite the roles the neural crest, cranial placodes and their derivatives play in several pathologies. The *in vitro* systems based on the differentiation of human pluripotent stem cells (hPSCs) strikingly recapitulate embryonic development in a dish. Extensively proved for the neurogenesis in the central nervous system (CNS) in the last 15 years, novel *in vitro* differentiation strategies were recently designed for the generation of peripheral nervous system (PNS)-related populations. It is the case of human neural crest, cranial placodes, cranial sensory and autonomic neurons, and enteric neurons. These novel models are equally important for enlightening the human early development and for developing new tools for the modern medicine. Better understanding of the programs for specification and maturation of the multitude of peripheral neurons is a major challenge confronting developmental and stem cell researchers in years to come.

Keywords: neural crest, cranial placodes, sensory ganglia, sympathetic ganglia, enteric ganglia, human embryo, pluripotent stem cells, peripheral neuropathies

1. Introduction

Human pluripotent stem cells (hPSCs), whether derived from the inner cell mass of an embryo at the blastocyst stage (human embryonic stem cells, hESCs) [1], or generated *via* reprogramming of differentiated somatic cells (human induced PSC, hiPSCs) [2], can self-renew indefinitely in culture while maintaining the ability to generate all the cell types in an organism.

The potential of using hPSCs for generating neural cells has been extensively exploited during the last 15 years. The derivation of central nervous system (CNS) cells was among the first differentiation protocols developed in the hESC field [3,4]; nowadays, efficient protocols are available for generating neuroepithelial cells [5], spinal cord motor neurons [6], midbrain dopaminergic neurons [7], and cortical neurons [8,9], among many others, valid from both hESCs and hiPSCs.

The effort to generate specific cell types was guided by the knowledge gained from studies on model organisms, as well as from the previous mouse embryonic stem cell differentiation studies. The cell biology of neurogenesis has been studied in most detail in CNS, but much progress has been achieved also for the peripheral nervous system (PNS) development (reviewed in [10–13]). PNS neurons are categorized into two anatomical and functional classes: sensory and autonomic neurons.

The *sensory neurons* are afferent neurons that relay information from a number of specific structures (including Merkel's discs, Meissner's, Pacini's and Ruffini's corpuscles, Golgi tendon organs, and muscle spindles) to the CNS. Accordingly, proprioceptive neurons provide spatial information regarding position and movement, mechanoceptive neurons mediate touch, nociceptive neurons respond to painful stimuli or itch, and thermoceptive neurons relay information regarding temperature. These neurons form *cranial ganglia* (trigeminal, geniculate, petrosal, nodose, and others), and *dorsal root ganglia* (DRG), a metameric series of ganglia next to the spinal cord.

The *autonomic neurons* are largely efferent motor neurons that carry information from the CNS to the various organs of the body, providing involuntary control of the visceral organs. The autonomic PNS consists of three major components: the sympathetic (SyNS) and parasympathetic (PSNS) nervous systems, which function together to maintain body homeostasis, and the enteric nervous system (ENS), which controls gut motility. SyNS controls the involuntary activities that occur under stressful "fight or flight" conditions, while PSNS, in general, promotes the visceral activities characteristic of periods of peace and relaxation. Axons from CNS neurons situated in vegetative centers in brainstem and spinal cord project and synapse with the ganglionic cell bodies located paravertebrally, along the body axis, for SyNS, in *sympathetic ganglia* (SG) or inside the target organs (for PSNS) as *parasympathetic ganglia*, distributed all over the body. ENS is the largest component of the autonomic nervous system and comprises local neural circuits in the gastrointestinal tract consisting of sensory, inter-, and motor neurons, capable of an autonomous control, without CNS inputs. Sympathetic neurons are noradrenergic (the most) or cholinergic (few, e.g., sudomotor). Parasympathetic neurons are cholinergic (the most) or adrenergic (few). Enteric neurons, whose number surpasses that of spinal cord neurons, have remarkable cytoarchitecture and neurotransmitter diversity, including serotonin (5-hydroxytryptamine, 5-HT), GABA (γ-aminobutyric-acid-positive), and nitric-oxide-synthase (NOS)-positive neurons.

The peripheral neurons originate from neural crest and cranial placodes, two transient embryonic epithelia generated in early embryogenesis at the junction between CNS primordia and non-neural ectoderm. These ectodermal structures are notable for their ability to transform into mesenchymal-like cells, migrate extensively along highly stereotypic pathways and differentiate into numerous derivatives, according to the environmental influences encountered during their journey and at their final sites. In addition to sensory and autonomic neurons of the PNS, they form many other derivatives, including, from the neural crest, Schwann cells and satellite glia, melanocytes, endocrine cells, chondroblasts, osteoblasts, odontoblasts and smooth muscle cells, among others [10–12], and from placodes, sense organ cells and neuroendocrine cells in the head [13].

Neural crest and placode development are multistep processes whose main features are conserved across all vertebrate groups. Section 2 summarizes, in a pan-vertebrate model, the current understanding of neural crest and placode induction, specification, migration, and diversification, pointing on the major signaling molecules and transcription factors that act in each stage.

During the past ten years, much progress has been made in elucidating the mechanisms that orchestrate the differentiation of neural crest and placode cells toward the multitude of neurons forming the PNS. It is the case of sensory neurogenesis and autonomic neurogenesis (sympathetic, parasympathetic, and enteric) in various vertebrate models, which are briefly reviewed in Section 3.

Understanding the developmental ontogeny of the diverse peripheral neural populations provides an essential framework for designing rational approaches towards hPSC directed differentiation. Recently, protocols for efficient differentiation of hPSCs toward neural crest cells (NCCs), cranial placode cells (CPCs), neural crest-derived sensory neurons, neural crest-derived enteric neurons, and placode-derived sensory neurons have been reported, and are briefly reviewed in Section 4.

The expression profiles reported from the *in vitro* differentiation protocols and confirmed in human embryos are combined in an *in vitro* model proposed in Section 5. As a lot little is known about the specification mechanisms in human embryo comparing with animal models, these *in vitro* approaches can gain further insights into the early human development. As the iPSC technology has opened unprecedented opportunities for medicine, current challenges and future avenues in the development of novel therapeutic strategies for PNS diseases are also discussed.

2. Peripheral nervous system development in a pan-vertebrate model

2.1. Origins: neural border, neural crest, and cranial placodes

PNS originates early in development from epiblast, in close connectivity with the CNS and epidermal development. At the beginning of gastrulation, a median domain named *neural ectoderm* starts to form in the anterior part of the embryo, which eventually generates the neural plate, the neural tube, and the entire CNS; a new territory is formed between the future neural and non-neural ectodermal domains, named the *neural border* domain (reviewed in [11–16]).

The induction of both neural and neural border domains is established in all vertebrates by a complex interplay of signaling pathways such as fibroblast growth factors (FGFs), bone morphogenic proteins (BMPs), and Wingless and Int proteins (WNTs), although the sources of these signals and their inhibitors may vary between species. At the anterior end of the embryo, the inhibitors DKK1 (dickkopf 1, a WNT inhibitor) and noggin (a BMP inhibitor) suppress posterior signals and pattern the neural ectoderm, leading to the formation of the anterior neural plate and tube. In the lateral part of the embryo, increased activity of WNT and BMP specifies the non-neural ectoderm. The neural border originates within a zone exposed to intermediate levels of FGFs and WNTs, as well as to BMP inhibitors. The neural border cells turn on the expression of a new set of transcription factors, named the *neural border specifiers*, which include *Tfap2, Msx, Zic, Gbx2, Pax3/7, Dlx5/6, Gata2/3, Foxi1/2*, and *Hairy2* (reviewed in [12–16]) (**Figure 1**).

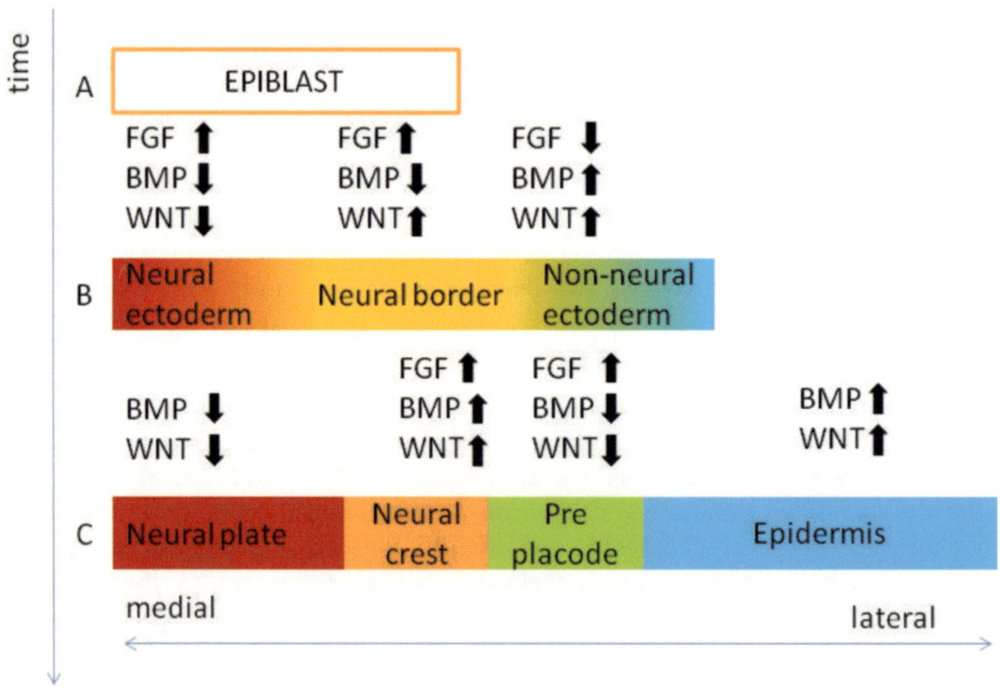

Figure 1. Induction of neural crest and preplacodal domain. (A and B) Prior and during gastrulation, FGF, BMP, and WNT initiate the differentiation of epiblast (A) medially into neural ectoderm and laterally into non-neural ectoderm. The neural border is induced between the non-neural and neural ectoderm (B). Prior and during neurulation, FGF, BMP, and WNT signaling at the neural border domain induces the neural crest (C), while attenuation of BMP and WNT signaling in the presence of FGF initiates the preplacode induction. In the medial part, the inhibition of BMP and WNT signaling in the neural ectoderm domain defines clear borders of the neural plate (red), which starts to form the neural tube and future central nervous system. In the lateral part, high levels of BMP and WNT define the epidermal domain (blue). BMP, bone morphogenic protein; FGF, fibroblast growth factor; NC ↑, activation; ↓, inhibition.

http://dx.doi.org/10.5772/63674

At subsequent stages, the neural plate closes to form the neural tube (a process named *neurulation*). As a consequence, the border territory elevates, forming the dorsal aspect of the opposing neural folds and of the new forming neural tube. Different signaling pathways together with the neural plate border specifiers act to segregate the neural border domain medially into the neural crest and laterally into the preplacodal domain. Neural crest is induced by the presence of FGF, BMP, and WNT activity, while FGF activity and inhibition of BMP and WNT signals in this stage induce the formation of the preplacodal domain. The emergence of the neural crest is marked by the expression of a new set of genes named *neural crest specifiers: FoxD3, Ets1* and *Snai1/2, Twist* and *Sox8/9/10*. The emergence of the preplacodal domain is marked by the expression of the *placodal specifiers Six1, Eya1/2*, and *Irx1* (reviewed in [12–16]) (**Figure 1**).

During neurulation and somitogenesis, the anterior-posterior (A/P) patterning of the neural plate/tube and of the associated neural crest and preplacodal domains takes place. As in the neural tube, the A/P patterning of the neural crest and placode domains is regulated by the major signaling pathways: FGFs, BMPs, WNTs, sonic hedgehog (SHH), Notch, and retinoic acid (RA). Their combined action further segregates both neural crest and preplacodal domains (reviewed in [12–16], **Figure 2**).

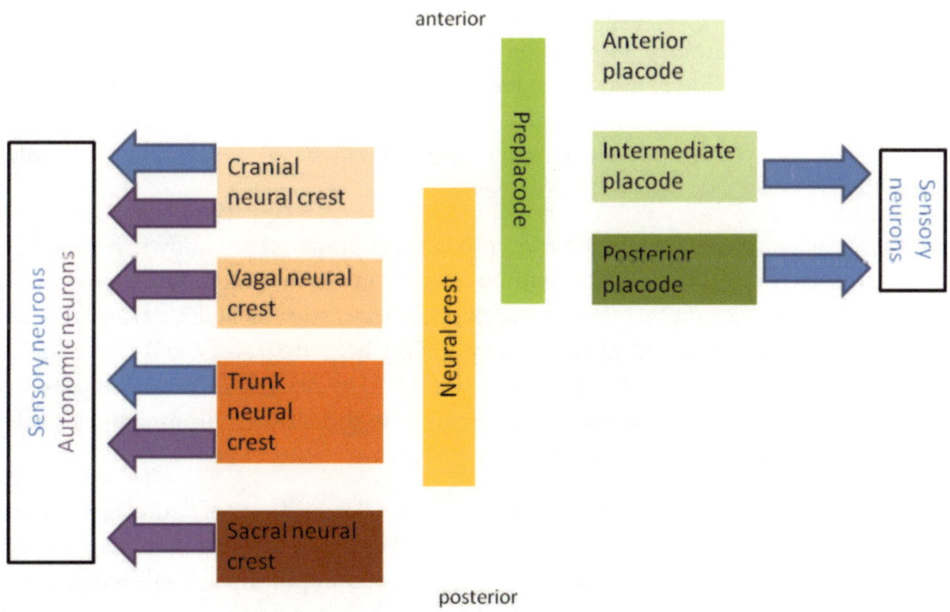

Figure 2. Regionalization and specification of neural crest and preplacode domains.

Cranial neural crest anterior to somite 1 contributes to the formation of the head (including bone, cartilage, connective tissue, teeth, and cranial ganglia (the last together with cranial placodes). Posterior to somite 1, vagal, trunk, and sacral neural crests are formed. Between somite 1–7, vagal neural crest gives rise to the ENS and cardiac outflow tracts. Between somite 7–28, trunk neural crest differentiates into DRG, SG, and norepinephrine-producing cells in the adrenal gland. Posterior to somite 28, sacral neural crest provides part of the ENS of the distal gut. Melanocytes and Schwann cells are generated from neural crest of all A/P levels (**Figure 2**).

Following the A/P patterning in the developing embryonic head, the preplacodal domain is divided into The anterior, posterior, and intermediate domains (**Figure 2**). Anterior domain comprises the adenohypophyseal, olfactory, and lens placodes. The posterior domain comprises the otic and epibranchial placodes. Between the anterior and posterior placodes develops the trigeminal placode. With the exception of the adenohypophyseal and lens placodes, all other cranial placodes give rise to sensory neurons of their associated sensory structures. The otic placode generates the sensory neurons of its associated vestibulocochlear (VIIIth) ganglion, and epibranchial placode gives rise to the sensory neurons of the geniculate, petrosal, and nodose ganglia. The trigeminal placode gives rise to the sensory neurons of the ophthalmic and maxillo-mandibular divisions of the trigeminal ganglion.

2.2. Migration: epithelial-to-mesenchymal transition and stereotypical pathways

After the neural crest and placodes are specified and patterned, their cells acquire mesenchymal properties and migratory morphology and commence migration. The epithelial-to-mesenchymal transition (EMT) was intensively studied in neural crest. Most of the work has focused on the adhesive changes that enable cells to delaminate. Cadherins are thought to be central to this process, a switch between type 1 cadherins (*Ecad* and *Ncad*) and type 2 cadherins (*Cad7* and *Cad11*) taking place in migratory cells.

The neural crest specifiers activate the EMT effector program, which allows the neural crest cells to delaminate from the neural folds/tube and become a migratory cell type, but also to maintain them in a proliferative state. Extracellular signals such as FGF2, neuregulin, neurotrophin 3 (NT3), and epithelial growth factor (EGF) help migratory cell survival and/or proliferation and may act as instructive cues [16,17]. Migratory NCCs are complex from a regulatory viewpoint, as they are constantly exposed to different environmental signals and are also starting to differentiate into diverse derivatives.

Additionally, different signaling systems are involved in the guidance of neural crest migration and the establishment of the correct migratory pathways [17]. For example, the melanocyte precursors follow a dorsolateral migratory stream, under the epidermis, while the sensory and autonomic precursors from trunk neural crest follow a ventromedial pathway, between the neural tube and the developing somites. Sympathetic progenitors express the chemokine receptor CXCR4, which is responsible for the directed migration toward the dorsal aorta, where the chemoattractant SDF-1 is produced [18].

Next section briefly describes the differentiation steps of the NCCs and CPCs that are fated to become sensory neurons, as well as of the NCCs fated for autonomic neurons. However, understanding the programs of neural crest diversification is a major challenge confronting development biologists in years to come.

3. Peripheral neurogenesis

3.1. Sensory neurogenesis

Cranial sensory neurons are derived mainly from trigeminal and epibranchial placodes and fewer from cranial neural crest, while the neurons forming the DRGs are derived solely from trunk neural crest. Specification of sensory neurogenic lineages has been better documented for the trunk neural crest (reviewed in [10, 11]). Each functional type of sensory neuron is characterized by its own unique set of receptors and ion channels, and their differentiation depends on unique sets of transcription factors.

A crucial molecular handle for analyses of sensory neuron development was provided by the discovery that functionally related neuronal subtypes require specific neurotrophic factors. Trophic factor signaling has long been recognized, particularly the neurotrophin ligand/receptor components: nerve growth factor (NGF)/TrkA, brain derived neurotrophic factor (BDNF)/TrkB, and NT-3/TrkC [19,20]. Premigratory NCCs delaminated at different time points, and sensory neurons are generated in a number of waves that derive from temporally distinct NCCs that enter in the ventromedial pathway and arrest migration adjacent to the neural tube, generating DRGs [21]. Large-diameter *TrkC/TrkB+* proprio- and mechanoreceptive neurons are produced first, while small-diameter *TrkA+* nociceptive neurons are subsequently generated. Later, boundary cap cells also generate a small population of *TrkA+* nociceptive sensory neurons.

The first two waves of NCC differentiation are regulated by *Neurogenin1* (*Ngn1*) and *Neurogenin2* (*Ngn2*). *Neurogenin-1* and *Neurogenin-2* are potent promoters of sensory specification and can be detected in a subset of proliferative migrating cells [22]. *Ngn1* appears to play a predominant role in formation of small-diameter nociceptive (*TrkA*$^+$) neurons with a minor requirement in the formation of large-diameter mechanoreceptor (*TrkB*$^+$) and proprioceptive (*TrkC*$^+$) neurons. In contrast, *Ngn2* plays a transient role in formation of large-diameter mechanoreceptor (*TrkB*$^+$) and proprioceptive (*TrkC*$^+$) neurons; *Ngn2*$^+$ cells also contribute to a small but significant fraction of nociceptive (*TrkA*$^+$) neurons.

The subsequent transition of neurogenic progenitors into post-mitotic neurons involves up-regulation of *Brn3a* and *Islet1,* as well as down-regulation of factors that maintain NCC multipotency, such as *Sox10* and *FoxD3* [23–25]. *Brn3a* and *Islet1* also direct the expression of transcription factors important for sensory neuron maturation, such as the Runx family. Runx1 is critical for the continuing differentiation of nociceptive neurons, while Runx3 primarily regulates proprioceptive maturation [21,26,27].

The signaling pathways and transcriptional changes that occur during sensory neuron specification have been well documented [28]. Many of the extrinsic and intrinsic cues that act in the neural crest specification act also in sensory neurogenesis. WNT/β-catenin signaling acts on sensory specification at the premigratory stage and during later stages of neuronal development [29].

During their early maturations stages, the sensory neurons begin to grow neurites that fasciculate with outgrowing spinal motor axons in the forming ventral root *en route* to the periphery or project centrally into the spinal cord via the dorsal root and innervate CNS targets. Somite-derived patterning cues direct the stereotyped position of early sensorimotor projections into the periphery and coordinate alignment with the developing vertebrae [30]. At further stages of maturation, peripheral innervation targets, central neurons, and associated glia produce neurotrophic factors that direct the development of receptive neuronal subtypes. Proper neurotrophic signaling is crucial also for sensory neuron survival and synapse formation [30].

3.2. Sympathetic neurogenesis

Sympathetic neurons are derived from trunk neural crest (**Figure 2**). Major differentiation steps and gene regulators acting on the way between neural crest progenitors and mature sympathetic neurons have been identified in several vertebrate species (reviewed in [31]).

The initial sympathetic differentiation depends on signals from the ventral neural tube and The notochord, as well as the environment surrounding the dorsal aorta. Sympathetic progenitors migrate along the ventro-medial path and reach the dorsal aorta, initially forming a continuous sympathetic chain that subsequently segregates into discrete ganglia. The ventro-medial path is shared with the DRG precursors, and the sympathetic precursors pass through the DRG primordia to reach the sites of primary sympathetic ganglia. The aorta-derived signals have been identified as members of the family of BMPs, which have an essential role in sympathetic neuron development [32]. Accordingly, migrating sympathetic progenitors express the chemokine receptor CXCR4 [18]. The cells then undergo a second migration to para-aortic sites, where secondary sympathetic ganglia are formed. Some of these cells from the primary sympathetic ganglia migrate deeper into the embryo toward the kidney, where they differentiate into predominantly neuro-endocrine cells (chromaffin cells) of the adrenal gland.

The BMP-induced differentiation involves the sequential onset of expression of *Phox2b*, a master gene of sympathetic neuron development, followed by *Ascl1, Insm1, Hand2, Phox2a, Gata2*, and *Gata3*. AP-2β is also important for sympathetic progenitor survival, while Hox genes for the patterning of the sympathetic chain [33]. Noradrenergic differentiation of sympathetic progenitors starts after stopping the migration in the vicinity of the dorsal aorta and is reflected by the expression of tyrosine hydroxylase (TH), followed by dopamine-beta-hydroxylase (DBH) [34] and pan-neuronal markers. However, these sympathetic neurons are immature and continue to proliferate in the sympathetic ganglia primordia under the control of a number of extrinsic signals, such as insulin-like growth factor I (IGF-1), artemin, WNTs, midkine, and BDNF (reviewed in [31]).

The vast majority of postmitotic sympathetic neurons are generated by asymmetric divisions of immature neurons, leading to a postmitotic neuron and an immature neuron that is able to

http://dx.doi.org/10.5772/63674

divide again [35]. Transcription factor codes that underlie the specification and sequential generation of different sympathetic neuron classes are still not known. However, a large proportion of transcription factors involved in the initial specification of sympathetic neurons are expressed up to the adult stage, such as Phox2b, and even selectively expressed in adult noradrenergic neurons, such as Hand2 [36]. These results demonstrate that several members of the gene regulatory network that controls initial sympathetic neuron development are also essential for the maintenance of differentiated neurons.

Survival and terminal differentiation of sympathetic neuron subtypes, as like of sensory neurons, is controlled by target-derived neurotrophic factors. Their terminal differentiation is controlled by specific retrograde signaling, which can result in a change of neuropeptide and neurotransmitter phenotype. For example, sweat glands and periosteum are innervated postnatally by noradrenergic sympathetic neurons that transdifferentiate to functional cholinergic sympathetic neurons [37,38].

3.3. Parasympathetic neurogenesis

Different parasympathetic neurons are generated from cranial, vagal, and sacral neural crest (**Figure 2**). The development of parasympathetic neurons is controlled by mechanisms that are closely related to those acting in the sympathetic neuron lineage, but also shows some interesting differences [31,39,40]. Signals and molecular pathways controlling the specification and differentiation of parasympathetic neurons have been characterized in detail for the ciliary ganglion derived from cranial neural crest and for the enteric neurons arriving from the vagal neural crest.

For the cranial neural crest, BMP5 and BMP7 are expressed at the site of autonomic ganglion formation and are shown to be essential and sufficient for ciliary neuron development. The ciliary ganglion and other parasympathetic ganglia, like the sympathetic ganglia, depend on *Ascl1* and *Phox2b* expression [41,42]. Unlike in the sympathetic lineage, parasympathetic neuron precursors do not express *Gata3*, *AP-2β* and *Hox* genes. A transient expression of noradrenergic characteristics (like TH and DBH) is observed, which is maintained in a very small subpopulation of cells. The majority of cells acquire a cholinergic phenotype, characterized by the expression of choline acetyltransferase (ChAT) and vesicular acetylcholine transporter (VAChT). For the cranial parasympathetic ganglia, as for the DRGs and in contrast to sympathetic ganglia, neuron differentiation starts after withdrawal from the cell cycle [31,41]. However, some postmitotic neurons in the ciliary ganglion do not start to differentiate after cell cycle exit but rather remain as quiescent postmitotic neuron progenitors that later differentiate to mature neurons [42].

Recent exciting work has shown that Schwann cell progenitors in the developing cranial nerve can generate parasympathetic neurons [43,44]. Thus, the developing cranial nerve appears to serve as both a guide and source of progenitors for the parasympathetic ganglia it will eventually innervate.

Detailed analysis revealed that both parasympathetic ganglia arise from cells that accompany the cranial nerve fibers as they grow towards the site of parasympathetic ganglion formation.

These migratory cells express SOX10, which indicates their crest origin, and additionally is a marker for Schwann cells [45–47]. During their migration, however, they turn on the expression of Phox2b, characteristic for autonomic neurons. These be-fated precursor cells are not restricted to cranial nerves but are also found in developing limb nerves where they generate small ganglia composed of Phox2b neurons. It is yet unclear which signals control the fate switch in the migration bi-fated precursor cells and how progenitors become restricted to a parasympathetic fate rather than to a sympathetic neuronal fate.

3.4. Enteric neurogenesis

The ENS develops from both the vagal and sacral neural crest. Vagal neural crest generates most of the ENS and migrates caudally to colonize the entire length of the bowel. Vagal neural crest acquires the ENS formation ability after further programming during migration to the foregut. This involves signaling by Hedgehog and retinoic acid systems which upregulate *Ret* expression of vagal neural crest. Enteric neuron precursors express Phox2a and TRKC. The mechanisms generating the large diversity of neurotransmitter phenotypes, including serotonin (5-hydroxytryptamine, 5-HT), GABA (γ-aminobutyric-acid-positive) and nitric-oxide-synthase (NOS)-positive neurons, are not yet known [39, 40].

4. Peripheral neurogenesis *in vitro*

4.1. Neural crest and placode induction protocols

During the last 15 years, many protocols for generating different neural progenitors and neurons from hESCs were developed. The early methods to differentiate human PSCs towards neural lineages were focused on producing neural stem cells and some CNS derivatives, whereby neural crest cells were observed as a by-product of neural differentiation. Some of these protocols involved dense self-organizing cell conglomerates, such as embryoid bodies, neural rosettes, or confluent cultures, that produce undefined signaling [48,49], while other protocols relied on the use of stromal feeder co-cultures [50,51]; however, these undefined protocols were not able to generate more than 10% of cells with neural crest identity.

A step forward in the increasing efficiency of neural induction in hPSCs was the introduction of the "dual SMAD inhibition" protocol, which generated high yields of neuroectodermal cells and CNS fates. This protocol uses the concomitant inhibition of BMP pathway with Noggin and of TGFβ/activin/Nodal signaling pathway with SB431542 (SB) [5]. Based on this protocol, different strategies were developed to direct the neuroectoderm cell towards neural crest cells, using dual SMAD inhibition and WNT signaling [52–54]. Most of them used the small molecule CHIR99021 (CHIR) to enhance WNT/β-catenin signaling by inhibiting $GSK3_{\beta}$. Other protocols were able to generate more neural crest cells by using the WNT agonist CHIR and only the TGFβ/activin/Nodal receptor inhibitor, but not BMP inhibitors [55, 56]. The efficiency of these protocols was variable, and additional separation methods were implemented to increase the neural crest yield, relying on surface markers such as HNK1 (human natural killer-1) and p75NTR (NGFR) [52–56].

A very recent protocol [57] dramatically improved the yield of neural crest cells in defined serum free-media under WNT activation by simply eliminating both the BMP inhibitor noggin and the TGFβ/activin/Nodal receptor inhibitor SB. This study established a rapid (5 days) and efficient protocol for generating a high yield of neural crest cells. This approach significantly shortens the total length of time required for the neural crest induction. At day 5, the differentiating cells expressed in high proportion (more than 80%) the neural crest markers SOX10, PAX7, and TFAP2A. Additionally, it was clearly demonstrated with this protocol that blocking the FGF and BMP pathways or increasing FGF and BMP activity dramatically decreased the neural crest induction (below 5% of the total cell population).

Another recent report [58] demonstrated that de-repression of endogenous BMP signaling during dual SMAD inhibition is sufficient for the selective induction of *human cranial placode cells*. Gene expression analysis at day 11 revealed, upon withdrawal of Noggin at day 2 or 3 of differentiation, a robust induction of the preplacodal markers *SIX1, EYA1,* and *DLX3*, and more than 70% of cells adopting a SIX1+ cranial placode fate.

Leung et al. [57] also investigated the cranial placode induction in the same defined culture conditions as for neural crest induction, but without CHIR. They found that E-cadherin, and preplacodal markers such as *EYA2, FOXC1* and *ISL1* were highly induced in day 5 cultures. Upon CHIR administration, however, these surface ectoderm and preplacodal transcripts were repressed.

The capacity of human neural crest and placode-like cells generated with different protocols to contribute to terminal derivatives associated with neural crest and placode development was tested in different approaches. Most of the neural crest-like cells were tested for their multipotency and were subjected to terminal differentiation into different derivatives, including sensory neurons, sympathetic neurons, enteric neurons, Schwann cells, melanocytes, chondrocytes, and osteoblasts. The placode-like cells were tested for their competence to generate sensory neurons, lens cells, and hormone-producing anterior pituitary cells [58].

Early protocols were based mainly on spontaneous differentiation, identifying different proportion of specific derivatives such as sensory neurons (Brn3a+/periferin+), sympathetic progenitors (TH+/periferin+), glial precursors ($S100\beta^{+}$ $GFAP^{+/-}$), and melanoblasts ($MITF^{+}$ $SOX10^{+}$). Some recent protocols were designed for the generation of sensory and enteric neurons, which are briefly presented below.

4.2. Sensory neuronal differentiation protocols

Human NCCs generated peripheral sensory neurons as well as other neural crest derivatives applying different protocols. For peripheral neuronal differentiation, NCCs were plated on a special coated surface in order to facilitate neuronal growth (such as Matrigel or Geltrex-coated plates, lanimin/fibronectin or collagen), followed by a treatment with a cocktail of neurotrophic molecules to sustain sensory neuronal differentiation and maturation: brain-derived neurotrophic factor (BDNF), neurotrophin-3 (NT-3), and nerve growth factor (NGF). Markers associated with peripheral sensory neurons (PRPH BRN3a, and ISL1,) and glial progenitors (S100β) were detected after 7–14 days [48–53, 56–61].

Chambers et al. [59] applied another protocol which resulted in the highly efficient derivation of human nociceptive neurons from PSCs within a short period of time. Specification into nociceptive neurons was achieved by adding an inhibitor of tyrosine kinase signaling (SU5402), a blocker of the NOTCH signaling (such as M N-[N-(3,5-difluorophenacetyl)-L-alanyl]-S-phenylglycine t-butyl ester, DAPT, which blocks γ-secretase) and a GSK-3β inhibitor (CHIR), which promotes WNT/β-catenin signaling [59–61].

The differentiation of neural crest cells toward a sensory phenotype was marked by the expression of the pro-neural transcription factors NGN2 and NGN1. Also consistent with early sensory neurogenesis, the expression levels of *PRPH, BRN3A, ISL1* were found to be highly up-regulated. The majority of neurons derived from all iPSC lines co-expressed BRN3A and PRPH, a combination that had been shown to be specific for peripheral sensory neurons *in vivo*. Most of the PRPH positive cells co-expressed ISL1, another recognized markers of sensory specification [23–25].

The increasing expression of the glutamatergic markers *VGLUT1* and *VGLUT2* as well as very low expression levels of markers for non-sensory neuronal phenotypes (such as *MASH1, HB9, DBH, GAD*, and *VACHT*) additionally supported an efficient sensory specification. The increasing expression of the synaptic marker *SYNAPTOPHYSIN* (*SYP*) indicated that the neurons were maturing to become functional. These newly differentiated neurons up-regulated markers of proprioceptive (TRK C, RUNX3), nociceptive (TRK A, RUNX1), or mechanoceptive (TRK B) subtypes.

Optimization of long-term differentiation will be necessary in order to sustain the maturation and subtype specification *in vitro*.

4.3. Autonomic neuronal differentiation protocols

A very recent study [62] describes an efficient strategy to derive enteric neurons with vagal origin and cranial parasympathetic neurons from human ESCs, based on CHIR and the dual SMAD neural crest induction protocol. Enteric neuron specification involves additional treatment with retinoic acid from day 6 to day 11. The differentiated cells were sorted for CD49D at day 11. ENC cells from the 11 day induction protocol were aggregated into 3D spheroids and cultured in defined medium containing CHIR and FGF2. After 4 days of suspension culture, the spheroids were plated on poly-ornithine/laminin/fibronectin-coated dishes in medium containing GDNF. The enteric neuron precursors migrated out of the plated spheroids and differentiated into neurons in 1–2 weeks, further maturing for up to 60 days.

Purified $CD49D^+$ precursors, derived in the presence of retinoic acid, expressed *HOXB2–HOXB5* indicative of vagal identity, but not more caudal HOX transcripts such as *HOXB9*. In further agreement with vagal identity, $CD49D^+$, retinoic-acid-treated neural crest precursors expressed markers of early enteric lineages including PAX3, EDNRB and RET. Replating 3D spheroids under differentiation conditions yielded immature neurons expressing TUJ1 and the enteric precursor marker PHOX2A and ASCL1 (day 20). Most $PHOX2A^+$ cells were positive for TRKC a surface marker expressed in enteric neuron precursors. Temporal expression analyses showed maintenance of ENC neuronal precursor marker expression by day 40 of differentiation, followed by an increase in the percentage of mature neurons by day 60. In agreement with enteric neuron identity, a broad range of neurotransmitter phenotypes was observed, including serotonin-positive (5-hydroxytryptamine$^+$, 5-HT$^+$), GABA$^+$ (γ-aminobutyric-acid-positive), and nitric-oxide-synthase-positive (NOS)$^+$ neurons.

The CD49D$^+$ crest progenitors not treated with RA were HOX-negative, expressing a cranial autonomic phenotype. CNC-derived precursors differentiated into tyrosine-hydroxylase-expressing neurons and gave rise to TRKB-positive rather than TRKC-positive precursors, and no 5-HT$^+$ neurons were observed [62].

5. From *in vitro* differentiation of PSCs to human PNS development and diseases

5.1. An in vitro neuronal differentiation model

Based on the recent reports presented above, a model for neuronal differentiation is proposed here (**Figure 3**). HESCs resemble epiblast cells in terms of signaling requirements and gene regulatory network for self-renewal and maintenance. Combining various chemically defined conditions, cell plating strategies, and activation of signaling pathways, such as activin, Nodal, FGF, and WNT, can direct the differentiation of hESCs into a number of cell

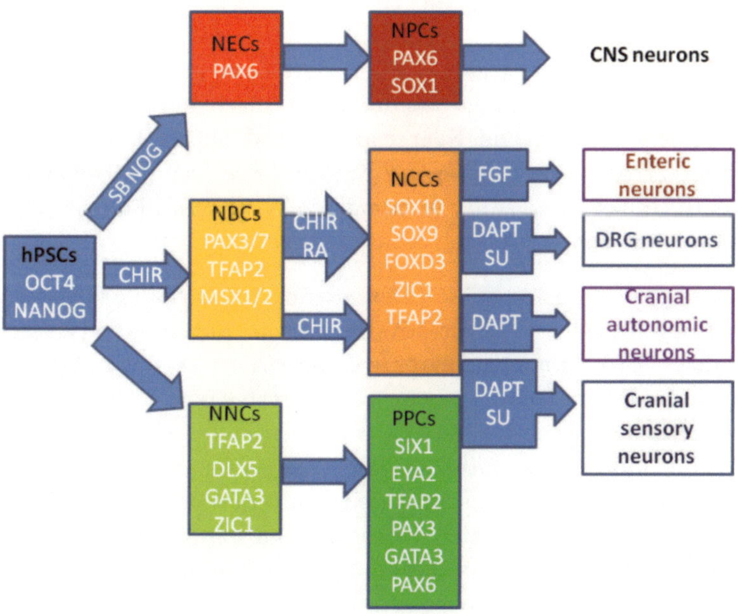

Figure 3. An *in vitro* model of human pluripotent stem cell (hPSC) neuronal differentiation. This model is based on recent PSC-derived protocols and *in vivo* expression data and depicts the stage-wise *in vitro* differentiation of hPSC toward neural progenitor cells, neural crest cells, preplacodal cells, and different classes of neurons. NCCs, neural crest cells; NNCs, non-neural cells; PPCs, preplacodal cells; NBCs, neural border cells; NECs, neuroepithelial cells; NPCs, neural progenitor cells.

types, including the ones generating neural plate, neural crest and placodes. As in development, different peripheral neurons can be generated from regionally patterned neural plate, neural crest, and cranial placode-like cells.

Through the blocking of BMP, activin, and Nodal (via dual SMAD inhibition), hPSCs loose stemness/pluripotency markers and acquire characteristics of neuroepithelial cells (NECs) and further neural progenitor cells (NPCs) and CNS neuronal types. Through the WNT activation (via GSK3 inhibition with small molecules such as CHIR), hPSCs acquire characteristics of non-neural ectodermal cells, neural border cells and neural crest cells. In the absence of WNT activity, markers associated with non-neural ectoderm, placodal and epidermal tissues arise. All these steps are characterized by batteries of transcription factors which are in accordance with the pan-vertebral model. This model relies on milestone-markers which have been identified in a specific fashion in embryonic human NC [63].

In the specification and differentiation steps, including further treatment with CHIR, the NCCs acquire a cranial phenotype, generating mainly cranial parasympathetic neurons. Additional patterning of NCCs with CHIR and RA allows the generation of enteric neurons and DRG sensory neurons. For sensory differentiation, FGF signaling should be blocked (with the inhibitor SU). Placode precursors generate mainly sensory neurons with a trigeminal phenotype. For all the neuronal differentiation steps, blocking NOTCH signals with DAPT dramatically increases the neuronal yield.

5.2. Modeling human early development

For obvious reasons, neural crest and placode development, as well as the sensory and autonomic neurogenesis, are difficult to be studied in human embryos. Access to gastrula stage embryos is extremely limited, and their experimental manipulation almost impossible. Nevertheless, neural crest marker expression patterns have been studied at various developmental stages [63]. However, no study in human embryos related the earliest events occurring during gastrulation and at the onset of neurulation.

The newly devised protocol using hPSC differentiation will certainly facilitate future studies on human PNS development. It is already the case in the same very recent study [57], demonstrating that in early human development, the earliest NC populations arise independently of neural and mesodermal tissues, from precursors that precede the standard neural plate border characteristics. By examining the expression dynamics of a number of neural-related transcripts, no correlation for the induction of PAX6+ NECs and SOX10+ NCCs was found. Furthermore, knockdown assays also suggested that reduced PAX6 function did not negatively interfere with NC induction. These recent data substantiate the notion that human NC forms independently of PAX6+ neuroectoderm. An early origin of human NCCs independent of neural progenitors has been suspected, in accordance with the recent embryological evidence has also suggested the capacity of non-neural ectoderm to form NC without acquiring a neural character (see section 2).

Neural induction in hESCs requires inhibition of either BMP or activin, or both [5]. The fact that administration of noggin blocked expression of NC and/or neural border markers suggests that human NC induction requires BMP activation, which is distinct from neural induction. It is also worth noting that transcripts for BMP ligands can be readily detected under CHIR treatment, supporting the suggested contribution of this pathway during later events in NC development, such as sensory and autonomic differentiation.

5.3. Human peripheral neurons and their application in medicine

The availability of hPSCs can provide not only a source for investigating the early human development, but also a source for generating human neurons for *in vitro* studies. Obviously, human neurons are difficult or impossible to be obtained from adult tissue, in primary culture, as it is the case for many other cells (e.g., blood cells, fibroblasts, skin cells). Many functional approaches can be performed with hESC-derived peripheral neurons, such as the electrophysiological investigation of different human neurons *in vitro* [64]. Many compounds can be tested for their neurotoxicity using assays based on human peripheral neurons, together with other central neurons [60].

In particular, specific neuronal subtypes generated from the patient iPSCs have become a source for studying disease mechanisms underlying different neurological disorders from which the affected neurons were not possible to be available before. The patient-derived *in vitro* models can recapitulate the disease in a dish, for determining its etiology and progression and could develop into a key drug discovery platform. During recent years, the generation of iPSC lines from human material has become routine. Even more, high-quality iPSCs obtained with non-integrative methods and the complete epigenetic resetting were reported [65]. This brings the iPSCs more close to the clinics, increasing their potential for use in regenerative medicine.

Protocols for sub-types of neurons with different degrees of heterogeneity are currently being used for different peripheral neuropathies (PNPs). Recent studies proved the utility of iPSC-derived neural crest cells and peripheral neurons for neurocristopathies (NCPs) and neurodegenerative diseases (NDDs).

NCPs are a major group of human congenital disorders caused by neural crest developmental deficiencies. To better understand the etiology of NCPs and perhaps identify targets for therapeutic intervention, it is critical to understand the detailed mechanisms of neural crest specification, migration, and differentiation in human embryos. The most common NCPs in which the peripheral neurons are affected are familial dysautonomia (FD) and Hirschsprung disease (HiD).

A first example is provided in a study using hiPSCs derived from FD patients, a pathology characterized by the degeneration of sensory and autonomic neurons [66]. Most FD patients carry a point mutation in the IKBKAP gene (I-k-B kinase complex associated protein), leading to a tissue-specific splicing defect, resulting in reduced levels of normal IKBKAP protein. Neural crest precursors obtained *in vitro* from FD hiPSCs show low levels of normal IKBKAP transcript, defective neuronal migration, and differentiation. These characteristics were used to monitor the effects of various drugs on NC-induced hiPSCs and have validated a compound restoring the IKBKAP splicing defect in these cells. These compounds rescue IKAP protein expression and the disease-specific loss of autonomic neuron marker expression [66].

Another example is HiD, in which the enteric neuron progenitors fail to colonize the full length of the bowel during early development. To investigate possible therapies for the HiD, a very recent study used hPSC *in vitro* under conditions that encouraged them to differentiate into cells resembling ENS precursors [62]. These progenitors were further transplanted into the colons of mice with a genetic mutation that causes a Hi-like disease. The transplant prevented premature death in these mice, although how the cells achieved this condition is not clear. Finally, hiPSC-derived ENS precursors enable the identification of candidate therapeutic targets [62]. In this case, the cell-based therapy can transform the future of patients with severe phenotype such as total intestinal aganglionosis.

In another class of PNP, specific peripheral neurons are affected by a neurodegenerative process. It is the case of Friedrich Ataxia (FRDA), an autosomal-recessive disease caused by a GAA triplet expansion in the first intron of the Frataxin (*FXN*) gene [67]. Major neuropathologic findings comprise a degeneration of DRGs, with loss of large sensory neurons, followed by cerebellar and cardiomyocytes degeneration. FRDA-iPSC lines have been established and were successfully differentiated into sensory neurons [67], as well as cardiomyocytes [68]. So far, no overt phenotype was observed in iPSC-derived neurons in FRDA in contrast to the reported mitochondrial phenotype in FRDA-iPSC-derived cardiomyocytes [67,68]. However, during the differentiation process of FRDA iPSCs to peripheral neurons *via* generation of neural crest cells, a differential expression of the frataxin protein was observed between control and FRDA iPSCs, with the FRDA sensory neurons lacking the up regulation found in control neurons. One limitation of this model is related to the maturity of neurons derived from iPSCs. Most iPSC-derived neurons seem to more closely resemble embryonic neurons than mature and aged neurons, and long-time cultures are required to model non-congenital disorders; additionally, it may be necessary to devise protocols that favor aging and degeneration-associated features [69].

To fully tap into the potential of the iPS technology and to progress toward a fundamental understanding of the causes of disease selectivity in the loss of neuron subtypes, it is necessary to establish reproducible and tailored protocols for differentiation of iPSCs specifically into these neuronal subtypes *in vitro*.

Better identification of a dysfunctional pathway in patients suffering from complex PNPs is the primary requirement for rational therapeutic drug development. The human iPSC-derived models could impact positively the screening of compound libraries and the drug safeties screens and in the same time reduce the animal dependency of the current drug development pipeline. iPSC technology is seen as an important driver of personalized medicine. Prior to treatment, patient-derived iPSCs or differentiated progenies can be used to tailor a particular drug type and dose according to the genetic and cellular profile. Based on all these facts, the hPSC-based PNS platform will allow the development of cell- and drug-based strategies for the treatment of different PNS diseases.

The most promising aspect of patient-derived cellular models is the idea of curing genetic diseases *in vitro*. Hence, patient donor cells (like fibroblasts) could be genetically corrected, reprogrammed into iPSCs and further differentiated into the desired progenitor cell. Another possibility is to generate isogenic iPSC lines by genetic modifications through target specific CRISPR/Cas9 technologies [70].

6. Conclusions

hPSCs have provided a revolutionary impact on modeling human PNS development, pathophysiology of different PNPs as well as on novel therapies for them. In the last few years, major efforts have been made to use hPSC to bypass the difficulties linked to working on the human embryo. The knowledge accumulated by embryologists working on various vertebrate model organisms has allowed the development of several strategies to manipulate cell fate choices from stem cells grown *in vitro*. These new *in vitro* cellular models have provided insights into the steps underlying determination and differentiation of diverse neural crest and placode cell lineages. Additionally, recent studies showed that it is now possible to differentiate patient-derived iPSCs into disease-susceptible neuronal phenotypes. The experiments described here proved the presence of disease-inherent phenotypes in iPSCs and the iPSC-derived affected neurons and even revealed new pathophysiological insights and promising candidate substance testing in some PNPs.

However, a lot of work is still needed to optimize and integrate the specific differentiation protocols in the disease-related iPSC models. More importantly, these need to be translated *in vivo* for all the PNPs, and a better appreciation of the precise signals and switches that dictate the survival, proliferation, and differentiation of neural crest cells into their distinct derivatives will facilitate their application in therapeutic and regenerative medicine. Nevertheless, patient iPSC-derived neurons provide a unique opportunity to gain insights into the pathophysiology of PNPs, as well as for drug screening, cell therapy, and personalized medicine.

As presented above, the *in vitro* derivation of PNS lineages from human PSCs did not achieve the same tremendous progress as of CNS lineages, despite the great medical needs also for peripheral neurons. However, the very recent progress in deriving human NCCs, placode cells, and enteric neurons marked a new stage in the human PNS development and diseases. However, understanding the programs of peripheral neurogenesis, especially in the later developmental stages, is a major challenge confronting development biologists in years to come.

Acknowledgements

This work was supported by Austrian FWF-Project P26886-B19.

Author details

Roxana Nat

Address all correspondence to: Irina-Roxana.Nat@i-med.ac.at

Institute for Neuroscience, Medical University of Innsbruck, Innsbruck, Austria

References

[1] Thomson J. A., Itskovitz-Eldor J., Shapiro S. S., Waknitz M. A., Swiergiel J. J., Marshall V. S., Jones J. M.. Embryonic stem cell lines derived from human blastocysts. Science. 1998. 282. 1145–1147.

[2] Takahashi K., Tanabe K., Ohnuki M., Narita M., Ichisaka T., Tomoda K., Yamanaka S. Induction of pluripotent stem cells from adult human fibroblasts by defined factors. Cell. 2007. 131(5). 861–872.

[3] Reubinoff B. E., Itsykson P., Turetsky T., Pera M. F., Reinhartz E., Itzik A., Ben-Hur T. Neural progenitors from human embryonic stem cells. Nature Biotechnology. 2001. 19. 1134–1140.

[4] Zhang S. C., Wernig M., Duncan I. D., Brüstle O., Thomson J. A. In vitro differentiation of transplantable neural precursors from human embryonic stem cells. Nature Biotechnology. 2001. 19(12). 1129–1133.

[5] Chambers S. M., Fasano C. A., Papapetrou E. P., Tomishima M., Sadelain M., Studer Lorenz. Highly efficient neural conversion of human ES and iPS cells by dual inhibition of SMAD signalling. Nature Biotechnology. 2009. 27. 275–280.

[6] Li X., Hu B., Jones S. A., Zhang Y., LaVaute T., Du Z., Zhang S. Directed differentiation of ventral spinal progenitors and motor neurons from human embryonic stem cells by small molecules. Stem Cells. 2008. 26. 886–893.

[7] Kriks S., Shim J., Piao J., Ganat Y. M., Wakeman D. R., Xie Z., Carrillo-Reid L., Auyeung G., Antonacci C., Buch A., Yang L., Beal M. F., Surmeier, Kordower J. H., Tabar V., Studer L. Dopamine neurons derived from human ES cells efficiently engraft in animal models of Parkinson's disease. Nature. 2011. 480. 547–551.

[8] Shi Y., Kirwan P., Smith J., Robinson H. P. C., Livesey F. J. Human cerebral cortex development from pluripotent stem cells to functional excitatory synapses. Nature Neuroscience. 2012. 15. 477–486.

[9] Nat R., Dechant G. Milestones of directed differentiation of mouse and human embryonic stem cells into telencephalic neurons based on neural development in vivo. Stem Cells and Development. 2011. 20. 947–958.

[10] Le Douarin N. M., Kalcheim C. The Neural Crest. 2nd ed. Cambridge University Press. 1999.

[11] Newbern J. Molecular control of the neural crest and peripheral nervous system development. Current Topics in Developmental Biology. 2015. 201–231.

[12] Simões-Costa M., Bronner M. E. Establishing neural crest identity: a gene regulatory recipe. Development. 2015. 142. 242–257.

[13] Singh S., Groves A. The molecular basis of craniofacial placode development. WIREs: Developmental Biology. 2016. doi:10.1002/wdev.226.

[14] Takaoka K., Yamamoto M., Hamada H. Origin of body axes in the mouse embryo. Current Opinion in Genetics & Development. 2007. 17. 344–350.

[15] Groves A. K., LaBonne C. Setting appropriate boundaries: fate, patterning and competence at the neural plate border. Developmental Biology. 2014. 389. 2–12.

[16] Sauka-Spengler T., Bronner-Fraser M. A. Gene regulatory network orchestrates neural crest formation. Nature Reviews Molecular Cell Biology. 2008. 9. 557–568.

[17] Takahashi Y., Sipp D., Enomoto H. Tissue interactions in neural crest cell development and disease. Science. 2013. 23. 860–863.

[18] Kasemeier-Kulesa J. C., McLennan R., Romine M. H., Kulesa P. M., Lefcort F. CXCR4 controls ventral migration of sympathetic precursor cells. Journal of Neuroscience. 2010. 30. 13078–13088.

[19] Ernsberg U. Role of neurotrophin signalling in the differentiation of neurons from dorsal root ganglia and sympathetic ganglia. Cell and Tissue Research. 2009. 336. 349–384.

[20] Usui N., Watanabe K., Ono K., Tomita K., Tamamaki N., Ikenaka K., Takebayashi H. Role of motoneuron-derived neurotrophin 3 in survival and axonal projection of sensory neurons during neural circuit formation. Development. 2012. 139. 1125–1132.

[21] Marmigère F., Montelius A., Wegner M., Groner Y., Reichardt L. F., Ernfors P. The Runx1/AML1 transcription factor selectively regulates development and survival of TrkA nociceptive sensory neurons. Nature Neuroscience. 2006. 9. 180–187.

[22] Marmigère F., Ernfors P. Specification and connectivity of neuronal subtypes in the sensory lineage. Nature Reviews Neuroscience. 2007. 8. 114–127.

[23] Dykes I. M., Tempest L., Lee S., Turner E. E. Brn3a and Islet1 act epistatically to regulate the gene expression program of sensory differentiation. The Journal of Neuroscience. 2011. 31. 9789–9799.

[24] Fedtsova N. G., Turner E. E. Brn-3.0 expression identifies early post-mitotic CNS neurons and sensory neural precursors. Mechanisms of Development. 1995. 53. 291–304.

[25] Sun Y., Dykes I. M., Liang X., Eng S. R., Evans S. M., Turner E. E. A central role for Islet in sensory neuron development linking sensory and spinal gene regulatory programs. Nature Neuroscience. 2008. 11. 1283–1293.

[26] Chen A. I., de Nooi J. C., Jessell T. M. Graded activity of transcription factor Runx3 specifies the laminar termination pattern of sensory axons in the developing spinal-cord. Neuron. 2006. 49. 395–408

[27] Inoue K., Ito K., Osato M., Lee B., 1 Bae S., Ito Y. The transcription factor Runx3 represses the neurotrophin receptor TrkB during lineage commitment of dorsal rootganglion neurons. The Journal of Biolgoical Chemistry. 2007. 282. 24175–24184.

[28] Lallemend F., Ernfors P. Molecular interactions underlying the specification of sensory neurons. Trends in Neuroscience. 2012. 35. 373–381.

[29] Hari L., Miescher I., Shakhova O., Suter U., Chin L., Taketo M., Richardson W. D., Kessaris N., Sommer L. Temporal control of neural crest lineage generation by Wnt/β-catenin signaling. Development and Stem Cells. 2012. 139. 2107–2117.

[30] Guo W., Robbins M. T., Wei F., Zou S., Dubner R., Ren K. Supraspinal brain-derived neurotrophic factor signaling: a novel mechanism for descending pain facilitation. The Journal of Neuroscience. 2006. 26. 126–137.

[31] Rohrer H. Transcriptional control of differentiation and neurogenesis in autonomic ganglia. European Journal of Neuroscience. 2011. 34. 1563–1573.

[32] Morikawa Y., Zehir A., Maska E., Deng C., Schneider M. D., Mishina Y., Cserjesi P. BMP signaling regulates sympathetic nervous system development through Smad4-dependent and -independent pathways. Development. 2009. 136. 3575–3584.

[33] Schmidt M., Huber L., Majdazari A., Schütz G., Williams T., Rohrer H. The transcription factors AP-2β and AP-2α are required for survival of sympathetic progenitors and differentiated sympathetic neurons. Developmental Biology. 2011. 355. 89–100.

[34] Ernsberger U., Patzke H., Tissier-Seta J. P., Reh T., Goridis C., Rohrer H. The expression of tyrosine hydroxylase and the transcription factors cPhox-2 and Cash-1: evidence for distinct inductive steps in the differentiation of chick sympathetic precursor cells.Mechanisms of Development. 1995. 52. 125–136.

[35] Tsarovina K., Pattyn A., Stubbusch J., Müller F., van der Wees J., Schneider C., Brunet JF., Rohrer H. Essential role of Gata transcription factors in sympathetic neuron development. Development. 2004. 131. 4775–4786.

[36] Hendershot T. J., Liu H., Clouthier D. E., Shepherd I. T., Coppola E., Studer M., Firulli A. B., Pittman D. L., Howard M. J. Conditional deletion of Hand2 reveals critical functions in neurogenesis and cell type-specific gene expression for development of neural crest-derived noradrenergic sympathetic ganglion neurons. Developmental Biology. 2008. 319. 179–191.

[37] Apostolova G., Dechant G. Development of neurotransmitter phenotypes in sympathetic neurons. Autonomic Neuroscience. 2009. 151. 30–38.

[38] Stanke M., Duong C. V., Pape M., Geissen M., Burbach G., Deller T., Gascan H., Parlato R., Schütz G., Rohrer H. Target-dependent specification of the neurotransmitter phenotype: cholinergic differentiation of sympathetic neurons is mediated in vivo by gp130 signaling. Development. 2006. 133. 141–150

[39] Reichenbach B., Delalande JM., 1 Kolmogorova E., Prier A., Nguyen T., Smith C. M., Holzschuh J., Shepherd I. T. Endoderm-derived Sonic hedgehog and mesoderm Hand2 expression are required for enteric nervous system development in zebrafish. Developmental Biology. 2008. 318. 52–64.

[40] Simkin J. E., Zhang D., Rollo B. N., Newgreen D. F. Retinoic acid upregulates ret and induces chain migration and population expansion in vagal neural crest cells to colonise the embryonic gut. PLOS. 2013. 8(5). e64077

[41] Lee V. M., Sechrist J. W., Bronner-Fraser M., Nishi R. Neuronal differentiation from postmitotic precursors in the ciliary ganglion. Developmental Biology. 2002. 252. 312–323.

[42] Müller F., Rohrer H. Molecular control of ciliary neuron development: BMPs and downstream transcriptional control in the parasympathetic lineage. Developmental. 2002. 129. 5707–5717.

[43] Dyachuk V., Furlan A., Shahidi M. K., Giovenco M., Kaukua N., Konstantinidou C., Pachnis V., Memic F., Marklund U., Müller T., Birchmeier C., Fried K., Ernfors P., Adameyko I. Parasympathetic neurons originate from nerve-associated peripheral glial progenitors. Science. 2014. 345(6192). 82–87.

[44] Espinosa-Medina, Outin E., Picard C. A., Chettouh Z., Dymecki S., Consalez G. G., Coppola E., Brunet J.-F. Parasympathetic ganglia derive from Schwann cell precursors. Science. 2014. 345. 87–90.

[45] Jessen K. R., Mirsky R. The origin and development of glial cells in peripheral nerves. Nature Reviews Neuroscience. 2005. 6. 671–682.

[46] Bremer M., Fröb F., Kichko T., Reeh P., Tamm E. R., Suter U., Wegner M. Sox10 is required for Schwann-cell homeostasis and myelin maintenance in the adult peripheral nerve. Glia. 2011. 59. 1022–1032.

[47] Britsch S., Goerich D. E., Riethmacher D., Peirano R. I., Rossner M., Nave K., Birchmeier C., Wegner M. The transcription factor Sox10 is a key regulator of peripheral glial development. Genes Development. 2015. 15. 66–78.

[48] Mizuseki K., Sakamoto T., Watanabe K., Muguruma K., Ikeya M., Nishiyama A., Arakawa A., Suemori H., Nakatsuji N., Kawasaki H., Murakami F., Sasai Y. Generation of neural crest-derived peripheral neurons and floor plate cells from mouse and primate embryonic stem cells. PNAS. 2003. 100 (10). 5828–5833.

[49] Lee G., Kim L., Elkabetz Y., Al Shamy G., Panagiotakos G., Barberi T., Tabar V., Studer L. Isolation and directed differentiation of neural crest stem cells derived from human embryonic stem cells. Nature Biotechnology. 2007. 25. 1468–1475

[50] Pomp O., Brokhman I., Ben-Dor I., Reubinoff B., Goldstein R. S. Generation of peripheral sensory and sympathetic neurons and neural crest cells from human embryonic stem cells. Stem Cells. 2005. 23 (7). 923–930

[51] Pomp O., Brokhman I., Ziegler L., Almog M., Korngreen A., Tavian M., Goldstein R. S.PA6-induced human embryonic stem cell-derived neurospheres: a new source of human peripheral sensory neurons and neural crest cells. Brain Research. 2008. 1230.50–60.

[52] Menendez L., Kulik M. J., Page A. T., Park S. S., Lauderdale J. D., Cunningham M. L.,Dalton S. Directed differentiation of human pluripotent cells to neural crest stem cells. Nature Protocols. 2013. 8 (1). 203–212.

[53] Menendez L., Yatskievych T. A., Antin P. B., Dalton S. Wnt signaling and a Smad pathway blockade direct the differentiation of human pluripotent stem cells to multipotent neural crest cells. PNAS. 2011. 108(48). 19240–19245

[54] Fukuta M., Nakai Y., Kirino K., Najagawa M., Sekiguchi K., Nagata S., Matsumoto Y., Yamamoto T., Umeda K., Heike T., Okumura N., Koizumi N., Sato T., Otsuka T., Kinoshita S., Ueno M., Ikeya M., Toguchida J. Derivation of mesenchymal stromal cellsfrom pluripotent stem cells through a neural crest lineage using small molecule compounds with defined media. PLoS One. 2014. 9(12). e112291

[55] Avery J., Dalton S. Methods for derivation of multipotent neural crest cells derived from human pluripotent stem cells. Embryonic Stem Cell Protocols. 2015. 1341. 197–208.

[56] Chambers S. M., Mica Y., Lee G., Studer L., Tomishima M. J. Dual-SMAD inhibition/WNT activation-based methods to induce neural crest and derivatives from human pluripotent stem cells. Methods in Molecular Biology. 2016. 1307. 329–343

[57] Leung A. W., Murdoch B., Salem A. F., Prasad M. S., Gomez G. A., Garcia-Castro M. I. Wnt/β-catenin signaling mediates human neural crest induction via a pre-neural border intermediate. Development. 2016. 143. 398–410.

[58] Dincer Z., Piao J., Niu L., Ganat Y., Kriks S., Zimmer B., Shi S., Tabar V., Studer l. Specification of functional cranial placode derivatives from human pluripotent stem cells. Cell Reports. 2013. 5(5). 1387–1402.

[59] Chambers SM., Qi Y., Mica Y. et al Combined small molecule inhibition accelerates developmental timing and converts human pluripotnet stem cells into nociceptors. Nat Biotechnol 2012; 30: 715-720

[60] Hoelting L., Klima S., Karreman C. et al Peripheral Neurons for Drug Safety Testing. Stem Cells Translational Medicine 2016 5: 464-475

[61] Young G. T., Gutteridge A., Fox H. D. Wilbrey A. L., Cao L., Cho L. T., Brown A. R., Benn C. L., Kammonen L. R., Friedman J. H., Bictash M., Whiting P., Bilsland J. G., Stevens E. B. Characterizing human stem cell-derived sensory neurons at the single cell level reveals their ion channel expression and utility in pain research. Molecular Therapy. 2014. 22(8). 1530–1543.

[62] Fattahi F., Steinbeck J., Kiks S., Tchieu J., Zimmer B., Kishinevsky S., Zeltner N., Mica Y., El-Nachef W., Zhao H., de Stanchina E., Gershon M. D., Grikscheit T. C., Chen S., Studer L. Deriving human ENS lineages for cell therapy and drug discovery in Hirschsprung disease. Nature. 2016. 531. 105–109.

[63] Betters E., Liu Y., Kjaeldgaard A., Sundstroem E., Garcia-Castro M. Analysis of early human neural crest development. Developmental Biology 2010. 344(2). 578–592.

[64] Prè D., Nestor M., Sproul A., Jacob S., Koppensteiner P., Chinchalongporn V., Zimmer M., Yamamoto A., Scott A., Noggle S., Arancio O. A Time Course analysis of the electrophysiological properties of neurons differentiated from human induced pluripotent stem cells (iPSCs). PLoS One. 2014. 9(7). e103418.

[65] Ma H., Morey R., O'Neil R. C., He Y., Daughtry B., Schultz M. D., et al. Abnormalities in human pluripotent cells due to reprogramming mechanisms. Nature 2014. 511.177–183.

[66] Lee G., Kim L., Elkabetz Y., Al Shamy G., Panagiotakos G., Barberi T., Tabar V., Studer L Modeling pathogenesis and treatment of familial dysautonimia patients using specific iPS cells Nature. 2009. 461. 4002–4406.

[67] Eigentler A., Boesch S., Schneider R., Dechant G., Nat R. Induced pluripotent stem cells from friedreich ataxia patients fail to upregulate frataxin during in vitro differentiation to peripheral sensory neurons. Stem Cells and Development. 2013. 15. 3271–3282.

[68] Hick A., Wattenhofer-Donze M., Chintawar S., Tropel P., Simard J. P., Vaucamps N. Et al. Neurons and cardiomyocytes derived from induced pluripotent stem cells as a model for mitochondrial defects in Friedreich's ataxia. Disease Models and Mechanisms 2013. 6. 608–621.

[69] Miller J. D., Ganat Y. M., Kishinevsky S., Bowman R. L., Liu B., Tu E. Y., et al. Human iPSC-based modeling of late-onset disease via progerin-induced aging. Cell Stem Cell. 2013; 13: 691–705.

[70] Doudna J., Charpentier E. The new frontier of genome engineering with CRISPR-Cas9.Science. 2014. 28(346). 6213. doi: 10.1126/science.1258096.

Chapter 16

Temporal Dynamics of Spontaneous Ca^{2+} Transients, ERBB4, vGLUT1, GAD1, Connexin, and Pannexin Genes in Early Stages of Human Stem Cell Neurodifferentiation

Pallavi V. Limaye, Michele L. McGovern,
Mandakini B. Singh, Katerina D. Oikonomou,
Glenn S. Belinsky, Erika Pedrosa,
Herbert M. Lachman and Srdjan D. Antic

Additional information is available at the end of the chapter

http://dx.doi.org/10.5772/62769

Abstract

Spontaneous Ca^{2+} transients drive stem cell proliferation and neurodifferentiation. Deciphering the relationship between neuronal and glial human genes on one side and spontaneous Ca^{2+} activity on the other side is essential for our understanding of normal brain development, and for insights into the pathogenesis of neurodegenerative and neurodevelopmental disorders. In the present study, forebrain neurons were derived from human embryonic and induced pluripotent stem cells (hESC-H9 and iPSC-15; 22q11.2 deletion) over a period of 21 days in vitro (DIV). Every 1–2 days, multisite optical imaging technique was applied to detect populations of cells with spontaneous Ca^{2+} transients. The expression levels of 14 genes of interest were analyzed by quantitative polymerase chain reaction (qPCR) on the same biological samples where physiological recordings were performed. The genes analyzed include: the schizophrenia candidate gene *ERBB4*, connexin (*Cx*) genes *Cx26*, *Cx36*, *Cx43*, *Cx45*, *Cx47*, pannexin-1 (*PNX1*), neuronal markers *PAX6*, *vGLUT1*, *GAD1*, *TUBB3*, glial lineage markers *BLBP*, *GFAP*, and housekeeping gene *ACTB*. We found that Ca^{2+} signals decrease in amplitude, decrease in duration, and increase in frequency during the first 21 days of human neurodifferentiation. The expression levels of *ERBB4*, *PAX6*, *GAD1*, *vGLUT1*, *BLBP*, *Cx36*, *Cx45*, and *PNX1* were found to be strongly positively correlated with the percentage of cells exhibiting spontaneous Ca^{2+} transients ("Active Cells"). While expression of *BLBP*, *Cx45*, *ERBB4*, *GAD1*, *PAX6*, *PNX1*, and *vGLUT1* were correlated with short-duration and long-amplitude Ca^{2+} transients, *Cx43*, *TUBB3*, and *Cx47* were better correlated with long-duration and short-amplitude transients. The expression dynamics of *Cx26* was unrelated to any aspect of spontaneous Ca^{2+} activity. Four genes showed an exponential time course with a distinct onset on a given DIV. The onset of *PNX1, ERBB4,* and *vGLUT1* occurred before, while the onset of *Cx36* occurred after the first action potentials were detected in early differentiating human neurons.

Keywords: embryonic, induced pluripotent, spontaneous activity, Ca^{2+} imaging, schizophrenia, action potential

1. Introduction

Deciphering human brain development requires the understanding of physiological processes of neuron progenitors and differentiating neurons [1–3]. Although electrical and optical recordings from postmitotic human neurons can be achieved using acute human fetal brain slices [4,5], the overwhelming majority of studies on the physiology of postmitotic human neurons use human embryonic or induced pluripotent stem cells [6–12]. The implementation of human stem cell technologies brings two advantages:

1. Culturing of human neurons alleviates severe ethical and technical limitations associated with physiological recording in fetal brain slices.
2. In vitro differentiation of human neurons underlies state-of-the-art experimental models of neurodegenerative disorders [13–15].

Much of the current research effort on stem cell-derived human neurons is focused on the gene expression levels, but it is also necessary to elucidate physiological signals that either drive genes, or are driven by genes. Among these signals are transient changes in intracellular Ca^{2+} concentration [16–19].

Spontaneous Ca^{2+} transients are detected in the developing cortex before the input of sensory information [20]. In the earliest stages of neurogenesis, Ca^{2+} transients control fundamental cell processes such as proliferation, migration, and differentiation of neural progenitors into neurons [2,21,22]. Spontaneous Ca^{2+} transients differ in characteristics such as duration, frequency, and amplitude, encoding information that can be decoded by proteins within the cell, thus allowing the latter to exert control over various processes. High-frequency Ca^{2+} oscillations trigger events such as the release of neurotransmitters, whereas low-frequency activity is indicative of slower processes such as gene transcription [23].

Given the interdependence between Ca^{2+} signals and the gene transcription process, we hypothesized that the expression levels of a particular human gene may correlate with the onset of a particular type of Ca^{2+} activity in differentiating cells. This working hypothesis is bidirectional. On one hand, a gene may code for proteins that bring Ca^{2+} into the cells, in which case an increase in gene expression changes the pattern of ongoing Ca^{2+} transients. On the other hand, if Ca^{2+} activity activates gene transcription, mRNA levels will rise in the cytosol in response to a certain amplitude, duration, or frequency of spontaneous Ca^{2+} transients. In order to test for possible correlations between Ca^{2+} activity and gene expression levels, forebrain neurons were derived from human stem cells over 21 days in vitro (DIV). At regular time intervals (every 2 days), the patterns of spontaneous Ca^{2+} signaling were recorded optically using a Ca^{2+}-sensitive dye. Multisite Ca^{2+} imaging allowed us to determine the duration, frequency, and amplitude of spontaneous Ca^{2+} transients at each DIV. The physiological measurements were followed by gene expression analysis on the same in vitro samples (cell cultures) in which physiological recordings were made. We experimented with hESC and iPSC technologies in order to study the physiological activity and gene expression during the earliest stages of human neurodifferentiation, when pluripotent cells transition into neuron progenitors, which in turn transition further to postmitotic neurons. The following 14 genes were examined: the schizophrenia candidate gene *ERBB4;* neuronal markers: *PAX6, vGLUT1, GAD1, TUBB3;* the glial lineage markers *BLBP* and *GFAP;* several connexins (*Cx26, Cx36, Cx43, Cx45, Cx47*); a pannexin (*PNX1*); and the housekeeping gene *ACTB*. We found that Ca^{2+} transients with specific temporal properties are significantly correlated to the expression of a select group of genes during these early stages of neuronal development.

2. Materials and method

2.1. Cell lines

Human embryonic stem cell line (hESC-H9) was obtained from the University of Connecticut Stem Cell Core. The induced pluripotent stem cell line (iPSC-15, 22q11.2 deletion, passages 11–14, in [24] called SZ22del15-6) was created at the Albert Einstein College of Medicine (Bronx, New York). Both cell lines have also been used in a previous report [8], where it has been demonstrated that there is no difference in spontaneous activity or expression of selected genes between these two lines.

2.1.1. Differentiation protocol

Stem cell lines were differentiated using a five-stage protocol (**Figure 1A**) with defined media constituents [6, 25]. This protocol does not use typical brain region morphogens in Stages IV–V (besides those present in B27), and so is associated with the default production of forebrain neurons [26]. Briefly, stem cell colonies (Stage I) were dissociated by collagenase, and stem cell aggregates were incubated for 4 days in stem cell (SC) media without bFGF on ultralow adherence plates (Costar, Wilkes Barre, PA) (Stage II). 5 μM dorsomorphin (Chemdea, Ridgewood, NJ) and 5 μM SB431542 (Ascent, Princeton, NJ) were added to the media from days 1–8. Stem cell aggregates were then seeded on dishes coated with 1:100 Geltrex and allowed to expand for 4–8 days in NEP (neuroepithelial)–basal medium (Stage III) until neuroepithelial colonies appeared. NEP-basal medium consisted of DMEM/F12, 1 mg/ml BSA (Sigma), 1× N2, 1× B27 supplements without retinoic acid, and 1× penicillin/streptomycin/antimycotic. Thirty minutes before selecting the resultant neuroepithelial colonies, 0.66 mg/ml ROCK inhibitor (Y27632, Wako USA, Richmond, VA) was added to the media. Colonies with neuroepithelial morphology were removed by trituration and seeded with 0.66 mg/ml ROCK inhibitor on glass cover slips coated with 1:100 Geltrex. Cells were grown in NEP-basal medium with 20 ng/ml basic fibroblast growth factor (bFGF) for 7 days (Stage IV). Cells were then maintained in NEP-basal medium in the presence of 10 ng/ml BDNF (brain-derived neurotrophic factor) (Peprotech, Rocky Hill, NJ) and 10 ng/ml GDNF (glial cell line-derived neurotrophic factor) (Peprotech) (Differentiation Stage V and maintenance). Media were changed every other day.

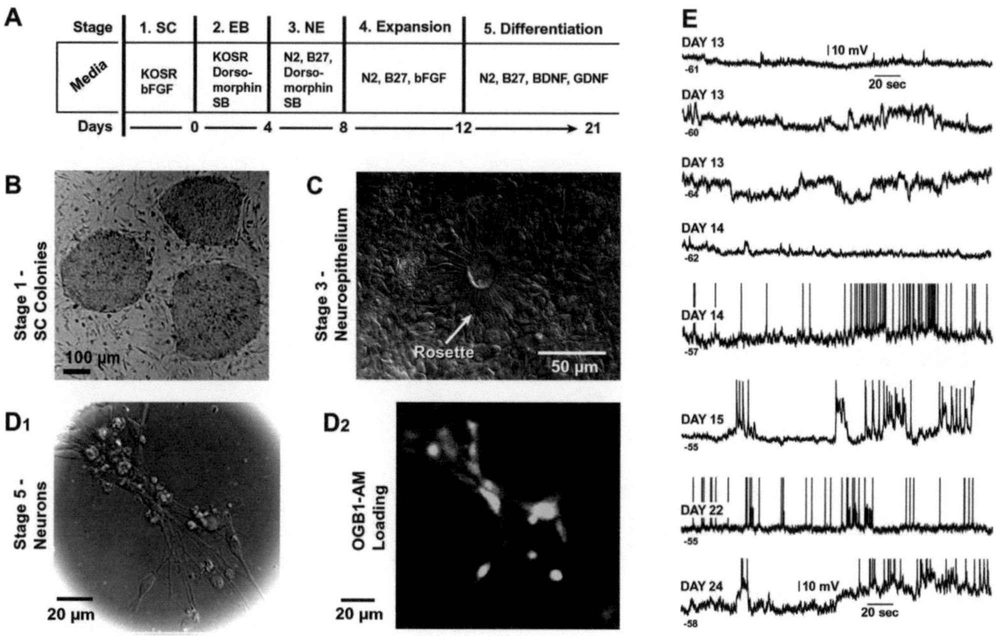

Figure 1. Human stem cell neurodifferentiation. (**A**) Timeline of the differentiation protocol comprising five stages. SC, induced pluripotent or embryonic stem cells; EB, embryoid bodies; NE, neuroepithelial colonies. (**B**) Dark-field photograph of undifferentiated stem cell colonies. (**C**) IR DIC photograph of the neuroepithelium with formed rosettes Stage III. (**D_1**) IR DIC photograph of postmitotic neurons—Stage V. (**D_2**) The same neurons as in D_1, loaded with OGB1-AM and photographed in the fluorescence channel (excitation, 470 nm). (**E**) Whole-cell recordings of spontaneous electrical activity from cultured cells on DIV-13 to DIV-24. Resting membrane potential shown under the trace. Action potentials are truncated for clarity.

2.1.2. Electrophysiology

Whole-cell patch recordings were performed as previously described [6]. Briefly, before recordings, cells were transferred to an Olympus (Tokyo, Japan) BX51WI upright microscope (equipped with infrared video microscopy) and incubated in DMEM/F12, 0.075% BSA, 0.25× B27, Pen/Strep/Antimycotic, and 14 mM HEPES, pH 7.4. Individual cells were selected for recordings based on a small round or ovoid cell body (diameters, 5–10 μm) and typically two or more extended processes. Pipettes (10 MΩ) were filled with an intracellular solution containing (in mM) 135 K-glutamate, 10 HEPES, 2 $MgCl_2$, 3 ATP-Na_2, 0.3 GTPNa_2, and 10 P-creatine Na_2, pH 7.3. Recordings were performed using Multiclamp 700B and Clampex9.2 (Molecular Devices, Union City, CA). In the current clamp configuration, we measured the resting membrane potential, and then a negative holding direct current (in the range −2 pA to

−15 pA) was applied to bring the membrane potential to approximately −60 mV. This was done to compare neurons under identical conditions during spontaneous activity. Spontaneous activity was monitored for 5 min continuously. Electrical traces were analyzed using Clampfit 9.2 (Molecular Devices).

2.2. Ca^{2+} imaging

Multisite Ca^{2+} imaging was used to record spontaneous activity in cells cultured on glass cover slips (12 mm). The cells were loaded with a Ca^{2+}-sensitive fluorescent dye by incubating the cover slips for 30 min with membrane-permeable Oregon Green 488 Bapta-1-AM (OGB1-AM, Invitrogen 06807) at 37°C. The cells were then washed with warmed media and allowed to recover for additional 30 min in the incubator (37°C). Cover slips were placed in the recording chamber and perfused with modified Krebs-Ringer solution. Images were projected onto an 80 × 80 pixel CCD camera (NeuroCCD-SMQ, RedShirtImaging, Decatur, GA) and sampled at 800 ms/frame. One to four nonoverlapping visual fields were sampled per cover slip. The cells were then frozen and kept at −80°C until RNA extraction. Analysis of optical signals was done off-line using Neuroplex software (RedShirtImaging). The following definitions were applied for the analysis of data obtained by multisite Ca^{2+} imaging:

Active Cells: In each visual field, the number of cells showing at least one Ca^{2+} transient was divided by the total number of cells in that field (%).

Repetitive Activity: The number of cells in the visual field exhibiting four or more Ca^{2+} transients per 5 min of optical recording, divided by the number of "Active Cells" (cells showing at least one Ca^{2+} transient) in that field (%).

Amplitude of a Ca^{2+} transient was measured from the baseline before the signal onset to the signal peak, and expressed as $\Delta F/F$ (%).

Duration of the Ca^{2+} transient was measured at half amplitude (half width) and expressed in seconds. The median of duration of all recorded Ca^{2+} signals ($n = 4744$) was determined to be at 8 s. Ca^{2+} signals were then separated into two groups, one consisting of "wide" and the other of "narrow" events. *Wide*: An event endowed with a half-width longer than 8 s. *Narrow*: An event was considered to be narrow if the duration (measured at half amplitude) was less than or equal to 8 s.

2.3. RNA extraction

The samples were removed from −80°C, and 300 μl of Trizol was added to each well containing a cover slip. Cells were detached from the cover slips by rinsing with Trizol. The cover slips were then discarded and 1 μl of yeast tRNA 10 μg/μl (Sigma) was added to each well. The cells were allowed to sit for 2 min at room temperature; then they were triturated, and the contents from each well were transferred to a separate 0.5 ml microcentrifuge tube. A volume of 60 μl of chloroform was added to each sample and shaken vigorously for 15 s. The tubes were incubated at room temperature for 15 min and were then centrifuged at 12,000 g for 15 min at 4°C. They were then placed on ice, and the clear, aqueous phase from each tube was removed and transferred to a new 0.5 ml microcentrifuge tube.150 µl of 100% isopropanol was then added to each sample. The contents of each tube were mixed, allowed to rest for 10 min at room temperature, and then centrifuged at 12,000 g for 10 min at 4°C. An RNA pellet was located in each tube and the supernatant removed. 0.3 ml of cold 75% ethanol was added to each tube, which was then vortexed and centrifuged at 7,500 g for 5 min at 4°C.The supernatant was again removed and the pellets allowed to air-dry. The pellets were then resuspended in 7 µl of RNase-free water and incubated at 55°C for 10 min, using the PCR machine (Eppendorf Mastercycler Gradient, Hauppauge, NY). In order to remove any genomic DNA contamination, samples were subjected to treatment with DNase according to the manufacturer's protocol (New England Biolabs).

2.4. cDNA synthesis

The following were mixed: 5 µl RNA, 0.75 µl of mixed OligodT and Random Hexamers 1:1 (Qiagen, Valencia, CA), 1 µl (10 mM each) dNTPs (Promega, Madision, WI), and 3.25 µl RNase-free water (Invitrogen). There was also an NTC (no template control) tube which contained 5 µl of RNase-free water in place of RNA. The reagents in each tube were mixed thoroughly and placed in the PCR machine with a preheated lid of 80°C. The samples were incubated at 65°C for 5 min and then immediately placed on ice (for at least 1 min). 4 µl of 5× SuperScriptIII Buffer (Invitrogen), 1 µl DTT (Invitrogen), 1 µl RNaseOut (Invitrogen), 3 µl DNase-free water, and 1 µl SuperScriptIII (Invitrogen) were added to the 10 µl already present in each tube. 4 µl 5× SuperScriptIII buffer, 1 µl DTT, 1 µl RNaseOut, and 4 µl DNase-free water were added to −RT (No Reverse Transcriptase) tubes. The reagents in each tube were mixed thoroughly and placed in the PCR machine at 25°C for 5 min, 37°C for 10 min, 50°C for 80 min, and 70°C for 15 min. The samples were then stored at −20°C. For positive controls, total RNA of human fetal brain (Clontech, Mountain View, CA) was used.

2.5. Preamplification

Primer pools of long outer primers were made in water, 1 µM each for up to 14 genes in total. The following reagents were mixed for 80 µl pre-amp reactions: 16 µl 5× GoTaq Flexi Buffer Colorless (Promega), 4.8 µl of 25 mM $MgCl_2$, 1.6 µl of 10 mM dNTPs, 3.2 µl primer pool, 1.28 µl GoTaq DNA polymerase (Promega), 48.12 µl water, and 5 µl cDNA. The lid of the Eppendorf Mastercycler Gradient was preheated to 104°C, and the pre-amplification PCR was run at 94°C—4 min [94°C—30 s, 55°C—30 s, 72°C—40 s] for 10 cycles, and 72°C—10 min.

2.6. Quantitative polymerase chain reaction (qPCR)

qPCR was performed on the genes of interest by mixing the following reagents: 3 µl 5× PCR buffer, 0.9 µl 25 mM $MgCl_2$, 0.5 µl 10mM dNTPs, 1 µl 10 µM mixed forward and reverse primers (**Table 1**), 0.75 µl 10 µM FAM BHQ Probe (Sigma), 0.4 µl GoTaq, 5 µl preamplified cDNA, and 13.45 µl water. The final volume in each well was 25 µl. qPCR was performed using the Bio-Rad CFX Connect Real-Time PCR Detection System. Samples were run at 94°C—3 min, [94°C—30 s, 55°C—30 s, 72°C—30 s] for 40 cycles, with the annealing temperature dependent on the primers. **Table 1** shows screened nested primers, probe sets, and the annealing temperatures. Gene expression data were normalized using housekeeping gene *ACTB* as a reference, using the 2-ΔΔCt method.

2.6.1. Data sets

In the present study, spontaneous Ca^{2+} activity was recorded in 25 cultures containing developing human neurons at various ages (differentiation days: 7, 8, 9, 10, 11, 14, 15,16, 17, 18, and 21). Ten cultures were from iPSC-15 and 15 cultures contained hESC-H9 cells (**Table 2A**). The total number of spontaneous Ca^{2+} signals analyzed from Day 7 to Day 21 is 4744. All 25 cultures were saved after the recording session and subjected to gene expression analysis. In addition to these 25 cultures, the identical gene analysis was performed on 23 cultures in which recordings were not performed (**Table 2B**). No significant differences in gene expression levels of 14 genes (see list above) were detected between recorded and nonrecorded cultures (data not shown); hence, the gene expression data from recorded and nonrecorded cultures were combined. The gene expression data from DIV 8 to DIV 21 were used in the "gene-to-gene" correlation analyses ($N = 45$ cultures/cover slips). The physiological and gene expression data from DIV 7 to DIV 21 were used in the activity-to-gene correlation analyses ($n = 4744$ calcium transients, $N = 48$ cultures).

Gene	Outer primers 5′–3′ forward; reverse (annealing temperature)
	Inner primers 5′–3′, forward; reverse (annealing temperature)
	Probe
ACTB	CCTCGCCTTTGCCGATCC; GATGCCGTGCTCGATGGGGT (55°C)
	CCTCGCCTTTGCCGATCC; GCGAAGCCGGCCTTGCACAT (55°C)
	famATGATATCGCCGCGCTCGTCGTCGAbhq
TUBB3	ACTCCCTTGAACAGGGACAGGGA; GTTCCGGGTTCCAGGTCCACCA (55°C)
	CACAGGCGTCCACAGTTCT; CGAGTCGCCCACGTAGTTG (55°C)
	famCATCAGTGATGAGCATGGCATCGACCbhq
ERBB4	GGAGTACTTGGTCCCTCAGGCT; GGGTGCCACTGGCTTGCGTA (55°C)
	GCAAGAATTGACTCGAATA; CTGGAATTGTGCTAGTTG (50°C)
	famCAGACACTCCTTGTTCAGCAGCbhq
GAD1	GATGATGGGCGTGCTGTTGC; GCGGCTGGGTTGGAGATGAC (55°C)
	CGACACCGGGGACAAGGCAATT; CCGTCGTTGAGGGCTGTCTGG (55°C)
	famCAATGGCGAGCCTGAGCACACAAACGTCTGbhq
vGLUT1	CGCTACATTATCGCCATCATGAG; GGTGGGGCCCATTTGCTCCA (55°C)
	CAGGAGGATTTATCTGTCAAAAAT; GGGTATGTGACCCCCTCTACCAAC (55°C)
	famATGCTGATCCCCTCAGCTGCCCGCbhq
PAX6	AGCCCAGTATAAGCGGGAGTGC; TCCCCCTCCTTCCTGTTGCTGG (55°C)

	TCTTTGCTTGGGAAATCCG; CTGCCCGTTCAACATCCTTAG (55°C)
	famTCATACATGCCGTCTGCGCCCATCTGbhq
PNX1	CGCTGTTTGTTCCATTCCGAC; TCTCTGTCGGGCATTCTTCTC (55°C)
	GGAGAATATTAAGAGCAGTGGTC; CTGTCTATGTTCATACCTTGGAG (45°C)
	famCGACCCAATGCTACTCCTGACAAACCTbhq
CX26	TCCCGACGCAGAGCAAAC; GCAGGGTGTTGCAGACAAAG (55°C)
	TCCCGACGCAGAGCAAAC; AGGGTGTTGCAGACAAAGTC (55°C)
	famCCCACACCTCCTTTGCAGCCACAbhq
CX36	GCAGCACTCCACTATGATCGG; GGAAGACCCAGTAACGTATGTG (55°C)
	ACTCCACTATGATCGGGAGGATCC; CTGCAGGGTGTTGCACAC (53°C)
	famCTGCTCATCATCGTACACCGTCTCCCCbhq
CX43	AGGAGTTCAATCACTTGGCG; GTACTGACAGCCACACCTTC (55°C)
	CTTGGCGTGACTTCACTAC; GCAGTTGAGTAGGCTTGAAC (55°C)
	famCGCTCCAGTCACCCATGTTGCCTGbhq
CX45	TGGACAACAGGGCATACC; GCAAACGCATCATAACAGAC (55°C)
	TGGACAACAGGGCATACC; CGGAAGACAATCAGAACAGTG (55°C)
	famAGTTGGAGCTTCCTGACTCGCCTGCbhq
CX47	CCCAGGACAATCAGGGATTC; CAGGCCAGAAAATTCCAGTC (55°C)
	CAGGACAATCAGGGATTC; CCAGAAAATTCCAGTCTCC (51°C)
	famCTGCTTGGATGCTGGTCCTCCbhq
BLBP	GGTGGGAAATGTGACCAAAC; TGTCTCCATCCAGGCTAACA (55°C)
	CAACGGTAATTATCAGTCAAG; GGCTAACAACAGACTTACA (51°C)
	famTCATCAGGACTCTCAGCACATTCAAbhq
GFAP	CTGCGGCTCGATCAACTCA; GTAGTCGTTGGCTTCGTGCT (55°C)
	GCCTATAGACAGGAAGCA; GTGGATCTTCCTCAAGAAC (49°C)
	famCCTCCAGCGACTCAATCTTCCTbhq

Annealing temperatures for standard three-step PCR are included.

Table 1. Screened nested primer and probe sets designed to amplify mRNA without amplifying genomic DNA.

2.7. Statistical analysis

Statistical analysis was performed using SigmaStat 2.03 (Systat, San Jose, CA). Data are presented as mean ± SEM. Statistical differences between the groups were evaluated with Student's *t*-test for paired data. Pearson's product–moment correlation coefficient was used to determine the correlation between gene expression levels and spontaneous activity. Unavailable values from DIV 7 to DIV 21 were interpolated from the original data set using the method of Least Squares Fit ($n = 15$, where n = number of DIV; $df = 13$, where $df = n - 2$). Only values with $p < 0.001$ ($r \geq 0.76$) were considered statistically significant. For "gene-to-gene" correlation analyses, $n = 12$, where n = number of DIV, and $df = 10$, where $df = n - 2$). DIV 7 was not included in these correlation analyses.

3. Results

3.1. Stem cell-derived human neurons

Two human stem cell lines (hESC-H9 and iPSC-15) were subjected to a five-stage neurodifferentiation protocol over a period of 21 days (**Figure 1A**). The first stage (Stage I) was used to grow and maintain undifferentiated stem cell colonies (**Figure 1B**). Day 0 (DIV 0) marks the Seeding of embryoid bodies. Upon seeding, the cells from the embryoid bodies begin to form neuroepithelial structures (**Figure 1C**, Rosettes). The neuroepithelial cells were expanded in a bFGF-rich media for 4 days (Stage IV). The expanded NEs were lifted on DIV 12 and re-seeded on glass cover slips in BDNF/GDNF-based neurodifferentiation media (Stage V). This protocol produced cells with neuronal morphologies (**Figure 1D$_1$**) and neuronal electrical properties, namely the ability to fire action potentials. Whole-cell recordings were performed on cells with a neuronal appearance—cells with processes visible in IR DIC microscopy (**Figure 1D$_1$**). That is to say that sampling of cells for electrical recordings was biased by visual appearance. The first action potentials were detected on DIV 14, $n = 3$ (n = number of cells) (**Figure 1E**). During the fifth stage (Stage V), whole-cell recordings revealed that some, not all, stem cell-derived neurons were spontaneously producing bursts of action potentials.

In our previous study, stem cell-derived human neurons were grown for 88 days, characterized electrophysiologically, and individual cells were captured for gene expression analysis [8]. The previous study, which used the same two stem cell lines under an identical five-stage differentiation protocol and systematic whole-cell recordings ($n = 229$ human cells from Day 13 to Day 88) produced detailed distributions of electrophysiological properties [8, supplemental figure S1]. In the present study, the focus is on the first 21 days of stem cell neurodifferentiation, when pluripotent stem cells transition to neural progenitors and when first postmitotic neurons are born. Also, in the present study, the gene analysis was performed on the whole population of cells, unbiased by visual appearance or physiological recordings; hence, all cell types present in the culture were included in the RNA samples.

3.2. Spontaneous Ca^{2+} transients

Every 1–2 days, glass cover slips containing cultured human cells were removed from the incubator, loaded with a Ca^{2+}-sensitive dye OGB1-AM, and positioned in the recording chamber of the electrophysiological imaging station (**Figure 1D$_2$**). Multisite optical recordings of unprovoked (spontaneous) Ca^{2+} activity were performed from DIV 7 to DIV 21 in 136 locations (visual fields, **Figure 2A$_1$**) distributed across 25 cover slips (one visual field approximately 380 × 380 μm^2). Regions of interest (ROIs) were placed on all OGB1-stained cells, indiscriminately (**Figure 2A$_2$**). Therefore, the ROIs include both neurons and nonneurons present in the culture. From each visual field, the Ca^{2+} signals were recorded for at least 5 min (**Figure 2B**). In the majority of visual fields, three successive optical recording trials were performed, extending the total monitoring to 15 min. Long recording sessions were needed, because spontaneous outbursts of electrical activity in developing human neurons are interspersed by silent periods that may last for several minutes [5]. Cells were deemed "active" if at least one Ca^{2+} transient was detected (**Figure 2B**, upper traces). Some cells showed no activity throughout the entire recording sessions (**Figure 2B**, lower traces). The percentage of

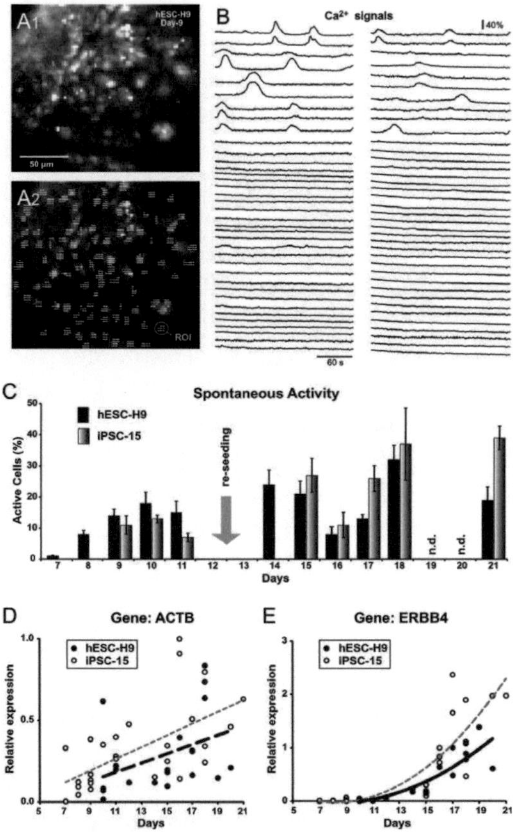

Figure 2. Two human lines (hESC-H9 and iPSC-15) produce cells with similar Ca^{2+} activity and gene expression. (**A_1**) hESC-H9 cells on DIV-9 were bulk-loaded with OGB1-AM. Image captured by fast data acquisition camera (80 × 80 pixels). (**A_2**) Same as in A_1, except gray dashes mark the actual pixels used to generate regions of interest (ROI). (**B**) Spontaneous Ca^{2+} transients from 75 ROIs are shown. Each trace represents 3.5 min of continuous optical monitoring. Cells with spontaneous Ca^{2+} activity, "Active Cells," are grouped on the top. The majority of ROIs have no spontaneous Ca^{2+} transients. (**C**) The average "Active Cells" per day of differentiation for two stem cell lines, hESC-H9 (black) and iPSC-15 (gray). "Active Cells": the number of ROIs with Ca^{2+} transients divided by the total number of ROIs in the visual field. Error bars are s.e.m. The number of visual fields studied by multisite Ca^{2+} imaging in the two stem cell lines on each DIV are listed in **Table 2A**. On DIV-12, the expanding colonies were lifted and reseeded on fresh cover slips in a new type of a differentiation media (addition of BDNF and GDNF). (**D**) Gene expression of ACTB measured in the hESC-H9 (black) and iPSC-15 line (white). Each data point represents one cover slip (one culture). Black and gray trend lines are polynomial fits through the hESC-H9 and iPSC-15 data sets, respectively. (**E**) ERBB4 expression in two stem cell lines normalized to that of ACTB.

cells displaying spontaneous Ca^{2+} activity ("Active Cells") was calculated for each visual field by dividing the number of active ROIs (cells with Ca^{2+} signals) with the total number of ROIs in the visual field (**Figure 2A$_2$**). The average percentage of cells displaying spontaneous Ca^{2+} activity did not significantly differ between hESC-H9 and iPSC cells (**Figure 2C**). Therefore, in the remaining analysis (figures), the physiological data from the two cell lines were combined.

(A)	**Physiology**			
	hESC-H9		**iPSC-15**	
	***n* (visual fields)**	***N* (cover slips)**	***n* (visual fields)**	***N* (cover slips)**
DIV-07	4	2	0	0
DIV-08	20	2	0	0
DIV-09	8	1	6	1
DIV-10	8	2	13	2
DIV-11	7	1	3	1
DIV-14	7	1	0	0
DIV-15	6	1	8	1
DIV-16	2	1	5	1
DIV-17	6	1	9	1
DIV-18	12	2	5	2
DIV-21	2	1	5	1
Total	82	15	54	10
(B)	Gene expression analysis			
	hESC-H9	iPSC-15	Combined	
	N (cover slips)	*N* (cover slips)	*N* (cover slips)	
DIV-07	3	0	3	
DIV-08	3	0	3	
DIV-09	5	0	5	
DIV-10	1	4	5	
DIV-11	3	2	5	
DIV-12	0	2	2	
DIV-14	1	1	2	
DIV-15	2	3	5	
DIV-16	3	1	4	
DIV-17	2	2	4	
DIV-18	3	3	6	
DIV-20	1	2	3	
DIV-21	1	0	1	
Total	28	20	48	

n indicates number of visual fields used for optical recordings. Upon physiological recordings cover slips were saved for gene analysis. (B) Experimental samples used for gene analysis. These samples include cover slips saved after physiological recordings (25 cover slips in A), as well as 23 additional cover slips that were saved without any optical recordings.

Table 2. (A) Experimental samples used for physiological recordings. *N* indicates number of cover slips with cultured cells.

3.2.1. Active cells

The average percentage of cells displaying spontaneous Ca^{2+} activity increased steadily from DIV-7 through DIV-21 (**Figure 3A**). On DIV-7, the average number of active cells was 1 ± 0.1%. By DIV-14, the average number of active cells was 24 ± 4.64%, and by DIV-21, approximately 40% of cells were active (40 ± 6.3%). There was a statistically significant increase in the average percentage of active cells after the introduction of neurodifferentiation media (DIV-14 to DIV-21) as compared to that before its introduction (DIV-7 to DIV-11) ($p < 0.05$). As human neurons mature, they fire repetitive action potentials important to initiate contact with the surrounding cells [27,28]. Belinsky et al. [8] have demonstrated that repetitive activity of immature neurons can be used as a measure of their maturity.

3.2.2. Amplitude

The amplitude of spontaneous Ca^{2+} transients was analyzed over the first 3 weeks of human stem cell neurodifferentiation. Only cells with Ca^{2+} transients were included in the analysis ("Active Cells") (**Figure 2B**, upper traces). The average peak amplitude of spontaneous Ca^{2+} transients was calculated for each day in vitro (**Figure 3B**), and there was a statistically significant decrease in the average percentage of active cells after the introduction of neuro-differentiation media (DIV-14 to DIV-21) as compared to that before its introduction (DIV-8 to DIV-11) ($p < 0.05$, one-tailed t-test). The difference is significant when the data series DIV-8 to DIV-10 is compared to the data series DIV-11 to DIV-21 ($p < 0.05$, two-tailed t-test). Larger signal amplitudes were predominantly distributed in the first half of the in vitro neurodifferentiation protocol, when cell cultures primarily consisted of undifferentiated cells and neuron progenitors (**Figure 3B**).

3.2.3. Repetitive activity

Among active cells, we found a group of cells with more frequent Ca^{2+} signals (repetitive activity). The criterion for repetitive activity was set to four or more spontaneous Ca^{2+} transients in the period of time required to complete one optical recording trial (5 min). The number of cells with repetitive activity was divided by the number of active cells in the same visual field, and expressed as a percentage (**Figure** 3C). The percentage of cells exhibiting repetitive activity significantly increased with the addition of brain and glial cell-derived neurotrophic factors (BDNF and GDNF) into the cell culture media (DIV-14 to DIV- 21, as compared to DIV-7 to DIV-11; $p < 0.005$).

3.2.4. Duration

We measured the half-width (duration measured at half amplitude) of all recorded Ca^{2+} signals ($n = 4726$ Ca^{2+} transients obtained from 132 visual fields in 23 cover slips; DIV-7 not considered) over the first 3 weeks of human stem cell neurodifferentiation. The duration of spontaneous Ca^{2+} transients was calculated for each day of differentiation and plotted as a function of time, *day in vitro* on the x-axis (**Figure 3D**). The signal duration in the time period between DIV-8 to DIV-11 was significantly longer compared to the time period DIV-15 to

DIV-21 ($p < 0.001$). The median duration of spontaneous Ca^{2+} transients was determined to be 8 s. The Spitzer laboratory had previously reported rapid (narrow) and slow (wide) Ca^{2+} transients, but the significance of these Ca^{2+} transients at very early stages of human neurodifferentiation has as yet to be unraveled [29,30]. Based on a median duration of 8 s, we separated Ca^{2+} transients into two groups, one consisting of wide signals with duration of more than 8 s (**Figure 3E**, Wide) and the other of narrow signals with duration equal to, or less than, 8 s (**Figure 3E**, Narrow). This allowed us to calculate the percentage of narrow and wide signals at each day of differentiation (**Figure 3F**). The proportion of narrow Ca^{2+} transients increased from DIV-8 to DIV-21 (**Figure** 3F, black bars). However, at the beginning of the neurodifferentiation protocol, during Stages III and IV, the number of narrow and wide signals was about 1:1, with nearly half of the total number of Ca^{2+} transients falling into the narrow category and another half being categorized as wide. The difference between narrow and wide counts in this period (DIV-8 to DIV-11) was not statistically significant ($p < 0.05$). From DIV-15 onward, this initial balance was thrown off with narrow Ca^{2+} transients greatly outnumbering wide transients (**Figure 3F**, DIV-14 to DIV-21). There was a statistically significant difference between wide and narrow signal counts after the introduction of the differentiation factors in the culture medium, the cultures predominantly containing postmitotic neurons (DIV-14 to DIV-21), as compared to DIV-7 to DIV-11 ($p < 0.01$). Thus, the ratio of wide and narrow (less than 8 s) events can be attributed to the maturation of stem cell-derived neurons.

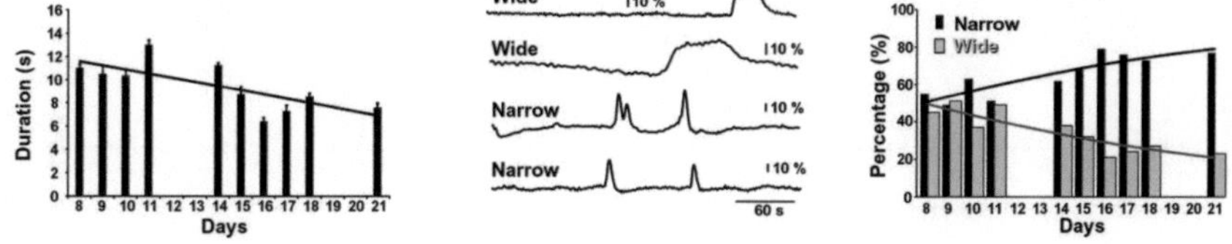

Figure 3. Spontaneous Ca^{2+}transients during early stages of neurodifferentiation. (**A**) In this and all remaining figures, the data from two cell lines are pooled together. Average number of "Active Cells" per visual field from DIV-7 to DIV-21. "Active Cells": the number of ROIs with Ca^{2+} transients divided by the total number of ROIs in that visual field (%). (**B**) Average amplitude of Ca^{2+} signals $\Delta F/F$ per day. The total number of spontaneous Ca^{2+} signals analyzed from DIV-7 to DIV-21 is 4744. (**C**) Cells with repetitive activity (4 or more Ca^{2+} transients per 5 min period) divided by the number of active cells in that visual field. An active cell is any cell with 1 or more Ca^{2+} transients per 5 min period. (**D**) Durations (half-widths) of all Ca^{2+} transients recorded in this project are plotted as average values per day. Same data set as in ***B***. (**E**) Optical traces showing wide and narrow spontaneous Ca^{2+} transients. "Wide" is defined as a signal half-width (width at half amplitude) greater than 8 s. "Narrow" is defined as a signal with half-width ≤ 8 s. (**F**) Distributions of Wide (gray) and Narrow (black) spontaneous Ca^{2+} transients per day, expressed as percentage of all transients on that day in vitro. Same data set as in ***D***.

Ca^{2+} release from internal stores has been reported to be especially sensitive to Ca^{2+} influx at early stages of differentiation, leading to the generation of Ca^{2+} transients of large amplitude and duration, with the sensitivity to Ca^{2+} influx decreasing as cells mature [29]. We found that larger amplitudes were predominantly distributed in the first half of our in vitro neurodifferentiation protocol, when cell cultures primarily consisted of undifferentiated cells and neuron progenitors. The amplitude of wide Ca^{2+} transients was not significantly correlated to the expression of any of the genes studied. Our results are consistent with those of Gu et al. [29]. We also observed a statistically significant positive correlation between percentage repetitive activity and the number of narrow events (action potentials; $r = 0.874$, $p < 0.001$).

The multisite Ca^{2+}imaging data indicate that spontaneous Ca^{2+} transients were present during the proliferation and differentiation phases of the in vitro neurodifferentiation protocol. The

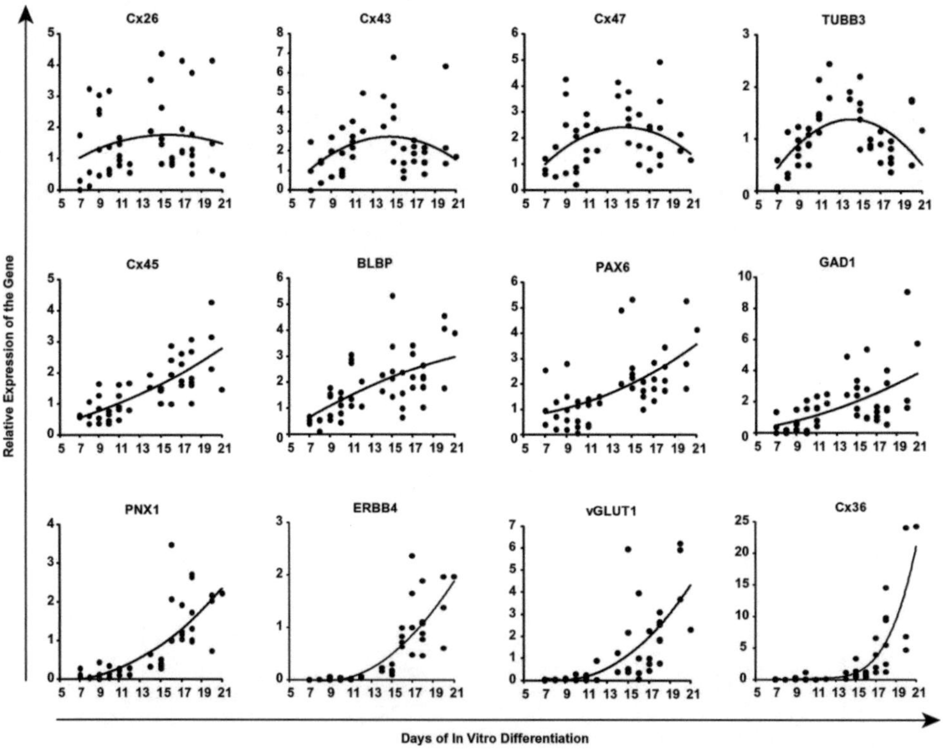

Figure 4. Gene expression levels during early stages of stem cell neurodifferentiation. The expression levels of 12 human genes (*Cx26, Cx43, Cx47, TUBB3, Cx45, BLBP, PAX6, GAD1, PNX1, ERBB4, vGLUT1,* and *Cx36*) normalized to the expression of *ACTB* over a period of 21 DIVs. Genes are organized in three rows based on the temporal patterns of expression. Top row: genes consistently expressed throughout the 21 DIVs, or with expression slowly rising and then slowly declining. Middle row: gene expression increases steadily throughout the study period (21 days). Bottom row: genes with expression starting very low and then rising exponentially as differentiation progresses. Each of the 48 data points shown in the graph represents one cover slip with cultured cells used in a qPCR reaction. N = number of cover slips. For DIV-7 ($N = 3$), DIV-8 ($N = 3$), DIV-9 ($N = 5$), DIV-10 ($N = 5$), DIV-11 ($N = 5$), DIV-12 ($N = 2$), DIV-14 ($N = 2$), DIV-15 ($N = 5$), DIV-16 ($N = 4$), DIV-17 ($N = 4$), DIV-18 ($N = 6$), DIV-20 ($N = 3$), and DIV-21 ($N = 1$) (**Table 2B**). The trend lines are second order polynomials, except gene *Cx36* which is fitted with a fourth order polynomial.

temporal patterns of three distinct parameters of Ca^{2+} transients: (a) increasing frequency, (b) decreasing peak amplitude, and (c) decreasing duration, are consistent with the gradual transition between the Ca^{2+} activities of undifferentiated cells (large amplitude–long duration) driven by a release from internal stores [17,31,32] to the spontaneous Ca^{2+} activities of postmitotic human nerve cells driven by spontaneous electrical activity [30,33,34]. The relation between electrical activity and Ca^{2+} transients in human postmitotic neurons was previously determined by simultaneous whole-cell and Ca^{2+} imaging from the same cell [8, **Figure 5**].

3.3. Gene expression during neuronal differentiation

Gene expression analysis was performed on the same biological samples in which physiological recordings were made (**Table 2A**), in order to determine if there are any correlations between gene onset, gene dynamics, and spontaneous Ca^{2+} activity. Connexin and pannexin gene isoforms are expressed in the human fetal brain at the development stages when spontaneous Ca^{2+} activity is also found [35]; these genes were analyzed in the present study along with several neuronal, glial, and human genes. More precisely, the samples were tested for four classes of genes: (1) connexins (*Cx26, Cx36, Cx43, Cx45, Cx47*, pannexin-1 (*PNX1*); (2) neuronal marker genes *PAX6, vGLUT1, GAD1, TUBB3;* (3) glial lineage markers *BLBP, GFAP;* (4) schizophrenia-associated gene (*ERBB4*). Housekeeping gene (*ACTB)* was used to normalize the gene expression values.

We compared the average expression for each of the genes, in the two cell lines, hESC-H9 and iPSC-15, over the time period DIV-7 to DIV-21 (**Table 2B**). Our gene expression results failed to detect any principal differences between the hESC-H9 and iPSC-15 cells ($p > 0.05$; $N = 28$ cover slips containing hESC-H9, and $N = 20$ cover slips with iPSC-15 cells). Our results are consistent with the single-cell PCR analysis of stem cell-derived neurons, which also found similar gene expression patterns between HESC-H9 and iPSC-15 lines [8]. Based on similar physiological (**Figure 2**C) and gene expression results (**Figure 2D**, E) obtained in two human cell lines, the data from the two cell lines were combined in the remainder of the study.

With the exception of *GFAP*, all tested genes (total number of genes tested = 14) were detected at some point between DIV-7 and DIV-21. The expression of *Cx26* was fairly constant (**Figure 4**, *Cx26*). While *Cx43, Cx47*, and *TUBB3* were upregulated in the presence of bFGF (cell proliferation stage, DIV-7 to DIV-12), their levels fell when bFGF was withdrawn on DIV-12 and differentiation factors BDNF and GDNF were added to the culture (**Figure 4**). The expression of genes *Cx45, BLBP, PAX6*, and *GAD1* did not seem to be influenced by the addition or removal of growth factors from the cell culture media (**Figure 4**). The expression of *ERBB4, Cx36*, and *vGLUT1* was negligible in the presence of bFGF (**Figure 4**), but sharply increased after the introduction of differentiation factors BDNF and GDNF to the culture medium on DIV-12.

The temporal patterns of gene expression during the course of early differentiation of human neurons were divided into four general categories (**a** – **d**) and characterized as follows:

a. Unchanging, constant expression (**Figure 5**, line "a"). The example of unchanging constant expression is the temporal pattern of *Cx26* (**Figure 4**).

b. Hyperbolic increase (**Figure 5**, line "b"). The examples of hyperbolic increase are the temporal patterns of *Cx43, Cx47,* and *TUBB3* (**Figure 4**).

c. Monotonously rising, linear increase (**Figure 5**, line "c"). The examples of linearly increasing pattern of expression are *PAX6, BLBP, Cx45,* and *GAD1* (**Figure 4**).

d. Exponential increase (**Figure 5**, "d" lines). The examples of exponentially increasing gene levels can be found in the expression patterns of *PNX1, vGLUT1, ERBB4,* and *Cx36* (**Figure 4**). An interesting aspect of the exponential increase pattern is the day of onset when gene expression becomes detectable. Our measurements showed that this day of onset varied between the genes. For example, the expression onset of *Cx36* (~DIV-17) was at least 3 days later than the onset of *ERBB4* and *vGLUT1* (~DIV-13) or 5 days later than the onset of *PNX1* (~DIV-11). Three genes, *ERBB4, vGLUT1,* and *Cx36,* exhibited their onsets after DIV-12, followed by a sharp rise in the gene expression level, suggesting that these genes responded to the addition of growth factors (GDNF and BDNF) on DIV-12.

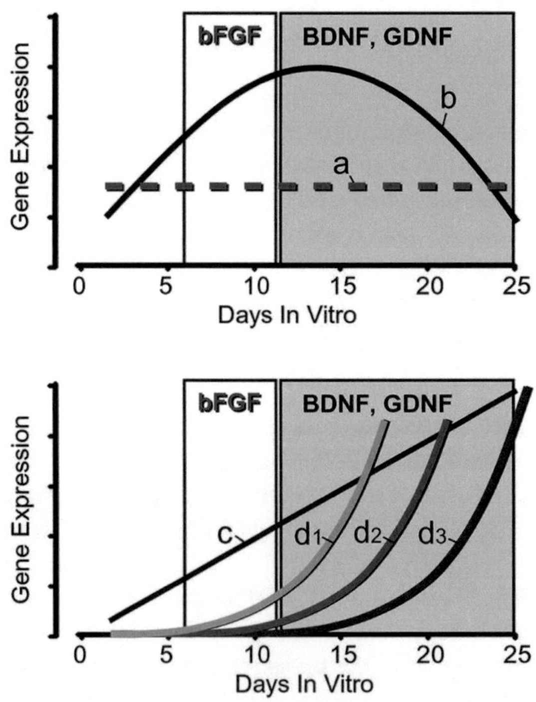

Figure 5. Characteristic patterns of gene dynamics during early stages of stem cell neurodifferentiation. Differences in differentiation media before and after DIV-12 are indicated by rectangles in the background. The four temporal patterns of gene expression levels (**a**–**d**) are based on the polynomial fits obtained from the raw data points shown in (Figure 4 (**a**) A relatively constant expression level over the entire course of differentiation for *Cx26*. (**b**) Hyperbolic pattern—a slow rise followed by a slow decline for genes *Cx43, Cx47,* and *TUBB3*. (**c**) Monotonous linear increase in the expression of *PAX6, BLBP, Cx45,* and *GAD1*. (**d_1, d_2,** and **d_3**) An exponential increase with a delayed onset. The onset of the exponential rise begins on DIV-11 for *PNX1*; on DIV-13 for *ERBB4* and *vGLUT1*; and on DIV-17 for *Cx36*.

Next, we examined gene expression levels in relation to one significant physiological event: the onset of AP (Action Potential) firing capacity. The first APs were detected in young human neurons on DIV-14 using patch clamp recordings (**Figure 1**E). This event divides the time course of our experiment into two periods: (a) before AP firing activity was present and (b) after AP firing was established in developing human neurons. The experimental data reveals three gene expression trends in relation to the AP firing onset: (i) Negatively correlated; (ii) Unperturbed/Unrelated; and (iii) Positively correlated. In the first group "Negatively correlated", the gene expression levels were rising before AP firing was present. Upon the onset of AP firing, the gene expression trend reversed from "rising" to "falling", Three genes (*Cx43, Cx47*, and *TUBB3*) showed an apparent decline after the AP firing activity was established on DIV-14 (**Figure 4**, Top row). In the second group "Unperturbed/Unrelated", the gene expression trend did not react to the appearance of the first APs in young neurons on DIV-14. These genes are *Cx45, BLBP, PAX6, GAD1*, and *PNX1* (**Figure 4**). In the third group "Positively correlated," the gene expression was not detected before the appearance of the first APs. Following the appearance of APs on DIV-14, three genes, *ERBB4, vGLUT1*, and *Cx36*, exhibited an exponential increase in expression levels (**Figure 4**, Bottom row).

3.4. Activity to gene correlations

We next examined if there are any significant correlations between a cell's physiology, manifested through spontaneous Ca^{2+} activity, and gene expression patterns during early neuronal differentiation. Recall that the multicell calcium imaging data (**Figure 2AB**) were used to quantify the population of spontaneously active cells ("Active Cells") on a given DIV (**Figure 3A**). The same cells, in which calcium imaging was performed, were also subjected to gene analysis (**Figure 4**). This allowed us to plot physiological data and gene expression analysis data on the same graph.

3.4.1. Active cells

For example, **Figure 6A** shows similar trends of "Active Cells" (fraction of cells exhibiting spontaneous Ca^{2+} activity) and relative expression levels of the *PNX1* gene. The time course of *Cx26* expression, on the other hand, was unrelated to the fraction of cells exhibiting spontaneous Ca^{2+} activity (**Figure 6B**, "Active Cells"). We used Pearson's product–moment correlation coefficient test to study the relationship between spontaneous Ca^{2+} activity and gene expression. Two groups of genes emerged from the numerical analysis of correlation. One group of genes possessed expression patterns highly correlated with the percentage of spontaneously active cells (**Table 3**, Row: "Active Cells"; bold numbers, $p < 0.001$). This group of genes included *Cx36, BLBP, Cx45, ERBB4, GAD1, PAX6, PNX1*, and *vGLUT1*. All aforementioned genes had a positive correlation coefficient (CC), indicating that gene expression levels increased with an increase in the number of spontaneously active cells ("Active Cells"). The other group of genes exhibited expression profiles unrelated to the "Active Cells" parameter (**Table 3**, Row: "Active Cells"; pale numbers; $p > 0.001$). This group of genes included *Cx26, Cx43, TUBB3*, and *Cx47*.

(A)		*Cx26*	*Cx43*	*Cx36*	*BLBP*	*TUBB3*	*Cx45*
1	Active cells (%)	0.5651	0.1844	**0.8624**	**0.9727**	−0.0570	**0.9998**
2	Repetitive activity (%)	0.5477	0.1638	**0.8715**	**0.9677**	−0.0779	**0.9991**
3	Narrow: signal duration < 8 s; all signal amplitudes included	0.7372	0.2599	0.7244	**0.9987**	0.1152	**0.9862**
4	Moderate duration: signal duration between 8 and 20 s; signal amplitude >10%	−0.5069	**−0.8902**	0.7758	0.1590	**−0.9474**	0.3678
5	Moderate duration: signal duration between 8 and 20 s; signal amplitude > 5%	0.5534	0.0203	**0.8566**	**0.9574**	−0.1268	**0.9971**
6	Wide: signal duration > 20 s; signal amplitude < 10%	**0.8996**	**0.9935**	−0.3271	0.4146	**0.9660**	0.2090
7	Wide: signal duration > 20 s; signal amplitude > 20%	**0.9106**	**0.9902**	−0.3034	0.4380	**0.9590**	0.2343
(B)		*Cx47*	*ERBB4*	*GAD1*	*PAX6*	*PNX1*	*vGLUT1*
1	Active cells (%)	0.1641	**0.9776**	**0.99998**	**0.9994**	**0.9958**	**0.9753**
2	Repetitive activity (%)	0.1434	**0.9818**	**0.9996**	**0.9999**	**0.9975**	**0.9797**
3	Narrow: signal duration <8 s; all signal amplitudes included	0.3490	**0.8972**	**0.9634**	**0.9591**	**0.9448**	**0.9071**
4	Moderate duration: signal duration between 8 and 20 s; signal amplitude >10%	**−0.8437**	0.6190	0.4631	0.4770	0.5175	0.6008
5	Moderate duration: signal duration between 8 and 20 s; signal amplit. >5%	0.1137	**0.9770**	**0.9996**	**0.9990**	**0.9958**	**0.9817**
6	Wide: signal duration >20 s; signal amplitude <10%	**0.9998**	−0.0802	0.1055	0.0899	0.0433	−0.0573
7	Wide: signal duration >20 s; signal amplitude >20%	**0.9989**	−0.0543	0.1313	0.1157	0.0692	−0.0314

All numerical values on the right side of the table are correlation coefficients (*CC*), calculated using Pearson Product–Moment Correlation test. Bold *CC* values are statistically significant ($p < 0.001$). Pale *CC* values are not statistically significant ($p > 0.001$).

Table 3. Correlation between parameter of spontaneous calcium activity (e.g. Active Cells [%]) and normalized gene expression level (e.g. *Cx36*).

3.4.2. Repetitive activity

We evaluated the relation between repetitive activity and gene expression. "Repetitive Activity" is a parameter extracted from the multicell calcium imaging data. This parameter describes the number of cells in the visual field exhibiting four or more Ca^{2+} transients per 5 min of optical recording, divided by the number of "Active Cells" (cells showing at least one Ca^{2+} transient) in that field (%). We found that the same set of genes positively correlated to "Active Cells" was also positively correlated with "Repetitive Activity" (**Table 3**, Row: "Repetitive Activity").

3.4.3. Ca^{2+} signal properties

As many as 4744 spontaneous Ca^{2+} transients were documented in 25 cover slips over the period of DIV-7 to DIV-21. Each spontaneous Ca^{2+} transient was measured in terms of signal peak amplitude and signal duration (half-width), which allowed us to group data points based on signal amplitude and signal duration (**Figure 3**).

3.5. Narrow signals (<8 s)

3.5.1. GAD1 and vGLUT1

Gu and Spitzer [30] observed that the expression of the neurotransmitter GABA in developing Xenopus spinal neurons is regulated by spontaneous Ca^{2+} spikes (<5 s duration) [30]. Based on the median duration of 8 s, we categorized data points into "Narrow" and "Wide" Ca^{2+} events (**Figure 3F**). The number of *Narrow* events (<8 s) from DIV-7 to DIV-21 was significantly correlated to the expression of *GAD1* and *vGLUT1* (**Table 3**, Row 3). Since GAD1 is involved in the synthesis of GABA [17,30], these data suggest that in developing human neurons, there is a close relationship between spontaneous Ca^{2+} activity and the differentiation of neurons into GABAergic and glutamatergic neurons.

3.5.2. Cx45

The number of narrow events (with duration <8 s) was strongly ($p < 0.001$) correlated to the expression of *Cx45* (**Figure 6D**; **Table 3**, Row 3). *Cx45* is known to influence the spontaneous firing patterns in the developing retina [2].

3.5.3. BLBP

The number of narrow events (duration < 8 s) from DIV-7 to DIV-21 was significantly ($p < 0.001$) correlated to the expression of *BLBP* (**Table 3**, Row 3). BLBP expression has been reported to increase during neuronal differentiation [36], being required to induce morphological changes in radial glia in response to neuronal cues [37,38].

3.6. Wide signals (>20 s)

3.6.1. TUBB3, Cx47, and Cx26

We found that Ca^{2+} transients with duration >20 s and peak amplitude ≤10% $\Delta F/F$ not only had a strong positive correlation to the expression of the neuronal marker *TUBB3*, but also to three connexin isoforms *Cx43, Cx47*, and *Cx26* ($p < 0.001$, **Table 3**, Row 6). Connexin isoforms *Cx43* and *Cx47* were previously found in astrocytes [39,40] and oligodendrocytes [41], respectively. *Cx26* has been reported to be located on neurons and astrocytes of the developing brain as well as the leptomeninges (Nagy *et al*., 2001).

The count of Ca^{2+} transients with duration >20 s and amplitude >20% $\Delta F/F$ showed a significant positive correlation ($p < 0.001$) to the expression levels of four genes: *Cx26, Cx43, TUBB3*, and *Cx47* (**Table 3**, Rows 7). Interestingly, these were the only four genes that showed no correlation to the parameters "Active Cells" and "Repetitive Activity" (**Table 3**, Rows 1 and 2). These subsets of Ca^{2+} transients with long duration and large amplitude (duration > 20 s and amplitude > 20% $\Delta F/F$, as well as duration > 20 s and amplitude ≤ 10% $\Delta F/F$) had no significant correlation to the expression of eight genes: *Cx36, BLBP, Cx45, ERBB4, GAD1, PAX6, PNX1*, and *vGLUT1*. These are exactly the same eight genes that are highly correlated to the parameters "Active Cells" and "Repetitive Activity" (**Table 3**, Rows 1 and 2).

Overall, the correlation analyses revealed potential links between the number of cells with spontaneous calcium transients ("Active Cells" and "Repetitive Activity") and expression patterns of two connexins (*Cx36, Cx45*), one pannexin (*PNX1*), and several general neuronal genes, including *ERBB4, GAD1, PAX6*, and *vGLUT1* (**Table 3**, Rows 1 and 2). These results indicate one of two possibilities: (a) that specific subclasses of spontaneous Ca^{2+} transients could be driven by the expression of specific human genes, (b) specific amplitude and duration of Ca^{2+} transients may selectively promote expression of some human genes at early stages of neuronal differentiation. In the current study, however, we did not perform experiments to explore if expression of a certain human gene affects the incidence of spontaneous Ca^{2+} transients, with certain combinations of amplitudes and durations.

3.7. Gene-to-gene correlations

The time courses of gene expression levels shown in **Figure 4** suggest that the expression of all genes in the bottom row—*PNX1, ERBB4, vGLUT1*, and *Cx36* correlate to each other. In addition to this, we found a strong correlation ($r = 0.91$, $p < 0.001$) between *Cx43* and *TUBB3* expression (**Figure 4**, Top row).

We found that there was a statistically significant correlation between *Cx36* and *GAD1* (GABA expression; $r = 0.79$, $p < 0.01$), as well as a statistically significant correlation between *Cx36* and *vGLUT1* ($r = 0.63$, $p < 0.05$).

These data indicate that systematic sampling of gene expression levels from DIV-7 to DIV-21 during human in vitro neurodifferentiation can be used to detect groups of genes that rise and fall in parallel to each other, suggesting that their expression is driven by the same sequence of cellular and molecular events [42].

4. Discussion

Abnormal fetal brain development is linked to a myriad of neurological and psychiatric diseases. Human fetal gestation is a particularly sensitive stage of brain development, during which aberrant genetic programs or external factors (viruses, pollutants, toxins, and other agents) exert the most devastating impact [43,44]. The etiopathogenesis of mental illnesses, such as autism and schizophrenia, is not exclusively determined by genetic variation, but rather by the interaction of genes with the environment [45,46]. Variations in the duration and amplitude of Ca^{2+} transients and their time of occurrence affect neuronal development and plasticity [3], but these physiological aspects of human neurodifferentiation are largely absent in clinical and laboratory research. Unsurmountable ethical and technical difficulties preclude experiments on human fetuses. One possibility for studying the physiology of young human neurons is to use hESC and iPSC technologies and mimic human neurodifferentiation in vitro [6,11,47,48].

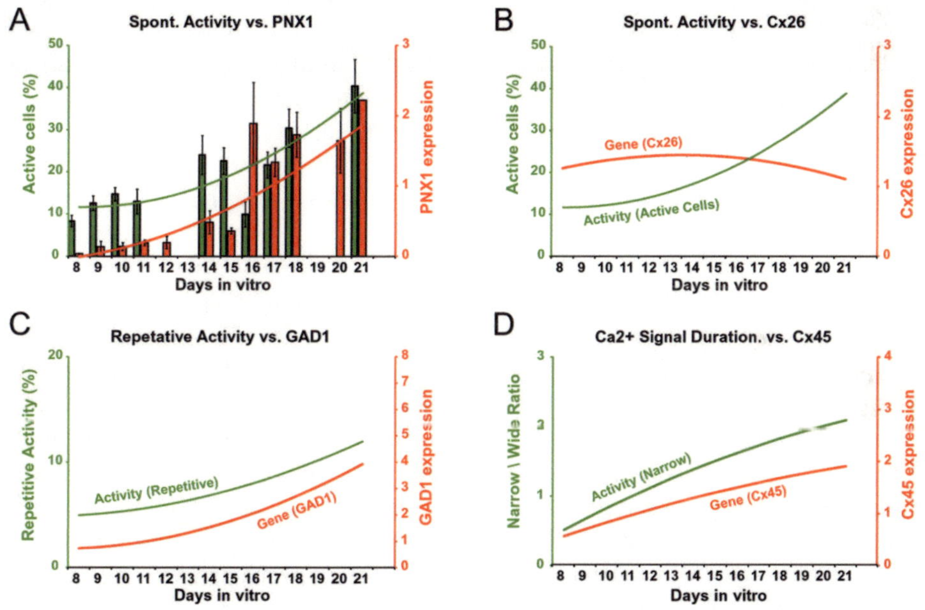

Figure 6. Correlations between different parameters of spontaneous Ca^{2+}transients and gene expression levels. (**A**) In this and all the following panels, a specific parameter of the spontaneous Ca^{2+} activity is plotted using the left ordinate axis (green), while the relative gene expression levels are plotted using the right ordinate axis (red). "Active Cells": the number of ROIs with Ca^{2+} transients divided by the total number of ROIs in that visual field (%). Each green bar represents an average value obtained on a given DIV. Each red bar represents an average relative expression of *PNX1* on a given DIV. The trend lines are polynomial fits through spontaneous Ca^{2+} activity (green) or the gene expression levels (red). The two trend lines are strongly correlated; correlation coefficient, $CC = 0.99$. (**B**) Same physiological parameter ("Active Cells") as in *A* is compared against the relative expression of *Cx26*. The two trend lines, physiological activity (green) and gene expression (red), are not correlated; $CC = 0.56$. (**C**) The percent of cells with repetitive Ca^{2+} activity on each day of in vitro differentiation (green) is in good correlation to the expression levels of *GAD1* (red); $CC = 0.97$. (**D**) The fraction of narrow Ca^{2+} transients on each day of in vitro differentiation (green) is positively correlated with the expression levels of *Cx45* (red); $CC = 0.98$.

In this experimental project, we have applied embryonic and pluriopotent stem cell technology to study physiological activity and gene expression during the earliest stages of human neurodifferentiation. Using multisite Ca^{2+} imaging and qPCR, we have compared time courses of spontaneous physiological Ca^{2+} activity and gene expression in two human cell lines (hESC-H9 and iPSC-15), as these cells transitioned from undifferentiated stem cells to postmitotic neurons. The iPSC-15 cell line was derived from a patient with schizo-affective disorder with a deletion on 22q11.2 that results in haploinsufficiency of ~ 40 genes, which is one of the most common causes of psychotic disorders [49–51]. We did not observe any significant differences in electrical properties, action potential frequencies, and the expression of genes important for neural development between the two cell lines, demonstrating the presence of intact cellular machinery for basic neuronal function [8].

Spontaneous Ca^{2+} transients were observed in both cell lines from the first day of recording (DIV-7), and increased in occurrence throughout the entire in vitro differentiation period. Ca^{2+} transients act via a variety of signal types, with amplitude, duration, and frequency, each encoding a specific piece of information [52]. The multisite optical imaging method used in the present study provided means for monitoring the critical parameters of Ca^{2+} transients: amplitude, duration, and frequency in up to 80 cells simultaneously. The results suggest that (a) lower amplitude, (b) greater ratio of narrow to wide signals, and (c) increased frequency are all characteristics of maturing nerve cells (**Figure 3E, F**).

4.1. Correlating spontaneous electrical activity with gene expression

Interestingly, as human cells differentiated from DIV-7 to DIV-21, there was a strong positive correlation between the number of cells exhibiting repetitive spontaneous Ca^{2+} transients and the expression levels of *GAD1* and *ERBB4* ($p < 0.001$; **Table 3**, Row 2). Also, there was a strong positive correlation ($p < 0.001$) between the number of narrow events and the expression levels of either *GAD1* or *ERBB4* (**Table 3**, Row 3). These parallel trends of *ERBB4* and *GAD1* are not surprising as GABAergic interneurons have been suggested to employ *ERBB4* receptor signaling early in the development of the cerebral cortex [53]. In accordance with our present observations, a previous study employing patch clamp recording followed by single-cell PCR revealed that individual neurons expressing *GAD1* and *ERBB4* are more likely to engage in spontaneous electrical activity than their counterparts which do not express these two genes [8]. As human cells differentiated into neurons and matured, the expression of *GAD1* and *ERBB4* increased at the same rate as the number of cells showing spontaneous Ca^{2+} activity (**Table 3**, Row 1) or the number of cells showing repetitive Ca^{2+} transients (**Figure 6C**). The regulation of the GABAergic neuronal phenotype (*GAD1* expression and production of GABA) appears to occur via an activity-dependent mechanism and is Ca^{2+}-mediated, as shown by removing extracellular Ca^{2+} or stimulating cultured neurons with different frequencies of Ca^{2+} spikes [30,54].

Narrow calcium events (Ca^{2+} transients with duration < 8 s) fit the description of AP-induced Ca^{2+} signals in postmitotic human neurons obtained by simultaneous patch clamp and optical recordings [8], and as such these short-duration Ca^{2+} signals mark the arrival of newborn neurons in stages IV and V of the current neurodifferentiation protocol (**Figure** 3F).

We, therefore, explored the possibility that the expression of all the other genes that correlated with repetitive activity also had a significant correlation to the number of narrow events. We found that except for *Cx36*, which is significantly correlated ($p < 0.001$) to percent repetitive activity, but not to the number of narrow events, the rest of the genes (*BLBP, Cx45, ERBB4, GAD1, PAX6, PNX1,* and *vGLUT1*) had a significant positive correlation to both repetitive activity and number of narrow events (**Table 3**, Rows 2 and 3). *Cx36* has been reported to be expressed during the development of retinal ganglion cells. It is thought that *Cx36* is capable of influencing spontaneous firing patterns, but is not required for spontaneous retinal wave generation [2].

The addition of growth factors BDNF and GDNF to the cell culture media appeared to have an effect on the expression of *ERBB4, vGLUT1,* and *Cx36*. These three genes were present at minimal levels, until the addition of growth factors on DIV-12 caused a sharp increase in their level of expression (**Figure 4**). Interestingly, the three genes showing exponential rise after DIV-14 are also highly correlated with Ca^{2+} signals of *Moderate* duration of 8–20 s and amplitude greater than 5 % $\Delta F/F$ (**Table 3**, Row 5). *Cx36* is thought to participate in neuronal gap junction formation in vivo. An increased expression of *Cx36* has been demonstrated to promote neuronal differentiation by enhancing cell–cell contact between progenitors [55]. In situ hybridization techniques have shown high levels of *Cx36* mRNA in olfactory bulbs, pineal gland, inferior olive, hippocampal pyramidal neurons, and in the retina. We found a significant correlation between *Cx36* and *GAD1* expression patterns ($r = 0.79$, $p < 0.01$), suggesting some functional link between these two genes. *Cx36* has been reported to be present on GABAergic interneurons in the neocortex and hippocampus [56]. In *Cx36*-deficient mice, gap junction-coupling among neocortical inhibitory interneurons has been shown to be nearly absent. GAD1 (GAD67) is an enzyme that synthesizes GABA and is itself influenced by Ca^{2+} signaling [17,30]. Thus, *Cx36* and *GAD1* have a close relationship, which may be a part of the regulatory mechanisms for expression of GABA and/or GABAergic interneuron differentiation.

In addition to having a significant correlation to *GAD1*, we found that there was a significant correlation between *Cx36* and the *vGLUT1* expression patterns ($r = 0.63$, $p < 0.05$) over the period DIV-8 to DIV-21. The *Cx36* gene has been reported to be induced at the time of the surge of the transcription factors that determine β-cell differentiation [57]. Thus, during neuronal differentiation, *Cx36* could potentially influence transcription factors that determine the terminal differentiation of neurons to either glutamatergic or GABAergic, with the *Cx36* channels forming intercellular pathways for transmission of developmentally relevant molecules. Upregulation of *Cx36* expression has also been previously associated with the sequential expression of specific ligand-gated responses (GABA and glutamate), thus heralding terminal differentiation of neurons, with the peak expression of *Cx36* being associated with a developmental window in which the neuronal network connections are being developed [58].

In the present study, we found that *Cx43, Cx47,* and *TUBB3* showed a hyperbolic pattern of expression, wherein there was an upregulation of these genes during the cell proliferation stage (in the presence of the proliferation factor bFGF) and a subsequent decrease in their expression. Both the increase and the decrease of *Cx43, Cx47,* and *TUBB3,* relative to the characteristic time periods of the differentiation protocol (e.g. replacement of proliferation media on DIV-12; appearance of first APs on DIV-14; sharp onset of *Cx36* expression on DIV-17), may become useful for understanding the roles of these three genes in human neurodifferentiation.

We found that the decline of *TUBB3* after DIV-14 was paralleled by an upregulation of *GAD1* (putative GABAergic neurons) and *vGLUT1* (putative glutamatergic neurons) during the differentiation phase (**Figure 4**, compare *TUBB3, GAD1,* and *vGLUT1*). Similar to our results, a decrease in the expression of the neuronal progenitor marker *TUBB3* along with a simultaneous increase in GABA and glutamatergic neurons during neuronal differentiation was also observed by [42]. We found a strong correlation between *Cx43* and *TUBB3* expression ($r = 0.91$, $p < 0.001$), which suggests a close relationship between *Cx43* expression on astrocytes and *TUBB3* expression on neuronal progenitors. *Cx43* is essential for the maintenance of neural progenitors in a proliferative state [59, 60] and forms communication channels, between migrating neurons and the glial cells, during neurogenesis [61].

By sampling data points on consecutive days (DIV-7 to DIV-21), we observed that the expression of human *GAD1* occurred before the expression of human *vGLUT1* (**Figure 4**), as has been reported in the literature, where the formation of GABAergic synapses in the CNS preceded that of glutamatergic synapses [62]. In summary, as cells exit from the cell cycle and undergo lineage commitment, as shown by the expression of *ERBB4, GAD1,* and *vGLUT1,* they demonstrate a changing profile of connexin expression with *Cx43* and *Cx47* declining, *Cx45* and *PNX1* increasing, and *Cx36* being activated.

Although *BLBP* was expressed in our cell culture, *GFAP* was not expressed at detectable levels. Both *GFAP* and *BLBP* are expressed on radial glia during early neuronal development, but their expression has been reported to be nonoverlapping [63]. In line with our experimental results, *GFAP* was observed to be expressed only after the expression of *BLBP* during in vitro culture of human embryonic stem cells [42]. Thus, culturing cells for a period longer than 21 days is necessary for observing *GFAP* expression.

When correlation analysis was performed between Ca^{2+} transients with specific characteristics of duration and amplitude, and the expression of specific genes involved in neurodifferentiation, we observed that specific subclasses of Ca^{2+} transients are linked to specific genes. Two possible scenarios can explain the observed positive correlations between spontaneous physiological activity and gene expression. First, specific calcium transients could be driven by the expression of specific human genes. Second, one type of calcium transients (e.g., short duration and small amplitude) may selectively influence the expression of some but not all human genes during early neuronal development.

Spike stimulation mimicking the frequency of spontaneous transients has been shown to be most effective in replicating the effects of spontaneous transients on neurotransmitter expression, for example, GABA expression [30,54]. Thus, it would be quite useful to find calcium spike properties (amplitude, duration, and frequencies) that promote the expression

http://dx.doi.org/10.5772/62769

of genes involved in early neuronal differentiation in order to improve current methods for modeling human brain development in culture.

Acknowledgements

This study was supported by Connecticut Innovations grant 09-SCA-UCHC-13 to SDA, NIH grants MH063503 and MH104775 to SDA, HCRAC Institutional grant to SDA, and NIH grants MH087840 and MH099427 to HML. Electrophysiological and optical measurements were performed in the *Stem Cell Physiology and Chemistry Core*, which is supported by Connecticut Innovations grant SCD-01-2009 to SDA. The LED-based epi-illumination system was designed and built by Dr. Peter Lee, Essel R&D Inc., Toronto, Canada. The authors are grateful to Drs. Steven Crocker and Richard Mains (UConn Health) for help with the qPCR apparatus.

Author contributions

Designed the study: GSB and SDA. Produced iPSCs: EP and HML. Stem cell neurodifferentiation: GSB, PVL, and MLM. Patch electrode recordings and multisite calcium imaging: MBS, KDO, and SDA. Gene expression levels: PVL, GSB, and MLM. Analyzed the experimental data: PVL, MLM, MBS, and SDA. Wrote the paper: PVL, HML, and SDA.

Competing interests

The authors have declared that no competing interests exist.

Author details

Pallavi V. Limaye[1], Michele L. McGovern[1], Mandakini B. Singh[1], Katerina D. Oikonomou[1], Glenn S. Belinsky[1], Erika Pedrosa[2], Herbert M. Lachman[2] and Srdjan D. Antic[1*]

*Address all correspondence to: antic@uchc.edu

1 Stem Cell Institute, Institute for Systems Genomics, UConn Health, Farmington, CT, USA

2 Department of Psychiatry and Behavioral Sciences, Albert Einstein College of Medicine, Bronx, New York, USA

References

[1] Moody, W.J. and M.M. Bosma (2005) *Ion channel development, spontaneous activity, and activity-dependent development in nerve and muscle cells*. Physiol Rev, 85(3): p. 883–941.

[2] Blankenship, A.G. and M.B. Feller (2011) *Mechanisms underlying spontaneous patterned activity in developing neural circuits*. Nat Rev Neurosci, 11(1): p. 18–29.

[3] Rosenberg, S.S. and N.C. Spitzer (2011) *Calcium signaling in neuronal development*. Cold Spring Harb Perspect Biol, 3(10): p. a004259.

[4] Moore, A.R., R. Filipovic, Z. Mo, M.N. Rasband, N. Zecevic, and S.D. Antic (2009) *Electrical excitability of early neurons in the human cerebral cortex during the second trimester of gestation*. Cereb Cortex, 19(8): p. 1795–805.

[5] Moore, A.R., W.L. Zhou, I. Jakovcevski, N. Zecevic, and S.D. Antic (2011) *Spontaneous electrical activity in the human fetal cortex in vitro*. J Neurosci, 31(7): p. 2391–8.

[6] Belinsky, G.S., A.R. Moore, S.M. Short, M.T. Rich, and S.D. Antic (2011) *Physiological properties of neurons derived from human embryonic stem cells using a dibutyryl cyclic AMP-based protocol*. Stem Cells Dev, 20(10): p. 1733–46.

[7] Chatzidaki, A., A. Fouillet, J. Li, J. Dage, N.S. Millar, E. Sher, and D. Ursu (2015) *Pharmacological characterisation of nicotinic acetylcholine receptors expressed in human iPSC-derived neurons*. PLoS One, 10(4): p. e0125116.

[8] Belinsky, G.S., M.T. Rich, C.L. Sirois, S.M. Short, E. Pedrosa, H.M. Lachman, and S.D. Antic (2014) *Patch-clamp recordings and calcium imaging followed by single-cell PCR reveal the developmental profile of 13 genes in iPSC-derived human neurons*. Stem Cell Res, 12(1): p. 101–18.

[9] Preynat-Seauve, O., D.M. Suter, D. Tirefort, L. Turchi, T. Virolle, H. Chneiweiss, M. Foti, J.A. Lobrinus, L. Stoppini, A. Feki, M. Dubois-Dauphin, and K.H. Krause (2009) *Development of human nervous tissue upon differentiation of embryonic stem cells in three-dimensional culture*. Stem Cells, 27(3): p. 509–20.

[10] Ma, L., Y. Liu, and S.C. Zhang (2011) *Directed differentiation of dopamine neurons from human pluripotent stem cells*. Methods Mol Biol, 767: p. 411–8.

[11] Liu, H., J. Lu, H. Chen, Z. Du, X.J. Li, and S.C. Zhang (2015) *Spinal muscular atrophy patient-derived motor neurons exhibit hyperexcitability*. Sci Rep, 5: p. 12189.

[12] Pre, D., M.W. Nestor, A.A. Sproul, S. Jacob, P. Koppensteiner, V. Chinchalongporn, M. Zimmer, A. Yamamoto, S.A. Noggle, and O. Arancio (2014) *A time course analysis of the electrophysiological properties of neurons differentiated from human induced pluripotent stem cells (iPSCs)*. PLoS One, 9(7): p. e103418.

[13] Brennand, K.J. and F.H. Gage (2011) *Concise review: the promise of human induced pluripotent stem cell-based studies of schizophrenia*. Stem Cells, 29(12): p. 1915–22.

[14] Chamberlain, S.J., P.F. Chen, K.Y. Ng, F. Bourgois-Rocha, F. Lemtiri-Chlieh, E.S. Levine, and M. Lalande (2010) *Induced pluripotent stem cell models of the genomic imprinting disorders Angelman and Prader–Willi syndromes*. Proc Natl Acad Sci U S A, 107(41): p. 17668–73.

[15] Zhao, D., M. Lin, J. Chen, E. Pedrosa, A. Hrabovsky, H.M. Fourcade, D. Zheng, and H.M. Lachman (2015) *MicroRNA profiling of neurons generated using induced pluripotent stem cells derived from patients with schizophrenia and schizoaffective disorder, and 22q11.2 del*. PLoS One, 10(7): p. e0132387.

[16] Rowe, E.W., D.M. Jeftinija, K. Jeftinija, and S. Jeftinija (2005) *Development of functional neurons from postnatal stem cells in vitro*. Stem Cells, 23(8): p. 1044–9.

[17] Holliday, J., R.J. Adams, T.J. Sejnowski, and N.C. Spitzer (1991) *Calcium-induced release of calcium regulates differentiation of cultured spinal neurons*. Neuron, 7(5): p. 787–96.

[18] Hill, E.J., C. Jimenez-Gonzalez, M. Tarczyluk, D.A. Nagel, M.D. Coleman, and H.R. Parri (2012) *NT2 derived neuronal and astrocytic network signalling*. PLoS One, 7(5): p. e36098.

[19] Tonelli, F.M., A.K. Santos, D.A. Gomes, S.L. da Silva, K.N. Gomes, L.O. Ladeira, and R.R. Resende (2012) *Stem cells and calcium signaling*. Adv Exp Med Biol, 740: p. 891–916.

[20] Konur, S. and A. Ghosh (2005) *Calcium signaling and the control of dendritic development*. Neuron, 46(3): p. 401–5.

[21] Rosenberg, S.S. and N.C. Spitzer (2011) *Calcium signaling in neuronal development*. Cold Spring Harb Perspect Biol., 3(10): p. a004259.

[22] Yamamoto, N. and G. Lopez-Bendito (2012) *Shaping brain connections through spontaneous neural activity*. Eur J Neurosci, 35(10): p. 1595–604.

[23] Uhlen, P. and N. Fritz (2010) *Biochemistry of calcium oscillations*. Biochem Biophys Res Commun, 396(1): p. 28–32.

[24] Pedrosa, E., V. Sandler, A. Shah, R. Carroll, C. Chang, S. Rockowitz, X. Guo, D. Zheng, and H.M. Lachman (2011) *Development of patient-specific neurons in schizophrenia using induced pluripotent stem cells*. J Neurogenet, 25(3): p. 88–103.

[25] Iacovitti, L., A.E. Donaldson, C.E. Marshall, S. Suon, and M. Yang (2007) *A protocol for the differentiation of human embryonic stem cells into dopaminergic neurons using only chemically defined human additives: studies in vitro and in vivo*. Brain Res, 1127(1): p. 19–25.

[26] Zeng, H., M. Guo, K. Martins-Taylor, X.F. Wang, Z. Zhang, J.W. Park, S.N. Zhan, M.S. Kronenberg, A. Lichtler, H.X. Liu, F.P. Chen, L.X. Yue, X.J. Li, and R.H. Xu (2010) *Specification of region-specific neurons including forebrain glutamatergic neurons from human induced pluripotent stem cells*. PLos One, 5(7): p. e11853.

[27] Price, D.J., H. Kennedy, C. Dehay, L. Zhou, M. Mercier, Y. Jossin, A.M. Goffinet, F. Tissir, D. Blakey, and Z. Molnar (2006) *The development of cortical connections*. Eur J Neurosci, 23(4): p. 910–20.

[28] Bando, Y., K. Irie, T. Shimomura, H. Umeshima, Y. Kushida, M. Kengaku, Y. Fujiyoshi, T. Hirano, and Y. Tagawa (2014) *Control of spontaneous Ca^{2+} transients is critical for neuronal maturation in the developing neocortex*. Cereb Cortex, 26(1): p. 106–17.

[29] Gu, X., E.C. Olson, and N.C. Spitzer (1994) *Spontaneous neuronal calcium spikes and waves during early differentiation*. J Neurosci, 14(11 Pt 1): p. 6325–35.

[30] Gu, X. and N.C. Spitzer (1995) *Distinct aspects of neuronal differentiation encoded by frequency of spontaneous Ca^{2+} transients*. Nature, 375(6534): p. 784–7.

[31] Leclerc, C., I. Neant, S.E. Webb, A.L. Miller, and M. Moreau (2006) *Calcium transients and calcium signalling during early neurogenesis in the amphibian embryo Xenopus laevis*. Biochim Biophys Acta, 1763(11): p. 1184–91.

[32] Resende, R.R., J.L. da Costa, A.H. Kihara, A. Adhikari, and E. Lorencon (2010) *Intracellular Ca^{2+} regulation during neuronal differentiation of murine embryonal carcinoma and mesenchymal stem cells*. Stem Cells Dev, 19(3): p. 379–94.

[33] Malmersjo, S., P. Rebellato, E. Smedler, H. Planert, S. Kanatani, I. Liste, E. Nanou, H. Sunner, S. Abdelhady, S. Zhang, M. Andang, A. El Manira, G. Silberberg, E. Arenas, and P. Uhlen (2013) *Neural progenitors organize in small-world networks to promote cell proliferation*. Proc Natl Acad Sci U S A, 110(16): p. E1524–32.

[34] Conhaim, J., C.R. Easton, M.I. Becker, M. Barahimi, E.R. Cedarbaum, J.G. Moore, L.F. Mather, S. Dabagh, D.J. Minter, S.P. Moen, and W.J. Moody (2011) *Developmental changes in propagation patterns and transmitter dependence of waves of spontaneous activity in the mouse cerebral cortex*. J Physiol, 589(Pt 10): p. 2529–41.

[35] Moore, A.R., W.L. Zhou, C.L. Sirois, G.S. Belinsky, N. Zecevic, and S.D. Antic (2014) *Connexin hemichannels contribute to spontaneous electrical activity in the human fetal cortex*. Proc Natl Acad Sci U S A, 111(37): p. E3919–28.

[36] Retrosi, G., N.J. Sebire, M. Bishay, E.M. Kiely, J. Anderson, P. De Coppi, E. Resca, D. Rampling, N. Bier, K. Mills, S. Eaton, and A. Pierro (2011) *Brain lipid-binding protein: a marker of differentiation in neuroblastic tumors*. J Pediatr Surg, 46(6): p. 1197–200.

[37] Feng, L., M.E. Hatten, and N. Heintz (1994) *Brain lipid-binding protein (BLBP): a novel signaling system in the developing mammalian CNS*. Neuron, 12(4): p. 895–908.

[38] Anton, E.S., M.A. Marchionni, K.F. Lee, and P. Rakic (1997) *Role of GGF/neuregulin signaling in interactions between migrating neurons and radial glia in the developing cerebral cortex*. Development, 124(18): p. 3501–10.

[39] Nagy, J.I., D. Patel, P.A. Ochalski, and G.L. Stelmack (1999) *Connexin30 in rodent, cat and human brain: selective expression in gray matter astrocytes, co-localization with connexin43 at gap junctions and late developmental appearance*. Neuroscience, 88(2): p. 447–68.

[40] Nadarajah, B., A.M. Jones, W.H. Evans, and J.G. Parnavelas (1997) *Differential expression of connexins during neocortical development and neuronal circuit formation*. J Neurosci, 17(9): p. 3096–111.

[41] Odermatt, B., K. Wellershaus, A. Wallraff, G. Seifert, J. Degen, C. Euwens, B. Fuss, H. Bussow, K. Schilling, C. Steinhauser, and K. Willecke (2003) *Connexin 47 (Cx47)-deficient mice with enhanced green fluorescent protein reporter gene reveal predominant oligodendrocytic expression of Cx47 and display vacuolized myelin in the CNS*. J Neurosci, 23(11): p. 4549–59.

[42] Taléns-Visconti, R., I. Sanchez-Vera, J. Kostic, M.A. Perez-Arago, S. Erceg, M. Stojkovic, and C. Guerri (2011) *Neural differentiation from human embryonic stem cells as a tool to study early brain development and the neuroteratogenic effects of ethanol*. Stem Cells Dev, 20(2): p. 327–39.

[43] Lewis, D.A. and P. Levitt (2002) *Schizophrenia as a disorder of neurodevelopment*. Annu Rev Neurosci, 25: p. 409–32.

[44] Rapoport, J.L., A.M. Addington, S. Frangou, and M.R. Psych (2005) *The neurodevelopmental model of schizophrenia: update 2005*. Mol Psychiatry, 10(5): p. 434–49.

[45] Matelski, L. and J. Van de Water (2015) *Risk factors in autism: thinking outside the brain*. J Autoimmun, 67: p. 1–7.

[46] Owen, M.J., A. Sawa, and P.B. Mortensen (2016) *Schizophrenia*. Lancet, pii: S0140℃6736(15): p. 01121–6.

[47] Brennand, K.J., A. Simone, J. Jou, C. Gelboin-Burkhart, N. Tran, S. Sangar, Y. Li, Y. Mu, G. Chen, D. Yu, S. McCarthy, J. Sebat, and F.H. Gage (2011) *Modelling schizophrenia using human induced pluripotent stem cells*. Nature, 473(7346): p. 221–5.

[48] Suzuki, M., A.D. Nelson, J.B. Eickstaedt, K. Wallace, L.S. Wright, and C.N. Svendsen (2006) *Glutamate enhances proliferation and neurogenesis in human neural progenitor cell cultures derived from the fetal cortex*. Eur J Neurosci, 24(3): p. 645–53.

[49] Lin, M., H.M. Lachman, and D. Zheng (2015) *Transcriptomics analysis of iPSC-derived neurons and modeling of neuropsychiatric disorders*. Mol Cell Neurosci, pii: S1044–7431(15): p. 30035–X.

[50] Meechan, D.W., T.M. Maynard, E.S. Tucker, and A.S. LaMantia (2011) *Three phases of DiGeorge/22q11 deletion syndrome pathogenesis during brain development: patterning, proliferation, and mitochondrial functions of 22q11 genes*. Int J Dev Neurosci, 29(3): p. 283–94.

[51] Shprintzen, R.J., A.M. Higgins, K. Antshel, W. Fremont, N. Roizen, and W. Kates (2005) *Velo-cardio-facial syndrome*. Curr Opin Pediatr, 17(6): p. 725–30.

[52] Parekh, A.B. and S. Muallem (2011) *Ca(2+) signalling and gene regulation*. Cell Calcium, 49(5): p. 279.

[53] Mei, L. and W.C. Xiong (2008) *Neuregulin 1 in neural development, synaptic plasticity and schizophrenia*. Nat Rev Neurosci, 9(6): p. 437–52.

[54] Watt, S.D., X. Gu, R.D. Smith, and N.C. Spitzer (2000) *Specific frequencies of spontaneous Ca^{2+} transients upregulate GAD 67 transcripts in embryonic spinal neurons*. Mol Cell Neurosci, 16(4): p. 376–87.

[55] Hartfield, E.M., F. Rinaldi, C.P. Glover, L.F. Wong, M.A. Caldwell, and J.B. Uney (2011) *Connexin 36 expression regulates neuronal differentiation from neural progenitor cells*. PLoS One, 6(3): p. e14746.

[56] Thompson, R.J. and B.A. Macvicar (2008) *Connexin and pannexin hemichannels of neurons and astrocytes*. Channels (Austin), 2(2): p. 81–6.

[57] Nlend, R.N., A. Ait-Lounis, F. Allagnat, V. Cigliola, A. Charollais, W. Reith, J.A. Haefliger, and P. Meda (2012) *Cx36 is a target of Beta2/NeuroD1, which associates with prenatal differentiation of insulin-producing beta cells*. J Membr Biol, 245(5–6): p. 263–73.

[58] Rozental, R., C. Giaume, and D.C. Spray (2000) *Gap junctions in the nervous system*. Brain Res Brain Res Rev, 32(1): p. 11–15.

[59] Duval, N., D. Gomes, V. Calaora, A. Calabrese, P. Meda, and R. Bruzzone (2002) *Cell coupling and Cx43 expression in embryonic mouse neural progenitor cells*. J Cell Sci, 115(Pt 16): p. 3241–51.

[60] Cheng, A., H. Tang, J. Cai, M. Zhu, X. Zhang, M. Rao, and M.P. Mattson (2004) *Gap junctional communication is required to maintain mouse cortical neural progenitor cells in a proliferative state*. Dev Biol, 272(1): p. 203–16.

[61] Montoro, R.J. and R. Yuste (2004) *Gap junctions in developing neocortex: a review*. Brain Res Brain Res Rev, 47(1–3): p. 216–26.

[62] Ben-Ari, Y. (2002) *Excitatory actions of GABA during development: the nature of the nurture*. Nat Rev Neurosci, 3(9): p. 728–39.

[63] Howard, B., Y. Chen, and N. Zecevic (2006) *Cortical progenitor cells in the developing human telencephalon*. Glia, 53(1): p. 57–66.

Section 6

Regenerative medicine

Chapter 17

Pluripotent Stem Cells: Differentiation Potential and Therapeutic Efficacy for Cartilage Repair

Wei Seong Toh

Additional information is available at the end of the chapter

http://dx.doi.org/10.5772/62818

Abstract

Articular cartilage injuries, often caused by trauma, have a limited potential to heal, which over time, may lead to osteoarthritis, an inflammatory and degenerative joint disease frequently associated with activity-related pain, swelling and impaired mobility. Many treatment modalities have been introduced but with limited success due to the formation of inferior fibrocartilage at the damaged area/injured site during the repair process. Pluripotent stem cells (human embryonic stem cells and induced pluripotent stem cells) provide a valuable cell source for critical understanding of pluripotency and lineage-specific differentiation, as well as for derivation of therapeutic mesenchymal cells and chondrocytes for articular cartilage repair. Here, we discuss the characteristics of pluripotent stem cells, their differentiation pathways to mesenchymal progenitors and chondrocytes, and their emerging roles in cartilage regeneration.

Keywords: pluripotent stem cells, mesenchymal stem cells, chondrocytes, cartilage, chondrogenesis

1. Introduction

Articular cartilage is a unique hypocellular, aneural, an avascular load-bearing tissue, supported by the underlying subchondral bone [1]. The extracellular matrix (ECM) is composed of a hydrated network of type II collagen fibrils, which are specifically arranged architecturally, and enforced with water-retaining aggrecan molecules linked to hyaluronic acid. The type II collagen network is further stabilized by other collagen types, IX and XI, in the territorial/inter-territorial matrix and biglycan, decorin, matrilins, perlecan, type IV collagen, laminins, type VI and

XVIII/endostatin collagens in the pericellular matrix surrounding the chondrocytes [2–5]. This combination of molecules gives the articular cartilage its unique ability to withstand the repetitive compressive loading in daily activities [1, 5, 6].

Articular cartilage injuries, commonly caused by excessive sports activities, have high incidence. When left untreated, cartilage lesions can lead to osteoarthritis, a degenerative joint disease characterized by the progressive degradation of the articular cartilage, subchondral bone, meniscus and ligaments, and the formation of painful osteophytes [7]. Osteoarthritis is the most common form of arthritis that affects 27 million people in United States, resulting in over 50% of total joint replacements, and the number of cases is increasing due to aging and obesity [6].

Current treatment modalities for articular cartilage injuries include microfracture, abrasion, drilling, osteochondral grafting, and more recently autologous chondrocyte implantation (ACI) [8]. Many of these methods help to repair the tissue and reduce pain to some degree. However, there are pertinent issues of inferior fibrocartilage repair, donor site morbidity and loss of chondrocytic phenotype upon expansion that necessitates the need for an alternative approach.

In recent years, stem cells have emerged as a promising cell source for treatment of cartilage lesions and osteoarthritis [9, 10]. Major advantages of applying stem cells for cartilage repair stem from their availability, proliferative capacity and multi-lineage differentiation potency [9, 10]. Major sources of stem cells include the adult and embryonic stem cells. Adult mesenchymal stem cells (MSCs), commonly isolated from bone marrow and adipose tissue, are currently evaluated in clinical trials for treatment of cartilage defects and osteoarthritis and have yielded promising results [11, 12]. However, one major limiting factor in use of adult MSCs relates to the impaired cellular proliferation and differentiation with increasing donor age [13, 14]. Furthermore, it has been established that cartilage differentiated from adult bone marrow MSCs is not phenotypically stable and has the propensity for hypertrophy development [15–17].

Pluripotent stem cells (PSCs) have the capacity for unlimited self-renewal and ability to differentiate into all cell types present in the body, thus representing an immortal cell source that could potentially provide unlimited supply of cells for cell-based therapies [18, 19]. However, one of the greatest challenges in pluripotent stem cell research is to understand, control, and develop an efficient and stable culture milieu for directing differentiation to a particular lineage, in this instance, chondrocytes [9]. In order to advance pluripotent stem cell therapies to the clinic, several biological challenges need to be overcome that include as follows: (1) directing lineage-specific differentiation, (2) defining the functional phenotype of derived chondrocytes, (3) eliminating the risk of teratoma formation, (4) designing adjuvant and tissue-engineering strategies to enhance the efficacy of cartilage repair and (5) deciphering the underlying mechanisms of cartilage repair. This chapter aims to discuss the latest advances in use of human PSCs in cartilage tissue engineering and regeneration.

http://dx.doi.org/10.5772/62818

2. Pluripotent stem cells

Pluripotent stem cells (PSCs) possess unlimited self-renewal and differentiation capacity, and include embryonic stem cells (ESCs) and induced pluripotent stem cells (iPSCs). ESCs are isolated from the inner cell mass of embryos [18], while iPSCs are created from somatic cells through reprogramming with defined gene and protein factors [19]. Human PSCs are well-poised as an alternative cell source to adult stem cells for transplantation. These cells proliferate in culture without loss of differentiation potential, thus enabling the generation of specific desired cell types in large numbers [9]. Furthermore, these pluripotent cells are capable of differentiating into the various cellular intermediates and committed cell types of a specific lineage by recapitulating the developmental embryogenesis. Specifically, in cartilage therapy, pluripotent stem cells enable the generation of early-staged chondroprogenitors and chondrocytes for cartilage-specific transplantation (i.e. hyaline, fibrocartilage and elastic cartilage). For repair of articular cartilage of the knee joint, hyaline cartilage repair is desired, so as to provide the long-term durability and ability to withstand repetitive mechanical loading in daily activities [1].

Major hurdles to clinical use of human PSCs include the risk of teratoma formation for human ESCs and iPSCs, and the risk of immune rejection in the case of human ESCs. Although nuclear reprogramming technology has been established that enables the generation of patient-specific PSCs, there is still safety issue of chromosomal variation as a result of genomic alteration [20]. These genetic alterations may in a long term lead to tumours, as observed in cases of gene therapy [21, 22]. On this note, several groups are exploring non-integrative methods such as transient viral transfections and small molecules to generate iPSCs [23, 24]. Such methods may enable generation of iPSCs that are safe for transplantation in near future.

3. Differentiation of PSCs into chondrogenic lineage

Development of an efficient and reproducible culture protocol for directing the differentiation of human PSCs into a defined specific chondrogenic lineage prior to transplantation is critical, as any remnants of PSCs could result in teratoma formation [9, 25]. On this note, as few as 245 human ESCs were reported to induce teratoma formation in 10 weeks' time [26]. Establishing the functional phenotype and competence of PSC-derived chondrogenic cells for transplantation is also a challenge. Hence, understanding the factors and underlying signalling pathways involved in lineage-specific differentiation of human PSCs to chondroprogenitors and chondrocytes, and establishing the functionality of these derived cells are necessary. We will discuss the various strategies utilizing approaches of growth factors, co-culture, conditioned medium and morphogenetic factors, gene modification, as well as derivation of mesenchymal progenitor cells. Additionally, tissue-engineering approaches incorporating three-dimensional (3-D) scaffolds and sustained release of soluble factors to influence stem cell chondrogenesis will be discussed.

3.1. Growth factors, chemicals and inductive medium

Early studies in PSC chondrogenesis involved differentiation *via* embryoid body (EB) formation, an *in vitro* culture system designed to mimic developmental embryogenesis with the formation of three germ layers, namely mesoderm, ectoderm and endoderm, and their tissue derivatives [27, 28]. In these studies, EBs were formed in suspension culture, followed by direct plating or dissociation into single cells and further differentiation into chondrocytic cells [27, 29]. Specific growth factors and chemicals may be added in certain fashion and combinations to induce specific cellular signalling and differentiation. Accordingly, Kramer et al. [27] investigated the differentiation of mouse ESCs *via* 5 days of EB formation followed by culture of EB outgrowth for 30 days. In that study, addition of bone morphogenetic protein (BMP)-2 (2 ng/ml) or BMP-4 (10 ng/ml), but not TGF-β1 (2 ng/ml), increased the number of cartilaginous nodules when applied for the entire culture period. It was further observed that application of BMP-2 from day 2 to 5 of EB formation had the optimal stimulatory effect on chondrogenesis. It was further demonstrated that co-stimulation with BMP-2 (10 ng/ml) and TGF-β1 (2 ng/ml) during day 3–5 of EB formation, followed by BMP-2 stimulation in EB outgrowth had a synergistic enhancement on chondrogenesis [29].

These findings somewhat agree with the study performed on human ESCs, where early and continuous application of TGF-β1 (10 ng/ml) had an inhibitory effect on chondrogenesis [30, 31], and only enhanced chondrogenesis when applied at a later stage, following 5 days of EB formation [30]. The early inhibitory effect of TGF-β1 on human ESC chondrogenesis relates to the role of TGF-β/activin/nodal signalling in the maintenance of undifferentiated state of human ESCs [30, 32]. Similarly, Toh et al. [33] showed that BMP-2 (100 ng/ml) was capable of inducing chondrogenic differentiation of dissociated human EB cells seeded as high-density micromass over a 21-day culture period. Notably, dissociation of the human EBs and replating them as high-density micromass enhanced the kinetics of chondrogenesis with early upregulation of type II collagen and glycosaminoglycan (GAG). This finding is in agreement with the high-density culture of adult MSCs in pellet and hydrogel cultures that cellular condensation is required for chondrogenesis [5, 34].

These studies subsequently prompted the investigation of growth factors and molecules involved in early germ layer induction of PSCs prior to chondrogenic differentiation. Kawaguchi et al. [35] showed that retinoic acid (10^{-7} M) added during EB formation enhanced early induction of pre-somitic mesoderm and neural crest markers including Mox1 and FoxD3. Subsequent stimulation with TGF-β3 (10 ng/ml) in EB outgrowth resulted in heightened type II collagen expression. Recently, Yamashita et al. [36] demonstrated a direct differentiation approach without the need of cell sorting and expansion. In that study, mesendodermal differentiation of human iPSCs was induced by Wnt3a (10 ng/ml) and Activin A (10 ng/ml), followed by chondrogenic differentiation under co-stimulation of BMP-2 (10 ng/ml), TGF-β1 (10 ng/ml) and GDF-5 (10 ng/ml) over a period of 42 days was able to produce homogenous hyaline cartilaginous particles at high efficiency. Further subcutaneous transplantation of the human iPSC-derived cartilaginous particles generated hyaline cartilage that expressed type II collagen, but not type I collagen, in immunodeficiency mice. Collectively, the TGF-β and BMP signalling have distinct roles during embryonic chondrogenesis, and the interplay of these factors, in addition to the co-factors and culture systems, have a significant impact on PSC chondrogenesis.

3.2. Co-culture, conditioned medium and morphogenetic factors

The co-culture approach represents one of the earliest strategies used in chondrogenic differentiation of PSCs [37–39]. In this approach, the PSCs are co-cultured with chondrocytes in cell–cell contact or *via* a transwell system that allows only exchange of the morphogenetic factors and diffusible signals.

Bigdeli et al. [39] performed co-culture of human ESCs with irradiated neonatal or adult articular chondrocytes in pellet cultures in the presence of TGF-β1 for 14 days, prior to dissociation. In that study, isolated mesenchymal cells were expandable and demonstrated chondrogenic potential with deposition of GAG upon differentiation in pellet culture in the presence of TGF-β3. In another study, Vats et al. [38] showed that co-culture with chondrocytes over transwell inserts induced chondrogenic differentiation of human ESCs with expression of type II collagen and GAG by the end of 28 days, without the need of exogenous growth factors. When transplanted subcutaneously into the backs of severe combined immunodeficient (SCID) mice, these co-cultured human ESCs formed cartilaginous tissue that stained positive for type II collagen and GAG. Further studies showed that co-culture with bovine chondrocytes induced chondrogenic commitment of human ESCs and iPSCs, generating expandable chondrogenic cells that expressed type II collagen at early passage, and maintained the chondrogenic potential for cartilage formation [40, 41].

In those studies, the morphogenetic factors secreted by the chondrocytes were sufficient to induce PSC chondrogenesis, without the need of cell–cell contact [40, 41]. However, the morphogenetic factors are poorly defined, and the exact factor(s) responsible for differentiation of PSCs remains a question.

3.3. Gene modification

Gene transfer strategies to induce chondrogenesis of dedifferentiated chondrocytes and MSCs have been widely studied [42, 43]. To date, gene transfection of the Sox Trio (Sox5, Sox6 and Sox9) seems to be the most promising approach in generating stable hyaline cartilage, with the suppression of hypertrophic maturation and osteogenesis [42]. However, this strategy has yet to be explored in chondrogenic induction of human PSCs.

To date, the use of gene transfer strategy in induction of PSC chondrogenesis is limited. Wei et al. [44] showed that iPSCs can be generated from human osteoarthritic chondrocytes and subsequently induced to differentiate into chondrocytes. In that study, lentiviral transfection of TGF-β1 was applied singly or in combination with primary chondrocytes in co-culture to induce differentiation of human iPSCs in alginate matrix. Notably, overexpression of TGF-β1 and chondrocyte co-culture demonstrated synergistic enhancement of iPSC chondrogenesis *in vitro* and ectopic cartilage formation *in vivo*. Recently, Diekman et al. (2015) demonstrated that inhibition of cell cycle inhibitor p21 in mouse iPSCs enhanced monolayer proliferation and subsequent chondrogenic differentiation with increased synthesis of GAG in pellet cultures over the control cells [45].

3.4. Derivation of mesenchymal progenitor cells from PSCs

Adult mesenchymal stem cells (MSCs) are present in several adult tissues, with the ability for lineage differentiation at least to bone, cartilage and adipose tissue [46–48]. Developmentally, MSCs that are involved in cartilage formation are derived mainly from mesoderm for limb joint and rib chondrocytes, and cranial neural crest for craniofacial chondrocytes. According to the Mesenchymal and Tissue Stem Cell Committee of the International Society for Cellular Therapy, the minimal criteria to define human MSCs include adherence to plastic surface, specific antigen expression ($CD73^+$ $CD90^+$ $CD105^+$ $CD34^-$ $CD11b^-$ $CD14^-$ $CD19^-$ $CD79a^-$ $HLA\text{-}DR^-$) and multi-lineage differentiation potential under standard *in vitro* differentiation conditions [49]. Due to the expression of major histocompatibility complex (MHC) class I, but not MHC II, adult MSCs are considered non-immunogenic, making them a valuable cell source for allogenic transplantation without the need of immunosuppression [50].

Several strategies have been reported for the derivation of mesenchymal stem/progenitor cells from PSCs [51–53]. Earlier studies have employed co-culture with stroma cells or growth factor induction with or without sorting for an enriched cell population [51, 52]. Barberi et al. [51] was the first to report the derivation of MSCs from human ESCs. In that study, mesenchymal differentiation of human ESCs was induced by co-culture with OP9 stromal cells followed by sorting of $CD73^+$ mesenchymal precursor cells. These mesenchymal precursor cells expressed MSC markers including CD44, CD73, CD105, CD166, VCAM, ICAM1 and STRO-1 and demonstrated multi-lineage differentiation to cartilage, bone, fat and skeletal muscle cells under inductive conditions. Lian et al. [52] subsequently reported a clinical compliant protocol that involved direct culture of human ESCs in culture medium supplemented with 10% serum replacement medium, 5 ng/ml of fibroblast growth factor (FGF)-2 and 5 ng/ml platelet-derived growth factor (PDGF)-AB. This was followed by sorting of $CD105^+$ CD24- mesenchymal precursor cells that displayed a MSC surface antigen profile ($CD29^+$, $CD44^+$, $CD49a^+$, $CD49e^+$, $CD105^+$, $CD166^+$, CD34- and CD45-) and tri-lineage (osteogenic, chondrogenic, adipogenic) differentiation potential. Others have employed simple expansion of PSCs or EB outgrowth cells in culture medium conditions, typically used for culture of MSCs [53–55]. In those studies, the initial cultures consisted of mixed cell populations. After a few passages over 3–5 weeks, a homogenous population of spindle-shaped MSC-like cells was obtained. These cells expressed key MSC surface markers and displayed the capacity to differentiate into osteoblasts, adipocytes and chondrocytes. When injected into thigh muscles of SCID mouse, these human PSC-derived mesenchymal cells did not form teratoma [52].

The chondrogenic response of mesenchymal cells depends on the developmental origin. Studies have shown that MSCs generated by different methods from various sources of iPSCs could display variability in their differentiation capacity [53, 56, 57]. In all the above-mentioned protocols of deriving MSCs from human PSCs, the developmental origin of derived MSCs is a question. On this note, Umeda et al. [58] reported the isolation of paraxial mesoderm-like cells from human iPSCs under chemically defined medium conditions. These cells expressed platelet-derived growth factor receptor (PDGFR)-α, but not vascular endothelial growth factor receptor (KDR). When cultured in chondrogenic conditions sequentially supplemented with PDGF-BB, TGF-β3 and BMP-4, these mesodermal mesenchymal cells were able to form hyaline-like cartilage that resembles the hyaline cartilage of the knee joint articular cartilage.

To advance the clinical use of human PSC-derived MSCs, the immunogenicity and immunological properties of these cells are important considerations in order to rule out potential adverse immune rejection upon cell transplantation. To date, few studies have studied the immunological characteristics of human PSC-derived MSCs [59–61]. Similar to human bone marrow MSCs, human ESC-derived MSCs express cell surface HLA class I (HLA-ABC), but not HLA class II (HLA-DR) molecules, and suppress T lymphocyte proliferation induced by allogenic cells or mitogenic stimuli [59–61]. Importantly, hESC-derived MSCs demonstrated therapeutic efficacy with marked increase in survival of lupus-prone mice and a reduction of symptoms in autoimmune model of uveitis [60]. These findings support the use of human PSC-derived MSCs as a potential therapeutic candidate for further clinical development and application.

4. Cartilage tissue engineering

Utilizing the combination of cells, scaffolds and/or biomolecules, tissue engineering has emerged as a potential alternative to tissue/organ transplantation in tissue/organ replacement, and a strategy to enhance the overall therapeutic efficacy in treatment of injuries and diseases. Due to its limited ability to self-repair, cartilage is an ideal candidate for tissue engineering [62]. The principles of tissue engineering not only apply to cell seeding in scaffolds for implantation, but also extend to modulation of various cellular processes and cell–cell–matrix interactions that are critical in tissue repair and regeneration [63]. Here, we provide an overview of the scaffolds (matrices) and stimulating factors that have been used to influence PSC chondrogenesis.

4.1. Scaffolds

Different biomaterials have been investigated in their ability to support PSC chondrogenesis in 2-D and 3-D. This ranges from natural polymers such as type I collagen [64], type II collagen [65], agarose [66, 67]), alginate [44, 68]), fibrin [69], fibronectin [69], hyaluronic acid [70] and gelatin [69, 71] to synthetic polymers including polycaprolactone (PCL) [71, 72], polylactic-co-glycolic acid (PLGA) [73, 74], poly(L-lactic acid) (PLLA) [73] and polyethylene glycol diacrylate (PEGDA) [40, 75, 76], and may be fabricated in the forms of sponges, meshes and hydrogels. Several parameters need to be considered in scaffold design for cartilage tissue engineering. These include the scaffold architecture (e.g. shape, porosity and pore size); mechanical properties (e.g. elasticity); biochemical properties (e.g. ligands for cell adhesion); biocompatibility and biodegradability (to enable cell survival, matrix deposition and tissue remodelling). Scaffold design parameters for cartilage tissue engineering have been extensively reviewed [63]. Biomaterials that have been used in PSC-based cartilage tissue engineering are summarized in **Table 1**.

Biomaterial	**References**
Natural polymers	
Agarose	[66, 67]
Alginate	[44, 68]
Gelatin	[69, 71]
Fibrin	[69]
Fibronectin	[69, 73]
Hyaluronic acid	[70]
Matrigel™	[73]
Type I collagen	[64]
Type II collagen	[65]
Synthetic polymers	
Polycaprolactone (PCL)	[71, 72]
Polylactic-co-glycolic acid (PLGA)	[73, 74]
Poly(L-lactic acid) (PLLA)	[73]
Polyethylene glycol (PEG)	[40, 75, 76, 98]

Table 1. Types of biomaterials used in cartilage tissue engineering.

4.1.1. ***Natural biopolymers***

Natural polymers resemble the natural extracellular matrix (ECM) found in the body and can interact with cells through surface receptors to influence cell fate and functions [77]. To recapitulate the native cartilage ECM environment, several earlier studies have investigated the use of type I and II collagens, and GAG as the major ECM components in fabrication of 2-D substrates and 3-D scaffolds for chondrocytes [78] and stem cell culture and chondrogenesis [79]. When cultured on type II collagen substrates, isolated human EB-derived mesenchymal cells retained the chondrogenic potential and displayed enhanced chondrogenesis with increased matrix deposition of type II collagen and GAG in subsequent pellet culture in the presence of TGF-β1 [65]. Other ECM proteins such as fibronectin and basement membrane molecules (e.g. type IV collagen, laminin, nidogens) have also been reported to regulate progressive stages of chondrogenesis [80–82]. Notably, substrate combination of fibronectin and gelatin was reported to enhance chondrogenic differentiation of human iPSCs, compared to gelatin alone [69].

However, the use of natural polymers has disadvantages of batch variability and difficulty in control of material properties including degradation, mechanics and bioactivity [77]. To overcome these limitations, blended combination of natural and synthetic polymers, chemical and physical modifications, or use of pure synthetic polymers such as PEG may be considered [83].

Levenberg et al. [73] was the first to study the role of 3-D PLGA/PLLA scaffolds coated with either Matrigel™ or fibronectin for directed differentiation of human EB cells in the presence of various factors including retinoic acid, activin-A, TGF-β1 or insulin growth factor (IGF)-I. It was concluded that TGF-β1 was needed for human EB chondrogenesis in 3-D PLGA/PLLA scaffolds, despite the presence of other lineage structures. Bai et al. [74] further demonstrated differentiation of human EB-derived cells in monolayer chondrogenic conditions supplemented with TGF-β3 and BMP-2, before dissociating the cells for suspension in alginate and seeding in PLGA scaffolds. Following 8-week subcutaneous implantation, cell-alginate-PLGA complexes formed cartilaginous tissues with lacunar cell morphology and stained positive for Sox9, type II collagen and GAG, amid at a low efficiency and the presence of fibrous tissue-like structures. On this note, the synthetic PLGA scaffold provides the mechanical framework [74], while alginate hydrogel enables the encapsulation of cells and maintains them in spherical cellular morphology that is supportive of chondrogenesis [68, 74]. Liu et al. [71] further showed that electrospun PCL/gelatin nanofibrous scaffolds enhanced chondrogenic differentiation of mouse iPSC-derived EB cells over plate controls. When implanted into the full-thickness defects created in the rabbit model, cell-seeded PCL/gelatin scaffolds restored the defects with hyaline-like repair tissue that stained positive for GAG and type II collagen.

4.1.2. ***Modifications of natural polymers***

Modification of natural polymers aims to improve the overall composition, physical and mechanical properties of the biomaterial scaffold. Modifications of hyaluronic acid [84, 85], gelatin [86, 87] and chondroitin sulphate [88, 89] to include chemical functionalities (e.g. thiols, methacrylates, tyramines) have been frequently performed to improve the properties of these natural polymers for various biomedical applications. Some of these semi-synthetic matrices have been used as scaffolds for PSCs and PSC-derived cells in cartilage tissue engineering. Of note, HA has been reported to influence chondrogenesis through interactions with surface receptor CD44, making this natural biopolymer an ideal candidate for chemical modifications for application in cartilage tissue engineering [85, 90]. Prestwich's group first reported thiol-modification of hyaluronic acid and gelatin that allow crosslinking by addition of PEGDA to form hydrogels for biomedical applications including cartilage tissue engineering [84, 91]. When used to encapsulate human EB-derived mesenchymal cells and culture in chondrogenic medium supplemented with TGF-β1 and BMP-7, these cell-seeded constructs displayed hyaline cartilaginous tissue characteristics including chondrocytic lacunar morphology, and deposition of high amounts of type II collagen and GAG, and minimal amount of type I collagen (**Figure 1**) [70]. When further implanted into rat osteochondral defects, these cell-seeded constructs regenerated the defects by undergoing an orderly remodelling process to form a hyaline cartilage layer and underlying subchondral bone by the end of 12 weeks. Similarly, thiol-modified dextran/PEG hybrid hydrogels have been reported for encapsulation of chondrocytes and mouse ESCs, and both cell types produced cartilaginous tissue [92].

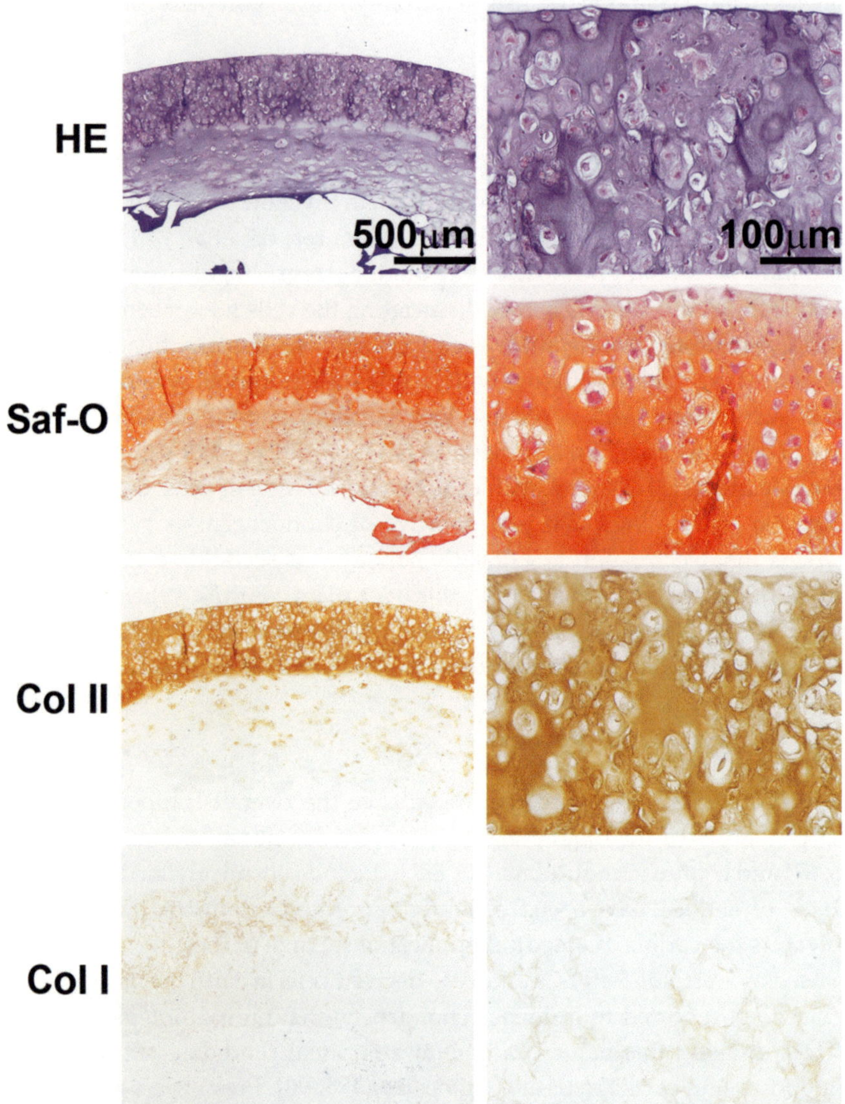

Figure 1. Tissue-engineered cartilage construct based on human ESC-derived mesenchymal cells. Histological staining by haematoxylin and eosin (HE) and safranin O/fast green (Saf-O) staining, as well as immunohistochemical staining for type II collagen (Col II) and type I collagen (Col I), was performed. Analysis revealed hyaline cartilaginous tissue characteristics of the tissue-engineered cartilage construct including chondrocytic lacunar morphology, abundant deposition of GAG and Col II, and minimal deposition of Col I.

4.1.3. *Synthetic polymers*

Synthetic polymer-based scaffolds with the defined composition and chemistry have evolved to become increasingly attractive as cell culture platforms to gain insights on effects of specific biophysical and/or biochemical cues on cell behaviour and function [93]. However, these synthetic polymers usually require modifications with peptides or blending with natural ECM proteins to confer bio-functionality to support various cellular functions including cell adhesion, proliferation and differentiation [40, 73, 75]. Among these synthetic polymers, PEG is widely used due in part to its high hydrophilicity, bio-inert property and versatility. The hydroxyl groups of PEG are easily modified with other chemical functionalities (e.g. acrylates, methacrylates, thiols, azides) that can react to form a 3D network [94, 95]. Furthermore, peptide sequences such as Arg-Gly-Asp (RGD) can be tethered to PEG to facilitate cell adhesion and spreading [95]. On this note, Hwang et al. [75] showed that mesenchymal cells derived from human EB outgrowth displayed heightened cartilage-specific gene expression and matrix production, when encapsulated in RGD-modified PEG hydrogels compared to PEG hydrogel alone. Similarly, expanded mesenchymal chondrogenic cells derived from co-culture of human ESCs and bovine chondrocytes, encapsulated in RGD-modified PEG hydrogels formed homogenous cartilaginous tissue *in vitro* and *in vivo* [40].

4.2. Stimulating factors

Apart from growth factors and cytokines that have been described above to influence chondrogenesis, other stimulating factors have also been commonly employed to induce, accelerate and/or enhance cartilage formation, as part of tissue-engineering triad strategy. These additional stimulating factors include the use of small molecules and application of mechanical stimulation (e.g. hydrostatic pressure, dynamic compression, bioreactors).

Small molecules including kartogenin, TD-198946, prostaglandin E2, glucosamine, dexamethasone and cytochalasin D have been reported to promote stem cell chondrogenesis [96]. Notably, cytochalasin D treatment has been shown to enhance chondrogenesis of mouse EB-derived cells through actin filament reorganization [97]. Glucosamine, an amino monosaccharide present in GAG, was also reported to increase the levels of cartilage-specific gene expression and matrix accumulation when applied to hydrogel cultures of mouse EBs in chondrogenic conditions supplemented with TGF-β1 [98].

The influence of biomechanical signals on stem cell chondrogenesis is widely reported [62, 99, 100]. Specifically, Terraciano et al. [101] showed that mesenchymal cells derived from human EBs responded differently to mechanical stimulation, depending on the extent of chondrogenic differentiation. In the absence of TGF-β1, EB-derived mesenchymal cells exhibited a downregulation of chondrogenic genes including Sox9, aggrecan and type II collagen. However, after TGF-β1 pre-conditioning, mechanical compression enhanced chondrogenic differentiation of EB-derived mesenchymal cells with increased chondrogenic gene expression and matrix deposition.

5. Cartilage regeneration using PSCs and cell derivatives

To date, several studies have applied PSCs and PSC-derived cells in cartilage repair in ectopic and orthotropic models *in vivo* [40, 69, 70, 76]. The effective application of PSCs and the associated outcomes in cartilage repair depends on several factors including 1) differentiation efficiency and homogeneity of cartilage formation; 2) delivery strategy; 3) site of transplantation, as previously reviewed [9]. Here, we discuss the latest developments of PSCs, and underlying mechanisms in cartilage regeneration from animal studies.

Although a number of studies have demonstrated the functionality of PSCs in cartilage repair in orthotropic models, few have investigated the survival of the implanted PSCs over the course of cartilage tissue repair [69, 70, 76]. In a study by Toh et al. [70], tissue engineered constructs based on human EB-derived mesenchymal cells were implanted into full-thickness cartilage defects in an immunosuppressed rat model. In that study [70], it was observed that human chimerism of human EB-derived mesenchymal cells declined drastically over a period of 12 weeks. The percentage human cell chimerism and survival declined from ~70% at 2 weeks to ~30% at 6 weeks and <5% at 12 weeks. Despite the rapid decline in cell survival, there was full cartilage regeneration composed of mainly the host rat chondrocytes populating the defect site at the end of 12 weeks, implicating the role of paracrine functions of human EB-derived mesenchymal cells in cartilage repair. It was further reported in another study [69] that significant number of implanted human ESC-derived mesenchymal cells could still be detected at the repair site by the end of 12 weeks, although the exact percentage human cell chimerism was not determined. It therefore remains unclear if extended period of cell survival may promote a better and long-lasting cartilage repair. There are, however, good reasons to hypothesize that human PSC-derived cells mediate tissue repair and regeneration *via* multiple mechanisms that require thorough investigations.

Based on the current understanding, it is well-accepted that MSCs mediate tissue repair and regeneration by mechanisms of direct differentiation into target cells to replace the damaged tissue, and secretion of trophic factors to orchestrate endogenous cell response to mediate tissue regeneration [10]. Proteomic analysis of MSC conditioned medium (secretome) revealed a broad range of bioactive molecules secreted by MSCs, including growth factors, cytokines, chemokines, extracellular matrix molecules and vesicles [102], some of which reported to play important roles in chondrogenesis immunomodulation, and tissue repair [103, 104]. Similarly, human PSC-derived MSCs have been reported to produce a broad range of these trophic factors that largely constitute the biological significance of these cells in tissue repair [105]. Among these bioactive molecules, extracellular vesicles (EVs) produced by human ESC-derived MSCs have been demonstrated in their efficacy to reduce the infarct size in a mouse model of myocardial ischemia/reperfusion injury [106], promote hepatic regeneration in a mouse model of drug-induced liver injury [107], and more recently to promote cartilage regeneration in a rat model of osteochondral defect [108].

6. Conclusions and perspectives

The discovery of human PSCs, including human ESCs and iPSCs, has opened new avenues and possibilities for treatment of cartilage injuries and diseases such as osteoarthritis. The field is progressing rapidly to develop protocols for differentiation of human PSCs into MSCs and chondrogenic cells for cartilage tissue engineering and repair.

Looking ahead, the therapeutic potential of these cells will depend largely on identifying and fine-tuning the strategies, conditions, and factors required to induce PSC chondrogenesis in a defined, reproducible, and clinically compliant manner. Interdisciplinary efforts of culture technologies, biomaterials and tissue engineering are converging to develop effective PSC therapies, including intra-articular cell injection and implantation of tissue engineered cartilage graft, for cartilage repair.

The application of PSCs in cartilage regeneration not only restricts to cellular therapies, but extends to secretome components including growth factors, cytokines, extracellular matrix molecules and vesicles that likely underpin the mechanistic roles of PSC-derived cells in tissue repair and regeneration. Deciphering the composition and the components of the secretome is going to shed light on the paracrine functions of PSC-derived cells in cartilage repair, as well as development of novel therapeutics and strategies for cartilage regeneration.

Acknowledgements

This work was supported by grants from the National University Healthcare System, National University of Singapore (R221000083112 and R221000081133) and National Medical Research Council Singapore (R221000080511).

Author details

Wei Seong Toh

Address all correspondence to: dentohws@nus.edu.sg

Faculty of Dentistry, National University of Singapore, Singapore

Tissue Engineering Program, Life Sciences Institute, National University of Singapore, Singapore

References

[1] Becerra J, Andrades JA, Guerado E, Zamora-Navas P, López-Puertas JM, Reddi AH. Articular cartilage: structure and regeneration. Tissue Engineering Part B: Reviews. 2010;16(6):617–27.

[2] Kvist AJ, Nyström A, Hultenby K, Sasaki T, Talts JF, Aspberg A. The major basement membrane components localize to the chondrocyte pericellular matrix—a cartilage basement membrane equivalent? Matrix Biology. 2008;27(1):22–33.

[3] Foldager CB, Toh WS, Gomoll AH, Olsen BR, Spector M. Distribution of basement membrane molecules, laminin and collagen type IV, in normal and degenerated cartilage tissues. Cartilage. 2014;5(2):123–32.

[4] Zhang Z. Chondrons and the pericellular matrix of chondrocytes. Tissue Engineering Part B: Reviews. 2015;21(3):267–77.

[5] Toh WS, Foldager CB, Hui JH, Olsen BR, Spector M. Exploiting stem cell-extracellular matrix interactions for cartilage regeneration: a focus on basement membrane molecules. Current Stem Cell Research & Therapy. 2016; doi: 10.2174/1574888X10666150723150525

[6] Ge Z, Hu Y, Heng BC, Yang Z, Ouyang H, Lee EH, et al. Osteoarthritis and therapy. Arthritis Care & Research. 2006;55(3):493–500.

[7] Goldring MB, Goldring SR. Osteoarthritis. Journal of Cellular Physiology. 2007;213(3): 626–34.

[8] Brittberg M, Lindahl A, Nilsson A, Ohlsson C, Isaksson O, Peterson L. Treatment of deep cartilage defects in the knee with autologous chondrocyte transplantation. New England Journal of Medicine. 1994;331(14):889–95.

[9] Toh WS, Lee EH, Cao T. Potential of human embryonic stem cells in cartilage tissue engineering and regenerative medicine. Stem Cell Reviews and Reports. 2010;7(3):544–59.

[10] Toh WS, Foldager CB, Pei M, Hui JHP. Advances in mesenchymal stem cell-based strategies for cartilage repair and regeneration. Stem Cell Reviews and Reports. 2014;10(5):686–96.

[11] Wakitani S, Imoto K, Yamamoto T, Saito M, Murata N, Yoneda M. Human autologous culture expanded bone marrow mesenchymal cell transplantation for repair of cartilage defects in osteoarthritic knees. Osteoarthritis and Cartilage. 2002;10(3):199–206.

[12] Nejadnik H, Hui JH, Feng Choong EP, Tai B-C, Lee EH. Autologous bone marrow-derived mesenchymal stem cells versus autologous chondrocyte implantation: an observational cohort study. The American Journal of Sports Medicine. 2010;38(6):1110–6.

[13] Stolzing A, Jones E, McGonagle D, Scutt A. Age-related changes in human bone marrow-derived mesenchymal stem cells: consequences for cell therapies. Mechanisms of Ageing and Development. 2008;129(3):163–73.

[14] Choudhery M, Badowski M, Muise A, Pierce J, Harris D. Donor age negatively impacts adipose tissue-derived mesenchymal stem cell expansion and differentiation. Journal of Translational Medicine. 2014;12(1):8.

[15] Pelttari K, Winter A, Steck E, Goetzke K, Hennig T, Ochs BG, et al. Premature induction of hypertrophy during in vitro chondrogenesis of human mesenchymal stem cells

correlates with calcification and vascular invasion after ectopic transplantation in SCID mice. Arthritis & Rheumatism. 2006;54(10):3254–66.

[16] Toh WS. Recent progress in stem cell chondrogenesis. Progress in Stem Cell. 2014;1(1): 7–17.

[17] Chen S, Fu P, Cong R, Wu H, Pei M. Strategies to minimize hypertrophy in cartilage engineering and regeneration. Genes & Diseases. 2015;2(1):76–95.

[18] Thomson JA, Itskovitz-Eldor J, Shapiro SS, Waknitz MA, Swiergiel JJ, Marshall VS, et al. Embryonic stem cell lines derived from human blastocysts. Science. 1998;282(5391): 1145–7.

[19] Takahashi K, Tanabe K, Ohnuki M, Narita M, Ichisaka T, Tomoda K, et al. Induction of pluripotent stem cells from adult human fibroblasts by defined factors. Cell. 2007;131(5):861–72.

[20] Okano H, Nakamura M, Yoshida K, Okada Y, Tsuji O, Nori S, et al. Steps toward safe cell therapy using induced pluripotent stem cells. Circulation Research. 2013;112(3): 523–33.

[21] Nienhuis AW, Dunbar CE, Sorrentino BP. Genotoxicity of retroviral integration in hematopoietic cells. Molecular Therapy. 2006;13(6):1031–49.

[22] Seggewiss R, Pittaluga S, Adler RL, Guenaga FJ, Ferguson C, Pilz IH, et al. Acute myeloid leukemia is associated with retroviral gene transfer to hematopoietic progenitor cells in a rhesus macaque. Blood. 2006;107(10):3865–7.

[23] Shi Y, Desponts C, Do JT, Hahm HS, Schöler HR, Ding S. Induction of pluripotent stem cells from mouse embryonic fibroblasts by Oct4 and Klf4 with small-molecule compounds. Cell Stem Cell. 2008;3(5):568–74.

[24] Warren L, Manos PD, Ahfeldt T, Loh Y-H, Li H, Lau F, et al. Highly efficient reprogramming to pluripotency and directed differentiation of human cells with synthetic modified mRNA. Cell Stem Cell. 2010;7(5):618–30.

[25] Heng BC, Cao T, Lee EH. Directing stem cell differentiation into the chondrogenic lineage in vitro. Stem Cells. 2004;22(7):1152–67.

[26] Hentze H, Soong PL, Wang ST, Phillips BW, Putti TC, Dunn NR. Teratoma formation by human embryonic stem cells: evaluation of essential parameters for future safety studies. Stem Cell Research. 2009;2(3):198–210.

[27] Kramer J, Hegert C, Guan K, Wobus AM, Müller PK, Rohwedel J. Embryonic stem cell-derived chondrogenic differentiation in vitro: activation by BMP-2 and BMP-4. Mechanisms of Development. 2000;92(2):193–205.

[28] Carpenter MK, Inokuma MS, Denham J, Mujtaba T, Chiu C-P, Rao MS. Enrichment of neurons and neural precursors from human embryonic stem cells. Experimental Neurology. 2001;172(2):383–97.

[29] zur Nieden NI, Kempka G, Rancourt DE, Ahr H-J. Induction of chondro-, osteo- and adipogenesis in embryonic stem cells by bone morphogenetic protein-2: effect of cofactors on differentiating lineages. BMC Developmental Biology. 2005;5(1):1–15.

[30] Yang Z, Sui L, Toh WS, Lee EH, Cao T. Stage-dependent effect of TGF-β1 on chondrogenic differentiation of human embryonic stem cells. Stem Cells and Development. 2008;18(6):929–40.

[31] Gong G, Ferrari D, Dealy CN, Kosher RA. Direct and progressive differentiation of human embryonic stem cells into the chondrogenic lineage. Journal of Cellular Physiology. 2010;224(3):664–71.

[32] James D, Levine AJ, Besser D, Hemmati-Brivanlou A. TGFβ/activin/nodal signaling is necessary for the maintenance of pluripotency in human embryonic stem cells. Development. 2005;132(6):1273–82.

[33] Toh WS, Yang Z, Liu H, Heng BC, Lee EH, Cao T. Effects of culture conditions and bone morphogenetic protein 2 on extent of chondrogenesis from human embryonic stem cells. Stem Cells. 2007;25(4):950–60.

[34] Toh WS, Lim TC, Kurisawa M, Spector M. Modulation of mesenchymal stem cell chondrogenesis in a tunable hyaluronic acid hydrogel microenvironment. Biomaterials. 2012;33(15):3835–45.

[35] Kawaguchi J, Mee PJ, Smith AG. Osteogenic and chondrogenic differentiation of embryonic stem cells in response to specific growth factors. Bone. 2005;36(5):758–69.

[36] Yamashita A, Morioka M, Yahara Y, Okada M, Kobayashi T, Kuriyama S, et al. Generation of scaffoldless hyaline cartilaginous tissue from human iPSCs. Stem Cell Reports. 2015;4(3):404–18.

[37] Sui Y, Clarke T, Singh Khillan J. Limb bud progenitor cells induce differentiation of pluripotent embryonic stem cells into chondrogenic lineage. Differentiation. 2003;71(9–10):578–85.

[38] Vats A, Bielby RC, Tolley N, Dickinson SC, Boccaccini AR, Hollander AP, et al. Chondrogenic differentiation of human embryonic stem cells: the effect of the microenvironment. Tissue Engineering. 2006;12(6):1687–97.

[39] Bigdeli N, Karlsson C, Strehl R, Concaro S, Hyllner J, Lindahl A. Coculture of human embryonic stem cells and human articular chondrocytes results in significantly altered phenotype and improved chondrogenic differentiation. Stem Cells. 2009;27(8):1812–21.

[40] Hwang NS, Varghese S, Elisseeff J. Derivation of Chondrogenically-committed cells from human embryonic cells for cartilage tissue regeneration. Plos One. 2008;3(6):e2498.

[41] Qu C, Puttonen KA, Lindeberg H, Ruponen M, Hovatta O, Koistinaho J, et al. Chondrogenic differentiation of human pluripotent stem cells in chondrocyte co-culture. The International Journal of Biochemistry & Cell Biology. 2013;45(8):1802–12.

[42] Ikeda T, Kamekura S, Mabuchi A, Kou I, Seki S, Takato T, et al. The combination of SOX5, SOX6, and SOX9 (the SOX trio) provides signals sufficient for induction of permanent cartilage. Arthritis & Rheumatism. 2004;50(11):3561–73.

[43] Yang HN, Park JS, Woo DG, Jeon SY, Do H-J, Lim H-Y, et al. Chondrogenesis of mesenchymal stem cells and dedifferentiated chondrocytes by transfection with SOX Trio genes. Biomaterials. 2011;32(30):7695–704.

[44] Wei Y, Zeng W, Wan R, Wang J, Zhou Q, Qiu S, et al. Chondrogenic differentiation of induced pluripotent stem cells from osteoarthritic chondrocytes in alginate matrix. European Cells & Materials. 2012;23:1–12.

[45] Diekman BO, Thakore PI, O'Connor SK, Willard VP, Brunger JM, Christoforou N, et al. Knockdown of the cell cycle inhibitor p21 enhances cartilage formation by induced pluripotent stem cells. Tissue Engineering Part A. 2014;21(7–8):1261–74.

[46] Pittenger MF, Mackay AM, Beck SC, Jaiswal RK, Douglas R, Mosca JD, et al. Multilineage potential of adult human mesenchymal stem cells. Science. 1999;284(5411):143–7.

[47] Toh WS, Liu H, Heng BC, Rufaihah AJ, Ye CP, Cao T. Combined effects of TGFβ1 and BMP2 in serum-free chondrogenic differentiation of mesenchymal stem cells induced hyaline-like cartilage formation. Growth Factors. 2005;23(4):313–21.

[48] Liu H, Toh WS, Lu K, MacAry PA, Kemeny DM, Cao T. A subpopulation of mesenchymal stromal cells with high osteogenic potential. Journal of Cellular and Molecular Medicine. 2009;13(8b):2436–47.

[49] Dominici M, Le Blanc K, Mueller I, Slaper-Cortenbach I, Marini F, Krause D, et al. Minimal criteria for defining multipotent mesenchymal stromal cells. The International Society for Cellular Therapy position statement. Cytotherapy. 2006;8(4):315–7.

[50] Mafi P, Hindocha S, Mafi R, Griffin M, Khan WS. Adult mesenchymal stem cells and cell surface characterization—a systematic review of the literature. The Open Orthopaedics Journal. 2011;5:253–60.

[51] Barberi T, Willis LM, Socci ND, Studer L. Derivation of multipotent mesenchymal precursors from human embryonic stem cells. Plos Medicine. 2005;2(6):e161.

[52] Lian Q, Lye E, Suan Yeo K, Khia Way Tan E, Salto-Tellez M, Liu TM, et al. Derivation of clinically compliant MSCs from CD105+, CD24− differentiated human ESCs. Stem Cells. 2007;25(2):425–36.

[53] Hynes K, Menicanin D, Mrozik K, Gronthos S, Bartold PM. Generation of functional mesenchymal stem cells from different induced pluripotent stem cell lines. Stem Cells and Development. 2014;23(10):1084–96.

[54] Chen X, Song X-H, Yin Z, Zou X-H, Wang L-L, Hu H, et al. Stepwise differentiation of human embryonic stem cells promotes tendon regeneration by secreting fetal tendon matrix and differentiation factors. Stem Cells. 2009;27(6):1276–87.

[55] Koyama N, Miura M, Nakao K, Kondo E, Fujii T, Taura D, et al. Human induced pluripotent stem cells differentiated into chondrogenic lineage via generation of mesenchymal progenitor cells. Stem Cells and Development. 2012;22(1):102–13.

[56] Diederichs S, Tuan RS. Functional comparison of human-induced pluripotent stem cell-derived mesenchymal cells and bone marrow-derived mesenchymal stromal cells from the same donor. Stem Cells and Development. 2014;23(14):1594–610.

[57] Kang R, Zhou Y, Tan S, Zhou G, Aagaard L, Xie L, et al. Mesenchymal stem cells derived from human induced pluripotent stem cells retain adequate osteogenicity and chondrogenicity but less adipogenicity. Stem Cell Research & Therapy. 2015;6(1):1–14.

[58] Umeda K, Zhao J, Simmons P, Stanley E, Elefanty A, Nakayama N. Human chondrogenic paraxial mesoderm, directed specification and prospective isolation from pluripotent stem cells. Scientific Reports. 2012;2:455.

[59] Trivedi P, Hematti P. Derivation and immunological characterization of mesenchymal stromal cells from human embryonic stem cells. Experimental Hematology. 2008;36(3):350–9.

[60] Kimbrel EA, Kouris NA, Yavanian GJ, Chu J, Qin Y, Chan A, et al. Mesenchymal stem cell population derived from human pluripotent stem cells displays potent immunomodulatory and therapeutic properties. Stem Cells and Development. 2014;23(14): 1611–24.

[61] Fu X, Chen Y, Xie F-N, Dong P, Liu W, Cao Y, et al. Comparison of immunological characteristics of mesenchymal stem cells derived from human embryonic stem cells and bone marrow. Tissue Engineering Part A. 2015;21(3–4):616–26.

[62] Chung C, Burdick JA. Engineering cartilage tissue. Advanced Drug Delivery Reviews. 2008;60(2):243–62.

[63] Toh WS, Spector M, Lee EH, Cao T. Biomaterial-mediated delivery of microenvironmental cues for repair and regeneration of articular cartilage. Molecular Pharmaceutics. 2011;8(4):994–1001.

[64] Zhang S, Jiang YZ, Zhang W, Chen L, Tong T, Liu W, et al. Neonatal desensitization supports long-term survival and functional integration of human embryonic stem cell-derived mesenchymal stem cells in rat joint cartilage without immunosuppression. Stem Cells and Development. 2013;22(1):90–101.

[65] Toh WS, Guo X-M, Choo AB, Lu K, Lee EH, Cao T. Differentiation and enrichment of expandable chondrogenic cells from human embryonic stem cells in vitro. Journal of Cellular and Molecular Medicine. 2009;13(9b):3570–90.

[66] Koay EJ, Hoben GMB, Athanasiou KA. Tissue engineering with chondrogenically differentiated human embryonic stem cells. Stem Cells. 2007;25(9):2183–90.

[67] Hoben GM, Willard VP, Athanasiou KA. Fibrochondrogenesis of hESCs: growth factor combinations and cocultures. Stem Cells and Development. 2008;18(2):283–92.

[68] Ko J-Y, Kim K-I, Park S, Im G-I. In vitro chondrogenesis and in vivo repair of osteochondral defect with human induced pluripotent stem cells. Biomaterials. 2014;35(11):3571–81.

[69] Cheng A, Kapacee Z, Peng J, Lu S, Lucas RJ, Hardingham TE, et al. Cartilage repair using human embryonic stem cell-derived chondroprogenitors. Stem Cells Translational Medicine. 2014;3(11):1287–94.

[70] Toh WS, Lee EH, Guo X-M, Chan JKY, Yeow CH, Choo AB, et al. Cartilage repair using hyaluronan hydrogel-encapsulated human embryonic stem cell-derived chondrogenic cells. Biomaterials. 2010;31(27):6968–80.

[71] Liu J, Nie H, Xu Z, Niu X, Guo S, Yin J, et al. The effect of 3D nanofibrous scaffolds on the chondrogenesis of induced pluripotent stem cells and their application in restoration of cartilage defects. Plos One. 2014;9(11):e111566.

[72] Fecek C, Yao D, Kaçorri A, Vasquez A, Iqbal S, Sheikh H, et al. Chondrogenic derivatives of embryonic stem cells seeded into 3D polycaprolactone scaffolds generated cartilage tissue in vivo. Tissue Engineering Part A. 2008;14(8):1403–13.

[73] Levenberg S, Huang NF, Lavik E, Rogers AB, Itskovitz-Eldor J, Langer R. Differentiation of human embryonic stem cells on three-dimensional polymer scaffolds. Proceedings of the National Academy of Sciences. 2003;100(22):12741–6.

[74] Bai HY, Chen GA, Mao GH, Song TR, Wang YX. Three step derivation of cartilage like tissue from human embryonic stem cells by 2D–3D sequential culture in vitro and further implantation in vivo on alginate/PLGA scaffolds. Journal of Biomedical Materials Research Part A. 2010;94A(2):539–46.

[75] Hwang NS, Varghese S, Zhang Z, Elisseeff J. Chondrogenic Differentiation of Human Embryonic Stem Cell-Derived Cells in Arginine-Glycine-Aspartate-Modified Hydrogels. Tissue Engineering. 2006;12(9):2695–706.

[76] Hwang NS, Varghese S, Lee HJ, Zhang Z, Ye Z, Bae J, et al. In vivo commitment and functional tissue regeneration using human embryonic stem cell-derived mesenchymal cells. Proceedings of the National Academy of Sciences. 2008;105(52):20641–6.

[77] Toh WS, Toh Y-C, Loh XJ. Hydrogels for stem cell fate control and delivery in regenerative medicine. In: Loh JX, editor. In-situ gelling polymers: for biomedical applications. Singapore: Springer Singapore; 2015. pp. 187–214.

[78] Nehrer S, Breinan HA, Ramappa A, Shortkroff S, Young G, Minas T, et al. Canine chondrocytes seeded in type I and type II collagen implants investigated In Vitro. Journal of Biomedical Materials Research. 1997;38(2):95–104.

[79] Bosnakovski D, Mizuno M, Kim G, Takagi S, Okumura M, Fujinaga T. Chondrogenic differentiation of bovine bone marrow mesenchymal stem cells (MSCs) in different hydrogels: influence of collagen type II extracellular matrix on MSC chondrogenesis. Biotechnology and Bioengineering. 2006;93(6):1152–63.

[80] Toh WS, Foldager CB, Olsen BR, Spector M. Basement membrane molecule expression attendant to chondrogenesis by nucleus pulposus cells and mesenchymal stem cells. Journal of Orthopaedic Research. 2013;31(7):1136–43.

[81] Singh P, Schwarzbauer JE. Fibronectin and stem cell differentiation—lessons from chondrogenesis. Journal of Cell Science. 2012;125(Pt 16):3703–12.

[82] Schminke B, Frese J, Bode C, Goldring MB, Miosge N. Laminins and nidogens in the pericellular matrix of chondrocytes: their role in osteoarthritis and chondrogenic differentiation. The American Journal of Pathology. 2016;186(2):410–8.

[83] Toh WS, Loh XJ. Advances in hydrogel delivery systems for tissue regeneration. Materials Science and Engineering: C. 2014;45:690–7.

[84] Shu XZ, Ahmad S, Liu Y, Prestwich GD. Synthesis and evaluation of injectable, in situ crosslinkable synthetic extracellular matrices for tissue engineering. Journal of Biomedical Materials Research Part A. 2006;79A(4):902–12.

[85] Burdick JA, Prestwich GD. Hyaluronic acid hydrogels for biomedical applications. Advanced Materials. 2011;23(12):H41–H56.

[86] Wang L-S, Du C, Toh WS, Wan ACA, Gao SJ, Kurisawa M. Modulation of chondrocyte functions and stiffness-dependent cartilage repair using an injectable enzymatically crosslinked hydrogel with tunable mechanical properties. Biomaterials. 2014;35(7):2207–17.

[87] Yue K, Trujillo-de Santiago G, Alvarez MM, Tamayol A, Annabi N, Khademhosseini A. Synthesis, properties, and biomedical applications of gelatin methacryloyl (GelMA) hydrogels. Biomaterials. 2015;73:254–71.

[88] Guo Y, Yuan T, Xiao Z, Tang P, Xiao Y, Fan Y, et al. Hydrogels of collagen/chondroitin sulfate/hyaluronan interpenetrating polymer network for cartilage tissue engineering. Journal of Materials Science: Materials in Medicine. 2012;23(9):2267–79.

[89] Ni Y, Tang Z, Cao W, Lin H, Fan Y, Guo L, et al. Tough and elastic hydrogel of hyaluronic acid and chondroitin sulfate as potential cell scaffold materials. International Journal of Biological Macromolecules. 2015;74:367–75.

[90] Kim IL, Mauck RL, Burdick JA. Hydrogel design for cartilage tissue engineering: a case study with hyaluronic acid. Biomaterials. 2011;32(34):8771–82.

[91] Liu Y, Shu XZ, Prestwich GD. Osteochondral defect repair with autologous bone marrow-derived mesenchymal stem cells in an injectable, in situ, cross-linked synthetic extracellular matrix. Tissue Engineering. 2006;12(12):3405–16.

[92] Jukes JM, van der Aa LJ, Hiemstra C, van Veen T, Dijkstra PJ, Zhong Z, et al. A newly developed chemically crosslinked dextran–poly(ethylene glycol) hydrogel for cartilage tissue engineering. Tissue Engineering Part A. 2010;16(2):565–73.

[93] Lutolf MP, Hubbell JA. Synthetic biomaterials as instructive extracellular microenvironments for morphogenesis in tissue engineering. Nature Biotechnology. 2005;23(1): 47–55.

[94] Bryant SJ, Anseth KS. Hydrogel properties influence ECM production by chondrocytes photoencapsulated in poly(ethylene glycol) hydrogels. Journal of Biomedical Materials Research. 2002;59(1):63–72.

[95] Burdick JA, Anseth KS. Photoencapsulation of osteoblasts in injectable RGD-modified PEG hydrogels for bone tissue engineering. Biomaterials. 2002;23(22):4315–23.

[96] Wang Y, Zhu G, Li N, Song J, Wang L, Shi X. Small molecules and their controlled release that induce the osteogenic/chondrogenic commitment of stem cells. Biotechnology Advances. 2015;33(8):1626–40.

[97] Zhang Z, Messana J, Hwang NS, Elisseeff JH. Reorganization of actin filaments enhances chondrogenic differentiation of cells derived from murine embryonic stem cells. Biochemical and Biophysical Research Communications. 2006;348(2):421–7.

[98] Hwang NS, Varghese S, Theprungsirikul P, Canver A, Elisseeff J. Enhanced chondrogenic differentiation of murine embryonic stem cells in hydrogels with glucosamine. Biomaterials. 2006;27(36):6015–23.

[99] Kelly DJ, Jacobs CR. The role of mechanical signals in regulating chondrogenesis and osteogenesis of mesenchymal stem cells. Birth Defects Research Part C: Embryo Today: Reviews. 2010;90(1):75–85.

[100] Haugh MG, Meyer EG, Thorpe SD, Vinardell T, Duffy GP, Kelly DJ. Temporal and spatial changes in cartilage-matrix-specific gene expression in mesenchymal stem cells in response to dynamic compression. Tissue Engineering Part A. 2011;17(23–24):3085–93.

[101] Terraciano V, Hwang N, Moroni L, Park HB, Zhang Z, Mizrahi J, et al. Differential response of adult and embryonic mesenchymal progenitor cells to mechanical compression in hydrogels. Stem Cells. 2007;25(11):2730–8.

[102] da Silva Meirelles L, Fontes AM, Covas DT, Caplan AI. Mechanisms involved in the therapeutic properties of mesenchymal stem cells. Cytokine & Growth Factor Reviews. 2009;20(5–6):419–27.

[103] Liu H, Lu K, MacAry PA, Wong KL, Heng A, Cao T, et al. Soluble molecules are key in maintaining the immunomodulatory activity of murine mesenchymal stromal cells. Journal of Cell Science. 2012;125(1):200–8.

[104] Jeong SY, Kim DH, Ha J, Jin HJ, Kwon S-J, Chang JW, et al. Thrombospondin-2 secreted by human umbilical cord blood-derived mesenchymal stem cells promotes chondrogenic differentiation. Stem Cells. 2013;31(10):2136–48.

[105] Sze SK, de Kleijn DPV, Lai RC, Khia Way Tan E, Zhao H, Yeo KS, et al. Elucidating the secretion proteome of human embryonic stem cell-derived mesenchymal stem cells. Molecular & Cellular Proteomics. 2007;6(10):1680–9.

[106] Lai RC, Arslan F, Lee MM, Sze NSK, Choo A, Chen TS, et al. Exosome secreted by MSC reduces myocardial ischemia/reperfusion injury. Stem Cell Research. 2010;4(3):214–22.

[107] Tan CY, Lai RC, Wong W, Dan YY, Lim S-K, Ho HK. Mesenchymal stem cell-derived exosomes promote hepatic regeneration in drug-induced liver injury models. Stem Cell Research & Therapy. 2014;5(3):1–14.

[108] Zhang S, Chu WC, Lai RC, Hui JH, Lee EH, Lim SK, et al. Human mesenchymal stem cell-derived exosomes promote orderly cartilage regeneration in an immunocompetent rat osteochondral defect model. Cytotherapy. 2016 (in press)

Chapter 18

Cartilage Regeneration Using Pluripotent Stem Cell-Derived Chondroprogenitors: Promise and Challenges

Naoki Nakayama, John Y. Lee, Nadine Matthias, Katsutsugu Umeda, Qing Yan and Johnny Huard

Additional information is available at the end of the chapter

http://dx.doi.org/10.5772/63052

Abstract

The cartilage of joints is long-lasting (i.e., permanent) cartilage and is not spontaneously repaired after injury in humans. There has been considerable interest in the clinical application of stem cells to the repair of damaged cartilage; however, current cell therapies using adult chondrocytes and mesenchymal stromal cells face problems associated with the low yield of such cells. The expansion culture, needed before transplantation, leads to the formation of fibrocartilage or growth plate-like (i.e., bone-forming) cartilage *in vivo*. Both types of cartilage are unsuitable for the repair of joint cartilage such as meniscus and articular cartilage. Joints are formed during embryogenesis. Therefore, we hypothesize that embryonic progenitor cells responsible for the development of joint cartilage would be the best for regenerating joint cartilage in the adult. Pluripotent stem cells (PSCs) are expected to differentiate in culture into any somatic cell types through processes that mimic embryogenesis, making human (h)PSCs a promising source of embryonic cells for regenerative medicine. However, regardless of the cell system used, the major research goals leading to clinical application to cartilage regeneration are to (1) expand chondrogenic cells (chondroprogenitors) to sufficient numbers without loss of their chondrogenic activity, and (2) direct the differentiation of such cells *in vivo* or *in vitro* toward articular or other types of chondrocytes of interest. The overall aim of the current review was to provide the basis of a strategy for meeting the goals for cartilage regeneration by the use of hPSC-derived chondroprogenitor cells. We provide an overview on signaling mechanisms that are known to affect the expandability and chondrogenic activity of adult and embryonic chondroprogenitors, as well as their differentiation *in vivo* or *in vitro* toward a particular type of chondrocyte. We then discuss alternative types of progenitor cells that might replace or combine with the hPSC-derived chondroprogenitors to regenerate permanent cartilage. We also include our recent achievement of successfully expanding hPSC-derived neural crest to generate ectomesenchymal chondroprogenitors that can be maintained for a long term in culture without loss of chondrogenic activity. Finally, we provide information on the challenges that hPSC progeny-based regenerative medicine will face, and discuss the implications for such challenges for the future use of PSC progeny to regenerate cartilage.

Keywords: joint, cartilage, maturation, expansion, regeneration

1. Introduction

Damaged human joint articular cartilage does not heal spontaneously and eventually develops into osteoarthritis, probably due to a lack of proper "regenerative (i.e., stem/progenitor)" cells. Consequently, there has been considerable interest in the clinical application of chondrogenic stem/progenitor cells in the repair of damaged cartilage. For example, cell-based therapies have been developed that use the endogenous mobilization of marrow cells (i.e., microfracture method), or cells such as expanded (dedifferentiated) articular chondrocytes or mesenchymal stromal cells (MSCs) added exogenously via a periosteal patch or collagen membrane (i.e., autologous chondrocyte implantation method). Although such therapies relieve the major clinical symptom of pain, they do not promote the regeneration of long-lasting hyaline cartilage. Tissue engineering approaches that deliver a biodegradable matrix seeded with dedifferentiated chondrocytes or MSCs have been tested for the treatment of large cartilage lesions in animal models with similarly disappointing long-term results. Furthermore, the cells that persist at the repair site are often not the donor cells. Instead, they appear to be endogenous cells that have migrated from other sites in the body. The findings suggest that the transplanted chondrogenic cells may function as trophic mediators by stimulating endogenous repair functions. A trophic role would be problematic since there is no control for the selective migration of beneficial cell types such as articular chondrocytes and their precursors over unwanted cell types such as fibroblastic and osteogenic cells.

As to the cell type to be used for cartilage regeneration, chondrocytes are naturally suitable. However, their use requires *ex vivo* expansion since it is difficult to obtain sufficient chondrocytes for the treatment. During expansion culture, chondrocytes lose their characteristics. They dedifferentiate to fibroblastic cells that possess varying degrees of chondrogenic activity that quickly declines as culture continues. Furthermore, isolation of autologous chondrocytes requires an initial surgical removal of normal cartilage, which contributes to the increased risk of morbidity. MSCs isolated from adult bone marrow or fat tissue are defined *in vitro* according to both their differentiation capacity, typically to bone, cartilage and fat cell lineages, and their self-renewal activity [1, 2]. Adult muscle-derived stem cells (MDSCs) isolated from skeletal muscle can also be differentiated into bone and cartilage cells [3–5] in addition to other population of muscle progenitor cells [6]. Therefore, MSCs and MDSCs are expected to serve as precursors of chondrocytes (hereafter called "chondroprogenitors") and alternative sources of cells for cartilage repair. However, as in the case of chondrocytes, such chondroprogenitors also often require *ex vivo* expansion to achieve sufficient cells for the treatment. Furthermore, expansion of MSCs is known to result in losses in chondrogenic activity [7, 8] that can be decreased to some degrees by the inclusion of fibroblast growth factor (FGF) in the growth

http://dx.doi.org/10.5772/63052

media [9–11]. Another limitation is that the *ex vivo*-expanded adult MSCs tend to lose long-term viability and produce fibrocartilage tissue that is poorly integrated into the rest of the hyaline cartilage. Consequently, the repaired cartilaginous tissue is often lost from the site of damage, creating the need for subsequent surgery [12–15]. Attempts have been made to improve the *in vivo* chondrogenicity of adult MSCs and dedifferentiated articular chondrocytes by forcing the expression of genes that encode chondrogenic growth factors, for example, transforming growth factor (TGF)β and bone morphogenetic protein (BMP). While the attempts have led to short-term benefit, they have failed to prevent the developed chondrocytes from the hypertrophic differentiation, terminal maturation, and mineralization that promote osteophyte formation in the longer term [12]. So it seems that such *in vivo* factor-induced chondrocytes may be committed to form growth plate-like "unstable" cartilage [16]. Further attempts at improvement have aimed at controlling terminal maturation of chondrocytes and the ensuing osteogenesis process, such as by inhibiting angiogenesis using a soluble form of the vascular endothelial growth factor (VEGF) receptor 1 (VEGFR1/Flt1) (while sFlt1 also increases the chondrogenic potential the cells) [3–5]. As stated by Somoza et al. [15], neither exogenously administered nor endogenously mobilized MSCs have contributed to the generation of durable and phenotypically correct (i.e., anatomic site-specific) articular cartilage. Evidence to date thus casts doubt on the potential usefulness of MSCs in the regeneration of hyaline articular cartilage. A better source of chondrocytes and better knowledge of the critical environmental cues for the development of articular cartilage by chondrocytes are clearly needed [17]. Regardless of the type of chondrogenic cells used, the main challenges with cell-based, articular cartilage tissue-engineering remain: (1) quality and durability of the *de novo* generated cartilage tissue, (2) resistance of the tissue to an endochondral ossification program, and (3) its effective integration into the anatomic site-specific host tissue.

Joint formation is initiated during embryogenesis by a specialized cell type, the "joint progenitor," which is responsible for generating hyaline articular cartilage. Furthermore, surface injury that is introduced *in utero* into fetal joint cartilage is repaired spontaneously without scarring in a large animal model [18] indicating that embryonic epiphyseal chondrocytes possess spontaneous repair activity. It is thus conceivable that embryonic joint progenitors and/or chondroprogenitors may elicit better regenerative capacity than adult dedifferentiated chondrocytes or MSCs at the injured cartilage even in the adult. However, the only practical source of such specialized embryonic cells from humans would be pluripotent stem cells (PSCs). Whether derived from an embryo (embryonic stem cells or ESCs) or induced from adult tissue cells (induced pluripotent stem cells or iPSCs), PSCs can be maintained in culture without loss of their pluripotency and are expected to differentiate in culture into any type of somatic cell. Interestingly, both types of PSC represent inner cell mass or epiblast so that their *in vitro* differentiation mimics early embryogenic processes [19–21]. Therefore, it is expected that human embryonic progenitors involved in joint cartilage formation can be generated using the wealth of knowledge accumulated from studies on the developmental biology of bone and cartilage. In general, PSCs are induced to differentiate into mesenchymal chondroprogenitors prior to chondrocyte formation. However, most of the early studies employed spontaneous (i.e., uncontrolled) differentiation methods to generate such mesenchymal cells, followed by expansion culture using serum-containing MSC media to enrich them [22]. Unfortunately, such expansion was accompanied by the loss of chondrogenic activity [23, 24], as seen with adult human chondrocytes and MSCs [25, 26]. Furthermore, some reports appear to have overlooked knowledge of embryology and treated PSCs as if they were limb mesenchyme or MSCs by trying to induce chondrogenesis directly in a TGFβ-containing chondrogenesis medium designed for inducing chondrogenesis from adult MSCs [27]. Consequently, potential benefits for cartilage repair by human (h)PSC-derived chondroprogenitors over those by adult hMSCs, whether in quantity or quality, have not been demonstrated [22]. We proposed that the disappointing early observations may have been attributable to the use of mesenchymal hPSC progeny of unclear embryonic origins ["Origin matters" 28], and undefined conditions for their expansion. The choice of both cell type and medium failed to draw on the wealth of available information on the signaling mechanisms involved in embryonic skeletogenesis, thereby hampering progress toward the effective repair of cartilage. Recently, reports including ours [23, 29–35] have been based on knowledge of embryology to generate chondroprogenitors of a particular embryonic origin and the use defined culture conditions for their expansion.

Concerning the challenges with the use of MSCs as stated by Somoza et al. [15], it is conceivable that the successful repair of articular cartilage will depend on both the ability of chondrogenic cells with high expansion capacity to form hyaline cartilage and a microenvironment at the site of repair that promotes chondrogenesis, while preventing terminal maturation and mineralization of chondrocytes. Therefore, we aim to give an overview on what we know from genetic as well as cellular (e.g., MSCs and MDSCs) studies, (1) about ways to expand chondroprogenitors and maintain their chondrogenic activity, and (2) about the potential signaling mechanisms and cell populations that affect the type (e.g., articular chondrocytes or growth plate chondrocytes) and/or differentiation state (e.g., immature or mature) of chondrocytes generated from chondroprogenitors. We will also summarize relevant information obtained from our studies of hPSC-derived chondroprogenitors.

2. Long-term expansion of chondroprogenitor cells without loss of their chondrogenic activity: advantages of hPSC-derived progenitors

With any cell system, one of the major research goals for cellular therapy or tissue-engineering approach for cartilage regeneration is to expand chondroprogenitors to yield large numbers without the loss of chondrogenic activity. These goals are important because the quality and durability of the *de novo* generated cartilage tissue seem to be inversely related to the extent of expansion culture of articular chondrocytes and MSCs. Retention of chondrogenic activity even after long-term expansion in adult stem/progenitor cells is not a trivial outcome. Standard serum-containing growth conditions support chondrogenic activity of MSCs for only 2–4 passages. However, attempts to circumvent the problems by improving the MSC culture conditions have met with some success. Here, we will provide an overview of the studies done with MSCs and hPSC-derived progenitor cells. As expected, under optimal culture conditions, the proliferative potential of PSC-derived progeny far exceeds that of their adult counterparts.

2.1. FGF, TGFβ, hedgehog, and WNTs for the expansion of adult MSC

FGF signaling is known to upregulate expression of *Sox9*, a key regulator of chondrogenesis [36–38], in chondrogenic mesenchymal cell lines and primary chondrocytes [39]. FGFs are also known to preserve the ability of limb-bud mesenchymal cells to undergo chondrogenesis, when the WNT signal is removed [40]. On the other hand, FGF2 does not consistently affect *in vitro* chondrogenesis from MSCs [41, 42]. However, FGF2 supplementation during expansion culture in the presence of 10% (v/v) fetal bovine serum (FBS) reproducibly enhanced their growth [9–11] and maintained the chondrogenic activity up to 7 passages for hMSCs [11] and up to 9 for rabbit MSCs [9] as assessed by the standard TGFβ-induced *in vitro* chondrogenesis culture. Serum-free methods for the expansion culture have also been established. The StemPro-based expansion culture for hMSCs from adult bone marrow sustained their growth, morphology, and tri-lineage (bone, cartilage and fat) potential for at least 5 passages [43]. In addition to FGF2, the StemPro culture contains platelet-derived growth factor PDGF-BB, TGFβ1, and insulin. Therefore, inclusion of FGF2 in the growth medium for hMSCs will be beneficial for expanding hMSCs for cartilage repair.

Indian hedgehog (Ihh) is a key mitogen for chondrocytes in the growth plate [44]. Similarly, hedgehog signaling stimulates the growth of human bone marrow MSCs and enhances chondrogenesis in serum-containing culture conditions [45]. On the other hand, WNT signaling is known to play important roles on the growth, specification, movement, and organization of early precursor cells for osteochondrogenesis [46]. The roles of WNT signaling on the proliferation of embryonic osteochondrogenic mesenchymal cells [47] and their association with osteogenesis have been well established [48, 49]. Very recently, WNT3a has been shown to further stimulate the FGF2-stimulated growth of hMSCs and enhance the FGF2-enhanced maintenance of chondrogenic potency for at least 20 days (4 passages) of expansion culture in the presence of 10% FBS [50]. In spite of these positive results with hedgehogs and WNTs in expanding mesenchymal cells, both factors are "developmental factors" that induce osteogenesis rather than chondrogenesis during skeletogenesis. Through the canonical signaling pathway (involving β-catenin), WNTs reduce Sox9 levels and antagonize the functions of Sox9 in chondrocytes [51], and Ihh is placed upstream of the WNT/β-catenin signaling events, essential for osteoblastogenesis in the perichondrium of developing bone [52–55]. We have also found that WNTs stimulate the growth of hPSC-derived chondroprogenitor cells (data not shown). Although the effects may depend on the developmental/differentiation stage of the target cell, use of WNTs and Ihh on chondrogenic mesenchymal cells probably needs careful prior examination to verify that they do not cause loss of chondrogenic activity during expansion. Research into ways of improving expansion culture for hMSCs to preserve their chondrogenicity reproducibly is still ongoing.

2.2. Suppression of TGFβ signaling successfully expands endothelial (progenitor) cells derived from PSCs

Mesenchymal progeny of PSCs had the similar problems as adult MSC, namely the loss of original phenotypes and developmental potentials during expansion. In this regard, Miyazono and colleagues [56] published a pioneering work in 2003 in which they demonstrated that Nodal/Activin/TGFβ receptor kinase inhibitor, SB431542, enhanced the growth and integrity of mouse (m)ESC-derived endothelial cells, leading to a successful protocol for the expansion of hPSC-derived endothelial cells [57]. TGFβ is known to be a potent inhibitor for lymphohematopoietic progenitor cell proliferation [58]. However, as the name "transforming growth factor" indicates, it was originally found as a potent growth-promoting factor for many transformed (tumor/cancer) cells and untransformed (normal) cells, including human adult bone marrow MSCs [43]. PSC-derived endothelial cells may produce TGFβ-like activity that serves as a "brake" on their own proliferation in culture.

2.3. Long-term expansion of endoderm stem/progenitor cells derived from hPSCs without loss of their developmental potency

Recently, multipotential endodermal stem/progenitor cells, which can undergo long-term expansion without loss of their phenotypes and developmental potentials (i.e., can self-renew), have been developed from hPSCs [59, 60]. The foregut stem cells can be maintained in culture over 18 passages in RPMI + B27-based serum-free medium supplemented with FGF2, Activin A, BMP4, hepatocyte growth factor (HGF), and epidermal growth factor (EGF) [60]. The endodermal progenitor cells can be maintained over 24 passages in IMDM:F12 (3:1) + N2/B27-based serum-free medium containing FGF2, BMP4, VEGF, and EGF [59]. Nodal/Activin/TGFβ receptor kinase inhibitor is not necessary for the long-term expansion of endodermal stem/progenitor cells. Therefore, the inhibition of Nodal/Activin/TGFβ receptor kinase is not a common requirement for the expansion of hPSC-derived progeny.

2.4. Long-term maintenance of hPSC-derived chondrogenic ectomesenchymal cells without loss of their chondrogenic activity

In contrast, methods for the long-term expansion of hPSC-derived chondrogenic mesenchymal cells had not been explored extensively. Research has focused more on the "genesis" but not "expansion" of chondrogenic activity from hPSCs. Although a short-term expansion of chondrogenic activity as judged by the increase in the $SOX9^+$ cell number was reported [61], expansion of PSC-derived mesenchymal cells has not been accompanied by long-term maintenance of their chondrogenic activity in the conventional MSC medium, as in the case of adult MSCs [22]. Proliferation without loss of chondrogenic activity is thus not an intrinsic property of the hPSC-derived "embryonic" chondrogenic cells. Rather, it is a property that can emerge under certain defined culture conditions. In this respect, we have previously established and refined signaling requirements for the differentiation of mouse and human ESCs and iPSCs to chondrogenic lateral plate mesoderm, paraxial mesoderm, and cranial neural crest-like progeny in a serum-free defined medium [22, 31, 33–35, 62–64]. Our group was the first to define the MIXL1-green fluorescent protein $(GFP)^+$ VEGFR2 $(KDR)^-$ PDGF receptor (PDGFR)α^+ human paraxial mesoderm progeny and T-GFP$^+$KDR$^-$PDGFRα^+ mouse paraxial mesoderm progeny, from which chondrogenic mesenchymal cells were derived [33–35, 62]. We have also recently reported that MIXL1-GFP$^-$CD271^{hi}CD73$^-$ human neural crest-like progeny develops in a chemically defined medium [CDM: IMDM:F12 (1:1), 0.5% (w/v) fatty acid-free BSA, 1% (v/v) synthetic lipids, 10 μg/ml insulin, 300 μg/ml holo-transferrin, 0.17 mM ascorbic acid-2-phosphate, 0.3 mM monothyoglycerol] in the presence of Nodal/Activin/TGFβ receptor inhibitor (SB431542) [65–68] and that they quickly become CD271$^+$CD73$^+$ ectomesenchymal cells expressing *SOX9* (albeit at lower levels than those achieved during chondrogenesis) during expansion in CDM in the presence of FGF2 and SB431542 (FSB condition, **Figure 1**) [31]. Significantly, such cells maintain their normal karyotype and chondrogenic activity at least at passage 10 and passage 16, respectively. In contrast, removal of SB431542 slows down or stops the proliferation with enhanced expression of *TGFB1/2/3* genes and TGFβ target genes such as *TGFBI* and decreases their chondrogenic activity after 2–4 passages. Removal of FGF, the only growth factor in the medium, led to immediate cessation of growth. Thus, the neural crest-derived ectomesenchymal cells originating from hPSCs are similar to PSC-derived endothelial cells in that the autocrine or paracrine action of TGFβ serves as an intrinsic "brake" on their expansion in culture. In order to shed light on the molecular mechanism of such long-term growth and maintenance of

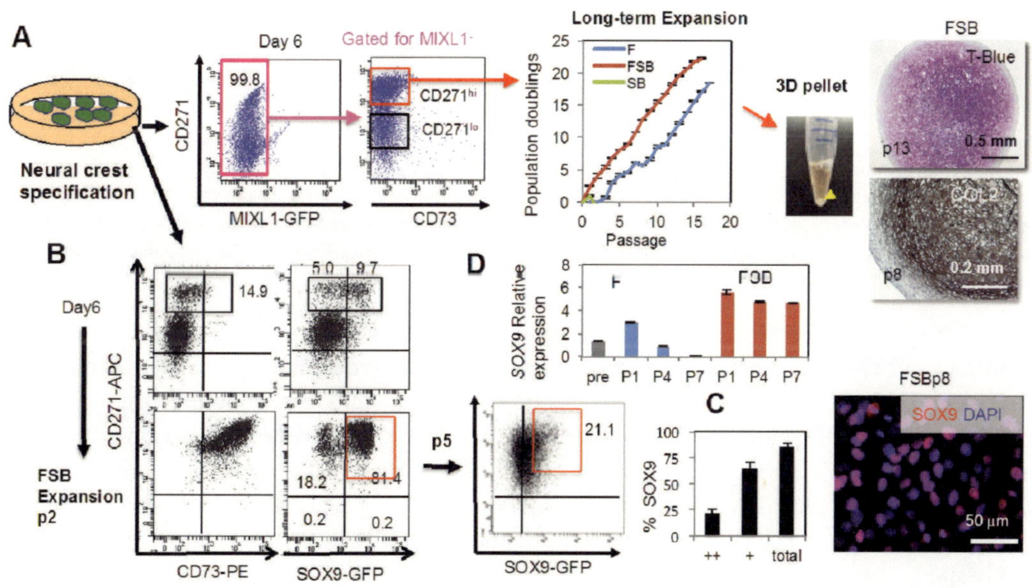

Figure 1. (A) Chondrogenic neural crest-ectomesenchymal cells. The MIXL1-GFP-CD271^{hi}CD73$^-$ cell fraction, enriched in the *SOX10* transcript, was isolated and expanded under FSB to generate SOX9-expressing chondroprogenitors. The formed cartilage particles were fixed, sectioned and stained with Toluidine blue (T-Blue) and anti-COL2 antibody. N: Noggin, SB: SB431542, F: FGF2, C: CHIR. (B) Ectomesenchymal cells from the SOX9-GFP hiPSC line. The iPSCs were differentiated into CD271^{hi}CD73$^-$ neural crest-like progeny: 2/3 of which (9.7%) are SOX9-GFP$^+$, were then expanded in FSB, leading to CD271$^+$CD73$^+$ cells that are mostly SOX9-GFP$^+$ (81.4%), but reduced to mostly SOX9-GFP^{lo} by passage (p) 5. (C) Immunostaining supported the FACS results. ++:high, +: low (D) *SOX9* mRNA expression was stably maintained by p7.

chondrogenic activity in culture, we performed genome-wide transcriptome analysis using RNA-seq technology to compare RNA profiles of ectomesenchymal cells cultured under FSB conditions with those of cells initially maintained in FSB but later exposed briefly to media without SB431542 (FSB → F) [31]. Bioinformatic analysis has revealed a list of protein-coding genes and non-coding RNA genes specifically expressed in cells cultured under either condition (**Figure 2**). Among them are growth-promoting genes, such as *LIN28A/B* and *MYCN*, neural crest/ectomesenchymal genes, such as *NGFR, TWIST1/2, ALX1/3/4,* and *SOX8,* and *SOX9* regulator genes, such as *MAF, ZBTB16,* and *LBH* [31].

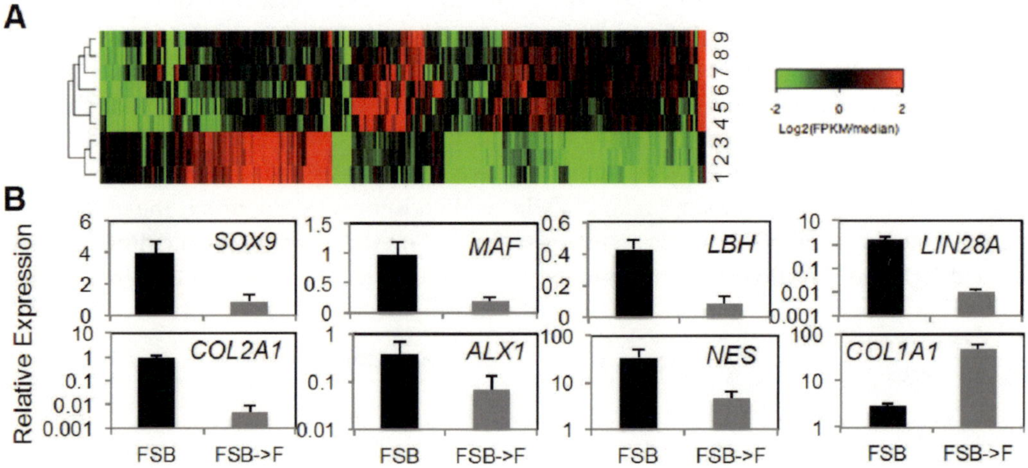

Figure 2. Comparative transcriptome analysis of ectomesenchymal cells using the RNA-seq technology. (A) Heat map of the top 250 genes, which are differentially expressed among the three groups. Lanes 1–3;"FSB", Lanes 4, 5, 8; "F", Lanes 6, 7, 9; "FSB→F". (B) RT-PCR confirmation of the relative gene expression levels predicted by the FPKM (Fragments Per Kilobase of transcript per Million mapped reads) values. F: FGF2, SB: SB431542, →: transition of culture conditions.

In contrast to PSC-derived endothelial cells and chondrogenic ectomesenchymal cells, the maintenance of human adult bone marrow MSCs in culture required TGFβ in a defined culture conditions [43]. Therefore, PSC-derived ectomesenchymal cells and bone marrow MSCs are distinct cell types, despite the proposal that both may share the common developmental origin of neural crest [69–71]. In support, application of SB431542 to mouse bone marrow MSC culture failed to support their growth and preferentially left adipocytes (data not shown), consistent with the notion that TGFβ suppresses adipocytic differentiation of mesenchymal progenitors [72].

3. Signaling mechanisms potentially leading to the genesis of long-lasting immature cartilage from hPSC-derived chondroprogenitor cells

The reproducible generation, either *in vivo* or *in vitro,* of articular chondrocytes that form immature, unmineralized cartilage that can be maintained in the long-term *in vivo* (designat-

http://dx.doi.org/10.5772/63052

ed hereafter, permanent cartilage) also remains a difficult task. For example, while articular chondrocytes, whether uncultured or briefly cultured, generate *in vitro* hyaline cartilage with no sign of hypertrophic differentiation and terminal maturation [73, 74], MSCs appear to be limited to an endochondral ossification program under the same condition—a program that will result in vascular penetration, marrow deposition, and ossification of the developed cartilaginous tissue [15, 49, 74, 75]. The critical environmental signals required for pathway switching toward the genesis of permanent hyaline cartilage have not been discovered. In this respect, it is worth noting that the differentiation state of chondrocytes generated *in vitro* from MSCs affects the stability of generated cartilage after transplantation. When MSCs are induced to generate mature, hypertrophic chondrocytes *in vitro*, ossification of the developed cartilage pellets *in vivo* is accelerated [16]. Therefore, controlling the hypertrophic differentiation of developed chondrocytes seems to be a promising first step toward the generation of permanent cartilage. Interestingly, even in the endochondral ossification process, there are signaling mechanisms that inhibit chondrocyte hypertrophic differentiation [49]. Such mechanisms have been defined via extensive genetic studies. The process of hypertrophic differentiation and terminal maturation is initiated by signaling events that eventually suppress chondrocyte determinants such as Sox9 and express determinants of the osteogenic program, such as Runx2 in chondrocytes. Sox9 is the master transcription factor for chondrocytes [36, 37, 76, 77]. Runx2/Cbfa1 is the master regulator of "mineralization" in both chondrocytes and osteoblasts [78–80] and facilitates hypertrophic differentiation of chondrocytes (e.g., stimulates the expression of *Col10a1* the marker for chondrocyte hypertrophic differentiation) in the growth plate [81–85]. Any of the signaling mechanisms that maintain Sox9 without inducing Runx2 or other osteogenic transcription factors in chondrocytes could be the pathway switcher for MSCs, which we believe should also help to generate permanent cartilage from hPSC-derived chondroprogenitors. We have summarized the roles of five well-known signaling mechanisms below that affect endochondral ossification *in vivo* and chondrogenesis from MSCs *in vitro* and the implications for each for chondrogenesis from hPSC-derived chondroprogenitors.

3.1. Indian hedgehiog (Ihh)-parathyroid hormone-related peptide (PTHrP) feedback loop

The Ihh-PTHrP feedback loop is one of the best-known mechanisms that controls chondrocyte hypertrophic differentiation in the growth plate [86–89]. After the cartilage primordium has formed during embryogenesis, perichondrial cells and chondrocytes at the ends of the cartilage produce PTHrP that signals through PTH receptors on chondrocytes. PTHrP action keeps chondrocytes in a proliferative state and delays their differentiation toward the post-mitotic hypertrophic chondrocytes [90]. The chondrocytes distant from the cartilage ends escape from such PTHrP effects and stop proliferating (pre-hypertrophic chondrocytes). Such chondrocytes synthesize Ihh, which in turn stimulates the synthesis of PTHrP from chondrocytes at the ends of the cartilage, potentially through the action of TGFβ [91, 92]. The pre-hypertrophic chondrocytes differentiate to hypertrophic chondrocytes and then to terminally matured, mineralized chondrocytes. The feedback loop between PTHrP and Ihh thus regulates the pace of differentiation of immature (proliferating) chondrocytes. The Ihh-PTHrP system may also function similarly during articular chondrogenesis [93].

Elucidation of the individual roles of hedgehog and PTHrP signaling is important if there is to be future application of the Ihh and PTHrP signaling mechanisms to control chondrocyte hypertrophic differentiation during chondrogenesis from MSCs- or hPSC-derived chondro-progenitors. To date, the role of hedgehog signaling on chondrocyte maturation appears contradictory. Ihh directly stimulates proliferation of (immature) chondrocyte but also stimulates osteogenesis from osteochondrogenic mesenchymal cells. In addition, in the absence of PTHrP, Ihh appears to stimulate hypertrophic differentiation of chondrocytes through activating canonical WNT signaling (analogous to the Ihh-induced osteogenesis process) [94, 95] and vascularization of the hypertrophic cartilage, leading to trabecular bone formation [44, 96]. However, the overproduction of Sonic hedgehog (Shh) in cartilage seems to upregulate Sox9 directly in chondrocytes and chondrogenic mesenchymal cells and prevents or slows the maturation of chondrocytes [97, 98].

In contrast, the effect of PTHrP on the suppression of chondrocyte hypertrophic differentiation seems clear. The critical mechanism of PTHrP in keeping chondrocytes proliferative and delaying their hypertrophic differentiation is the activation of Gs and adenylyl cyclase to increase the level of intracellular cyclic adenosine monophosphate (cAMP) [99]. The rise in cAMP leads to the suppression of Kip2 cyclin inhibitor expression [100] and activation of Sox9 by cAMP-dependent protein kinase (PKA)-mediated phosphorylation [101, 102]. PTHrP decreases the level of *Runx2* mRNA [103] and stimulates the degradation of Runx2 protein [104]. The activated Sox9 [101, 102] antagonizes Runx2 function via Nkx3.2 protein [105, 106] or facilitates its degradation [107]. The delay in chondrocyte hypertrophic differentiation by PTHrP is not only dependent on the inhibition of Runx2 function and expression [108]. Histone deacetylase 4 (Hdac4), when phosphorylated by Ca^{++}/calmodulin-dependent protein kinase II (CaMKII), stays in the cytoplasm of chondrocytes, where it is dephosphorylated by protein phosphatase 2A that is activated by PTHrP signaling through the cAMP–PKA pathway. Dephosphorylation of Hdac4 facilitates its translocation to the nucleus, which leads to the inhibition of Mef2c action and suppression of chondrocyte hypertrophic differentiation (e.g., suppression of *Col10a1* expression) [109]. Interestingly, *Pthrp* mutant mice show mineraliza-tion of nasal cartilage [93]. Nasal cartilage is permanent cartilage. Therefore, one of the mechanisms by which permanent cartilage is maintained is via the activation of PTHrP signaling. Thus, the mechanism of PKA-mediated suppression of hypertrophic differentia-tion of chondrocytes is well established.

Investigation of the effects of PTHrP on *in vitro* chondrogenesis from MSCs, especially on the suppression of hypertrophic differentiation of developed chondrocytes, has, however, yielded contradictory results. On the one hand, PTHrP was found to enhance hyaline chondrogene-sis by maintaining type II collagen (COL2) expression and suppressing the expression of type I (COL1) and type X collagens (COL10) [110]. On the other hand, PTHrP suppressed the gene expression of both *COL2A1* and *COL10A1* during *in vitro* chondrogenesis. The degree of post-transplantational suppression of mineralization of cartilage developed with PTHrP was not clear [75]. Intermittent administration of PTHrP(1–34) alleviated suppression of *COL2A1* [111]. However, the degree of suppression of *COL10A1* and *ALPL* expression by such treatment was marginal despite being statistically significant [111]. Since hedgehog stimulates the growth of hMSCs and enhances their chondrogenesis to the *COL10A1*-expressing hypertrophic state [45, 112], suppression of hedgehog signaling by the PTHrP-activated PKA [113] may be one of the mechanisms by which PTHrP suppresses chondrogenesis from MSCs *in vitro*.

3.2. TGFβ-BMP signaling

TGFβ was first identified as cartilage-inducing factor (CIF-A) purified from bovine demineralized bone [114] but was soon found to have a negative effect on chondrocyte terminal maturation during scaffold-free 3-dimensional (3D) chondrogenesis culture (called "pellet culture") [115]. In fact, canonical TGFβ signaling that goes through the TGFβ receptor (TGFBR)–Smads such as Smad3 inhibits chondrocyte hypertrophic differentiation in the growth plate and, interestingly, also during *articular* cartilage formation [92, 116–118]. Furthermore, the TGFβ/Smad3 signaling system is essential for the integrity of developed articular cartilage [117, 119–124]. Therefore, TGFβ is now a standard ingredient for inducing chondrogenesis from MSCs [27, 125, 126]. BMP signaling is known to be essential for chondrogenesis *in vivo* [127, 128] and to enhance chondrogenesis from mesenchymal cells. It also facilitates hypertrophic differentiation of chondrocytes, although it delays terminal maturation of the hypertrophic chondrocytes [127, 129–132]. In combination, TGFβ and BMP signaling synergistically enhanced chondrogenesis [7, 133–136] but also stimulated the expression of signs of hypertrophic differentiation such as *COL10A1* transcript [133, 135, 137]. The latter effect was dependent on the type of BMP used (e.g., compared to BMP2, BMP7 is less likely to induce hypertrophic differentiation [138]). In support, activation of the BMP receptor (BMPR)–Smads (Smad 1/5/8) is linked to chondrocyte hypertrophic differentiation [73].

Thus, one way to stimulate chondrogenesis from chondroprogenitor cells while preventing hypertrophic differentiation of developed chondrocytes is to avoid overt activation of BMP signaling. Another is to select the appropriate BMP to use. For example, GDF5, one of the BMP-family proteins isolated from articular cartilage [139, 140], is involved in joint morphogenesis in the mouse [139, 141–145]. GDF5 is also needed for proper maintenance of articular cartilage in adult humans [146–149]. *In vitro*, GDF5 induces weaker chondrogenesis and chondrocyte hypertrophic differentiation than BMP4 during micromass culture of mouse limb bud mesenchymes [150] and suppresses expression of MMP13, one of the markers of terminally matured chondrocytes, during pellet culture of human chondrocytes [151]. Unfortunately, however, the use of TGFβ alone (i.e., without exogenous BMPs) for the induction of chondrogenesis *in vitro* is not guaranteed to result in immature, non-hypertrophic chondrocytes and often leads to the expression of hypertrophic chondrocyte markers [16, 74, 75, 138]. This effect could be due to the TGFβ–BMP synergy established by the TGFβ-induced production of BMPs from differentiating MSC, since TGFβ was found to induce BMP expression during growth plate chondrogenesis [152]. In fact, our results from mESC-derived mesodermal cells indicate that TGFβ-induced *in vitro* chondrogenesis depends on endogenous BMP activity [34]. In addition, TGFβ can activate the BMP signaling mechanism in chondrocytes: TGFβ normally signals through ALK5 (the Type I TGFBR) and TGFBR-Smad, but in the presence of CD105 (Endoglin), TGFβ signals through ALK1 (a Type I BMPR) and thereby activates BMPR–Smads instead [122]. Such a shift in the TGFβ signaling mechanism seems to affect the stability and health of articular cartilage, which seems to be facilitated by WNT signaling [153].

Therefore, control of the strength and temporal action of BMP signaling via exogenously added BMPs such as GDF5 or BMP7 or by inhibitors for ALK1 or other BMPR–Smad activating receptors during chondrogenesis may lead to an optimal condition for hPSC-derived chondroprogenitors to form permanent cartilage preferentially. Alternatively, combinatory control with other signaling mechanisms such as WNT signaling may be needed.

3.3. WNT signaling

As noted, WNT signaling through the canonical pathway of stabilization and nuclear translocation of β-catenin stimulates osteogenesis and inhibits chondrogenesis by blocking *Sox9* expression [52] from osteochondrogenic mesenchymal cells during limb formation [154, 155]. However, canonical WNT signaling promotes hypertrophic differentiation of the chondrocytes typically in the growth plate [156, 157]. Genetic analysis has demonstrated that WNT/β-catenin signaling can initiate the hypertrophic differentiation process by inhibiting the PTHrP signaling mechanism [158]. One possible mechanism is via the direct binding of β-catenin to PTH1R, the major receptor for PTH and PTHrP [159]. Such physical interaction seems to switch the PTHrP–PTH1R signaling pathway from cAMP–PKA to Ca^{++}–CaMKII, which is known to support the hypertrophic differentiation of chondrocytes (e.g., Mef2c-dependent expression of *Col10a1*). When the PTH1R-β-catenin interaction is blocked in chondrocytes, their hypertrophic differentiation is also blocked. Furthermore, WNT/β-catenin signaling stimulates terminal maturation of hypertrophic chondrocytes in a PTHrP signaling-independent manner, while it also indirectly stimulates BMP signaling [158]. Another possible mechanism would involve direct stimulation in chondrocytes of transcription of the hypertrophic differentiation-inducing genes such as *Ihh* [160], *Col10a1,* and *Runx2* [161]. In support, limb cartilage of *Wnt9a*$^{-/-}$ mice express decreased levels of *Ihh* and *Col10a1* in hypertrophic chondrocytes [162, 163]. Interestingly, WNT/β-catenin signaling seems to help to establish the Ihh–PTHrP feedback loop. The *Wnt9a*$^{-/-}$ limb cartilage also shows decreased levels of *Pthrp* at the ends of cartilage anlagen (e.g., the prospective shoulder joint). Thus, the canonical WNT signaling mechanism has a negative effect in the early stage of chondrogenesis and a positive effect in later stages, especially in the chondrocyte maturation stage, during endochondral ossification.

On the other hand, WNTs are implicated in the commitment of MSCs to differentiation. For example, WNT signals promote osteogenesis from MSCs by inducing *RUNX2* expression [164], which in turn induces expression of the osteogenic gene *SP7/OSX* [165]. In contrast, as with TGFβ signaling, WNT signaling prevents MSCs from adopting an adipogenic fate [166, 167]. The same holds true for the chondrogenic commitment of MSCs: WNT signaling inhibits chondrogenesis from hMSCs by actively suppressing the induction of Sox9 [40]. These observations suggest that the inhibition of canonical WNT signaling would promote chondrogenesis from MSCs and suppress hypertrophic differentiation of the developed chondrocytes. Recently, Narcisi et al. [50] reported the interesting observation that the suppression of (all) WNT signaling during the last 3 weeks of the 5-week, serum-free, TGFβ-induced chondrogenesis from hMSCs not only suppressed, albeit marginally, signs of chondrocyte hypertrophic differentiation such as *COL10A1* and *ALPL* expression in the generated cartilage pellets but also maintained the pellets in an unmineralized state for 8 weeks at an ectopic site in nude mice.

3.4. Natriuretic peptide signaling

The C-type natriuretic peptide (CNP) signaling pathway is a major contributor to postnatal skeletal growth in humans [168, 169]. The CNP gene *Nppc* is highly expressed in chondrocytes in the growth plate [170, 171] and essential for bone formation [172–176]. The CNP receptor (membrane-bound guanylyl cyclase–coupled receptor B, *Npr2*) is a membrane-bound receptor carrying the intracellular guanylyl cyclase domain and is essential for eliciting skeletal phenotypes of CNP [177, 178]. CNP signaling thus results in upregulation of Cyclic guanosine monophosphate (cGMP) and activation of cGMP-dependent protein kinase (PKG). CNP stimulates hypertrophic differentiation of chondrocytes through activating the type II PKG (PKGII, *Prkg2*) [179, 180]. In fact, the phenotypes of mouse mutants of *Nppc*, *Npr2*, and *Prkg2* are similar [174, 177, 181]. Recently, glycogen synthase kinase (GSK 3β) was identified as the critical target of PKGII leading to the enhancement of β-catenin-dependent stimulation of hypertrophic differentiation of chondrocytes [182]. Thus, the CNP–NPR2–PKGII pathway can mimic canonical WNT signaling to stimulate hypertrophic differentiation of chondrocytes. However, manipulation of the CNP–NPR2–PKGII pathway during chondrogenesis from hMSCs or after the transplantation of hMSC-derived cartilage pellets for the generation and maintenance of permanent cartilage has not been tested extensively.

3.5. VEGF signaling

For tissue engineering approaches to cartilage repair, extensive improvements have been made to scaffold/hydrogel technology to provide a suitable chemical and physical environment for the embedded chondrocytes or chondroprogenitors. However, the growth factors tested to date have been selected for their potential to facilitate chondrogenesis (e.g., TGFβs or BMPs) rather than to prevent terminal maturation and mineralization. VEGF, the best known angiogenic factor, is produced in the growth plate specifically from hypertrophic chondrocytes, not from the resting or proliferating immature chondrocytes. Furthermore, VEGF-mediated blood vessel invasion is essential for coupling resorption of cartilage to bone formation [183]. Since the VEGFR2 (KDR/Flk1) is also expressed in hypertrophic chondrocytes, it is expected that the suppression of VEGF function may prevent terminal maturation and mineralization. The VEGF inhibition strategy has already been applied to stem cell-based cartilage regeneration therapy in animal models. Orthotopically transplanted MDSCs require BMP4 to differentiate effectively into chondrocytes to repair osteochondral defects in articular cartilage [4]. However, the suppression of VEGF function by forced expression of a soluble form of VEGFR1 (sFlt1) as well as BMP4 significantly enhanced the repair function of MDSCs [3, 5]. Incorporation of an inhibitor of VEGF signaling in a scafford embedded with human nasal chondrocytes allowed *in vivo* cartilage reformation after ectopic transplantation and increased the survival of the graft [184]. Thus, the manipulation of VEGF signaling may be useful in generating permanent cartilage. But thus far, any effect of suppression of VEGF action on the prevention of terminal maturation and mineralization of chondrocytes developed from hMSCs or (briefly cultured) chondrocytes *in vivo* has not been clearly demonstrated.

Bony cartilage recovered from transplanted mice, which originates from cartilage pellets generated by the hPSC-derived ectomesenchymal cells (**Figure 3** EctM), is usually highly vascular and becomes larger than the transplanted pellet (data not shown). In contrast, unmineralized or partially mineralized cartilage, which originates from cartilage pellets generated by uncultured paraxial mesoderm (**Figure 3** PM), is generally not vascular and slightly smaller than the transplanted pellet (data not shown). Suppression of production or function of VEGF from chondrocytes developed from adult- or hPSC-derived chondroprogenitors may thus lead to long-lasting (i.e., permanent) cartilage *in vivo*.

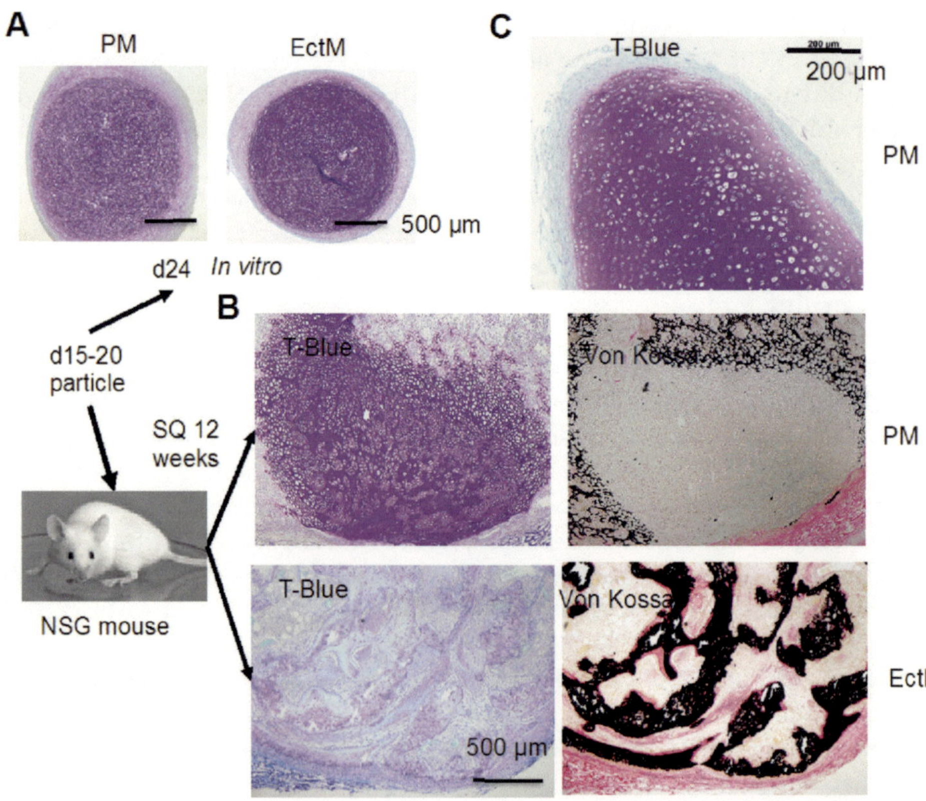

Figure 3. *In vivo* maturation of the cartilage particles developed with the FSB-expanded (p7) ectomesenchymal cells (EctM) and freshly isolated paraxial mesoderm (PM) derived from H9 hESCs. (A) The day-24 cartilage particles were directly fixed and stained with Toluidine Blue (T-Blue). (B) The day-20 cartilage pellets were transplanted into NSG mice subcutaneously (SQ) for 12 weeks, and were fixed, sectioned and stained with T-Blue and von Kossa. Paraxial

mesoderm-derived cartilage pellets constantly showed cartilaginous areas remain even after 12 weeks, and occasionally no sign of mineralization was detected (HES3 hESC-derived PM, C).

4. Cell types potentially leading to the genesis of permanent cartilage

The manipulation of signaling mechanisms may not be sufficient to generate permanent articular cartilage reproducibly *in vivo* or *in vitro*, even from hPSC-derived chondroprogenitor cells. Another strategy is to discover and use a novel population of cells derived from hPSCs that are themselves destined to form permanent joint cartilage or have the capacity to instruct other chondroprogenitors to form permanent articular cartilage. Permanent cartilage-forming cells have been sought in various adult tissues [185, 186]. The most relevant to articular cartilage regeneration would be synovial MSCs. However, thus far, no report has been published to indicate permanent cartilage formation from such MSCs [187]. Here, we will discuss potentially useful types of cells for articular cartilage regeneration that can be generated from hPSCs.

4.1. Nasal chondrocytes and their precursors (ectomesenchymal cells)-novel chondroprogenitor cells

Numerous attempts to regenerate permanent articular cartilage using chondrocytes from articular cartilage have had little success, mainly because of improper *ex vivo* expansion culture methods to compensate for their low cell yields. Nasal and auricular cartilage are permanent cartilage that develops from neural crest. Similarly to articular chondrocytes, nasal and auricular chondrocytes can be expanded in a medium containing 10% (v/v) FBS with TGFβ1, FGF2, and PDGF-BB, as demonstrated by Martin and colleagues [188]. Although inevitably dedifferentiated, the redifferentiation capacity of the cells was maintained at least for 2 passages, resulting in $COL2^{hi}COL1^{lo}$ hyaline cartilage pellets. The same group has explored the idea of using a tissue-engineering approach to repair articular cartilage by nasal chondrocytes [189]. Recently, they have demonstrated that human adult nasal chondrocytes can be clonally propagated as dedifferentiated cells with only a minimal loss of their re-differentiation capacity (i.e., ability to self-renew) under 5% (v/v) human serum with TGFβ1 and FGF2 [190]. The self-renewal activity is much stronger in nasal chondrocytes than in articular chondrocytes. Furthermore, cartilage constructs generated from dedifferentiated nasal chondrocytes showed a better capacity to repair condyle defects in goat articular cartilage than constructs made by dedifferentiated articular chondrocytes. Considering that adult stem cells generally require *ex vivo* expansion to a clinical scale for cartilage repair and regeneration, dedifferentiated nasal chondrocytes are a better cell source than articular chondrocytes and may even be better than MSCs.

On the other hand, the $CD271^{+}CD73^{+}$ ectomesenchymal cells expressing *SOX9* that can be expanded in CDM containing FGF2 and SB431542 for at least 16 passages develop from the $MIXL1\text{-}GFP^{-}CD271^{hi}CD73^{-}$ neural crest-like progeny of hPSCs (**Figure 1**) [31]. Therefore, $CD271^{+}CD73^{+}$*SOX9*$^{+}$ cells can be the developmental origin of nasal chondrocytes. However, the type of cartilage formed and the optimal chondrogenesis-promoting culture condition for the CD271⁺CD73⁺*SOX9*⁺ cells and dedifferentiated nasal chondrocytes are different (data not shown). Furthermore, cartilage pellets arising from the chondroprogenitors derived and expanded either from neural crest-like progeny (**Figure 3**) [31] or from paraxial mesoderm (data not shown) are unstable and readily mineralize in 8 weeks when transplanted subcutaneously into immunocompromized mice, as was found for adult MSCs [16, 74, 75]. By contrast, the unexpanded paraxial mesodermal progeny of hPSCs showed a greater capacity to generate hyaline cartilage particles *in vitro* than adult hMSCs [33], and the cartilage pellet made *in vitro* was stable for 12–19 weeks and showed only low levels of mineralization (**Figure 3**) [31]. Similarly, among adult cells, only unexpanded or briefly expanded articular chondrocytes retain the ability to regenerate long-lasting hyaline cartilage [74]. Such results suggest that current conditions for expanding chondrogenic mesenchymal cells whether from adult human tissues or hPSCs may skew the developmental fate of the cells upon induction of chondrogenesis toward growth plate-type chondrocytes. Since our protocol to generate chondrogenic ectomesenchymal cells involves expansion culture of neural crest-like progeny of hPSCs, we have been unable to determine whether expansion culture truly prevents the ectomesenchymal cells from generating nasal cartilage-like permanent cartilage.

When and how the decision is made during chondrogenesis to generate growth plate-like chondrocytes destined to form bone, or articular or nasal/auricular chondrocytes destined to form permanent cartilage are not yet fully understood. Research is undergoing to figure out the way to generate nasal chondrocytes from hPSC-neural crest-derived ectomesenchymal cells, which we believe will lead to answers to some of the questions.

4.2. Embryonic joint progenitor cells; articular chondroprogenitors or instructor cells for permanent cartilage formation

During embryogenesis, joint cartilage formation is initiated by the GDF5⁺ joint progenitor cells, which are distinct from the "general (SOX9⁺)" chondroprogenitors that can give rise to growth plate chondrocytes [191]. Lineage tracing experiments demonstrate that the GDF5⁺ mesenchymal cells accumulate as a band within a cartilage anlage, constituting the interzone, or at the edges of the independently formed cartilage anlagen during joint (e.g., hip) formation [192, 193]. The joint progenitor cells expressing Wnt9a, Erg, and collagen IIA eventually give rise to articular cartilage, ligament, synovial lining, and other joint tissues but contribute little if at all to the growth plate cartilage. Isolated joint progenitor cells are induced to form chondroprogenitors, albeit weakly compared with the standard chondrocytes, in response to GDF5, and forced expression of Wnt9a during chondrogenesis is inhibitory. However, conditional deletion of β-catenin inhibited the genesis of lubricin-expressing, flat superficial zone-like cell layer on articular cartilage [192], suggesting that canonical WNT signaling may change the chondrogenic fate of the joint progenitor from mid-deep zone chondrocytes to superficial zone chondrocytes.

These results points to the potential usefulness of the joint progenitor cells for the repair of articular cartilage. Since the joint progenitors are characterized mainly during embryogenesis [191], it is conceivable to aim to generate, isolate, and characterize joint progenitor cells from PSCs. There is an interesting observation suggesting that the suppression of hedgehog signaling might promote genesis of the GDF5⁺ cell during mESC differentiation [194]. As noted, hedgehog signaling, when not linked to the Ihh–PTHrP feedback mechanism, induces osteogenic gene expression in chondrocytes and facilitates endochondral ossification, which also causes osteoarthritic cartilage degradation in adult articular cartilage [195]. Unfortunately, simple suppression of hedgehog signaling does not promote the development of *in vivo* permanent cartilage from the chondrogenic mesoderm generated during ESC differentition even though at the level of gene expression it supports articular-type chondrocyte formation *in vitro*. [29, 30]. Continued efforts are warranted to identify and isolate (and expand as required) hPSC-derived GDF5⁺ cells to elucidate their role in developing articular cartilage and the repair of damaged articular cartilage.

4.3. Postnatal joint stem/progenitor cells?

Consistent with GDF5⁺ embryonic joint progenitor cells being involved in the formation of a superficial layer of articular cartilage [192], the superficial zone chondrocytes show a distinct pattern of expression of stem/progenitor cell markers (Notch1, Stro1, vascular cell adhesion-1, side population) [196–199]. Within the articular cartilage, Notch1 and Stro1 are expressed exclusively in the superficial zone, although Stro1 expression is not specific to articular cartilage, since the growth plate also abunduntly expresses Stro1 [200], and Notch1⁺ cells derived from the superficial zone are enriched in the colony-forming unit fibroblast (CFU-F) activity that is dependent on active Notch signaling [196]. Therefore, the superficial zone has been hypothesized to harbor stem/progenitor cell activity. However, it has not been demonstrated whether such stem/progenitor cells can behave as resident joint stem/progenitor cells and "regenerate or repair" articular cartilage more effectively than the conventional adult chondrogenic cells such as bone marrow MSCs.

Another area of potential interest with regard to joint stem/progenitor cells is the perichondrial groove of Ranvier (or the zone of Ranvier). The area is located at the periphery of growth plate and is enriched in proliferating cells. Studies by Karlsson et al. [200] on the knee of adult rabbits have proved the existence of different subpopulations of progenitor cells in articular cartilage and the perichondrial groove of Ranvier. As with the superficial zone chondrocytes of articular cartilage, the cells in the groove of Ranvier express markers associated with stem cells and their niche (e.g., Stro1 and Jagged1), whereas cells in the growth plate directly adjacent to it do not express many of such markers. Mice lacking Tgfbr2 signaling in developing limb bud mesenchyme fail to form interzone, resulting in the absence of interphalangeal joints [201] as well as tendons and ligaments [202]. The Tgfbr2-expressing cells are in fact first detected at embryonic day (E) 13.5 within the interphalangeal joint interzone. Interestingly, by E16.5, the Tgfbr2-expressing cells are enriched in the perichondrial groove of Ranvier, in part of the superficial layer of articular cartilage, in the synovium, and in the tendon's enthuses, and they remain in the same area postnatally [203]. Such Tgfbr2-expressing cells are slow-growing cells and exhibit stem cell traits in expressing joint progenitor markers.

The knowledge that joint progenitor-like cells, whether in the superficial zone of articular cartilage or the perichondrial groove of Ranvier, are present in postnatal joints, combined with the establishment of molecular tools to detect them are likely to inform future studies on the biology of postnatal joint progenitor cells. Isolation and functional characterization of these joint progenitor candidates will not only open the possibilities for an alternative cell source for regenerative therapy for cartilage, but also for the discovery of specific small or large molecule therapeutics that facilitate the regenerative activity of endogenous joint progenitor cells. However, as far as cellular therapy or tissue engineering therapy is concerned, the limitations noted with autologous chondrocyte/MSC implantation will also apply to these stem/progenitor cell systems—they will require expansion to clinically relevant levels without the loss of their developmental potential.

5. Conclusion and future perspective

As expected, PSC-derived chondrogenic progeny has shown the advantage over adult tissue-derived MSCs of "expansion without loss of chondrogenic potential" (**Figure 1**). However, the permanent(-like) cartilage-forming activity residing in the uncultured paraxial mesodermal cells (**Figure 3** PM) and the nasal cartilage-forming activity that the ectomesenchymal cells should possess, appear not to be readily maintained during expansion, even under the defined CDM + FGF2 + SB431542 (FSB) condition. In this sense, there is still room for the improvement in the expansion culture system. Controlling hypertrophic differentiation during chondrogenesis from such expanded hPSC-progeny by manipulating known signaling mechanisms or using novel cell populations may overcome the problem associated with expansion culture. To date, however, such an approach either has not been systematically examined, even for the widely used adult stem cells, or has failed to give positive results. Probably, mechanistic principles that direct the generation of articular or nasal cartilage from corresponding precursor cells need to be elucidated for the reproducible generation of permanent cartilage from MSCs or hPSC-derived, expanded chondroprogenitors.

For many years, the major challenges to making the hPSC system practical for therapeutic purposes have been how to direct differentiation to the specific cell lineage of interest and isolate the target cell population. But the important next challenge will be to control the aging or maturation of cell types generated from hPSCs so as to increase their functionality when transplanted into host adult tissues. Lastly, we will discuss two challenges associated with the use of embryonic/fetal cells for adult tissue repair/regeneration, and their potential relevance to the repair of articular cartilage by hPSC-derived progeny.

5.1. Overgrowth of embryonic/fetal progenitors

Human PSCs (both ESCs and iPSCs) are teratoma-forming cells by definition. Therefore, contamination of "undifferentiated" teratoma-forming cells in the population of "differentiated, functional" progeny has been the major concern with the use of hPSC-derived progenitor cells for therapeutic purposes. Even though the standard hPSCs are what is called "primed" PSCs, that is, they show less ability to self-renew as a single cell than mPSC-like "naïve" PSCs, many strategies have been devised to remove undifferentiated hPSCs. The use of iPSCs is associated with the additional concern that they are generated by transduction of adult tissue cells with a set of reprogramming genes that include *c-myc*—reactivation of the *c-myc* transgene after differentiation of iPSCs seems to cause spontaneous tumor formation [204]. New reprogramming methods that do not result in the integration of a stable transgene into the recipient chromosome were quickly developed [205, 206]. Thus, the technical issues associated with the PSC system are being resolved. However, there remains a definite "overgrowth" problem, common to primitive progenitor cells upon transplantation. Neural stem cells (NSCs) are the most frequently used PSC-derived progeny for transplantation experiments. The transplantation into the adult rodent brain of NSCs generated from hESCs and expanded in the presence of FGF2 and EGF resulted in long-term graft survival and neural stem-like behavior in the host striatum and subventricular zone [207]. However, NSCs expanded under Shh + FGF8 [208] as well as rosette-forming NSCs expandable under Shh + FGF8 or Shh + Notch-ligands [209], both of which originate from hESCs, have revealed signs of neural overgrowth upon transplantation as evidenced by large grafts composed of primitive neural progenitors/rosette structures. Thus far, the neural overgrowth problem has not been associated with simple contamination by undifferentiated PSCs in the NSC population. Furthermore, neural overgrowth directly from human basal forebrain precursor cells transplanted into rodent models of Huntington's disease has been reported [210]. Therefore, neural overgrowth seems to be associated generally with primitive neural progenitor cells, rather than PSC-derived progeny specifically, and the inhibition of Notch signaling seems to prevent such overgrowth [211].

High proliferative properties of embryonic/fetal cell types that are the main cell types derived *in vitro* from hPSCs may thus cause the problem of over-proliferation after transplantation. While the problem might be lineage/cell type dependent, we need to be aware of it when cartilage regeneration is performed using hPSC-derived chondroprogenitors and/or joint progenitors. Thus far, there have been no reports of cartilage overgrowth during the repair of articular cartilage by the use of PSC derived progenitors.

5.2. Development of "adult"-type cells

As noted, the process of differentiation of PSCs in culture can mimic many aspects of early embryogenesis. However, the terminally differentiated, functional cell types derived from PSCs often represent the embryonic counterpart of the adult cell type of interest, suggesting that the *in vitro* differentiation of PSCs mimics the developmental processes of embryonic/fetal cell types, not necessarily those of the adult cell type. Unfortunately, the functionality of embryonic cells often differs from that of their adult counterpart. For example, it is well known that embryonic hematopoietic progenitor cells (HPCs) that first appear in the yolk sac during embryogenesis give rise to primitive erythrocytes expressing embryonic globins and also demonstrate poor marrow repopulation activity upon transplantation into the adult, partly because of their poor homing capacity to the marrow and their poor viability in circulation. *In vitro* development of HPCs from PSCs has been extensively studied. The capacity of PSCs to give rise to hematopoietic cells in culture was first demonstrated by Doetschman et al. [212] using mESCs. However, detailed analyses on the hematopoietic progenitors were not reported until the early 1990s [213–218]. Since then, many have shown the development of myeloid cells, lymphocytes, and definitive erythrocytes expressing fetal (or sometimes adult) globins. Successful development of marrow-repopulating activity (MRA), the property of definitive hematopoietic stem cells (HSCs), from mPSCs has also been reported. However, the majority of such reports have not been followed up or reproduced. Thus far, the most reliable, reproducible, and widely used method of achieving the genesis of MRA involves forced expression of the HoxB4 [219–221] or Cdx4 [222, 223] transgene in differentiating mPSCs. No similar genetic methods have been reported for hPSCs [224]. Nevertheless, in order to make use of hPSCs for studying disorders of adult hematopoiesis or for marrow-reconstituting therapies, there is a need for a robust, reproducible way of generating from hPSCs HSCs that resemble those derived from cord blood and bone marrow, ideally without the need for genetic manipulation.

In addition, cardiac muscle cells and their precursors (i.e., cardiac stem/progenitor cells) have been developed from hPSCs [225, 226]. The integration of such cells into rodent heart muscle improved cardiac functions in rodent models of myocardial infarction [227–230]. However, as is the case for erythrocytes, embryonic and adult cardiomyocytes differ in the composition of their major functional protein, the myosin heavy chain (MHC). For example, in the mouse, the slow MHC (MHCβ or MYH6) is predominantly expressed in the ventricle of embryonic hearts, while fast MHC (MHCα or MYH7) is dominant in the adult heart. In humans, this switch in ventricular MHCs is less pronounced, and MHCβ persists in the adult ventricle. In addition, the spontaneous beating phenotype of the hPSC-derived cardiomyocytes resembles fetal but not adult cardiomyocytes, and the majority of the hPSC-derived cardiomyocytes fail to fully recapitulate the electrophysiological function of adult ventricular cardiomyocytes [225, 231]. The most critical limitation is their deficiency in I_{K1}, the potassium inwardly rectifying channel. Since the hPSC-derived, spontaneously beating, cardiomyocyte clusters can transiently serve as a pacemaker in a pig heart upon transplantation [227], their functional differences from adult cardiomyocytes may increase the risk of cardiac arrhythmias upon transplantation. Therefore, similar to the case of HSCs, a robust, reproducible method that generates from hPSCs cardiac stem/progenitor cells capable of giving rise to mature adult cardiomyocytes is needed if hPSCs are to be used for the repair of damaged heart muscle.

Where cartilage is concerned, no adult–fetal differences in the biochemical property of joint chondrocytes have been elucidated. Differences have only been observed in the cellularity and physical properties of child and adult articular cartilage. Since there appears to be no active replenishing of articular chondrocytes or spontaneous regeneration of damaged articular or meniscus cartilage in adult humans and large animals, we hypothesize that embryonic/fetal cells are likely to perform better in cartilage repair than adult cells.

Author details

Naoki Nakayama[1*], John Y. Lee[1], Nadine Matthias[1], Katsutsugu Umeda[2], Qing Yan[3] and Johnny Huard[1,4]

*Address all correspondence to: naoki.nakayama@uth.tmc.edu

1 Brown Foundation Institute of Molecular Medicine, The University of Texas Health Science Center at Houston Medical School, Houston TX, USA

2 Department of Pediatrics, Kyoto University School of Medicine, Kyoto, Japan

3 Houston Methodist Research Institute, Houston TX, USA

4 Department of Orthopaedic Surgery, The University of Texas Health Science Center at Houston Medical School, Houston TX, USA

References

[1] Caplan, AI. Mesenchymal stem cells. J Orthop Res 1991;9(5): 641–650.

[2] Pittenger, MF, Mackay, AM, Beck, SC, Jaiswal, RK, Douglas, R, Mosca, JD, Moorman, MA, Simonetti, DW, Craig, S, Marshak, DR. Multilineage potential of adult human mesenchymal stem cells. Science 1999;284(5411): 143–147.

[3] Kubo, S, Cooper, GM, Matsumoto, T, Phillippi, JA, Corsi, KA, Usas, A, Li, G, Fu, FH, Huard, J. Blocking vascular endothelial growth factor with soluble Flt-1 improves the chondrogenic potential of mouse skeletal muscle-derived stem cells. Arthritis Rheum 2009;60(1): 155–165.

[4] Kuroda, R, Usas, A, Kubo, S, Corsi, K, Peng, H, Rose, T, Cummins, J, Fu, FH, Huard, J. Cartilage repair using bone morphogenetic protein 4 and muscle-derived stem cells. Arthritis Rheum 2006;54(2): 433–442.

[5] Matsumoto, T, Cooper, GM, Gharaibeh, B, Meszaros, LB, Li, G, Usas, A, Fu, FH, Huard, J. Cartilage repair in a rat model of osteoarthritis through intraarticular transplantation of muscle-derived stem cells expressing bone morphogenetic protein 4 and soluble Flt-1. Arthritis Rheum 2009;60(5): 1390–1405.

[6] Peault, B, Rudnicki, M, Torrente, Y, Cossu, G, Tremblay, JP, Partridge, T, Gussoni, E, Kunkel, LM, Huard, J. Stem and progenitor cells in skeletal muscle development, maintenance, and therapy. Mol Ther 2007;15(5): 867–877.

[7] Sekiya, I, Colter, DC, Prockop, DJ. BMP-6 enhances chondrogenesis in a subpopulation of human marrow stromal cells. Biochem Biophys Res Commun 2001;284(2): 411–418.

[8] Pittenger, MF, Mbalaviele, G, Black, M, Mosca, JD, Marshak, DR. Mesenchymal stem cells. In: Koller, MR, Palsson, B, OMasters, JRW (eds.), Primary Mesenchymal Cells. Boston: Kluwer Academic Publishers; 2001. 189–207.

[9] Tsutsumi, S, Shimazu, A, Miyazaki, K, Pan, H, Koike, C, Yoshida, E, Takagishi, K, Kato, Y. Retention of multilineage differentiation potential of mesenchymal cells during proliferation in response to FGF. Biochem Biophys Res Commun 2001;288(2): 413–419.

[10] Bianchi, G, Banfi, A, Mastrogiacomo, M, Notaro, R, Luzzatto, L, Cancedda, R, Quarto, R. Ex vivo enrichment of mesenchymal cell progenitors by fibroblast growth factor 2. Exp Cell Res 2003;287(1): 98–105.

[11] Solchaga, LA, Penick, K, Goldberg, VM, Caplan, AI, Welter, JF. Fibroblast growth factor-2 enhances proliferation and delays loss of chondrogenic potential in human adult bone-marrow-derived mesenchymal stem cells. Tissue Eng Part A 2010;16(3): 1009–1019.

[12] Steinert, AF, Ghivizzani, SC, Rethwilm, A, Tuan, RS, Evans, CH, Noth, U. Major biological obstacles for persistent cell-based regeneration of articular cartilage. Arthritis Res Ther 2007;9(3): 213.

[13] Matsumoto, T, Okabe, T, Ikawa, T, Iida, T, Yasuda, H, Nakamura, H, Wakitani, S. Articular cartilage repair with autologous bone marrow mesenchymal cells. J Cell Physiol 2010;225(2): 291–295.

[14] Barry, F, Murphy, M. Mesenchymal stem cells in joint disease and repair. Nat Rev Rheumatol 2013;9(10): 584–594.

[15] Somoza, RA, Welter, JF, Correa, D, Caplan, AI. Chondrogenic differentiation of mesenchymal stem cells: Challenges and unfulfilled expectations. Tissue Eng Part B Rev 2014;20(6):596–608.

[16] Scotti, C, Tonnarelli, B, Papadimitropoulos, A, Scherberich, A, Schaeren, S, Schauerte, A, Lopez-Rios, J, Zeller, R, Barbero, A, Martin, I. Recapitulation of endochondral bone formation using human adult mesenchymal stem cells as a paradigm for developmental engineering. Proc Natl Acad Sci USA 2010;107(16): 7251–7256.

[17] Bianco, P, Robey, PG, Simmons, PJ. Mesenchymal stem cells: revisiting history, concepts, and assays. Cell Stem Cell 2008;2(4): 313–319.

[18] Namba, RS, Meuli, M, Sullivan, KM, Le, AX, Adzick, NS. Spontaneous repair of superficial defects in articular cartilage in a fetal lamb model. J Bone Joint Surg Am 1998;80(1): 4–10.

[19] Gadue, P, Huber, TL, Nostro, MC, Kattman, S, Keller, GM. Germ layer induction from embryonic stem cells. Exp Hematol 2005;33(9): 955–964.

[20] Keller, G. Embryonic stem cell differentiation: emergence of a new era in biology and medicine. Genes Dev 2005;19(10): 1129–1155.

[21] Nishikawa, S-I, Jakt, LM, Era, T. Embryonic stem-cell culture as a tool for developmental cell biology. Nat Rev Mol Cell Biol 2007;8(6): 502–507.

[22] Nakayama, N, Umeda, K. From pluripotent stem cells to lineage-specific chondrocytes: essential signalling and cellular intermediates. In: Atwood, C (ed.), Embryonic

Stem Cells: The Hormonal Regulation of Pluripotency and Embryogenesis. Wien: INTECH; 2011. 621–648.

[23] Diekman, BO, Christoforou, N, Willard, VP, Sun, H, Sanchez-Adams, J, Leong, KW, Guilak, F. Cartilage tissue engineering using differentiated and purified induced pluripotent stem cells. Proc Natl Acad Sci USA 2012;109(47): 19172–19177.

[24] Bakre, MM, Hoi, A, Mong, JC, Koh, YY, Wong, KY, Stanton, LW. Generation of multipotential mesendodermal progenitors from mouse embryonic stem cells via sustained Wnt pathway activation. J Biol Chem 2007;282(43): 31703–31712.

[25] Dell'Accio, F, De Bari, C, Luyten, FP. Molecular markers predictive of the capacity of expanded human articular chondrocytes to form stable cartilage in vivo. Arthritis Rheum 2001;44(7): 1608–1619.

[26] Koelling, S, Kruegel, J, Irmer, M, Path, JR, Sadowski, B, Miro, X, Miosge, N. Migratory chondrogenic progenitor cells from repair tissue during the later stages of human osteoarthritis. Cell Stem Cell 2009;4(4): 324–335.

[27] Mackay, AM, Beck, SC, Murphy, JM, Barry, FP, Chichester, CO, Pittenger, MF. Chondrogenic differentiation of cultured human mesenchymal stem cells from marrow. Tissue Eng 1998;4(4): 415–428.

[28] Quarto, N, Wan, DC, Kwan, MD, Panetta, NJ, Li, S, Longaker, MT. Origin matters: differences in embryonic tissue origin and Wnt signaling determine the osteogenic potential and healing capacity of frontal and parietal calvarial bones. J Bone Miner Res 2010;25(7): 1680–1694.

[29] Craft, AM, Ahmed, N, Rockel, JS, Baht, GS, Alman, BA, Kandel, RA, Grigoriadis, AE, Keller, GM. Specification of chondrocytes and cartilage tissues from embryonic stem cells. Development 2013;140(12): 2597–2610.

[30] Craft, AM, Rockel, JS, Nartiss, Y, Kandel, RA, Alman, BA, Keller, GM. Generation of articular chondrocytes from human pluripotent stem cells. Nat Biotechnol 2015;33(6): 638–645.

[31] Umeda, K, Oda, H, Yan, Q, Matthias, N, Zhao, J, Davis, BR, Nakayama, N. Long-term expandable SOX9(+) chondrogenic ectomesenchymal cells from human pluripotent stem cells. Stem Cell Reports 2015;4(4): 712–726.

[32] Toh, WS, Lee, EH, Guo, XM, Chan, JK, Yeow, CH, Choo, AB, Cao, T. Cartilage repair using hyaluronan hydrogel-encapsulated human embryonic stem cell-derived chondrogenic cells. Biomaterials 2010;31(27): 6968–6980.

[33] Umeda, K, Zhao, J, Simmons, P, Stanley, E, Elefanty, A, Nakayama, N. Human chondrogenic paraxial mesoderm, directed specification and prospective isolation from pluripotent stem cells. Sci Rep 2012;2:455.

[34] Nakayama, N, Duryea, D, Manoukian, R, Chow, G, Han, CY. Macroscopic cartilage formation with embryonic stem-cell-derived mesodermal progenitor cells. J Cell Sci 2003;116(Pt 10): 2015–2028.

[35] Zhao, J, Li, S, Trilok, S, Tanaka, M, Jokubaitis-Jameson, V, Wang, B, Niwa, H, Nakayama, N. Small molecule-directed specification of sclerotome-like chondroprogenitors and induction of a somitic chondrogenesis program from embryonic stem cells. Development 2014;141(20): 3848–3858.

[36] Mori-Akiyama, Y, Akiyama, H, Rowitch, DH, de Crombrugghe, B. Sox9 is required for determination of the chondrogenic cell lineage in the cranial neural crest. Proc Natl Acad Sci USA 2003;100(16): 9360–9365.

[37] Akiyama, H, Kim, JE, Nakashima, K, Balmes, G, Iwai, N, Deng, JM, Zhang, Z, Martin, JF, Behringer, RR, Nakamura, T, de Crombrugghe, B. Osteo-chondroprogenitor cells are derived from Sox9 expressing precursors. Proc Natl Acad Sci USA 2005;102(41): 14665–14670.

[38] Hargus, G, Kist, R, Kramer, J, Gerstel, D, Neitz, A, Scherer, G, Rohwedel, J. Loss of Sox9 function results in defective chondrocyte differentiation of mouse embryonic stem cells in vitro. Int J Dev Biol 2008;52(4): 323–332.

[39] Murakami, S, Kan, M, McKeehan, WL, de Crombrugghe, B. Up-regulation of the chondrogenic Sox9 gene by fibroblast growth factors is mediated by the mitogen-activated protein kinase pathway. Proc Natl Acad Sci USA 2000;97(3): 1113–1118.

[40] ten Berge, D, Brugmann, SA, Helms, JA, Nusse, R. Wnt and FGF signals interact to coordinate growth with cell fate specification during limb development. Development 2008;135(19): 3247–3257.

[41] Solchaga, LA, Penick, K, Porter, JD, Goldberg, VM, Caplan, AI, Welter, JF. FGF-2 enhances the mitotic and chondrogenic potentials of human adult bone marrow-derived mesenchymal stem cells. J Cell Physiol 2005;203(2): 398–409.

[42] Hellingman, CA, Koevoet, W, Kops, N, Farrell, E, Jahr, H, Liu, W, Baatenburg de Jong, RJ, Frenz, DA, van Osch, GJ. Fibroblast growth factor receptors in in vitro and in vivo chondrogenesis: relating tissue engineering using adult mesenchymal stem cells to embryonic development. Tissue Eng Part A 2010;16(2): 545–556.

[43] Ng, F, Boucher, S, Koh, S, Sastry, KS, Chase, L, Lakshmipathy, U, Choong, C, Yang, Z, Vemuri, MC, Rao, MS, Tanavde, V. PDGF, TGF-beta, and FGF signaling is important for differentiation and growth of mesenchymal stem cells (MSCs): transcriptional profiling can identify markers and signaling pathways important in differentiation of MSCs into adipogenic, chondrogenic, and osteogenic lineages. Blood 2008;112(2): 295–307.

[44] Long, F, Zhang, XM, Karp, S, Yang, Y, McMahon, AP. Genetic manipulation of hedgehog signaling in the endochondral skeleton reveals a direct role in the regulation of chondrocyte proliferation. Development 2001;128(24): 5099–5108.

[45] Warzecha, J, Gottig, S, Bruning, C, Lindhorst, E, Arabmothlagh, M, Kurth, A. Sonic hedgehog protein promotes proliferation and chondrogenic differentiation of bone marrow-derived mesenchymal stem cells in vitro. J Orthop Sci 2006;11(5): 491–496.

[46] Gordon, MD, Nusse, R. Wnt signaling: multiple pathways, multiple receptors, and multiple transcription factors. J Biol Chem 2006;281(32): 22429–22433.

[47] De Boer, J, Wang, HJ, Van Blitterswijk, C. Effects of Wnt signaling on proliferation and differentiation of human mesenchymal stem cells. Tissue Eng 2004;10(3–4): 393–401.

[48] Leucht, P, Minear, S, Ten Berge, D, Nusse, R, Helms, JA. Translating insights from development into regenerative medicine: the function of Wnts in bone biology. Semin Cell Dev Biol 2008;19(5): 434–443.

[49] Kozhemyakina, E, Lassar, AB, Zelzer, E. A pathway to bone: signaling molecules and transcription factors involved in chondrocyte development and maturation. Development 2015;142(5): 817–831.

[50] Narcisi, R, Cleary, MA, Brama, PA, Hoogduijn, MJ, Tuysuz, N, ten Berge, D, van Osch, GJ. Long-term expansion, enhanced chondrogenic potential, and suppression of endochondral ossification of adult human MSCs via WNT signaling modulation. Stem Cell Rep 2015;4(3): 459–472.

[51] Akiyama, H, Lyons, JP, Mori-Akiyama, Y, Yang, X, Zhang, R, Zhang, Z, Deng, JM, Taketo, MM, Nakamura, T, Behringer, RR, McCrea, PD, de Crombrugghe, B. Interactions between Sox9 and beta-catenin control chondrocyte differentiation. Genes Dev 2004;18(9): 1072–1087.

[52] Hill, TP, Spater, D, Taketo, MM, Birchmeier, W, Hartmann, C. Canonical Wnt/beta-catenin signaling prevents osteoblasts from differentiating into chondrocytes. Dev Cell 2005;8(5): 727–738.

[53] Hu, H, Hilton, MJ, Tu, X, Yu, K, Ornitz, DM, Long, F. Sequential roles of Hedgehog and Wnt signaling in osteoblast development. Development 2005;132(1): 49–60.

[54] Mak, KK, Chen, MH, Day, TF, Chuang, PT, Yang, Y. Wnt/beta-catenin signaling interacts differentially with Ihh signaling in controlling endochondral bone and synovial joint formation. Development 2006;133(18): 3695–3707.

[55] Rodda, SJ, McMahon, AP. Distinct roles for Hedgehog and canonical Wnt signaling in specification, differentiation and maintenance of osteoblast progenitors. Development 2006;133(16): 3231–3244.

[56] Watabe, T, Nishihara, A, Mishima, K, Yamashita, J, Shimizu, K, Miyazawa, K, Nishikawa, S, Miyazono, K. TGF-beta receptor kinase inhibitor enhances growth and

integrity of embryonic stem cell-derived endothelial cells. J Cell Biol 2003;163(6): 1303–1311.

[57] James, D, Nam, HS, Seandel, M, Nolan, D, Janovitz, T, Tomishima, M, Studer, L, Lee, G, Lyden, D, Benezra, R, Zaninovic, N, Rosenwaks, Z, Rabbany, SY, Rafii, S. Expansion and maintenance of human embryonic stem cell-derived endothelial cells by TGFbeta inhibition is Id1 dependent. Nat Biotechnol 2010;28(2): 161–166.

[58] Fortunel, NO, Hatzfeld, A, Hatzfeld, JA. Transforming growth factor-beta: pleiotropic role in the regulation of hematopoiesis. Blood 2000;96(6): 2022–2036.

[59] Cheng, X, Ying, L, Lu, L, Galvao, AM, Mills, JA, Lin, HC, Kotton, DN, Shen, SS, Nostro, MC, Choi, JK, Weiss, MJ, French, DL, Gadue, P. Self-renewing endodermal progenitor lines generated from human pluripotent stem cells. Cell Stem Cell 2012;10(4): 371–384.

[60] Hannan, NR, Fordham, RP, Syed, YA, Moignard, V, Berry, A, Bautista, R, Hanley, NA, Jensen, KB, Vallier, L. Generation of multipotent foregut stem cells from human pluripotent stem cells. Stem Cell Rep 2013;1(4): 293–306.

[61] Oldershaw, RA, Baxter, MA, Lowe, ET, Bates, N, Grady, LM, Soncin, F, Brison, DR, Hardingham, TE, Kimber, SJ. Directed differentiation of human embryonic stem cells toward chondrocytes. Nat Biotechnol 2010;28(11): 1187–1194.

[62] Tanaka, M, Jokubaitis, V, Wood, C, Wang, Y, Brouard, N, Pera, M, Hearn, M, Simmons, P, Nakayama, N. BMP inhibition stimulates WNT-dependent generation of chondrogenic mesoderm from embryonic stem cells. Stem Cell Res 2009;3(2–3): 126–141.

[63] Wang, Y, Nakayama, N. WNT and BMP signaling are both required for hematopoietic cell development from human ES cells. Stem Cell Res 2009;3(2–3): 113–125.

[64] Wang, Y, Umeda, K, Nakayama, N. Collaboration between WNT and BMP signaling promotes hemoangiogenic cell development from human fibroblast-derived iPS cells. Stem Cell Res 2010;4(3): 223–231.

[65] Pomp, O, Brokhman, I, Ben-Dor, I, Reubinoff, B, Goldstein, RS. Generation of peripheral sensory and sympathetic neurons and neural crest cells from human embryonic stem cells. Stem Cells 2005;23(7): 923–930.

[66] Lee, G, Kim, H, Elkabetz, Y, Al Shamy, G, Panagiotakos, G, Barberi, T, Tabar, V, Studer, L. Isolation and directed differentiation of neural crest stem cells derived from human embryonic stem cells. Nat Biotechnol 2007;25(12): 1468–1475.

[67] Jiang, X, Gwye, Y, McKeown, SJ, Bronner-Fraser, M, Lutzko, C, Lawlor, ER. Isolation and characterization of neural crest stem cells derived from in vitro differentiated human embryonic stem cells. Stem Cells Dev 2009;18(7): 1059–1070.

[68] Zhou, Y, Snead, ML. Derivation of cranial neural crest-like cells from human embryonic stem cells. Biochem Biophys Res Commun 2008;376(3): 542–547.

[69] Takashima, Y, Era, T, Nakao, K, Kondo, S, Kasuga, M, Smith, AG, Nishikawa, S. Neuroepithelial cells supply an initial transient wave of MSC differentiation. Cell 2007;129(7): 1377–1388.

[70] Nagoshi, N, Shibata, S, Kubota, Y, Nakamura, M, Nagai, Y, Satoh, E, Morikawa, S, Okada, Y, Mabuchi, Y, Katoh, H, Okada, S, Fukuda, K, Suda, T, Matsuzaki, Y, Toyama, Y, Okano, H. Ontogeny and multipotency of neural crest-derived stem cells in mouse bone marrow, dorsal root ganglia, and whisker pad. Cell Stem Cell 2008;2(4): 392–403.

[71] Morikawa, S, Mabuchi, Y, Niibe, K, Suzuki, S, Nagoshi, N, Sunabori, T, Shimmura, S, Nagai, Y, Nakagawa, T, Okano, H, Matsuzaki, Y. Development of mesenchymal stem cells partially originate from the neural crest. Biochem Biophys Res Commun 2009;379(4): 1114–1119.

[72] Choy, L, Derynck, R. Transforming growth factor-beta inhibits adipocyte differentiation by Smad3 interacting with CCAAT/enhancer-binding protein (C/EBP) and repressing C/EBP transactivation function. J Biol Chem 2003;278(11): 9609–9619.

[73] Hellingman, CA, Davidson, EN, Koevoet, W, Vitters, EL, van den Berg, WB, van Osch, GJ, van der Kraan, PM. Smad signaling determines chondrogenic differentiation of bone-marrow-derived mesenchymal stem cells: inhibition of Smad1/5/8P prevents terminal differentiation and calcification. Tissue Eng Part A 2011;17(7–8): 1157–1167.

[74] Pelttari, K, Winter, A, Steck, E, Goetzke, K, Hennig, T, Ochs, BG, Aigner, T, Richter, W. Premature induction of hypertrophy during in vitro chondrogenesis of human mesenchymal stem cells correlates with calcification and vascular invasion after ectopic transplantation in SCID mice. Arthritis Rheum 2006;54(10): 3254–3266.

[75] Weiss, S, Hennig, T, Bock, R, Steck, E, Richter, W. Impact of growth factors and PTHrP on early and late chondrogenic differentiation of human mesenchymal stem cells. J Cell Physiol 2010;223(1): 84–93.

[76] Bi, W, Deng, JM, Zhang, Z, Behringer, RR, de Crombrugghe, B. Sox9 is required for cartilage formation. Nat Genet 1999;22(1): 85–89.

[77] Akiyama, H, Chaboissier, MC, Martin, JF, Schedl, A, de Crombrugghe, B. The transcription factor Sox9 has essential roles in successive steps of the chondrocyte differentiation pathway and is required for expression of Sox5 and Sox6. Genes Dev 2002;16(21): 2813–2828.

[78] Komori, T, Yagi, H, Nomura, S, Yamaguchi, A, Sasaki, K, Deguchi, K, Shimizu, Y, Bronson, RT, Gao, YH, Inada, M, Sato, M, Okamoto, R, Kitamura, Y, Yoshiki, S, Kishimoto, T. Targeted disruption of Cbfa1 results in a complete lack of bone formation owing to maturational arrest of osteoblasts. Cell 1997;89(5): 755–764.

[79] Mundlos, S, Otto, F, Mundlos, C, Mulliken, JB, Aylsworth, AS, Albright, S, Lindhout, D, Cole, WG, Henn, W, Knoll, JH, Owen, MJ, Mertelsmann, R, Zabel, BU, Olsen, BR.

Mutations involving the transcription factor CBFA1 cause cleidocranial dysplasia. Cell 1997;89(5): 773–779.

[80] Otto, F, Thornell, AP, Crompton, T, Denzel, A, Gilmour, KC, Rosewell, IR, Stamp, GW, Beddington, RS, Mundlos, S, Olsen, BR, Selby, PB, Owen, MJ. Cbfa1, a candidate gene for cleidocranial dysplasia syndrome, is essential for osteoblast differentiation and bone development. Cell 1997;89(5): 765–771.

[81] Inada, M, Yasui, T, Nomura, S, Miyake, S, Deguchi, K, Himeno, M, Sato, M, Yamagiwa, H, Kimura, T, Yasui, N, Ochi, T, Endo, N, Kitamura, Y, Kishimoto, T, Komori, T. Maturational disturbance of chondrocytes in Cbfa1-deficient mice. Dev Dyn 1999;214(4): 279–290.

[82] Kim, IS, Otto, F, Zabel, B, Mundlos, S. Regulation of chondrocyte differentiation by Cbfa1. Mech Dev 1999;80(2): 159–170.

[83] Enomoto, H, Enomoto-Iwamoto, M, Iwamoto, M, Nomura, S, Himeno, M, Kitamura, Y, Kishimoto, T, Komori, T. Cbfa1 is a positive regulatory factor in chondrocyte maturation. J Biol Chem 2000;275(12): 8695–8702.

[84] Takeda, S, Bonnamy, JP, Owen, MJ, Ducy, P, Karsenty, G. Continuous expression of Cbfa1 in nonhypertrophic chondrocytes uncovers its ability to induce hypertrophic chondrocyte differentiation and partially rescues Cbfa1-deficient mice. Genes Dev 2001;15(4): 467–481.

[85] Ueta, C, Iwamoto, M, Kanatani, N, Yoshida, C, Liu, Y, Enomoto-Iwamoto, M, Ohmori, T, Enomoto, H, Nakata, K, Takada, K, Kurisu, K, Komori, T. Skeletal malformations caused by overexpression of Cbfa1 or its dominant negative form in chondrocytes. J Cell Biol 2001;153(1): 87–100.

[86] Lanske, B, Karaplis, AC, Lee, K, Luz, A, Vortkamp, A, Pirro, A, Karperien, M, Defize, LH, Ho, C, Mulligan, RC, Abou-Samra, AB, Juppner, H, Segre, GV, Kronenberg, HM. PTH/PTHrP receptor in early development and Indian hedgehog-regulated bone growth. Science 1996;273(5275): 663–666.

[87] Vortkamp, A, Lee, K, Lanske, B, Segre, GV, Kronenberg, HM, Tabin, CJ. Regulation of rate of cartilage differentiation by Indian hedgehog and PTH-related protein. Science 1996;273(5275): 613–622.

[88] Vortkamp, A. The Indian hedgehog—PTHrP system in bone development. Ernst Schering Res Found Workshop 2000;29: 191–209.

[89] Kronenberg, HM. PTHrP and skeletal development. Ann N Y Acad Sci 2006;1068:1–13.

[90] Chung, UI, Lanske, B, Lee, K, Li, E, Kronenberg, H. The parathyroid hormone/parathyroid hormone-related peptide receptor coordinates endochondral bone development by directly controlling chondrocyte differentiation. Proc Natl Acad Sci USA 1998;95(22): 13030–13035.

[91] Alvarez, J, Sohn, P, Zeng, X, Doetschman, T, Robbins, DJ, Serra, R. TGFbeta2 mediates the effects of hedgehog on hypertrophic differentiation and PTHrP expression. Development 2002;129(8): 1913–1924.

[92] Serra, R, Karaplis, A, Sohn, P. Parathyroid hormone-related peptide (PTHrP)-dependent and -independent effects of transforming growth factor beta (TGF-beta) on endochondral bone formation. J Cell Biol 1999;145(4): 783–794.

[93] Chen, X, Macica, CM, Nasiri, A, Broadus, AE. Regulation of articular chondrocyte proliferation and differentiation by Indian hedgehog and parathyroid hormone-related protein in mice. Arthritis Rheum 2008;58(12): 3788–3797.

[94] Kobayashi, T, Soegiarto, DW, Yang, Y, Lanske, B, Schipani, E, McMahon, AP, Kronenberg, HM. Indian hedgehog stimulates periarticular chondrocyte differentiation to regulate growth plate length independently of PTHrP. J Clin Investig 2005;115(7): 1734–1742.

[95] Mak, KK, Kronenberg, HM, Chuang, PT, Mackem, S, Yang, Y. Indian hedgehog signals independently of PTHrP to promote chondrocyte hypertrophy. Development 2008;135(11): 1947–1956.

[96] Long, F, Chung, UI, Ohba, S, McMahon, J, Kronenberg, HM, McMahon, AP. Ihh signaling is directly required for the osteoblast lineage in the endochondral skeleton. Development 2004;131(6): 1309–1318.

[97] Tavella, S, Biticchi, R, Schito, A, Minina, E, Di Martino, D, Pagano, A, Vortkamp, A, Horton, WA, Cancedda, R, Garofalo, S. Targeted expression of SHH affects chondrocyte differentiation, growth plate organization, and Sox9 expression. J Bone Miner Res 2004;19(10): 1678–1688.

[98] Bien-Willner, GA, Stankiewicz, P, Lupski, JR. SOX9cre1, a cis-acting regulatory element located 1.1 Mb upstream of SOX9, mediates its enhancement through the SHH pathway. Hum Mol Genet 2007;16(10): 1143–1156.

[99] Sakamoto, A, Chen, M, Kobayashi, T, Kronenberg, HM, Weinstein, LS. Chondrocyte-specific knockout of the G protein G(s)alpha leads to epiphyseal and growth plate abnormalities and ectopic chondrocyte formation. J Bone Miner Res 2005;20(4): 663–671.

[100] MacLean, HE, Guo, J, Knight, MC, Zhang, P, Cobrinik, D, Kronenberg, HM. The cyclin-dependent kinase inhibitor p57(Kip2) mediates proliferative actions of PTHrP in chondrocytes. J Clin Investig 2004;113(9): 1334–1343.

[101] Huang, W, Zhou, X, Lefebvre, V, de Crombrugghe, B. Phosphorylation of SOX9 by cyclic AMP-dependent protein kinase A enhances SOX9's ability to transactivate a Col2a1 chondrocyte-specific enhancer. Mol Cell Biol 2000;20(11): 4149–4158.

[102] Huang, W, Chung, UI, Kronenberg, HM, de Crombrugghe, B. The chondrogenic transcription factor Sox9 is a target of signaling by the parathyroid hormone-related

peptide in the growth plate of endochondral bones. Proc Natl Acad Sci USA 2001;98(1): 160–165.

[103] Li, TF, Dong, Y, Ionescu, AM, Rosier, RN, Zuscik, MJ, Schwarz, EM, O'Keefe, RJ, Drissi, H. Parathyroid hormone-related peptide (PTHrP) inhibits Runx2 expression through the PKA signaling pathway. Exp Cell Res 2004;299(1): 128–136.

[104] Zhang, M, Xie, R, Hou, W, Wang, B, Shen, R, Wang, X, Wang, Q, Zhu, T, Jonason, JH, Chen, D. PTHrP prevents chondrocyte premature hypertrophy by inducing cyclin-D1-dependent Runx2 and Runx3 phosphorylation, ubiquitylation and proteasomal degradation. J Cell Sci 2009;122(Pt 9): 1382–1389.

[105] Lengner, CJ, Hassan, MQ, Serra, RW, Lepper, C, van Wijnen, AJ, Stein, JL, Lian, JB, Stein, GS. Nkx3.2-mediated repression of Runx2 promotes chondrogenic differentiation. J Biol Chem 2005;280(16): 15872–15879.

[106] Provot, S, Kempf, H, Murtaugh, LC, Chung, UI, Kim, DW, Chyung, J, Kronenberg, HM, Lassar, AB. Nkx3.2/Bapx1 acts as a negative regulator of chondrocyte maturation. Development 2006;133(4): 651–662.

[107] Cheng, A, Genever, PG. SOX9 determines RUNX2 transactivity by directing intracellular degradation. J Bone Miner Res 2010;25(12): 2680–2689.

[108] Guo, J, Chung, UI, Yang, D, Karsenty, G, Bringhurst, FR, Kronenberg, HM. PTH/PTHrP receptor delays chondrocyte hypertrophy via both Runx2-dependent and -independent pathways. Dev Biol 2006;292(1): 116–128.

[109] Kozhemyakina, E, Cohen, T, Yao, TP, Lassar, AB. Parathyroid hormone-related peptide represses chondrocyte hypertrophy through a protein phosphatase 2A/histone deacetylase 4/MEF2 pathway. Mol Cell Biol 2009;29(21): 5751–5762.

[110] Kim, YJ, Kim, HJ, Im, GI. PTHrP promotes chondrogenesis and suppresses hypertrophy from both bone marrow-derived and adipose tissue-derived MSCs. Biochem Biophys Res Commun 2008;373(1): 104–108.

[111] Fischer, J, Aulmann, A, Dexheimer, V, Grossner, T, Richter, W. Intermittent PTHrP(1–34) exposure augments chondrogenesis and reduces hypertrophy of mesenchymal stromal cells. Stem Cells Dev 2014;23(20): 2513–2523.

[112] Steinert, AF, Weissenberger, M, Kunz, M, Gilbert, F, Ghivizzani, SC, Gobel, S, Jakob, F, Noth, U, Rudert, M. Indian hedgehog gene transfer is a chondrogenic inducer of human mesenchymal stem cells. Arthritis Res Ther 2012;14(4): R168.

[113] Hammerschmidt, M, Bitgood, MJ, McMahon, AP. Protein kinase A is a common negative regulator of Hedgehog signaling in the vertebrate embryo. Genes Dev 1996;10(6): 647–658.

[114] Seyedin, SM, Thompson, AY, Bentz, H, Rosen, DM, McPherson, JM, Conti, A, Siegel, NR, Galluppi, GR, Piez, KA. Cartilage-inducing factor-A. Apparent identity to transforming growth factor-beta. J Biol Chem 1986;261(13): 5693–5695.

[115] Kato, Y, Iwamoto, M, Koike, T, Suzuki, F, Takano, Y. Terminal differentiation and calcification in rabbit chondrocyte cultures grown in centrifuge tubes: regulation by transforming growth factor beta and serum factors. Proc Natl Acad Sci USA 1988;85(24): 9552–9556.

[116] Ferguson, CM, Schwarz, EM, Reynolds, PR, Puzas, JE, Rosier, RN, O'Keefe, RJ. Smad2 and 3 mediate transforming growth factor-beta1-induced inhibition of chondrocyte maturation. Endocrinology 2000;141(12): 4728–4735.

[117] Yang, X, Chen, L, Xu, X, Li, C, Huang, C, Deng, CX. TGF-beta/Smad3 signals repress chondrocyte hypertrophic differentiation and are required for maintaining articular cartilage. J Cell Biol 2001;153(1): 35–46.

[118] Sueyoshi, T, Yamamoto, K, Akiyama, H. Conditional deletion of Tgfbr2 in hypertrophic chondrocytes delays terminal chondrocyte differentiation. Matrix Biol 2012;31(6): 352–359.

[119] Serra, R, Johnson, M, Filvaroff, EH, LaBorde, J, Sheehan, DM, Derynck, R, Moses, HL. Expression of a truncated, kinase-defective TGF-beta type II receptor in mouse skeletal tissue promotes terminal chondrocyte differentiation and osteoarthritis. J Cell Biol 1997;139(2): 541–552.

[120] Yao, JY, Wang, Y, An, J, Mao, CM, Hou, N, Lv, YX, Wang, YL, Cui, F, Huang, M, Yang, X. Mutation analysis of the Smad3 gene in human osteoarthritis. Eur J Hum Genet 2003;11(9): 714–717.

[121] Wu, Q, Kim, KO, Sampson, ER, Chen, D, Awad, H, O'Brien, T, Puzas, JE, Drissi, H, Schwarz, EM, O'Keefe, RJ, Zuscik, MJ, Rosier, RN. Induction of an osteoarthritis-like phenotype and degradation of phosphorylated Smad3 by Smurf2 in transgenic mice. Arthritis Rheum 2008;58(10): 3132–3144.

[122] van der Kraan, PM, Blaney Davidson, EN, Blom, A, van den Berg, WB. TGF-beta signaling in chondrocyte terminal differentiation and osteoarthritis: modulation and integration of signaling pathways through receptor-Smads. Osteoarthr Cartil 2009;17(12): 1539–1545.

[123] Li, TF, Gao, L, Sheu, TJ, Sampson, ER, Flick, LM, Konttinen, YT, Chen, D, Schwarz, EM, Zuscik, MJ, Jonason, JH, O'Keefe, RJ. Aberrant hypertrophy in Smad3-deficient murine chondrocytes is rescued by restoring transforming growth factor beta-activated kinase 1/activating transcription factor 2 signaling: a potential clinical implication for osteoarthritis. Arthritis Rheum 2010;62(8): 2359–2369.

[124] Valdes, AM, Spector, TD, Tamm, A, Kisand, K, Doherty, SA, Dennison, EM, Mangino, M, Tamm, A, Kerna, I, Hart, DJ, Wheeler, M, Cooper, C, Lories, RJ, Arden, NK, Doherty,

M. Genetic variation in the SMAD3 gene is associated with hip and knee osteoarthritis. Arthritis Rheum 2010;62(8): 2347–2352.

[125] Johnstone, B, Hering, TM, Caplan, AI, Goldberg, VM, Yoo, JU. In vitro chondrogenesis of bone marrow-derived mesenchymal progenitor cells. Exp Cell Res 1998;238(1): 265–272.

[126] De Bari, C, Dell'Accio, F, Luyten, FP. Human periosteum-derived cells maintain phenotypic stability and chondrogenic potential throughout expansion regardless of donor age. Arthritis Rheum 2001;44(1): 85–95.

[127] Tsumaki, N, Nakase, T, Miyaji, T, Kakiuchi, M, Kimura, T, Ochi, T, Yoshikawa, H. Bone morphogenetic protein signals are required for cartilage formation and differently regulate joint development during skeletogenesis. J Bone Miner Res 2002;17(5): 898–906.

[128] Yoon, BS, Ovchinnikov, DA, Yoshii, I, Mishina, Y, Behringer, RR, Lyons, KM. Bmpr1a and Bmpr1b have overlapping functions and are essential for chondrogenesis in vivo. Proc Natl Acad Sci USA 2005;102(14): 5062–5067.

[129] Enomoto-Iwamoto, M, Iwamoto, M, Mukudai, Y, Kawakami, Y, Nohno, T, Higuchi, Y, Takemoto, S, Ohuchi, H, Noji, S, Kurisu, K. Bone morphogenetic protein signaling is required for maintenance of differentiated phenotype, control of proliferation, and hypertrophy in chondrocytes. J Cell Biol 1998;140(2): 409–148.

[130] Grimsrud, CD, Romano, PR, D'Souza, M, Puzas, JE, Reynolds, PR, Rosier, RN, O'Keefe, RJ. BMP-6 is an autocrine stimulator of chondrocyte differentiation. J Bone Miner Res 1999;14(4): 475–482.

[131] Grimsrud, CD, Romano, PR, D'Souza, M, Puzas, JE, Schwarz, EM, Reynolds, PR, Roiser, RN, O'Keefe, RJ. BMP signaling stimulates chondrocyte maturation and the expression of Indian hedgehog. J Orthop Res 2001;19(1): 18–25.

[132] Minina, E, Wenzel, HM, Kreschel, C, Karp, S, Gaffield, W, McMahon, AP, Vortkamp, A. BMP and Ihh/PTHrP signaling interact to coordinate chondrocyte proliferation and differentiation. Development 2001;128(22): 4523–4534.

[133] Sekiya, I, Larson, BL, Vuoristo, JT, Reger, RL, Prockop, DJ. Comparison of effect of BMP-2, -4, and -6 on in vitro cartilage formation of human adult stem cells from bone marrow stroma. Cell Tissue Res 2005;320(2): 269–276.

[134] Xu, D, Gechtman, Z, Hughes, A, Collins, A, Dodds, R, Cui, X, Jolliffe, L, Higgins, L, Murphy, A, Farrell, F. Potential involvement of BMP receptor type IB activation in a synergistic effect of chondrogenic promotion between rhTGFbeta3 and rhGDF5 or rhBMP7 in human mesenchymal stem cells. Growth factors (Chur, Switzerland) 2006;24(4): 268–278.

[135] Shen, B, Wei, A, Tao, H, Diwan, AD, Ma, DD. BMP-2 enhances TGF-beta3-mediated chondrogenic differentiation of human bone marrow multipotent mesenchymal stromal cells in alginate bead culture. Tissue Eng Part A 2009;15(6): 1311–1320.

[136] Murphy, MK, Huey, DJ, Hu, JC, Athanasiou, KA. TGF-beta1, GDF-5, and BMP-2 stimulation induces chondrogenesis in expanded human articular chondrocytes and marrow-derived stromal cells. Stem Cells 2015;33(3): 762–773.

[137] Sekiya, I, Vuoristo, JT, Larson, BL, Prockop, DJ. In vitro cartilage formation by human adult stem cells from bone marrow stroma defines the sequence of cellular and molecular events during chondrogenesis. Proc Natl Acad Sci USA 2002;99(7): 4397–4402.

[138] Caron, MM, Emans, PJ, Cremers, A, Surtel, DA, Coolsen, MM, van Rhijn, LW, Welting, TJ. Hypertrophic differentiation during chondrogenic differentiation of progenitor cells is stimulated by BMP-2 but suppressed by BMP-7. Osteoarthr Cartil 2013;21(4): 604–613.

[139] Storm, EE, Huynh, TV, Copeland, NG, Jenkins, NA, Kingsley, DM, Lee, SJ. Limb alterations in brachypodism mice due to mutations in a new member of the TGF beta-superfamily. Nature 1994;368(6472): 639–643.

[140] Chang, SC, Hoang, B, Thomas, JT, Vukicevic, S, Luyten, FP, Ryba, NJ, Kozak, CA, Reddi, AH, Moos, M, Jr. Cartilage-derived morphogenetic proteins. New members of the transforming growth factor-beta superfamily predominantly expressed in long bones during human embryonic development. J Biol Chem 1994;269(45): 28227–28234.

[141] Storm, EE, Kingsley, DM. Joint patterning defects caused by single and double mutations in members of the bone morphogenetic protein (BMP) family. Development 1996;122(12): 3969–3979.

[142] Storm, EE, Kingsley, DM. GDF5 coordinates bone and joint formation during digit development. Dev Biol 1999;209(1): 11–27.

[143] Francis-West, PH, Parish, J, Lee, K, Archer, CW. BMP/GDF-signalling interactions during synovial joint development. Cell Tissue Res 1999;296(1): 111–119.

[144] Tsumaki, N, Tanaka, K, Arikawa-Hirasawa, E, Nakase, T, Kimura, T, Thomas, JT, Ochi, T, Luyten, FP, Yamada, Y. Role of CDMP-1 in skeletal morphogenesis: promotion of mesenchymal cell recruitment and chondrocyte differentiation. J Cell Biol 1999;144(1): 161–173.

[145] Settle, SH, Jr., Rountree, RB, Sinha, A, Thacker, A, Higgins, K, Kingsley, DM. Multiple joint and skeletal patterning defects caused by single and double mutations in the mouse Gdf6 and Gdf5 genes. Dev Biol 2003;254(1): 116–130.

[146] Rountree, RB, Schoor, M, Chen, H, Marks, ME, Harley, V, Mishina, Y, Kingsley, DM. BMP receptor signaling is required for postnatal maintenance of articular cartilage. Plos Biol 2004;2(11): e355.

[147] Miyamoto, Y, Mabuchi, A, Shi, D, Kubo, T, Takatori, Y, Saito, S, Fujioka, M, Sudo, A, Uchida, A, Yamamoto, S, Ozaki, K, Takigawa, M, Tanaka, T, Nakamura, Y, Jiang, Q, Ikegawa, S. A functional polymorphism in the 5' UTR of GDF5 is associated with susceptibility to osteoarthritis. Nat Genet 2007;39(4): 529–533.

[148] Southam, L, Rodriguez-Lopez, J, Wilkins, JM, Pombo-Suarez, M, Snelling, S, Gomez-Reino, JJ, Chapman, K, Gonzalez, A, Loughlin, J. An SNP in the 5'-UTR of GDF5 is associated with osteoarthritis susceptibility in Europeans and with in vivo differences in allelic expression in articular cartilage. Hum Mol Genet 2007;16(18): 2226–2232.

[149] Chapman, K, Takahashi, A, Meulenbelt, I, Watson, C, Rodriguez-Lopez, J, Egli, R, Tsezou, A, Malizos, KN, Kloppenburg, M, Shi, D, Southam, L, van der Breggen, R, Donn, R, Qin, J, Doherty, M, Slagboom, PE, Wallis, G, Kamatani, N, Jiang, Q, Gonzalez, A, Loughlin, J, Ikegawa, S. A meta-analysis of European and Asian cohorts reveals a global role of a functional SNP in the 5' UTR of GDF5 with osteoarthritis susceptibility. Hum Mol Genet 2008;17(10): 1497–1504.

[150] Hatakeyama, Y, Tuan, RS, Shum, L. Distinct functions of BMP4 and GDF5 in the regulation of chondrogenesis. J Cell Biochem 2004;91(6): 1204–1217.

[151] Enochson, L, Stenberg, J, Brittberg, M, Lindahl, A. GDF5 reduces MMP13 expression in human chondrocytes via DKK1 mediated canonical Wnt signaling inhibition. Osteoarthr Cartil 2014;22(4): 566–577.

[152] Duneas, N, Crooks, J, Ripamonti, U. Transforming growth factor-beta 1: induction of bone morphogenetic protein genes expression during endochondral bone formation in the baboon, and synergistic interaction with osteogenic protein-1 (BMP-7). Growth Factors 1998;15(4): 259–277.

[153] van den Bosch, MH, Blom, AB, van Lent, PL, van Beuningen, HM, Blaney Davidson, EN, van der Kraan, PM, van den Berg, WB. Canonical Wnt signaling skews TGF-beta signaling in chondrocytes towards signaling via ALK1 and Smad 1/5/8. Cell Signal 2014;26(5): 951–958.

[154] Rudnicki, JA, Brown, AM. Inhibition of chondrogenesis by Wnt gene expression in vivo and in vitro. Dev Biol 1997;185(1): 104–118.

[155] Kawakami, Y, Wada, N, Nishimatsu, SI, Ishikawa, T, Noji, S, Nohno, T. Involvement of Wnt-5a in chondrogenic pattern formation in the chick limb bud. Dev Growth Differ 1999;41(1): 29–40.

[156] Hartmann, C, Tabin, CJ. Dual roles of Wnt signaling during chondrogenesis in the chicken limb. Development 2000;127(14): 3141–3159.

[157] Ryu, JH, Kim, SJ, Kim, SH, Oh, CD, Hwang, SG, Chun, CH, Oh, SH, Seong, JK, Huh, TL, Chun, JS. Regulation of the chondrocyte phenotype by beta-catenin. Development 2002;129(23): 5541–5550.

[158] Hartmann, C, Tabin, CJ. Wnt-14 plays a pivotal role in inducing synovial joint formation in the developing appendicular skeleton. Cell 2001;104(3): 341–351.

[159] Guo, X, Day, TF, Jiang, X, Garrett-Beal, L, Topol, L, Yang, Y. Wnt/beta-catenin signaling is sufficient and necessary for synovial joint formation. Genes Dev 2004;18(19): 2404–2417.

[160] Guo, X, Mak, KK, Taketo, MM, Yang, Y. The Wnt/beta-catenin pathway interacts differentially with PTHrP signaling to control chondrocyte hypertrophy and final maturation. Plos One 2009;4(6): e6067.

[161] Yano, F, Saito, T, Ogata, N, Yamazawa, T, Iino, M, Chung, UI, Kawaguchi, H. beta-catenin regulates parathyroid hormone/parathyroid hormone-related protein receptor signals and chondrocyte hypertrophy through binding to the intracellular C-terminal region of the receptor. Arthritis Rheum 2013;65(2): 429–435.

[162] Spater, D, Hill, TP, O'Sullivan, RJ, Gruber, M, Conner, DA, Hartmann, C. Wnt9a signaling is required for joint integrity and regulation of Ihh during chondrogenesis. Development 2006;133(15): 3039–3049.

[163] Dong, YF, Soung do, Y, Schwarz, EM, O'Keefe, RJ, Drissi, H. Wnt induction of chondrocyte hypertrophy through the Runx2 transcription factor. J Cell Physiol 2006;208(1): 77–86.

[164] Gaur, T, Lengner, CJ, Hovhannisyan, H, Bhat, RA, Bodine, PV, Komm, BS, Javed, A, van Wijnen, AJ, Stein, JL, Stein, GS, Lian, JB. Canonical WNT signaling promotes osteogenesis by directly stimulating Runx2 gene expression. J Biol Chem 2005;280(39): 33132–33140.

[165] Zhang, C, Cho, K, Huang, Y, Lyons, JP, Zhou, X, Sinha, K, McCrea, PD, de Crombrugghe, B. Inhibition of Wnt signaling by the osteoblast-specific transcription factor Osterix. Proc Natl Acad Sci USA 2008;105(19): 6936–6941.

[166] Shang, Y, Zhang, C, Wang, S, Xiong, F, Zhao, C, Peng, F, Feng, S, Yu, M, Li, M, Zhang, Y. Activated beta-catenin induces myogenesis and inhibits adipogenesis in BM-derived mesenchymal stromal cells. Cytotherapy 2007;9(7): 667–681.

[167] Li, HX, Luo, X, Liu, RX, Yang, YJ, Yang, GS. Roles of Wnt/beta-catenin signaling in adipogenic differentiation potential of adipose-derived mesenchymal stem cells. Mol Cell Endocrinol 2008;291(1–2): 116–124.

[168] Potter, LR, Abbey-Hosch, S, Dickey, DM. Natriuretic peptides, their receptors, and cyclic guanosine monophosphate-dependent signaling functions. Endocr Rev 2006;27(1): 47–72.

[169] Pejchalova, K, Krejci, P, Wilcox, WR. C-natriuretic peptide: an important regulator of cartilage. Mol Genet Metab 2007;92(3): 210–215.

[170] Hagiwara, H, Sakaguchi, H, Itakura, M, Yoshimoto, T, Furuya, M, Tanaka, S, Hirose, S. Autocrine regulation of rat chondrocyte proliferation by natriuretic peptide C and its receptor, natriuretic peptide receptor-B. J Biol Chem 1994;269(14): 10729–10733.

[171] Hagiwara, H, Sakaguchi, H, Lodhi, KM, Suda, K, Hirose, S. Subtype switching of natriuretic peptide receptors in rat chondrocytes during in vitro culture. J Biochem 1994;116(3): 606–609.

[172] Suda, M, Ogawa, Y, Tanaka, K, Tamura, N, Yasoda, A, Takigawa, T, Uehira, M, Nishimoto, H, Itoh, H, Saito, Y, Shiota, K, Nakao, K. Skeletal overgrowth in transgenic mice that overexpress brain natriuretic peptide. Proc Natl Acad Sci USA 1998;95(5): 2337–2342.

[173] Yasoda, A, Ogawa, Y, Suda, M, Tamura, N, Mori, K, Sakuma, Y, Chusho, H, Shiota, K, Tanaka, K, Nakao, K. Natriuretic peptide regulation of endochondral ossification. Evidence for possible roles of the C-type natriuretic peptide/guanylyl cyclase-B pathway. J Biol Chem 1998;273(19): 11695–11700.

[174] Chusho, H, Tamura, N, Ogawa, Y, Yasoda, A, Suda, M, Miyazawa, T, Nakamura, K, Nakao, K, Kurihara, T, Komatsu, Y, Itoh, H, Tanaka, K, Saito, Y, Katsuki, M, Nakao, K. Dwarfism and early death in mice lacking C-type natriuretic peptide. Proc Natl Acad Sci USA 2001;98(7): 4016–4021.

[175] Komatsu, Y, Chusho, H, Tamura, N, Yasoda, A, Miyazawa, T, Suda, M, Miura, M, Ogawa, Y, Nakao, K. Significance of C-type natriuretic peptide (CNP) in endochondral ossification: analysis of CNP knockout mice. J Bone Miner Metab 2002;20(6): 331–336.

[176] Yasoda, A, Komatsu, Y, Chusho, H, Miyazawa, T, Ozasa, A, Miura, M, Kurihara, T, Rogi, T, Tanaka, S, Suda, M, Tamura, N, Ogawa, Y, Nakao, K. Overexpression of CNP in chondrocytes rescues achondroplasia through a MAPK-dependent pathway. Nat Med 2004;10(1): 80–86.

[177] Tamura, N, Doolittle, LK, Hammer, RE, Shelton, JM, Richardson, JA, Garbers, DL. Critical roles of the guanylyl cyclase B receptor in endochondral ossification and development of female reproductive organs. Proc Natl Acad Sci USA 2004;101(49): 17300–17305.

[178] Tsuji, T, Kunieda, T. A loss-of-function mutation in natriuretic peptide receptor 2 (Npr2) gene is responsible for disproportionate dwarfism in cn/cn mouse. J Biol Chem 2005;280(14): 14288–14292.

[179] Miyazawa, T, Ogawa, Y, Chusho, H, Yasoda, A, Tamura, N, Komatsu, Y, Pfeifer, A, Hofmann, F, Nakao, K. Cyclic GMP-dependent protein kinase II plays a critical role in C-type natriuretic peptide-mediated endochondral ossification. Endocrinology 2002;143(9): 3604–3610.

[180] Chikuda, H, Kugimiya, F, Hoshi, K, Ikeda, T, Ogasawara, T, Shimoaka, T, Kawano, H, Kamekura, S, Tsuchida, A, Yokoi, N, Nakamura, K, Komeda, K, Chung, UI, Kawagu-

chi, H. Cyclic GMP-dependent protein kinase II is a molecular switch from proliferation to hypertrophic differentiation of chondrocytes. Genes Dev 2004;18(19): 2418–2429.

[181] Pfeifer, A, Aszodi, A, Seidler, U, Ruth, P, Hofmann, F, Fassler, R. Intestinal secretory defects and dwarfism in mice lacking cGMP-dependent protein kinase II. Science 1996;274(5295): 2082–2086.

[182] Kawasaki, Y, Kugimiya, F, Chikuda, H, Kamekura, S, Ikeda, T, Kawamura, N, Saito, T, Shinoda, Y, Higashikawa, A, Yano, F, Ogasawara, T, Ogata, N, Hoshi, K, Hofmann, F, Woodgett, JR, Nakamura, K, Chung, UI, Kawaguchi, H. Phosphorylation of GSK-3beta by cGMP-dependent protein kinase II promotes hypertrophic differentiation of murine chondrocytes. J Clin Investig 2008;118(7): 2506–2515.

[183] Gerber, HP, Vu, TH, Ryan, AM, Kowalski, J, Werb, Z, Ferrara, N. VEGF couples hypertrophic cartilage remodeling, ossification and angiogenesis during endochondral bone formation. Nat Med 1999;5(6): 623–628.

[184] Carlevaro, MF, Cermelli, S, Cancedda, R, Descalzi Cancedda, F. Vascular endothelial growth factor (VEGF) in cartilage neovascularization and chondrocyte differentiation: auto-paracrine role during endochondral bone formation. J Cell Sci 2000;113(Pt 1): 59–69.

[185] Atesok, K, Doral, MN, Bilge, O, Sekiya, I. Synovial stem cells in musculoskeletal regeneration. J Am Acad Orthop Surg 2013;21(4): 258–259.

[186] De Bari, C, Kurth, TB, Augello, A. Mesenchymal stem cells from development to postnatal joint homeostasis, aging, and disease. Birth Defects Res C Embryo Today 2010;90(4): 257–271.

[187] De Bari, C, Dell'Accio, F, Luyten, FP. Failure of in vitro-differentiated mesenchymal stem cells from the synovial membrane to form ectopic stable cartilage in vivo. Arthritis Rheum 2004;50(1): 142–150.

[188] Tay, AG, Farhadi, J, Suetterlin, R, Pierer, G, Heberer, M, Martin, I. Cell yield, proliferation, and postexpansion differentiation capacity of human ear, nasal, and rib chondrocytes. Tissue Eng 2004;10(5–6): 762–770.

[189] Candrian, C, Vonwil, D, Barbero, A, Bonacina, E, Miot, S, Farhadi, J, Wirz, D, Dickinson, S, Hollander, A, Jakob, M, Li, Z, Alini, M, Heberer, M, Martin, I. Engineered cartilage generated by nasal chondrocytes is responsive to physical forces resembling joint loading. Arthritis Rheum 2008;58(1): 197–208.

[190] Pelttari, K, Pippenger, B, Mumme, M, Feliciano, S, Scotti, C, Mainil-Varlet, P, Procino, A, von Rechenberg, B, Schwamborn, T, Jakob, M, Cillo, C, Barbero, A, Martin, I. Adult human neural crest-derived cells for articular cartilage repair. Sci Transl Med 2014;6(251): 251ra119.

[191] Decker, RS, Koyama, E, Pacifici, M. Genesis and morphogenesis of limb synovial joints and articular cartilage. Matrix Biol 2014;39: 5–10.

[192] Koyama, E, Shibukawa, Y, Nagayama, M, Sugito, H, Young, B, Yuasa, T, Okabe, T, Ochiai, T, Kamiya, N, Rountree, RB, Kingsley, DM, Iwamoto, M, Enomoto-Iwamoto, M, Pacifici, M. A distinct cohort of progenitor cells participates in synovial joint and articular cartilage formation during mouse limb skeletogenesis. Dev Biol 2008;316(1): 62–73.

[193] Nowlan, NC, Sharpe, J. Joint shape morphogenesis precedes cavitation of the developing hip joint. J Anat 2014;224(4): 482–489.

[194] Rockel, JS, Yu, C, Craft, A, Whetstone, H, Reilly, K, Alman, BA. Hedgehog signaling regulation is required for beta-catenin-mediated interzone cell differentiation and synovial joint morphogenesis. 60th Annual Meeting for Orthopaedic Research Society 2014:Poster 0072.

[195] Alman, BA. The role of hedgehog signalling in skeletal health and disease. Nat Rev Rheumatol 2015;11(9): 552–560.

[196] Dowthwaite, GP, Bishop, JC, Redman, SN, Khan, IM, Rooney, P, Evans, DJ, Haughton, L, Bayram, Z, Boyer, S, Thomson, B, Wolfe, MS, Archer, CW. The surface of articular cartilage contains a progenitor cell population. J Cell Sci 2004;117(Pt 6): 889–897.

[197] Hattori, S, Oxford, C, Reddi, AH. Identification of superficial zone articular chondrocyte stem/progenitor cells. Biochem Biophys Res Commun 2007;358(1): 99–103.

[198] Grogan, SP, Miyaki, S, Asahara, H, D'Lima, DD, Lotz, MK. Mesenchymal progenitor cell markers in human articular cartilage: normal distribution and changes in osteoarthritis. Arthritis Res Ther 2009;11(3): R85.

[199] Otsuki, S, Grogan, SP, Miyaki, S, Kinoshita, M, Asahara, H, Lotz, MK. Tissue neogenesis and STRO-1 expression in immature and mature articular cartilage. J Orthop Res 2010;28(1): 96–102.

[200] Karlsson, C, Thornemo, M, Henriksson, HB, Lindahl, A. Identification of a stem cell niche in the zone of Ranvier within the knee joint. J Anat 2009;215(3): 355–363.

[201] Spagnoli, A, O'Rear, L, Chandler, RL, Granero-Molto, F, Mortlock, DP, Gorska, AE, Weis, JA, Longobardi, L, Chytil, A, Shimer, K, Moses, HL. TGF-beta signaling is essential for joint morphogenesis. J Cell Biol 2007;177(6): 1105–1117.

[202] Pryce, BA, Watson, SS, Murchison, ND, Staverosky, JA, Dunker, N, Schweitzer, R. Recruitment and maintenance of tendon progenitors by TGFbeta signaling are essential for tendon formation. Development 2009;136(8): 1351–1361.

[203] Li, T, Longobardi, L, Myers, TJ, Temple, JD, Chandler, RL, Ozkan, H, Contaldo, C, Spagnoli, A. Joint TGF-beta type II receptor-expressing cells: ontogeny and characterization as joint progenitors. Stem Cells Dev 2013;22(9): 1342–1359.

[204] Nakagawa, M, Koyanagi, M, Tanabe, K, Takahashi, K, Ichisaka, T, Aoi, T, Okita, K, Mochiduki, Y, Takizawa, N, Yamanaka, S. Generation of induced pluripotent stem cells without Myc from mouse and human fibroblasts. Nat Biotechnol 2008;26(1): 101–106.

[205] Okita, K, Nakagawa, M, Hyenjong, H, Ichisaka, T, Yamanaka, S. Generation of mouse induced pluripotent stem cells without viral vectors. Science 2008;322(5903): 949–953.

[206] Yu, J, Hu, K, Smuga-Otto, K, Tian, S, Stewart, R, Slukvin, II, Thomson, JA. Human induced pluripotent stem cells free of vector and transgene sequences. Science 2009;324(5928): 797–801.

[207] Tabar, V, Panagiotakos, G, Greenberg, ED, Chan, BK, Sadelain, M, Gutin, PH, Studer, L. Migration and differentiation of neural precursors derived from human embryonic stem cells in the rat brain. Nat Biotechnol 2005;23(5): 601–606.

[208] Roy, NS, Cleren, C, Singh, SK, Yang, L, Beal, MF, Goldman, SA. Functional engraftment of human ES cell-derived dopaminergic neurons enriched by coculture with telomerase-immortalized midbrain astrocytes. Nat Med 2006;12(11): 1259–1268.

[209] Elkabetz, Y, Panagiotakos, G, Al Shamy, G, Socci, ND, Tabar, V, Studer, L. Human ES cell-derived neural rosettes reveal a functionally distinct early neural stem cell stage. Genes Dev 2008;22(2): 152–165.

[210] Geny, C, Naimi-Sadaoui, S, Jeny, R, Belkadi, AM, Juliano, SL, Peschanski, M. Long-term delayed vascularization of human neural transplants to the rat brain. J Neurosci 1994;14(12): 7553–7562.

[211] Ogura, A, Morizane, A, Nakajima, Y, Miyamoto, S, Takahashi, J. Gamma-secretase inhibitors prevent overgrowth of transplanted neural progenitors derived from human-induced pluripotent stem cells. Stem Cells Dev 2013;22(3): 374–382.

[212] Doetschman, TC, Eistetter, H, Katz, M, Schmidt, W, Kemler, R. The in vitro development of blastocyst-derived embryonic stem cell lines: formation of visceral yolk sac, blood islands and myocardium. J Embryol Exp Morphol 1985;87: 27–45.

[213] Wiles, MV. Embryonic stem cell differentiation in vitro. Methods Enzymol 1993;225: 900–918.

[214] Muller, AM, Dzierzak, EA. ES cells have only a limited lymphopoietic potential after adoptive transfer into mouse recipients. Development 1993;118(4): 1343–1351.

[215] Keller, G, Kennedy, M, Papayannopoulou, T, Wiles, MV. Hematopoietic commitment during embryonic stem cell differentiation in culture. Mol Cell Biol 1993;13(1): 473–486.

[216] Snodgrass, HR, Schmitt, RM, Bruyns, E. Embryonic stem cells and in vitro hematopoiesis. J Cell Biochem 1992;49(3): 225–230.

[217] Schmitt, RM, Bruyns, E, Snodgrass, HR. Hematopoietic development of embryonic stem cells in vitro: cytokine and receptor gene expression. Genes Dev 1991;5(5): 728–740.

[218] Burkert, U, von Ruden, T, Wagner, EF. Early fetal hematopoietic development from in vitro differentiated embryonic stem cells. New Biol 1991;3(7): 698–708.

[219] Kyba, M, Perlingeiro, RC, Daley, GQ. HoxB4 confers definitive lymphoid-myeloid engraftment potential on embryonic stem cell and yolk sac hematopoietic progenitors. Cell 2002;109(1): 29–37.

[220] Hanna, J, Wernig, M, Markoulaki, S, Sun, CW, Meissner, A, Cassady, JP, Beard, C, Brambrink, T, Wu, LC, Townes, TM, Jaenisch, R. Treatment of sickle cell anemia mouse model with iPS cells generated from autologous skin. Science 2007;318(5858): 1920–1923.

[221] Matsumoto, K, Isagawa, T, Nishimura, T, Ogaeri, T, Eto, K, Miyazaki, S, Miyazaki, J, Aburatani, H, Nakauchi, H, Ema, H. Stepwise development of hematopoietic stem cells from embryonic stem cells. Plos One 2009;4(3): e4820.

[222] Wang, Y, Yates, F, Naveiras, O, Ernst, P, Daley, GQ. Embryonic stem cell-derived hematopoietic stem cells. Proc Natl Acad Sci USA 2005;102(52): 19081–19086.

[223] McKinney-Freeman, SL, Lengerke, C, Jang, IH, Schmitt, S, Wang, Y, Philitas, M, Shea, J, Daley, GQ. Modulation of murine embryonic stem cell-derived CD41+c-kit+ hematopoietic progenitors by ectopic expression of Cdx genes. Blood 2008;111(10): 4944–4953.

[224] Wang, L, Menendez, P, Shojaei, F, Li, L, Mazurier, F, Dick, JE, Cerdan, C, Levac, K, Bhatia, M. Generation of hematopoietic repopulating cells from human embryonic stem cells independent of ectopic HOXB4 expression. J Exp Med 2005;201(10): 1603–1614.

[225] Rajala, K, Pekkanen-Mattila, M, Aalto-Setala, K. Cardiac differentiation of pluripotent stem cells. Stem Cells Int 2011;2011: 383709.

[226] Dambrot, C, Passier, R, Atsma, D, Mummery, CL. Cardiomyocyte differentiation of pluripotent stem cells and their use as cardiac disease models. Biochem J 2011;434(1): 25–35.

[227] Kehat, I, Khimovich, L, Caspi, O, Gepstein, A, Shofti, R, Arbel, G, Huber, I, Satin, J, Itskovitz-Eldor, J, Gepstein, L. Electromechanical integration of cardiomyocytes derived from human embryonic stem cells. Nat Biotechnol 2004;22(10): 1282–1289.

[228] Laflamme, MA, Chen, KY, Naumova, AV, Muskheli, V, Fugate, JA, Dupras, SK, Reinecke, H, Xu, C, Hassanipour, M, Police, S, O'Sullivan, C, Collins, L, Chen, Y, Minami, E, Gill, EA, Ueno, S, Yuan, C, Gold, J, Murry, CE. Cardiomyocytes derived from human embryonic stem cells in pro-survival factors enhance function of infarcted rat hearts. Nat Biotechnol 2007;25(9): 1015–1024.

[229] van Laake, LW, Passier, R, Monshouwer-Kloots, J, Verkleij, AJ, Lips, DJ, Freund, C, den Ouden, K, Ward-van Oostwaard, D, Korving, J, Tertoolen, LG, van Echteld, CJ, Doevendans, PA, Mummery, CL. Human embryonic stem cell-derived cardiomyocytes survive and mature in the mouse heart and transiently improve function after myocardial infarction. Stem Cell Res 2007;1(1): 9–24.

[230] Yang, L, Soonpaa, MH, Adler, ED, Roepke, TK, Kattman, SJ, Kennedy, M, Henckaerts, E, Bonham, K, Abbott, GW, Linden, RM, Field, LJ, Keller, GM. Human cardiovascular progenitor cells develop from a KDR+ embryonic-stem-cell-derived population. Nature 2008;453(7194): 524–528.

[231] Hoekstra, M, Mummery, CL, Wilde, AA, Bezzina, CR, Verkerk, AO. Induced pluripotent stem cell derived cardiomyocytes as models for cardiac arrhythmias. Front Physiol 2012;3: 346.

Chapter 19

Retinoic Acid Receptor Signaling in the Differentiation of Pluripotent Stem Cells into Skeletal Muscle Lineage

Jihong Chen and Qiao Li

Additional information is available at the end of the chapter

http://dx.doi.org/10.5772/62819

Abstract

Pluripotent stem cells have the capacity to differentiate into many types of cell lineages including skeletal myocytes. Nevertheless, the frequency of pluripotent stem cells generating skeletal myocytes in the absence of developmental cues is very low, and signaling molecules are required to commit them to muscle lineage. Thereby, *in vitro* stem cell differentiation has been used for decades to study molecular mechanisms of myogenic specification. Similar to human embryonic stem (ES) cells, various mouse pluripotent stem cells respond well to development cues *in vitro* to differentiate into cell types of all three primary germ layers. In tissue cultures, they can be induced into myogenic differentiation with an aggregation protocol which involves the formation of embryoid bodies (EBs). Their commitment into the skeletal muscle lineage recapitulates closely the cellular and molecular processes occurring in the early embryogenesis. Treatment of these stem cells with regulatory signals important for embryonic development, such as ligands of nuclear receptors, during EB formation markedly enhances the efficiency of myogenic differentiation. However, many challenges remain. Understanding on a molecular level, how different signaling pathways and chromatin dynamics converge during stem cell differentiation to specify the muscle lineage is imperative for identifying effective signaling molecules to generate sufficient amount of muscle progenitor cells for potential therapeutics. To this end, mouse stem cells will continue to serve as valuable model systems due to their close resemblance to skeletal myogenesis *in vivo,* and the ease of manipulation in experimental procedures. In this chapter, we will focus on recent research findings on nuclear receptor signaling in the specification of skeletal muscle lineage.

Keywords: Histone acetylation, Ligand, Myogenic specification, Nuclear receptor, Stem cells

1. Introduction

Pluripotent stem cells are valuable systems for delineating mechanisms of cellular differentiation due to their abilities to differentiate into virtually all cell types *in vitro*. Based on their derivative origins, there are ES cells, adult stem (AS) cells, and induced pluripotent stem (iPS) cells. ES cells are derived from the inner cell mass of blastocysts, have the capacity of unlimited self-renewal, and can give rise to derivatives of all three primary germ layers. AS cells, known as somatic stem cells, have limited ability to proliferate and can give rise to multiple cell lineages of the organ from which they are originated, but not all lineages of the three germ layers. iPS cells are generated from somatic cells by inducing conditions which reprogram the cells from a nonpluripotent state into a pluripotent, or ES cell-like state.

The first evidence for the pluripotent nature of embryonic cells was obtained from studies of mouse embryonal carcinoma (EC) cells, decades before the isolation of mouse ES cells [1]. Subcloned from teratocarcinomas, the EC cells grow as adherent cells in tissue-culture dishes and proliferate indefinitely [2]. When cultured in Petri dishes, they are unable to adhere and thereby form cell aggregates that contain a core of stem cells surrounded by epithelial cells. These cell aggregates are known as embryoid bodies (EBs), because they resemble the inner cell mass of embryo and develop extensive cavities and different cell types when subsequently grown as adhesive culture [3]. Since then, many EC cell lines have been generated and provided valuable experimental systems for studies of early development and cellular differentiation. More importantly, they paved the way for the isolation of mouse ES cells. Although pluripotent EC cells are much less used today, they remain an invaluable system for the studies of myogenic differentiation [4].

Mouse ES cells were first isolated in the early 1980s, from blastocysts grown on feeder-layer of division-incompetent mouse fibroblasts cells [5, 6]. These ES cells express all markers of EC cells and can differentiate extensively *in vivo and in vitro*. The requirement for ES cell to differentiate *in vitro* is, in essence, the same as for the EC cells, going through the stage of cell aggregation or EB formation [7]. However, ES cells need to be cultured in an inhibitory condition, such as in the presence of leukemia inhibitory factor (LIF), to retain the undifferentiated state, since they are prone to spontaneously differentiation [8, 9]. If grown in suspension without the LIF, the ES cells readily form EBs and differentiate as a result.

2. Ligand-enhanced myogenic differentiation

P19 pluripotent stem cell line, isolated from an experimental teratocarcinoma induced in the C3H/HC mice, exhibits typical EC morphology and normal karyotype [10]. Like other EC cell lines, the P19 cells can be grown as undifferentiated monolayer in tissue-culture dishes indefinitely and induced into differentiation to form cell lineages of all three germ layers [11]. In addition, they are amenable for genetic manipulation to incorporate and express ectopic genes and for selection of subclones or stable clones that retain their ability to differentiate [12].

http://dx.doi.org/10.5772/62819

All these characteristics have made them an excellent model system for mechanistic studies of early development.

One particular trait is the capacity of the P19 stem cells to generate skeletal myocytes in response to developmental cues. If grown in Petri dishes, the P19 cells readily form aggregates and develop into EBs [13]. Mesoderm specification takes place at the early stage of EB formation, coinciding with expression of Brachyury T, a member of the T-box family of transcription factors [14, 15]. However, EB formation *per se* does not result in myogenic differentiation, which requires additional regulatory signals. When cultured in the presence of signaling molecules, such as dimethyl sulfoxide (DMSO) or all-*trans* retinoic acid (RA), during EB formation, the P19 cells differentiate into skeletal myocytes in a relative low frequency [16, 17]. Treatment with combination of inducers, such as with both DMSO and RA, markedly enhances the myogenic conversion of P19 stem cells [18].

The differentiation of P19 stem cells is affected by the concentration of RA treatment. Cells exposed to high concentrations of RA (>10^{-7} M) develop into neurons and astrocytes, but fail to differentiate into skeletal myocytes [19–21]. On the other hand, EBs formed at low concentrations of RA (<10^{-7} M) develop into striated muscle, wherein the working concentration of RA is typically below the Kd, about 5–30 nM [4, 21–23]. The efficiency of P19 myogenic commitment is also affected by the timing and duration of RA treatments. As shown in **Figure 1**, cells treated with RA and DMSO during the full 4-day period of EB formation generated about 10% of myocytes by day 9 of differentiation, as determined by quantitative immnuofluorescence microscopy. However, if the EBs were allowed to form for 1 day in the absence of any treatment and then treated for the remaining 3-day period, the efficiency of myogenic differentiation increased by about twofold (**Figure 1**). When EBs were treated just for the last 2-day of EB formation, only about 5% of skeletal myocytes were generated (**Figure 1**). Finally, the ability of P19 cells to undergo myogenesis is also influenced by other factors in the serum, and EB formation is a prerequisite for myogenic differentiation of the pluripotent P19 EC cells [24].

Most interestingly, recent studies have identified bexarotene, a selective ligand of retinoid X receptor (RXR), to be an effective enhancer for the generation of skeletal myocytes by pluri-

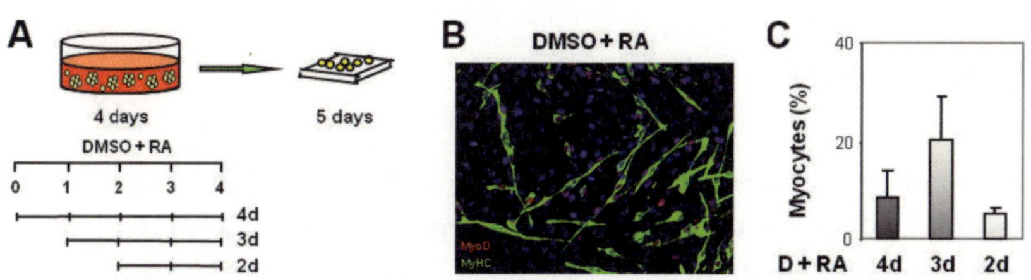

Figure 1. Effects of time course treatments on myogenic differentiation. (A) P19 stem cells were treated with DMSO and RA for the indicated times in Petri dishes during EB formation, maintained on coverslips for additional 5 days without any treatments and then stained for microscopic analysis. (B) Representative image of myosin heavy chain (green), MyoD (red), and DNA (blue) co-staining. (C) Quantification of myocytes is presented as the fractions of cells stained positively for myosin heavy chain in relation to the total cell populations. Error bars are the standard deviations of three independent experiments.

potent stem cells [4, 23]. Particularly, bexarotene enhances myogenic differentiation in a concentration dependent manner. The range of working concentration is wide (10–1000 nM) and fits the kinetics of its affinity to RXR as a ligand [4, 23]. More importantly, high concentrations of bexarotene do not inhibit the differentiation of P19 cells into skeletal myocytes [4, 23], which is in marked contrast with the dose effects of RA on myogenic conversion [21]. Nevertheless, the below Kd narrow concentration range also applies to the enhancement effect of arotinoid acid, a selective ligand for retinoic acid receptor (RAR), on myogenic specification [25]. In addition, the efficacy of bexarotene in P19 myogenic differentiation is comparable to RA and arotinoid acid [4, 23].

Early events of embryonic myogenesis are also closely recapitulated by the differentiation of ES cells into skeletal muscle lineage [26, 27]. RA is able to enhance myogenic differentiation of ES cells. More specifically, RA also affects the differentiation of ES cells into skeletal myocytes in a time- and concentration-dependant manner, similar manner as in pluripotent P19 EC cells. High concentrations of RA ($>10^{-7}$ M) induce neuronal differentiation in the ES cells, but inhibit myogenic commitment. Treatment of EBs with low concentrations of RA ($<10^{-7}$ M) at day 2–5 of differentiation leads to the induction of skeletal myogenesis, but the inhibition of cardiomyogenesis [28]. However, when low concentrations of RA are administered at day 5–7 of differentiation, skeletal myogenesis is inhibited, whereas cardiomyogenesis is induced [28].

Since ES cells respond poorly to RA regarding myogenic differentiation, the effect of bexarotene on the differentiation of ES cells into skeletal muscle lineage thus becomes critical [24]. A hanging-drop procedure was used to form the EBs which leads to ES cell differentiation. DMSO was omitted from the medium due to the toxicity to ES cells, and RA was administered in parallel as a comparison. Consistent with literature, RA had a low efficacy, about 3%, at converting the ES cells into skeletal muscle lineage [4]. However, bexarotene is about fivefold more efficient than RA and significantly increased the specification of skeletal muscle lineage [4]. Taken together, these data demonstrate that the RXR ligand is a more effective signaling molecule than RA to enhance the differentiation of ES cells into skeletal muscle lineage [4].

3. Ligand-inducible transcription factors

Vitamin A plays important roles in patterning and development during vertebrate embryogenesis [29]. Proper distribution and metabolism of vitamin A is fundamental for normal embryonic development and growth. Deficiency in vitamin A during early stage of embryogenesis results in congenital malformations affecting the patterning and development of many organ systems [30]. On the other hand, high concentrations of vitamin A or pharmacological concentration of RA, a potent derivative of vitamin A, have severe teratogenic consequences [31]. The diversified effects of RA are mediated by multiple levels of effectors, including enzymes that control the synthesis and degradation of RA, the cytoplasmic RA-binding proteins, and the nuclear receptors that are activated by RA [32].

http://dx.doi.org/10.5772/62819

The RARs are ligand-inducible transcription factors mediating the effects of RA on cellular activities [33]. There are three subtypes, namely RARα, RARβ, and RARγ, which can all be activated by both all-*trans*- and 9-*cis* RA [33]. Single subtype of RAR knockout mice is viable and appears normal, exhibiting few developmental defects [34, 35]. Nevertheless, double RAR knockout mice present a wide range of developmental abnormalities which resemble the vitamin A deficiency syndrome [36–39]. In fact, there appears to be a large degree of functional redundancy between RARs which play important roles in many distinct stages of embryonic patterning and organogenesis [33].

As a transcription factor, the RAR binds to RA-responsive element constitutively as a heterodimer with RXR (**Figure 2**). In the absence of ligand, the DNA-bound RAR/RXR heterodimer functions as a transcription repressor by associating with the NCoR co-repressor complex, but, upon RA induction, it acts as an activator by recruiting the p300 coactivator complexes to activate transcription [40, 41]. Often, RA-responsive promoters are classified as pre-set or poised promoters, as the TBP and Pol II complex associate with the TATA box constitutively [42, [43]. In this bimodal mode, ligand induction is through the RAR, wherein RXR is generally considered as a silent partner [44]. However, RXR is also amenable to RXR ligand induction and to form RXR homodimers or permissive heterodimers via dimerization with other nuclear receptors [45–47].

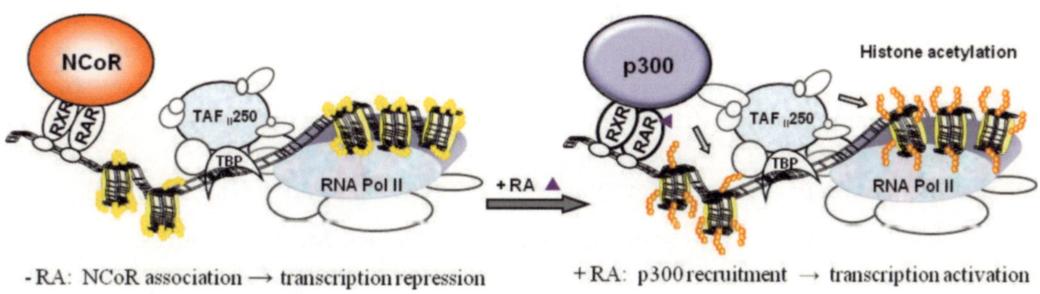

Figure 2. Schematic presentation of the bimodal function of RAR/RAR. In the absence of RA, the DNA bound RAR/RXR heterodimer associates with the NCoR corepressor complex to repress gene transcription. Upon RA induction, the corepressor complex is released, and the p300 coactivator complex is recruited by RAR/RXR to activate gene transcription.

The RXRs also consist of three subtypes, namely RXRα, RXRβ, and RXRγ, which bind to 9-*cis* RA [33]. Mice with subtypes of RXRs knocked out are also well characterized. RXRβ and RXRγ null mice are viable and mostly normal [48, [49]. In contrast, RXRα null mutants die in utero and display myocardial and ocular malformations [48]. Most interestingly, the RXRα null mutants exhibit developmental defects similar to the fetal vitamin A deficiency syndrome [50, [51]. Furthermore, the compound RXR and RAR knockout mice recapitulate most of the defects observed in RAR double mutants [48, [52]. Therefore, RXRα is the main subtype involved in embryonic development, and RXRα/RAR is the main functional unit to mediate RA signals during embryonic development [53, 54]. In addition, RXR is involved in an array of signaling cascades and has the capacity to converge multiple pathways as an liganded receptor [55, 56].

4. Transcription networks of myogenesis

Skeletal myogenesis is a complex process coordinated temporally by multiple myogenic regulatory factors including Myf5, MyoD, myogenin, and Mrf4 [57, 58]. While Myf5 and MyoD initiate the expression of muscle-specific genes and commit the progenitor cells into the muscle lineage [59–61], myogenin and Mrf4 mainly regulate the late stage of differentiation, such as the fusion of myoblasts into myotubes [62–65]. At the upstream, Wnt signaling and Shh from the dorsal neural tube and notochord act as the positive regulators of Myf5 gene expression, whereas the expression of MyoD depends on the function of progenitor factor Pax3 and Myf5 [66]. Although both mesoderm factor Meox1 and Pax3 are important for myogenesis, overexpression of Mexo1 *per se* is not sufficient to induce P19 myogenic differentiation [67, 68].

During P19 myogenic specification, Meox1 and Pax3 expression are upregulated by RA by day 4 of differentiation [4, 22, 23]. Similarly, Myf5 transcripts can also be detected by day 4 of differentiation following RA treatment [23, 69]. Interestingly, bexarotene increases the transcript level of Meox1 with a greater efficiency than RA (about twofold), whereas RA has a larger impact than bexarotene on gene expression of Pax3 and Myf5 [4, 23]. In addition, the temporal gene expression pattern induced by bexarotene during P19 myogenic differentiation is similar to during myogenesis *in vivo,* and RXR ligand acts as an effective enhancer for the specification of muscle lineage [4]. It worth noting that bexarotene and RA have comparable efficacies at enhancing P19 myogenic differentiation [4]. While RA enhances skeletal myogenesis by expanding the progenitor population [22], bexarotene may affect germ layer fate determinations, particularly promoting mesoderm differentiation [4].

Intriguingly, bexarotene is a more efficient enhancer than RA for myogenesis in the ES cell system [4, 23]. Similar as in the P19 stem cells, bexarotene augments Meox1 transcripts more potently than RA in ES cell system, whereas RA is more efficient at increasing Pax3 transcripts [4, 23]. Nonetheless, bexarotene alone is able to induce the expression of early differentiation marker meox1, whereas RA requires additional signaling molecules to induce Meox1 expression. Hence, bexarotene may enhance the commitment of skeletal muscle lineage by fine-tuning premyogenic transcriptional networks which then preferentially affect the downstream myogenic program. Comprehensive and condition-specific gene expression profiling will uncover additional early regulators activated by RXR selective signaling during mesoderm differentiation, identify novel regulators of myogenic differentiation, and determine why RXR agonist is an effective inducer for ES myogenic specification.

Genetic manipulation can also be used as an approach to induce myogenic differentiation in ES cells. The premyogenic factor Pax3 plays a critical role in embryonic muscle formation by acting upstream of the myogenic-specific program [70–72], whereas Pax7 is important for the maintenance of the muscle satellite cells [73–75]. Consequently, ectopic expression of Pax3 during EB differentiation enhances mesoderm formation and increases the myogenic poten-

http://dx.doi.org/10.5772/62819

tial of Pax3-induced ES cells [76]. Similarly, forced Pax7 expression promotes the expansion of myogenic progenitors which possess muscle regeneration potentials [77]. In any events, activating the myogenic signaling pathway with small molecular inducers, which can be easily supplemented into, or removed from differentiation media, to direct myogenic commitment remains a more practical and attractive approach in view of potential cell-based therapies. Most intriguingly, the premyogenic factor Pax3 is also an inhibitor of cardiac differentiation in lineage specification [23]. The activation of Pax3 by RA or bexarotene during myogenic differentiation coincides with inhibiting the expression of early cardiac factors including GATA4, Tbx5, and Nkx2.5, and the inhibitory effect of bexarotene or RA on cardiac differentiation depends on the function of Pax3 [23]. Thus, the premyogenic factor Pax3 plays dual roles in stem cell fate determination by regulating and integrating different signaling pathways.

5. Enhancer elements of myogenesis

In eukaryotic cells, the regulatory DNA elements, such as the enhancers and promoters, are organized with histones to form nucleosomes which are further packaged into a higher-order chromatin structure [78–81]. The organization of chromatin structure not only establishes hierarchical platforms, but also provides epigenetic information including histone acetylation, for the intricate interactions amongst regulatory-proteins in cell fate determination, and ultimately for the control of specific transcription networks [82]. Thus, chromatin signatures are valuable signatures to identify novel regulatory elements, in addition to the sequences of DNA binding elements [83–85].

Recent genome-wide analyses have revealed an apparent functional relationship between chromatin dynamics and transcriptional activation in lineage specification. For instance, active promoters are often associated with multiple histone modifications, whereas enhancers are generally associated with the transcriptional coactivator p300 and histone acetylation [86, 87]. Moreover, genetic evidence in mouse and ES cell model systems has demonstrated that the expression of Myf5 and MyoD genes specifically depends on the histone acetyltransferase (HAT) activity of p300 [88]. The question is, on a molecular level, how different signaling pathways and chromatin dynamics converge to direct cell fate determination. Understanding the molecular mechanisms of myogenic specification is imperative for manipulating stem cell fate determinations in view of cell-based therapies.

A long-range RAR-binding site has been identified within the *Pax3* locus [22]. Both RAR and RXR are found at this region at the early stage of myogenic specification regardless of RA treatment, as determined by a real-time PCR-based chromatin immunoprecipitation (ChIP-qPCR) analysis [4]. In addition, the association of transcriptional coactivator p300 to this RAR-binding region increased markedly following treatment with RA (**Figure 3**) or arotinoid acid, a RAR selective ligand [25]. More interestingly, the *Myf5* epaxial enhancer is also a direct target of p300, as determined by the ChIP-qPCR analysis (**Figure 3B**).

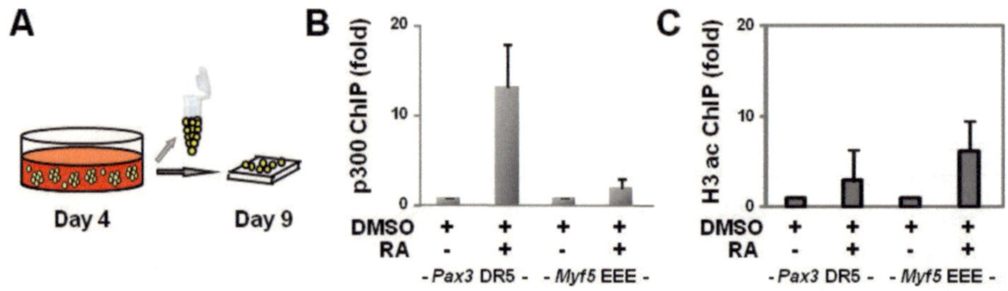

Figure 3. Occupancy of transcriptional coactivator p300 at myogenic loci. (A) Time line of the ChIP-qPCR procedure. Cells were cultivated in Petri dishes and treated with RA in the presence or absence of DMSO during EB formation. On day 4, aliquot of EBs was seeded on coverslips to assess the efficiency of myogenic differentiation on day 9. The remaining EBs was used for ChIP. (B) Association of p300 to the *Pax3* locus (*Pax3* DR5) and the *Myf5* early epaxial enhancer (*Myf5* EEE) was determined by ChIP-qPCR analysis, which is presented as the fold change in relation to DMSO controls. Input DNA was used as internal controls. Error bars are the standard deviations of three independent experiments. (C) The levels of H3K27 acetylation were also analyzed in parallel using the ChIP-qPCR assay.

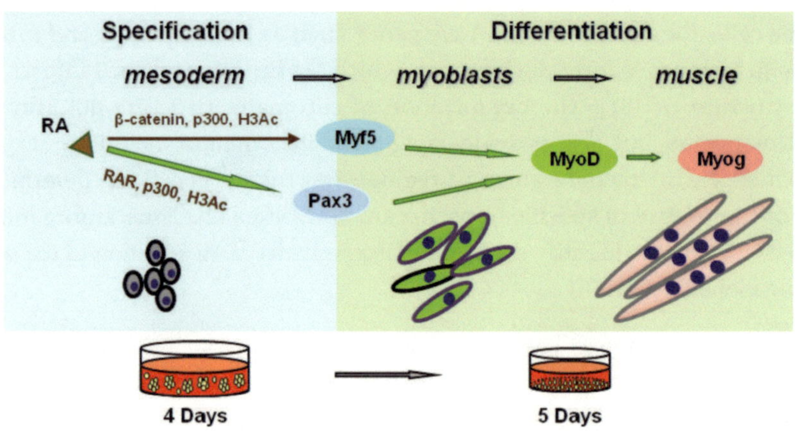

Figure 4. RA signaling in myogenic differentiation. RA enhances myogenic differentiation by augmenting Pax3 and Myf5 gene expression partly through promoting p300 occupancy and histone acetylation at the RA-responsive loci (solid brown and open green arrows).

The association of p300 to the *Pax3* locus is dependent on liganded RAR in an on-and-off mode [4] and increased about 15-fold following RA addition (**Figure 3B**). Intriguingly, the occupancy of p300 at the *Myf5* early enhancer, which does not harbor RAR binding sites, increased only about twofold following RA treatment (**Figure 3B**). Nevertheless, histones acetylation increased at both the *Pax3* locus and the *Myf5* early enhancer following RA treatment (**Figure 3C**). Therefore, RA regulates myogenic differentiation through p300-instigated histone acetylation in either RAR-binding dependent or independent manner (**Figure 4**). The relevance of histone acetylation during myogenic differentiation is additionally reinforced by the fact that valproic acid, a histone deacetylase inhibitor, is able to act in concert with RA to enhance the commitment of stem cells into the skeletal muscle lineage [89]. Nonetheless, chromatin signature for bexarotene to activate myogenic networks, particular-

http://dx.doi.org/10.5772/62819

ly how RXR selective signaling is transmitted during myogenic differentiation, remains to be defined. Systemic ChIP-seq analysis will identify additional p300-dependent myogenic enhancers and uncover novel regulatory elements to confer p300 function and histone acetylation in RXR-mediated stem cell differentiation.

6. Cell-based therapies for muscle-related diseases

Many diseases and conditions, including cancer, AIDS, muscular dystrophies, and chronological aging develop severe muscle wasting and would benefit enormously from muscle regeneration therapies. The unique architecture of skeletal muscle tissue makes it difficult to obtain differentiated skeletal muscle for tissue transplantation. Hence, muscle repair or regeneration may be best achieved through transplantation of the progenitor cells which are already committed into muscle lineage but not yet differentiated into skeletal myocytes. These progenitor cells will differentiate into functional skeletal muscle *in vivo* following transplantation. However, there are many challenges with respect to the effectiveness of myogenic differentiation, and the safety and long-term engraftment of transplanted progenitor cells. Some specific issues include what type of stem cells is best suited for production of progenitor cells and how to enrich the desired progenitor cells for clinical application.

Successful long-term therapeutics for skeletal muscle regeneration requires the contribution of transplanted progenitors to both the muscle fibers and the muscle stem cell pool. Muscle satellite cells may be an idea cell source of muscle regeneration, because they are not only able to generate muscle efficiently, but also able to establish the satellite cell pool following transplantation [90, 91]. However, their therapeutic potentials are greatly limited by their low abundance in the muscle. Another limitation is that the expansion of these cells *in vitro* reduces subsequently their regeneration capacity *in vivo* [92]. Similar limitation has also been found in murine and human hematopoietic stem cells [93, 94]. Finally, in the severe cases of muscular dystrophies, the regenerative source of satellite cells is unfortunately exhausted [95].

The ES cells can be expanded unlimitedly in tissue cultures, while maintaining their pluripotent differentiation potential. In addition, ES-derived myogenic progenitors have the ability to seed in the skeletal muscle stem cell compartment [76, 77]. Thus, ES cell-based muscle regeneration has unique advantage and presents immense promise. However, the use of ES cells in muscle-related disease is curtailed by the low frequency of myogenic differentiation in cultures and the difficulty in identifying and isolating the progenitor cells. The low frequency of ES cells to commit into skeletal muscle lineage is mostly attributed to the paucity of mesoderm formation during EB-differentiation in the absence of inducing signals.

To harness the full potential of ES cells in muscle regeneration, it is imperative that we identify small signaling molecules capable of efficiently directing ES cells into skeletal muscle lineage. Attempts at using RA in ES cell cultures have yielded moderate results, while RXR ligand appears to be a better inducer for myogenic differentiation. However, the molecular pathways involved have not yet been fully defined. A comprehensive deciphering of the differentiation cues in ES cultures and a better understanding of the regulation of the myogenic pathway *in vivo* will help us to identify additional small molecule inducers and devise optimal protocols to generate desirable myogenic progenitors for muscle regeneration or repair.

Small molecule inducers have been used to reprogram somatic cells, to maintain induced pluripotent states and to directly control lineage specification. They have the potential to enable the development of viable cell-based therapies and to control endogenous cell populations for supporting regeneration events. The impact of chemical biology on emerging regenerative medicine will only increase with time, with a greater comprehension of the signaling pathways regulating cell fate determination, and the molecular mechanisms promoting nascent cell survival and engraftment.

7. Perspectives

Pluripotent stem cells, regardless of their origin, possess a tremendous potential for use in the treatment of muscle-related disease, because of their capacities to differentiate into skeletal muscle lineage. However, small molecule inducers are required to specify myogenic differentiation *in vitro* with an efficacy suitable for cell-based therapies. Recent studies have uncovered the ability of RXR ligand to efficiently induce the commitment of ES cells into skeletal muscle lineage, but the molecular pathways involved remain to be determined. Concerted RNA-seq and ChIP-seq analysis with stem cell differentiation as a model system will uncover novel early regulators and epigenetic signatures important for myogenic differentiation. A systematic understanding on a molecular level, how different signaling pathways and chromatin dynamics converge to direct cell fate determination, is imperative for identifying additional small signaling molecules and developing nontoxic protocols with optimal combination of signaling molecules and treatment conditions to direct the specification of skeletal muscle lineage. The strategies and protocols devised in mouse stem cell systems can then be transferred to human ES cells and to other types of pluripotent stem cells in view of generating muscle progenitors for clinical applications.

Acknowledgements

This work was supported by grant from the Natural Sciences and Engineering Research Council of Canada. We thank our colleagues for a supporting research environment.

Author details

Jihong Chen and Qiao Li*

*Address all correspondence to: Qiao.Li@uottawa.ca

Faculty of Medicine, University of Ottawa, Ottawa, Ontario, Canada

References

[1] Ahuja, H.S., Szanto, A., Nagy, L. and Davies, P.J. (2003) The retinoid X receptor and its ligands: versatile regulators of metabolic function, cell differentiation and cell death. *J Biol Regul Homeost Agents*, 17, 29–45.

[2] Albella, B., Segovia, J.C., Guenechea, G., Pragnell, I.B. and Bueren, J.A. (1999) Preserved long-term repopulation and differentiation properties of hematopoietic grafts subjected to ex vivo expansion with stem cell factor and interleukin 11. *Transplantation*, 67, 1348–1357.

[3] Armour, C., Garson, K. and McBurney, M.W. (1999) Cell-cell interaction modulates myoD-induced skeletal myogenesis of pluripotent P19 cells in vitro. *Exp Cell Res*, 251, 79–91.

[4] Bajard, L., Relaix, F., Lagha, M., Rocancourt, D., Daubas, P. and Buckingham, M.E. (2006) A novel genetic hierarchy functions during hypaxial myogenesis: Pax3 directly activates Myf5 in muscle progenitor cells in the limb. *Genes Dev*, 20, 2450–2464.

[5] Bowles, J., Knight, D., Smith, C., Wilhelm, D., Richman, J., Mamiya, S., Yashiro, K., Chawengsaksophak, K., Wilson, M.J., Rossant, J., Hamada, H. and Koopman, P. (2006) Retinoid signaling determines germ cell fate in mice. *Science*, 312, 596–600.

[6] Braun, T., Bober, E., Rudnicki, M.A., Jaenisch, R. and Arnold, H.H. (1994) MyoD expression marks the onset of skeletal myogenesis in Myf-5 mutant mice. *Development*, 120, 3083–3092.

[7] Cerletti, M., Jurga, S., Witczak, C.A., Hirshman, M.F., Shadrach, J.L., Goodyear, L.J. and Wagers, A.J. (2008) Highly efficient, functional engraftment of skeletal muscle stem cells in dystrophic muscles. *Cell*, 134, 37–47.

[8] Chambon, P. (1996) A decade of molecular biology of retinoic acid receptors. *FASEB J*, 10, 940–954.

[9] Chambon, P. (2005) The nuclear receptor superfamily: a personal retrospect on the first two decades. *Mol Endocrinol*, 19, 1418–1428.

[10] Chen, J. and Li, Q. (2016) Implication of retinoic acid receptor selective signaling in myogenic differentiation. *Sci Rep*, 5, 18856.

[11] Chen, J., Wang, Y., Hamed, M., Lacroix, N. and Li, Q. (2015) Molecular Basis for the Regulation of transcriptional coactivator p300 in myogenic differentiation. *Sci Rep*, 5, 13727.

[12] Darabi, R., Gehlbach, K., Bachoo, R.M., Kamath, S., Osawa, M., Kamm, K.E., Kyba, M. and Perlingeiro, R.C. (2008) Functional skeletal muscle regeneration from differentiating embryonic stem cells. *Nat Med*, 14, 134–143.

[13] Darabi, R., Santos, F.N., Filareto, A., Pan, W., Koene, R., Rudnicki, M.A., Kyba, M. and Perlingeiro, R.C. (2011) Assessment of the myogenic stem cell compartment follow-

ing transplantation of Pax3/Pax7-induced embryonic stem cell-derived progenitors. *Stem Cells*, 29, 777–790.

[14] Doetschman, T.C., Eistetter, H., Katz, M., Schmidt, W. and Kemler, R. (1985) The in vitro development of blastocyst-derived embryonic stem cell lines: formation of visceral yolk sac, blood islands and myocardium. *J Embryol Exp Morphol*, 87, 27–45.

[15] Dorschner, M.O., Hawrylycz, M., Humbert, R., Wallace, J.C., Shafer, A., Kawamoto, J., Mack, J., Hall, R., Goldy, J., Sabo, P.J., Kohli, A., Li, Q., McArthur, M. and Stamatoyannopoulos, J.A. (2004) High-throughput localization of functional elements by quantitative chromatin profiling. *Nat Methods*, 1, 219–225.

[16] Edwards, M.K., Harris, J.F. and McBurney, M.W. (1983) Induced muscle differentiation in an embryonal carcinoma cell line. *Mol Cell Biol*, 3, 2280–2286.

[17] Edwards, M.K. and McBurney, M.W. (1983) The concentration of retinoic acid determines the differentiated cell types formed by a teratocarcinoma cell line. *Dev Biol*, 98, 187–191.

[18] Evans, M.J. and Kaufman, M.H. (1981) Establishment in culture of pluripotential cells from mouse embryos. *Nature*, 292, 154–156.

[19] Francetic, T., Le May, M., Hamed, M., Mach, H., Meyers, D., Cole, P.A., Chen, J. and Li, Q. (2012) Regulation of Myf5 early enhancer by histone acetyltransferase p300 during stem cell differentiation. *Mol Biol*, 1, 103. doi:110.4172/ 2168–9547.1000103.

[20] Francetic, T. and Li, Q. (2011) Skeletal myogenesis and Myf5 activation. *Transcription*, 2, 109–114.

[21] Gampe, R.T., Jr., Montana, V.G., Lambert, M.H., Wisely, G.B., Milburn, M.V. and Xu, H.E. (2000) Structural basis for autorepression of retinoid X receptor by tetramer formation and the AF-2 helix. *Genes Dev*, 14, 2229–2241.

[22] Grondona, J.M., Kastner, P., Gansmuller, A., Decimo, D., Chambon, P. and Mark, M. (1996) Retinal dysplasia and degeneration in RARbeta2/RARgamma2 compound mutant mice. *Development*, 122, 2173–2188.

[23] Guenther, M.G., Levine, S.S., Boyer, L.A., Jaenisch, R. and Young, R.A. (2007) A chromatin landmark and transcription initiation at most promoters in human cells. *Cell*, 130, 77–88.

[24] Hatzis, P. and Talianidis, I. (2002) Dynamics of enhancer-promoter communication during differentiation-induced gene activation. *Mol Cell*, 10, 1467–1477.

[25] Heintzman, N.D., Stuart, R.K., Hon, G., Fu, Y., Ching, C.W., Hawkins, R.D., Barrera, L.O., Van Calcar, S., Qu, C., Ching, K.A., Wang, W., Weng, Z., Green, R.D., Crawford, G.E. and Ren, B. (2007) Distinct and predictive chromatin signatures of transcriptional promoters and enhancers in the human genome. *Nat Genet*, 39, 311–318.

[26] Hermanson, O., Glass, C.K. and Rosenfeld, M.G. (2002) Nuclear receptor coregulators: multiple modes of modification. *Trends Endocrinol Metab*, 13, 55–60.

[27] Higazi, A., Abed, M., Chen, J. and Li, Q. (2011) Promoter context determines the role of proteasome in ligand-dependent occupancy of retinoic acid responsive elements. *Epigenetics*, 6, 202–211.

[28] Jones-Villeneuve, E.M., McBurney, M.W., Rogers, K.A. and Kalnins, V.I. (1982) Retinoic acid induces embryonal carcinoma cells to differentiate into neurons and glial cells. *J Cell Biol*, 94, 253–262.

[29] Jones-Villeneuve, E.M., Rudnicki, M.A., Harris, J.F. and McBurney, M.W. (1983) Retinoic acid-induced neural differentiation of embryonal carcinoma cells. *Mol Cell Biol*, 3, 2271–2279.

[30] Kassar-Duchossoy, L., Giacone, E., Gayraud-Morel, B., Jory, A., Gomes, D. and Tajbakhsh, S. (2005) Pax3/Pax7 mark a novel population of primitive myogenic cells during development. *Genes Dev*, 19, 1426–1431.

[31] Kastner, P., Grondona, J.M., Mark, M., Gansmuller, A., LeMeur, M., Decimo, D., Vonesch, J.L., Dolle, P. and Chambon, P. (1994) Genetic analysis of RXR alpha developmental function: convergence of RXR and RAR signaling pathways in heart and eye morphogenesis. *Cell*, 78, 987–1003.

[32] Kastner, P., Mark, M., Ghyselinck, N., Krezel, W., Dupe, V., Grondona, J.M. and Chambon, P. (1997) Genetic evidence that the retinoid signal is transduced by heterodimeric RXR/RAR functional units during mouse development. *Development*, 124, 313–326.

[33] Kennedy, K.A., Porter, T., Mehta, V., Ryan, S.D., Price, F., Peshdary, V., Karamboulas, C., Savage, J., Drysdale, T.A., Li, S.C., Bennett, S.A. and Skerjanc, I.S. (2009) Retinoic acid enhances skeletal muscle progenitor formation and bypasses inhibition by bone morphogenetic protein 4 but not dominant negative beta-catenin. *BMC Biol*, 7, 67.

[34] Kim, T.H., Abdullaev, Z.K., Smith, A.D., Ching, K.A., Loukinov, D.I., Green, R.D., Zhang, M.Q., Lobanenkov, V.V. and Ren, B. (2007) Analysis of the vertebrate insulator protein CTCF-binding sites in the human genome. *Cell*, 128, 1231–1245.

[35] Krezel, W., Dupe, V., Mark, M., Dierich, A., Kastner, P. and Chambon, P. (1996) RXR gamma null mice are apparently normal and compound RXR alpha +/−/RXR beta −/−/RXR gamma −/− mutant mice are viable. *Proc Natl Acad Sci USA*, 93, 9010–9014.

[36] Kurokawa, R., DiRenzo, J., Boehm, M., Sugarman, J., Gloss, B., Rosenfeld, M.G., Heyman, R.A. and Glass, C.K. (1994) Regulation of retinoid signalling by receptor polarity and allosteric control of ligand binding. *Nature*, 371, 528–531.

[37] Laperriere, D., Wang, T.T., White, J.H. and Mader, S. (2007) Widespread Alu repeat-driven expansion of consensus DR2 retinoic acid response elements during primate evolution. *BMC Genom*, 8, 23.

[38] Le May, M., Mach, H., Lacroix, N., Hou, C., Chen, J. and Li, Q. (2011) Contribution of Retinoid X Receptor Signaling to the Specification of Skeletal Muscle Lineage. *J Biol Chem*, 286, 26806–26812.

[39] Li, Q., Foote, M. and Chen, J. (2014) Effects of histone deacetylase inhibitor valproic acid on skeletal myocyte development. *Sci Rep*, 4, 7207.

[40] Li, Q., Le May, M., Lacroix, N. and Chen, J. (2013) Induction of Pax3 gene expression impedes cardiac differentiation. *Sci Rep*, 3, 2498.

[41] Lohnes, D., Kastner, P., Dierich, A., Mark, M., LeMeur, M. and Chambon, P. (1993) Function of retinoic acid receptor gamma in the mouse. *Cell*, 73, 643–658.

[42] Lohnes, D., Mark, M., Mendelsohn, C., Dolle, P., Dierich, A., Gorry, P., Gansmuller, A. and Chambon, P. (1994) Function of the retinoic acid receptors (RARs) during development (I). Craniofacial and skeletal abnormalities in RAR double mutants. *Development*, 120, 2723–2748.

[43] Lonard, D.M. and O'Malley B, W. (2007) Nuclear receptor coregulators: judges, juries, and executioners of cellular regulation. *Mol Cell*, 27, 691–700.

[44] Lufkin, T., Lohnes, D., Mark, M., Dierich, A., Gorry, P., Gaub, M.P., LeMeur, M. and Chambon, P. (1993) High postnatal lethality and testis degeneration in retinoic acid receptor alpha mutant mice. *Proc Natl Acad Sci USA*, 90, 7225–7229.

[45] Mangelsdorf, D.J., Thummel, C., Beato, M., Herrlich, P., Schutz, G., Umesono, K., Blumberg, B., Kastner, P., Mark, M., Chambon, P. and Evans, R.M. (1995) The nuclear receptor superfamily: the second decade. *Cell*, 83, 835–839.

[46] Mark, M., Ghyselinck, N.B. and Chambon, P. (2009) Function of retinoic acid receptors during embryonic development. *Nucl Recept Signal*, 7, e002.

[47] Maroto, M., Reshef, R., Munsterberg, A.E., Koester, S., Goulding, M. and Lassar, A.B. (1997) Ectopic Pax-3 activates MyoD and Myf-5 expression in embryonic mesoderm and neural tissue. *Cell*, 89, 139–148.

[48] Martin, G.R. (1981) Isolation of a pluripotent cell line from early mouse embryos cultured in medium conditioned by teratocarcinoma stem cells. *Proc Natl Acad Sci USA*, 78, 7634–7638.

[49] Martin, G.R. and Evans, M.J. (1974) The morphology and growth of a pluripotent teratocarcinoma cell line and its derivatives in tissue culture. *Cell*, 2, 163–172.

[50] Martin, G.R. and Evans, M.J. (1975) Differentiation of clonal lines of teratocarcinoma cells: formation of embryoid bodies in vitro. *Proc Natl Acad Sci USA*, 72, 1441–1445.

[51] McBurney, M.W. (1993) P19 embryonal carcinoma cells. *Int J Dev Biol*, 37, 135–140.

[52] McBurney, M.W., Jones-Villeneuve, E.M., Edwards, M.K. and Anderson, P.J. (1982) Control of muscle and neuronal differentiation in a cultured embryonal carcinoma cell line. *Nature*, 299, 165–167.

[53] McBurney, M.W. and Rogers, B.J. (1982) Isolation of male embryonal carcinoma cells and their chromosome replication patterns. *Dev Biol*, 89, 503–508.

[54] Montarras, D., Morgan, J., Collins, C., Relaix, F., Zaffran, S., Cumano, A., Partridge, T. and Buckingham, M. (2005) Direct isolation of satellite cells for skeletal muscle regeneration. *Science*, 309, 2064–2067.

[55] Nabeshima, Y., Hanaoka, K., Hayasaka, M., Esumi, E., Li, S. and Nonaka, I. (1993) Myogenin gene disruption results in perinatal lethality because of severe muscle defect. *Nature*, 364, 532–535.

[56] Niederreither, K. and Dolle, P. (2008) Retinoic acid in development: towards an integrated view. *Nat Rev Genet*, 9, 541–553.

[57] Orkin, S.H. and Hochedlinger, K. (2011) Chromatin connections to pluripotency and cellular reprogramming. *Cell*, 145, 835–850.

[58] Parker, M.H., Seale, P. and Rudnicki, M.A. (2003) Looking back to the embryo: defining transcriptional networks in adult myogenesis. *Nat Rev Genet*, 4, 497–507.

[59] Patapoutian, A., Yoon, J.K., Miner, J.H., Wang, S., Stark, K. and Wold, B. (1995) Disruption of the mouse MRF4 gene identifies multiple waves of myogenesis in the myotome. *Development*, 121, 3347–3358.

[60] Pavri, R., Lewis, B., Kim, T.K., Dilworth, F.J., Erdjument-Bromage, H., Tempst, P., de Murcia, G., Evans, R., Chambon, P. and Reinberg, D. (2005) PARP-1 determines specificity in a retinoid signaling pathway via direct modulation of mediator. *Mol Cell*, 18, 83–96.

[61] Petropoulos, H., Gianakopoulos, P.J., Ridgeway, A.G. and Skerjanc, I.S. (2004) Disruption of Meox or Gli activity ablates skeletal myogenesis in P19 cells. *J Biol Chem*, 279, 23874–23881.

[62] Rawls, A., Morris, J.H., Rudnicki, M., Braun, T., Arnold, H.H., Klein, W.H. and Olson, E.N. (1995) Myogenin's functions do not overlap with those of MyoD or Myf-5 during mouse embryogenesis. *Dev Biol*, 172, 37–50.

[63] Relaix, F., Rocancourt, D., Mansouri, A. and Buckingham, M. (2005) A Pax3/Pax7-dependent population of skeletal muscle progenitor cells. *Nature*, 435, 948–953.

[64] Ridgeway, A.G., Petropoulos, H., Wilton, S. and Skerjanc, I.S. (2000) Wnt signaling regulates the function of MyoD and myogenin. *J Biol Chem*, 275, 32398–32405.

[65] Ridgeway, A.G. and Skerjanc, I.S. (2001) Pax3 is essential for skeletal myogenesis and the expression of Six1 and Eya2. *J Biol Chem*, 276, 19033–19039.

[66] Rohwedel, J., Maltsev, V., Bober, E., Arnold, H.H., Hescheler, J. and Wobus, A.M. (1994) Muscle cell differentiation of embryonic stem cells reflects myogenesis in vivo: developmentally regulated expression of myogenic determination genes and functional expression of ionic currents. *Dev Biol*, 164, 87–101.

[67] Roth, J.F., Shikama, N., Henzen, C., Desbaillets, I., Lutz, W., Marino, S., Wittwer, J., Schorle, H., Gassmann, M. and Eckner, R. (2003) Differential role of p300 and CBP acetyltransferase during myogenesis: p300 acts upstream of MyoD and Myf5. *EMBO J*, 22, 5186–5196.

[68] Rudnicki, M.A., Schnegelsberg, P.N., Stead, R.H., Braun, T., Arnold, H.H. and Jaenisch, R. (1993) MyoD or Myf-5 is required for the formation of skeletal muscle. *Cell*, 75, 1351–1359.

[69] Sacco, A., Doyonnas, R., Kraft, P., Vitorovic, S. and Blau, H.M. (2008) Self-renewal and expansion of single transplanted muscle stem cells. *Nature*, 456, 502–506.

[70] Sapin, V., Dolle, P., Hindelang, C., Kastner, P. and Chambon, P. (1997) Defects of the chorioallantoic placenta in mouse RXRalpha null fetuses. *Dev Biol*, 191, 29–41.

[71] Seale, P., Sabourin, L.A., Girgis-Gabardo, A., Mansouri, A., Gruss, P. and Rudnicki, M.A. (2000) Pax7 is required for the specification of myogenic satellite cells. *Cell*, 102, 777–786.

[72] Shi, X. and Garry, D.J. (2006) Muscle stem cells in development, regeneration, and disease. *Genes Dev*, 20, 1692–1708.

[73] Smith, A.G., Heath, J.K., Donaldson, D.D., Wong, G.G., Moreau, J., Stahl, M. and Rogers, D. (1988) Inhibition of pluripotential embryonic stem cell differentiation by purified polypeptides. *Nature*, 336, 688–690.

[74] Smith, S.C., Reuhl, K.R., Craig, J. and McBurney, M.W. (1987) The role of aggregation in embryonal carcinoma cell differentiation. *J Cell Physiol*, 131, 74–84.

[75] Solter, D. (2006) From teratocarcinomas to embryonic stem cells and beyond: a history of embryonic stem cell research. *Nat Rev Genet*, 7, 319–327.

[76] Sucov, H.M., Dyson, E., Gumeringer, C.L., Price, J., Chien, K.R. and Evans, R.M. (1994) RXR alpha mutant mice establish a genetic basis for vitamin A signaling in heart morphogenesis. *Genes Dev*, 8, 1007–1018.

[77] Tajbakhsh, S., Rocancourt, D., Cossu, G. and Buckingham, M. (1997) Redefining the genetic hierarchies controlling skeletal myogenesis: Pax-3 and Myf-5 act upstream of MyoD. *Cell*, 89, 127–138.

[78] Tanaka, T. and De Luca, L.M. (2009) Therapeutic potential of "rexinoids" in cancer prevention and treatment. *Cancer Res*, 69, 4945–4947.

[79] Tapscott, S.J. (2005) The circuitry of a master switch: Myod and the regulation of skeletal muscle gene transcription. *Development*, 132, 2685–2695.

[80] Traycoff, C.M., Cornetta, K., Yoder, M.C., Davidson, A. and Srour, E.F. (1996) Ex vivo expansion of murine hematopoietic progenitor cells generates classes of expanded cells possessing different levels of bone marrow repopulating potential. *Exp Hematol*, 24, 299–306.

[81] van der Heyden, M.A. and Defize, L.H. (2003) Twenty one years of P19 cells: what an embryonal carcinoma cell line taught us about cardiomyocyte differentiation. *Cardiovasc Res*, 58, 292–302.

[82] Venuti, J.M., Morris, J.H., Vivian, J.L., Olson, E.N. and Klein, W.H. (1995) Myogenin is required for late but not early aspects of myogenesis during mouse development. *J Cell Biol*, 128, 563–576.

[83] Vidricaire, G., Jardine, K. and McBurney, M.W. (1994) Expression of the Brachyury gene during mesoderm development in differentiating embryonal carcinoma cell cultures. *Development*, 120, 115–122.

[84] Visel, A., Blow, M.J., Li, Z., Zhang, T., Akiyama, J.A., Holt, A., Plajzer-Frick, I., Shoukry, M., Wright, C., Chen, F., Afzal, V., Ren, B., Rubin, E.M. and Pennacchio, L.A. (2009) ChIP-seq accurately predicts tissue-specific activity of enhancers. *Nature*, 457, 854–858.

[85] Wang, Q., Carroll, J.S. and Brown, M. (2005) Spatial and temporal recruitment of androgen receptor and its coactivators involves chromosomal looping and polymerase tracking. *Mol Cell*, 19, 631–642.

[86] Weintraub, H., Tapscott, S.J., Davis, R.L., Thayer, M.J., Adam, M.A., Lassar, A.B. and Miller, A.D. (1989) Activation of muscle-specific genes in pigment, nerve, fat, liver, and fibroblast cell lines by forced expression of MyoD. *Proc Natl Acad Sci USA*, 86, 5434–5438.

[87] Weitzer, G., Milner, D.J., Kim, J.U., Bradley, A. and Capetanaki, Y. (1995) Cytoskeletal control of myogenesis: a desmin null mutation blocks the myogenic pathway during embryonic stem cell differentiation. *Dev Biol*, 172, 422–439.

[88] Wendling, O., Chambon, P. and Mark, M. (1999) Retinoid X receptors are essential for early mouse development and placentogenesis. *Proc Natl Acad Sci USA*, 96, 547–551.

[89] Wendling, O., Dennefeld, C., Chambon, P. and Mark, M. (2000) Retinoid signaling is essential for patterning the endoderm of the third and fourth pharyngeal arches. *Development*, 127, 1553–1562.

[90] Wendling, O., Ghyselinck, N.B., Chambon, P. and Mark, M. (2001) Roles of retinoic acid receptors in early embryonic morphogenesis and hindbrain patterning. *Development*, 128, 2031–2038.

[91] Williams, R.L., Hilton, D.J., Pease, S., Willson, T.A., Stewart, C.L., Gearing, D.P., Wagner, E.F., Metcalf, D., Nicola, N.A. and Gough, N.M. (1988) Myeloid leukaemia inhibitory factor maintains the developmental potential of embryonic stem cells. *Nature*, 336, 684–687.

[92] Wilson, J.G., Roth, C.B. and Warkany, J. (1953) An analysis of the syndrome of malformations induced by maternal vitamin A deficiency. Effects of restoration of vitamin A at various times during gestation. *Am J Anat*, 92, 189–217.

[93] Wobus, A.M., Rohwedel, J., Maltsev, V. and Hescheler, J. (1994) In vitro differentiation of embryonic stem cells into cardiomyocytes or skeletal muscle cells is specifically modulated by retinoic acid. *Roux's Arch Dev Biol*, 204, 36–45.

[94] Wolffe, A.P. (1997) Histones, nucleosomes and the roles of chromatin structure in transcriptional control. *Biochem Soc Trans*, 25, 354–358.

[95] Zhang, X.K., Lehmann, J., Hoffmann, B., Dawson, M.I., Cameron, J., Graupner, G., Hermann, T., Tran, P. and Pfahl, M. (1992) Homodimer formation of retinoid X receptor induced by 9-cis retinoic acid. *Nature*, 358, 587–591.

Chapter 20

Generating Self-Organizing Stomach Tissue from Mouse Embryonic Stem Cells

Taka-aki K. Noguchi

Additional information is available at the end of the chapter

http://dx.doi.org/10.5772/62648

Abstract

Advances in developmental research have allowed for the differentiation of pluripotent stem cells into various somatic cells in vitro. Recently, it was revealed that the aggregation of pluripotent stem cells or their derivatives during differentiation in three-dimensional (3D) cultures with collagen gel could mimic the process of spontaneous organogenesis in vitro as mimicking proper development in vivo. These methods are thought to be useful for monitoring the progress of organ formation and disease physiology, as could be done in an experimental animal. Here, we introduce a recently established method for stomach lineage differentiation from pluripotent stem cells with Matrigel-based 3D culture leading to stomach development. This method induces embryonic stem cell (ES cell) aggregates to spontaneously self-organize into stomach tissue; therefore, it has potential for modeling stomach organogenesis and development in vitro. We further discuss the ability of these in vitro stomach tissues to serve as a new model for gastric disease.

Keywords: Stomach, Pluripotent stem cells, Differentiation, Self-organization, Disease modeling

1. Introduction

Embryonic stem cells (ES cells) are derived from the early blastocyst and are in a pluripotent state with the ability to differentiate into all three germ layers and each type of somatic cell in vitro [1, 2]. Recently, defined transcription factors were found to be able to convert adult somatic cells into an ES cell-like state, termed induced pluripotent stem cells (iPS cells), providing great opportunities for developing in vitro developmental and disease models and future regenerative medicine including cell transplantation [3, 4]. To obtain somatic cells derived from ES or iPS cells in vitro, we have established directed differentiation methods for several somatic cell types following the in vivo developmental stages step-by-step. However, little is known about the differentiation of ES and iPS cells into stomach lineages in vitro.

The stomach consists of the gastric epithelium, which forms a functional gland structure for the production of digestive enzymes and gastric acid and is surrounded by a muscle layer derived from the mesoderm. Gastrointestinal organs, including the stomach, recruit splanchnic mesoderm to obtain each organ-specific cell type in their epithelium [5]. In developmental studies in chicken and mice, the early stomach mesenchyme regulates the adjacent stomach epithelium specification and differentiation into functional stomach cells from the embryonic gut endoderm [5]. In 2005, Kim *et al.* showed that mesenchymal Barx1 regulates stomach epithelium development by inhibition of Wnt signaling [6]. Disruption of Barx1 or blocking of Wnt inhibition causes intestinalization in the stomach. This phenomenon is rescued by ectopic Barx1 in the adjacent mesenchyme and by restored Wnt inhibition [6, 7]. For this reason, we hypothesized that stomach specification from ES cells in vitro might require not only the gut epithelium, but also Barx1+ mesenchyme.

In this chapter, we describe recent advances in our knowledge of stomach lineage specification with both gut epithelium and mesenchymal Barx1 from pluripotent stem cells [8]. We found that ES cell aggregates differentiate into endodermal and mesodermal lineages in basic medium and are specified into early stomach lineages by induction of SHH and inhibition of Wnt signaling. Of note, this culture condition induced the formation of self-organizing stomach primordium-like spheroids composed of Sox2+ anterior foregut endoderm and Barx1+ stomach mesenchyme. Furthermore, these stomach primordium-like spheroids could differentiate into a more mature state in 3D culture forming a gastric gland with functional secretion abilities, such as the ability to produce gastric acid and digestive enzymes. These results indicate that stomach tissue can be generated from ES cells using specific conditions in vitro and that these self-organized stomach tissues would possibly be useful for modeling development and disease.

2. Self-organization of stomach primordium-like spheroids from ES cell aggregates

Recently, we have established a method for directed differentiation of ES cells into stomach primordium-like spheroids comprised of both gut epithelium and stomach mesenchyme (**Figure 1** [8]). Previous reports indicated that the small number of ES cell aggregates in floating culture form a definitive endoderm at day six [9], and they differentiated into an intestinal gut-like structure with gut epithelium and stromal mesenchyme in serum containing basic medium [10]. With respect to the dual generation of the endodermal and mesodermal lineages, we postulated that these ES cell aggregates differentiate into both definitive endoderm and early mesoderm in floating culture at day six, and they then have the potential to develop into several gut lineages including the stomach depending on growth factor stimulation. Therefore, a modification in culture conditions after formation of ES cell aggregates at day six could direct

http://dx.doi.org/10.5772/62648

the development of stomach lineages, including mesenchymal Barx1, in gut-like structures from the intestinal fate. To date, we have screened culture conditions with several growth factors cocktails for the induction of stomach lineages in ES cell aggregates and found that SHH and DKK1 efficiently induce mesenchymal Barx1 with epithelial Sox2, but not intestinal Cdx2. This condition also resulted in Sox2+/Barx1+ stomach primordium-like spheroids, and they spontaneously grew bigger and differentiated into anterior to posterior embryonic stomachs as the culture progressed, which can be categorized as self-formation or self-organization [11].

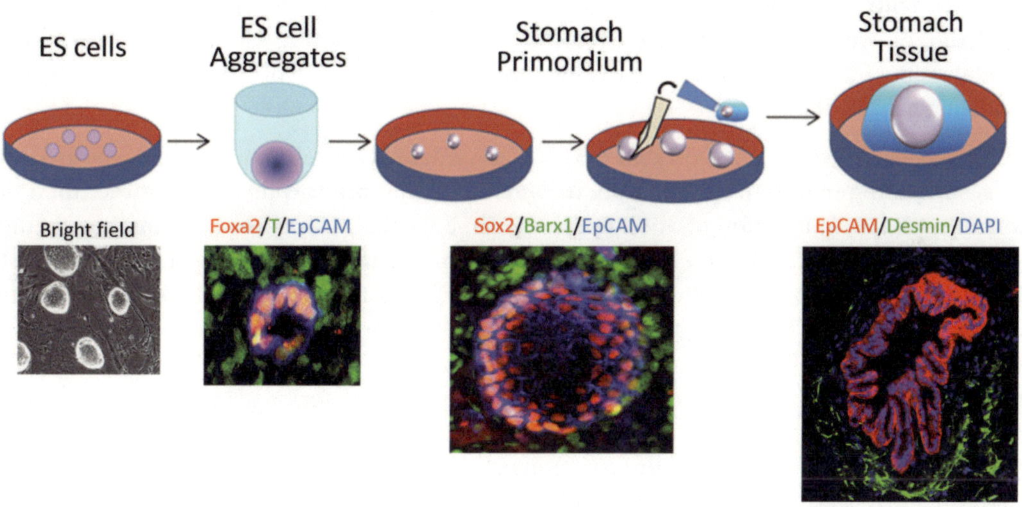

Figure 1. Self-organization of stomach tissue from mouse embryonic stem cells. Illustrated summary of differentiation and formation of stomach tissue from ES cells. Modified with permission from Noguchi *et al.* 2015 [8].

3. Differentiation of ES cell-derived stomach primordium to stomach tissue in vitro

Self-organization was observed for several differentiation methods by using aggregate cultures of ES cells or their derivatives, such as for the optic cup [12], pituitary gland [13], and kidney [14]. These methods induce aggregates of pluripotent stem cells or their derivatives to form embryonic primordium-like structures, and frequently induce them to differentiate terminally and functionally. Our stomach-like spheroids, which are close to embryonic stomach primordium, also recapitulate further in vivo stomach maturation. Developmental studies indicated that mesenchymal Barx1 expression covers the anterior to posterior regions of the embryonic stomach [15, 16], and inside the stomach the epithelium further differentiates into Sox2+ forestomach/corpus region and Pdx1+ antrum region, forming gland structures in the corpus and antrum regions. As well as showing stomach development, our ES cell-derived stomach primordium surrounded by Barx1+ mesenchyme further differentiates into the whole region of stomach and builds mature stomach tissue in a long-term 3D culture. In brief, ES cell-derived stomach primordium-like spheroids were cultured in growth factor-free medium and differentiated into a Sox2+ anterior stomach region and a Pdx1+ posterior region. Next, the spheroids were transferred to Matrigel-based 3D culture medium with mitogenic growth factors, as previously shown with gastric stem cell primary culture (Figure 1 [17]). These ES cell-derived stomach tissues, called e-ST, had similar gene expression profiles, morphology, and gastric cell zoning as a neonatal or adult stomach [8]. Of note, the ultrastructure of the gastric parietal cells and chief cells in the e-ST were very close to that of a neonatal or adult stomach, and the e-ST showed functional production of gastric acid in the presence of histamine and secretion of pepsinogen in the culture medium, which was not observed with previous stomach differentiation methods from pluripotent stem cells [18]. Accordingly, we suggest that the e-ST could be widely applicable as a new in vitro experimental model for mimicking and analyzing the in vivo physiology of stomach tissues, such as modeling gastric ulcers.

Our analysis, however, revealed that e-ST had abundant chief cells and low enteroendocrine cells, suggesting that they were mainly differentiated into the corpus region, rather than the antrum/pylorus. This differentiation tendency can be explained because our method included long-term SHH stimulation during stomach primordium formation [8]. During the early development of the stomach, the expression level of *Shh* is high in the endodermal forestomach, but low in the hindstomach [19], and *Shh* −/− embryos show a reduced size of the forestomach region [20]. Endodermal *Shh* expression is limited by FGF10, secreted from mesenchyme in the hindstomach, and maintains anterior-posterior balance during stomach development [21]. Consistent with this knowledge, our ES cell-derived stomach primordium might differentiate into mainly the forestomach region when cultured in SHH-containing medium during early-phase attachment culture, and then further differentiate into the corpus state during late-phase 3D culture. Thus, addition of recombinant FGF10 and reduction of SHH in medium at the early phase of e-ST culture could divert differentiation tendency of stomach primordium-like spheroids towards the hindstomach fate. These results indicated that our e-ST differentiation method could be useful for in vitro developmental studies, such as analyzing anterior-posterior patterning during stomach development.

4. Modeling diseases using e-ST

In terms of the possibility of modeling in vivo development using e-ST, as they self-organize into stomach tissue in vitro, their characteristics could be also explained as the recently defined “organoids” [22]. Organoids are 3D structures, in which each cell self-organizes into the proper tissue maintenance and differentiation state in vitro, similar to their in vivo tissue state, and are derived from either pluripotent stem cells, neonatal and adult stem cells, or their progenitors [22]. These organoids could recapitulate not only in vivo development and tissue homeostasis, but also disease states with gene manipulation by using gene-engineering tools including CRISPR/Cas9 [22].

As their definition and applications, we think that our e-ST can also be categorized as “organoids” and could mimic several diseases in vitro, as a Menetrier's disease model has shown [8]. Briefly, we introduced a tetracycline-inducible transforming growth factor alpha (TGF-α) overexpression cassette into mouse ES cells and then differentiated these ES cells into e-ST. This model could mimic chronic TGF-α overexpression in the stomach, and as a result successfully modeled a Menetrier's disease state with hypertrophic mucosa [8, 23]. Of note, TGF-α -induced e-ST showed reduced gastric acid, indicating that this e-ST could also model disease physiology. These finding indicate that e-ST could model in vitro gene-modified gastric diseases, and gene editing tools such as CRISPER/Cas9 may be beneficial for modeling cancers stemming from gene mutations.

5. Current limitations of e-ST

The method presented herein for the generation of stomach tissue from ES cells, however, remains challenging for developmental and clinical applications in the future. First, it is unclear how ES cell aggregates form definitive endoderm and mesoderm layers in serum-containing medium. We do not know how the growth factors in serum induce ES cell aggregates to differentiate into endoderm and mesoderm fates, and how these factors might affect applications such as in vitro analysis of stomach development in e-ST. Furthermore, serum is not categorized as a good manufacturing practices (GMP)-complaint material, and thus, it is not suitable for clinical applications. To be fully acceptable for developmental models, we have to generate chemically defined conditions for the differentiation of ES cells aggregates into e-ST. In addition, defined conditions could also help us to establish more efficient ways to generate e-ST.

Second, it is not clear whether e-ST can actually be established from human ES cells to model human development and disease. Currently, aggregates of human ES cells or their derivatives could recapitulate self-organizing human optic cup [24], and human pituitary gland [25] in vitro, and the methods are deeply related to mouse ES cell differentiation methods. With respect to those reports, we think it could be possible to recapitulate self-organization during human stomach development from human ES cells, as well as mouse ES cells, with some modifications. By establishing human e-ST from human ES cells, functional features of human stomachs could be also modeled as recently published for human pituitary tissue [25]. To achieve the differentiation of human stomach tissue from human ES cells, as well as e-ST from mouse ES cells, we have to consider the differences between "naïve" mouse ES cells and "primed" human ES cells [26]. Recently, naïve-like human ES cells were established by several groups [27–29]. Naive-like human ES cells possess similar characters to naïve mouse ES cells under specific conditions; therefore, conversion of human ES cells to a naïve-like state might have a potential to mimic the self-organization of the human stomach as naïve mouse ES cells did.

Acknowledgements

This work was supported in part by a Grant-in-Aid for Scientific Research from the Japan Society for the Promotion of Science (JSPS fellow: T. N.).

Author details

Taka-aki K. Noguchi*

Address all correspondence to: taka.noguchi@aist.go.jp

Graduate School of Life and Environmental Sciences, the University of Tsukuba, Tsukuba, Japan

References

[1] Evans MJ, Kaufman MH. Establishment in culture of pluripotential cells from mouse embryos. Nature. 1981;292(5819):154–6.

[2] Martin GR. Isolation of a pluripotent cell line from early mouse embryos cultured in medium conditioned by teratocarcinoma stem cells. Proc Natl Acad Sci U S A. 1981;78(12):7634–8.

[3] Takahashi K, Yamanaka S. Induction of pluripotent stem cells from mouse embryonic and adult fibroblast cultures by defined factors. Cell. 2006;126(4):663–76. 10.1016/j.cell.2006.07.024

[4] Takahashi K, Tanabe K, Ohnuki M, Narita M, Ichisaka T, Tomoda K, et al. Induction of pluripotent stem cells from adult human fibroblasts by defined factors. Cell. 2007;131(5):861–72. doi: 10.1016/j.cell.2007.11.019

[5] Fukuda K, Yasugi S. The molecular mechanisms of stomach development in vertebrates. Dev Growth Differ. 2005;47(6):375–82. doi: 10.1111/j.1440-169X.2005.00816.x

[6] Kim BM, Buchner G, Miletich I, Sharpe PT, Shivdasani RA. The stomach mesenchymal transcription factor Barx1 specifies gastric epithelial identity through inhibition of transient Wnt signaling. Dev Cell. 2005;8(4):611–22. doi: 10.1016/j.devcel.2005.01.015

[7] Zorn AM, Wells JM. Vertebrate endoderm development and organ formation. Annu Rev Cell Dev Biol. 2009;25:221–51. doi: 10.1146/annurev.cellbio.042308.113344

[8] Noguchi TK, Ninomiya N, Sekine M, Komazaki S, Wang PC, Asashima M, et al. Generation of stomach tissue from mouse embryonic stem cells. Nat Cell Biol. 2015;17(8):984–93. doi: 10.1038/ncb3200

[9] Matsuura R, Kogo H, Ogaeri T, Miwa T, Kuwahara M, Kanai Y, et al. Crucial transcription factors in endoderm and embryonic gut development are expressed in gut-like structures from mouse ES cells. Stem Cells. 2006;24(3):624–30. doi: 10.1634/stemcells.2005-0344

[10] Torihashi S, Kuwahara M, Ogaeri T, Zhu P, Kurahashi M, Fujimoto T. Gut-like structures from mouse embryonic stem cells as an in vitro model for gut organogene-

sis preserving developmental potential after transplantation. Stem Cells. 2006;24(12): 2618–26. doi: 10.1634/stemcells.2006-0148

[11] Sasai Y, Eiraku M, Suga H. In vitro organogenesis in three dimensions: self-organising stem cells. Development. 2012;139(22):4111–21. doi: 10.1242/dev.079590

[12] Eiraku M, Takata N, Ishibashi H, Kawada M, Sakakura E, Okuda S, et al. Self-organizing optic-cup morphogenesis in three-dimensional culture. Nature. 2011;472(7341): 51–6. doi: 10.1038/nature09941

[13] Suga H, Kadoshima T, Minaguchi M, Ohgushi M, Soen M, Nakano T, et al. Self-formation of functional adenohypophysis in three-dimensional culture. Nature. 2011;480(7375):57–62. doi: 10.1038/nature10637

[14] Takasato M, Er PX, Becroft M, Vanslambrouck JM, Stanley EG, Elefanty AG, et al. Directing human embryonic stem cell differentiation towards a renal lineage generates a self-organizing kidney. Nat Cell Biol. 2014;16(1):118–26. doi: 10.1038/ncb2894

[15] Verzi MP, Stanfel MN, Moses KA, Kim BM, Zhang Y, Schwartz RJ, et al. Role of the homeodomain transcription factor Bapx1 in mouse distal stomach development. Gastroenterology. 2009;136(5):1701–10. doi: 10.1053/j.gastro.2009.01.009

[16] Noguchi TK, Ishimine H, Nakajima Y, Watanabe-Susaki K, Shigeta N, Yamakawa N, et al. Novel cell surface genes expressed in the stomach primordium during gastrointestinal morphogenesis of mouse embryos. Gene Expr Patterns. 2012;12(3-4):154–63. doi: 10.1016/j.gep.2012.01.001

[17] Barker N, Huch M, Kujala P, van de Wetering M, Snippert HJ, van Es JH, et al. Lgr5(+ve) stem cells drive self-renewal in the stomach and build long-lived gastric units in vitro. Cell Stem Cell. 2010;6(1):25–36. doi: 10.1016/j.stem.2009.11.013

[18] Mc Cracken KW, Cata EM, Crawford CM, Sinagoga KL, Schumacher M, Rockich BE, et al. Modelling human development and disease in pluripotent stem-cell-derived gastric organoids. Nature. 2014;516(7531):400–4. doi: 10.1038/nature13863

[19] Bitgood MJ, Mc Mahon AP. Hedgehog and Bmp genes are coexpressed at many diverse sites of cell-cell interaction in the mouse embryo. Dev Biol. 1995;172(1):126–38. doi: 10.1006/dbio.1995.0010

[20] Ramalho-Santos M, Melton DA, Mc Mahon AP. Hedgehog signals regulate multiple aspects of gastrointestinal development. Development. 2000;127(12):2763–72.

[21] Spencer-Dene B, Sala FG, Bellusci S, Gschmeissner S, Stamp G, Dickson C. Stomach development is dependent on fibroblast growth factor 10/fibroblast growth factor receptor 2b-mediated signaling. Gastroenterology. 2006;130(4):1233–44. doi: 10.1053/j.gastro.2006.02.018

[22] Huch M, Koo BK. Modeling mouse and human development using organoid cultures. Development. 2015;142(18):3113–25. doi: 10.1242/dev.118570

[23] Takagi H, Jhappan C, Sharp R, Merlino G. Hypertrophic gastropathy resembling Menetrier's disease in transgenic mice overexpressing transforming growth factor alpha in the stomach. J Clin Invest. 1992;90(3):1161–7. doi: 10.1172/JCI115936

[24] Nakano T, Ando S, Takata N, Kawada M, Muguruma K, Sekiguchi K, et al. Self-formation of optic cups and storable stratified neural retina from human ESCs. Cell Stem Cell. 2012;10(6):771–85. doi: 10.1016/j.stem.2012.05.009

[25] Ozone C, Suga H, Eiraku M, Kadoshima T, Yonemura S, Takata N, et al. Functional anterior pituitary generated in self-organizing culture of human embryonic stem cells. Nat Commun. 2016;7:10351. doi: 10.1038/ncomms10351

[26] Nichols J, Smith A. Naive and primed pluripotent states. Cell Stem Cell. 2009;4(6):487–92. doi: 10.1016/j.stem.2009.05.015

[27] Gafni O, Weinberger L, Mansour AA, Manor YS, Chomsky E, Ben-Yosef D, et al. Derivation of novel human ground state naive pluripotent stem cells. Nature. 2013;504(7479):282–6. doi: 10.1038/nature12745

[28] Theunissen TW, Powell BE, Wang H, Mitalipova M, Faddah DA, Reddy J, et al. Systematic identification of culture conditions for induction and maintenance of naive human pluripotency. Cell Stem Cell. 2014;15(4):471–87. doi: 10.1016/j.stem.2014.07.002

[29] Takashima Y, Guo G, Loos R, Nichols J, Ficz G, Krueger F, et al. Resetting transcription factor control circuitry toward ground-state pluripotency in human. Cell. 2014;158(6):1254–69. doi: 10.1016/j.cell.2014.08.029

Chapter 21

Human Embryonic Stem Cells and Induced Pluripotent Stem Cells: The Promising Tools for Insulin-Producing Cell Generation

Wilairat Leeanansaksiri,
Piyaporn Rattananinsruang and
Chavaboon Dechsukhum

Additional information is available at the end of the chapter

http://dx.doi.org/10.5772/62850

Abstract

Diabetes mellitus, a disease with abnormally high level of blood glucose, can cause a wide range of chronic complications that affect almost every parts of the body. The major goal of diabetes treatments is to control elevated blood glucose without causing abnormally low levels of blood glucose. Despite islet transplantation provided endogenous insulin secretion in individuals with diabetes, the scarcity of cadaveric donors for pancreatic β-cell still remains a major obstacle. In this regard, the needs for an unlimited supply for cell replacement strategy have led to explore the way of generating insulin-producing cells to use in the disease treatment. Human embryonic stem cells (hESCs) offer a source to produce the desired kind of cell. Currently, several researchers achieved insulin-producing cells from hESCs using a multistep differentiation protocols, growth factors, and/or chemical compounds. In this review, we summarized the hESCs derivation, culture methods, and characteristics of hESCs. We also emphasized on the current methods for direct differentiation of hESCs into embryoid bodies (EBs) and toward insulin-producing cells, characterization of these insulin-producing cells, and the limitation of hESCs. Since the discovery of induced pluripotent stem cells (iPSCs), which have similar properties to hESCs but less ethical issues than hESCs, can be created directly from somatic cells that hold great promise as the therapeutic source for developing cell-based therapy. Herein, the methods to produce iPSC-derived insulin-producing cells are also discussed. Moreover, the encapsulation technology which is a powerful tool for accelerate hESCs and iPSCs applications in medicine which provide a new avenue for diabetes treatment in the future is also included in this review. Understanding the basic knowledge of hESCs and iPSCs, their differentiation capability toward insulin-producing cells will stimulate more therapeutic value of hESCs and iPSCs for diabetic treatments, drug screening, and regenerative medicine.

Keywords: diabetes mellitus, human embryonic stem cells, insulin-producing cells, insulin, encapsulation, induced pluripotent stem cells

1. Introduction

Diabetes mellitus is one of the most common chronic diseases that threaten the health and health economics worldwide [1–3]. The disease and cases management interventions improve short- and long-term health, and/or economic outcomes of patients are improving health and quality of their lives. General treatments include weight reduction, a diabetes diet, and exercise which are used to control diabetes [4, 5]. When these treatments fail to control the elevated blood glucose, oral medications and insulin injection therapy will be applied in the treatment strategy, respectively. Insulin therapy has the potential to improve symptoms, enhancing quality of life and provide a sense of well-being [6–8]. However, the intensive insulin treatment can cause hypoglycemia [9–11]. Therefore, transplantation of high effective insulin-producing cells containing physiological regulation of blood glucose level is critical choice of treatment. At present, there are three major cell-based therapy approaches to restore insulin-producing cells in diabetes patients: 1) direct transplantation of donor islet and pancreatic cells to patient 2) activation pancreatic progenitors residing in islet mass into insulin-producing cells (neogenesis) and transplantation, and 3) stem cell approach by differentiation pluripotent stem cells into insulin-secreted cells and then transplantation. The direct transplantation of islet cells and neogenesis methods have demonstrated normoglycemia in the absence of exogenous insulin therapy [12–14]. Nevertheless, the limitations of islet cell replacement are the following factors: non-functioning of isolated islets, the small number of transplanted islets, the immunogenicity of isolated islets lead to immune rejection, transplantation to inappropriate sites, recurrence of auto-immunity in the transplanted islets, and immunosuppression [15–18]. The third method, stem cell approach: Pluripotent stem cells including hESCs and iPSCs are the main targets for insulin-producing cell induction. These two sources of stem cells could provide limitless sources of cells for pancreatic β-cell replacement strategy. Therefore, these research areas have led researchers to explore the way of generating effective insulin-producing cells for diabetes treatment. In addition, the hESCs and iPSCs contain the higher differentiation capacity than adult stem cells [19–21].

Herein, we review the current knowledge of hESCs and iPSCs, followed by the directed differentiation of these cells toward the functional insulin-producing cells. In addition, the encapsulation technology, a powerful tool for accelerate hESCs and iPSCs applications in medicine, is also included in this review. We also summarize and discuss evidence that both hESCs and iPSCs are promising cell sources for future diabetes treatment.

http://dx.doi.org/10.5772/62850

2. Human embryonic stem cells

2.1. Derivation, culture, and characterization of hESCs

In the early era of hESCs research, the hESCs are isolated from the inner cell mass (ICM) of blastocyst-stage embryos by immunosurgery or mechanically methods. However, this process involves with animal-derived substances such as mouse antibodies and guinea pig complement [22, 23]. Alternatively, the use of Tyrode's acid for the removal of zona pellucida and mechanical isolation of ICM can serve as a potentially useful method for the establishment of hESCs line in the present time. Indeed, this technique also implies that the blastocyst could not contact with animal-derived pronase, antibodies, and complement factors [22]. The hESCs lines can be maintained in an undifferentiated or pluripotency state *in vitro* for prolonged periods of time. The potential of hESCs to differentiate into representing ectoderm, mesoderm, and endoderm derivatives has generated the possible use of hESCs in therapeutic applications [24]. The derivation process involves culturing of the ICM of blastocyst stage, induce to proliferate and differentiate into desired cell types [25]. The first successful derivation of hESCs was isolated from the ICM of human blastocyst and placed on mitotically inactivated murine feeder cells [26].

There are several methods have been reported for culture of undifferentiated hESCs *in vitro* such as culture the undifferentiated hESCs on feeder layers, for example, mouse embryonic fibroblast (MEF) or laminin- or Matrigel-coated-plastic surface with MEF conditioned medium. However, these methods possibly transfer harmful animal pathogens to human transplant recipients in clinical application [27]. Therefore, human feeder layers are used for hESCs culture instead, for example, human adult marrow cells, human fetal muscle (FM), human adult skin (AS), commercial human fetal skin (FS; D551/CCL-10, American Type Culture Collection [ATCC]), human adult uterine endometrial cells (hUECs), human adult breast parenchymal cells (hBPCs), and embryonic fibroblasts (hEFs). They are capable to support undifferentiated stage and proliferation state of hESCs [28–31]. Conditioned medium from hESCs-derived fibroblasts (hESC-dFs) also efficiently supports growth of hESCs in feeder-free culture systems [32]. Moreover, a three-dimensional (3D) porous natural polymer scaffolds (chitosan and alginate) effectively support self-renewal of hESCs without the need of feeder cells or conditioned medium [33]. Recent study has demonstrated that a defined engineered 3D microfiber system allows adequate propagation and cryopreservation of hESCs under feeder-free chemically defined conditions [34]. However, these culture conditions still have the ingredients from animal such as fetal bovine serum (FBS) and bovine albumin that contain in culture medium. For the clinical potential in cell replacement therapy, differentiated cells from hESCs should be cultured in xeno-free systems [27, 31]. Interestingly, Chen et al. reported that the suspension culture system under defined and serum-free conditions provides a powerful approach for scale-up expansion of hESCs. It was demonstrated that cell banks of several hESCs lines are generated from this system under current good manufacturing procedures (cGMP) or cGMP-equivalent conditions [35].

Characterization of hESCs lines can be achieved by both cellular and molecular analysis. Cellular characterizations can be determined by (1) morphologies of hESCs colonies: form flat and compact colonies with distinct cell borders, (2) morphologies of hESCs have a high ratio of nucleus to cytoplasm and have prominent nucleoli, (3) the hESCs exhibit high levels of telomerase activity and show normal karyotype, (4) most of the cells can be subcultured after freezing, thawing, and replating, (5) the cells can be differentiated into a variety of cell types both *in vitro* and *in vivo* conditions, and (6) the cells can generate teratoma in animal model or *in vivo*. The general molecular properties of hESCs lines can be examined by the expression of several transcripts, for example, the stage-specific antigens (SSEA-3 and SSEA-4), the glycoproteins tumor recognition antigen (TRA-1-60, TRA-1-81 and TRA-2-54), germ cell tumor marker (GCTM-2), trophoblast giant (TG343 and 30), cluster of differentiation (CD9 and 133), Octamer-4 (Oct4), Nanog, SRY-box-containing gene 2 (*Sox2*), teratocarcinoma-derived growth factor 1 (Tdgf1), left–right determination factor 2 (LeftyA), RNA exonuclease 1 (Rex-1), Stellar, Dazl, Nanos 1, pumilio gene (*Pum 1* and 2), growth differentiation factor-3 (Gdf3), thymus cell antigen 1 (Thy-1), and alkaline phosphatase [25, 36]. Other markers that are common to characterize hESCs are following: SSEA-3, SSEA-4, TRA-1-60, TRA-1-81, Oct4, and alkaline phosphatase. Nevertheless, there are differences between hESCs in their pluripotency or the genetic profile under the same conditions, their potential for large-scale culture and growth under feeder-free protocols, or their ability to form teratoma after injection into severe combined immunodeficiency (SCID) mice. Moreover, their capacity to differentiate into different cell types under *in vitro* conditions is variable [25, 36]. It is important to note that the difference in various hESCs lines is useful for the scientists to choose the appropriate hESCs line for their research.

2.2. Differentiation of hESCs into embryoid bodies (EBs) and definitive endoderm (DE)

To differentiate hESCs into particular cell lineages *in vitro*, the formation of embryoid bodies (EBs) represents an important step to generate three germ layers; ectoderm, mesoderm, and endoderm. The EBs will be then derive into tissue-specific progenitor cells and toward the desire final differentiated lineages. It should be noted that hESCs are able to differentiate through EBs parallels embryonic development due to the EBs recapitulates events during embryogenesis [37]. The use of EBs to produce a variety of desired cell types represents an exciting approach for therapeutic applications.

For the production of EBs, several methods have been designed. The first method: Methylcellulose (MC) method developed to form EB from single embryonic stem cell but has limitation to use EBs for medical application due to the contamination of methylcellulose. The second method, hanging drop (HD) method has been widely used to generate EBs, and further differentiate into a variety of cell types can be applied into other xenofree or chemically defined medium culture protocols that suitable for human therapeutic applications. However, it is a troublesome multiple steps methods. Moreover, it is hardly exchange that the medium for a drop and the observations of forming EBs in drops by direct microscopic is difficult [38]. Therefore, the third method has been developed called suspension culture method—the obtained EBs tends to be more heterogeneous in size and shape because their self-organized aggregation in culture. The heterogeneity of EBs structures may influence cell fate differentiation [38, 39].

In fact, improvement of the definitive endoderm (DE) layer development will increase the successful rate of insulin-producing cell generation from pluripotent stem cells. Therefore, there are several factors have been used to activation of DE formation, for example, activin A, GDF8, Wnt3a, bone morphogenetic proteins (BMPs), fibroblast growth factors (FGFs), inhibition of PI3K/Akt, and chemical inhibition of GSK3β [40]. In addition, collagen type 1 could improve the differentiation of hESCs into definitive endoderm [41].

2.3. Differentiation of hESCs into insulin-producing cells

At present time, cell-based therapy for diabetes is mostly target to type I diabetes. It has been described that islet-like clusters derive hESCs differentiation can be achieved in multistep procedures (nestin-positive protocol): (i) formation of EBs (stage 1), (ii) selective differentiation of cell populations expressing nestin using fetal calf serum depletion and culture with ITSF (stage 2), (iii) proliferation and maintenance of precursor cells (stage 3), and (iv) the differentiation induction and maintenance of insulin-positive cells (stage 4). The resultant cells are positive for dithizone (DTZ), a zinc-chelating agent known to selectively stain pancreatic β-cells, and are immunoreactive for antibodies against insulin, glucagons, and C-peptide. Insulin and other pancreatic β-cell-related genes such as glucagon, somatostatin, KIR6.2 and SUR1, IAPP, Isl1, PC1/3, PC2, GK, Nkx6.1, GLUT2, and Pax4 are expressed in the differentiating cells. The results indicated that differentiated cells can express genes involved in the β-cell differentiation pathway [42].

Moreover, insulin-producing islet-like clusters (ILCs) are generated from hESCs according to the method (definitive endoderm-based protocols) developed by Jiang and colleagues [43] who culture hESC lines under feeder-free conditions and direct differentiation toward ILCs by using a multistep, serum-free protocol. The 36-day differentiation protocol consists of four stages which included definitive endoderm induction (stage 1), pancreatic endoderm formation (stage 2), endocrine induction (stage 3), and islet-like clusters maturation (stage 4). The hESCs generate definitive endoderm coexpressing CXCR4 and Sox17, and CXCR4 and Foxa2 when treated with sodium butyrate and activin A. The Pdx1-expressing pancreatic endoderm is then induced by the addition of bFGF, EGF, noggin, and B27 supplement. Following withdrawal of bFGF, these cells are allowed to develop pancreatic endocrine cells. Gene expression analysis shows that pancreatic endoderm cells also start to express other pancreas-related genes such as HlxB9, Ptf1a, Ngn3, and Nkx6.1. Upon further differentiation of Pdx1-positive cell clusters to day 36, immunocytochemical-staining data demonstrated that the C-peptide-, glucagon-, and somatostatin-positive cells were predominantly localized in the small bud-like clusters as well as in some of the smaller ILCs. In addition, the ILCs generated by this protocol are able to secrete C-peptide in response to 20 mM glucose [43].

Based on the differentiation protocol, the nestin-positive progenitor-based and definitive endoderm-based protocols are successful in generating insulin-producing cells from hESCs. However, there is still some debate on the therapeutic potential between the cells obtained from these two protocols. Therefore, other approaches are developed. Bruin and colleagues implemented an approach in which hESCs were differentiated into fetal-like pancreatic cells *in vitro*. Taking an approach, this group set out to replace activinA/wnt3A with GDF8/GSK3β inhibitor to enhance efficiency of definitive endoderm production. This study achieved to generate insulin-producing cells and also revealed several key features of polyhormonal insulin-positive cells that differ from mature pancreatic β-cells, including defects in glucose transporter expression, KATP channel function, and prohormone-processing enzymes [44].

The strategies for differentiation of hESCs into insulin-producing cells also have been demonstrated by many research groups. They demonstrated that the insulin-producing cells are expressed the markers associated with pancreatic β-cell differentiation pathway, able to produce and secrete insulin in response to glucose concentration [42, 45–49]. The spontaneous differentiation of insulin-producing cells can also be observed with undifferentiated hESCs colonies when hESCs are propagated on a feeder layer of MEFs [45]. It has been demonstrated that *in vitro* differentiation in suspension culture results in the formation of discrete embryoid bodies (EBs) and a more consistent pattern of nestin-positive progenitors which possibly are insulin-producing progenitors [49] (**Figure 1**).

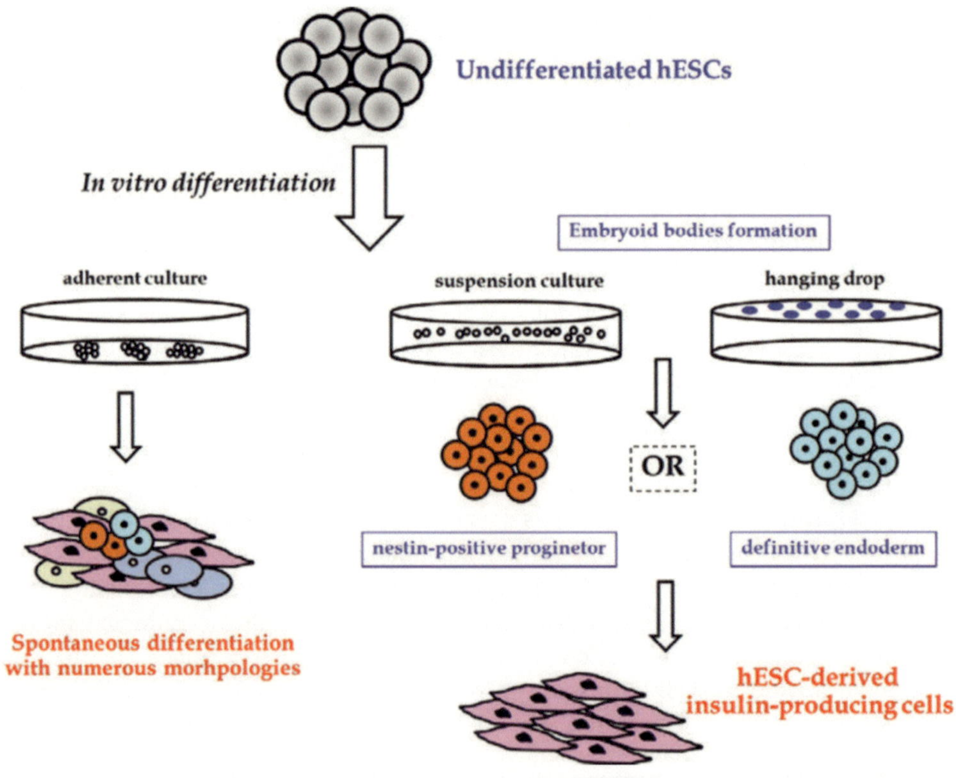

Figure 1. Summary of *in vitro* differentiation protocol for derivation of insulin-producing cells from hESCs. The directions of hESCs into insulin-producing cells are based on the variety of experimental approaches. During *in vitro* differentiation hESCs, the procedures were performed through a multistep including adherent culture, suspension culture, and hanging drop for generating the precursor cells. Cells grown under adherent conditions displayed a pleiotropic pattern with numerous morphologies. In addition, the differentiation of hESCs into insulin-producing cells was successfully induced by the nestin-positive progenitor-based and definitive endoderm-based protocols [43–46, 49].

Recently, it has been shown that maturation of stem cell into beta-cells can be driven by the expression of peptide hormone called urocortin 3 which also served as islet cell maturation marker [50]. In addition, the scalable differentiation protocol also able to generate several millions of glucose-responsive β-cells from hESCs [51]. Interestingly, it has been revealed that acceleration of Cdk4-cyclin D complex in the cell cycle of hESCs also play a potential key role in cell fate decision of hESCs differentiation into insulin-producing cell [52].

Although several authors achieved to demonstrate the rapid progress for possible treatment of diabetes using insulin-producing cells generated from hESCs, there are some crucial aspects need to be concerned. Based on this regard, the culture should be performed in xeno-free system in order to further develop potential medical applications. Moreover, the contamination of undifferentiated hESCs in the culture of insulin-producing cell induction needs to be removed before it can be applied for therapeutic purposes to eliminate the teratoma formation. The immune rejection aspect is also need to be assessed. For examples, during differentiation stage, the major histocompatibility complex (MHC) is up-regulated, leading to the non-self-proteins expression on the graft cells which may result in immune rejection of the graft in the absence of immunosuppressive therapy [25]. In addition, the expression of high level of telomerase activity can lead to teratoma formation after injection which should be noted [53]. Furthermore, prolonged growth of hESCs and differentiation of these cells into any stages of beta-cells development *in vitro* may also cause chromosomal aberrations. Therefore, complete characterization of hESC lines, insulin-producing cells, or other pancreatic cell stages that will be applied for therapeutic purposes need to be well analysis, for example, their molecular status, a continuous genetic, and chromosomal features.

2.4. Transplantation of hESC-derived insulin-producing cells

In order to gain insights into the human condition, the function of hESC-derived insulin-producing cells should be achieved *in vivo*. Also, many research groups have tried to transplant insulin-producing cells derived from hESCs into animal models of diabetes mellitus. For instance, the transplantation of hESC-derived ILC under the kidney capsule of streptozotocin (STZ)-induced diabetic immuno-incompetent mice exerted functional benefits. The grafted cells continued to contain cells that were shown the ability to secrete human C-peptide in response to an oral bolus of glucose. In addition, the transplanted ILCs could promote the mean survival of recipients as compared to controls, which were transplanted with human fibroblast cells [54]. Hua and colleagues reported the protocol for differentiating hESCs into pancreatic insulin-producing cells and transplanted the cells into severe combined immunodeficient (SCID)/non-obese diabetic (NOD) mice to assess graft survival and function. The terminally differentiated cells were glucose-responsive and expressed C-peptide, similar to pancreatic islets. When transplanted into the epididymal fat pad of SCID/NOD mice, these cells were capable of correcting hyperglycemia for ≥ 8 weeks. Notably, none of the treated animals developed tumors [55]. Alternatively, the function of pancreatic endoderm or pancreatic progenitors derived from hESCs has been observed following transplantation into diabetic models. Recent study has demonstrated that the engrafted hESC-derived pancreatic endoderm generated functional endocrine cells when implanted in the epididymal fat pad of

SCID/Beige mice. The grafted cells exhibited appropriate expression of pancreatic transcription factors, expressed prohormone-processing enzymes, and contained mature, homogeneous endocrine secretory granules. Moreover, the engrafted hESC-derived endocrine cells regulated glucose homeostasis in the host by synthesized and released insulin in response to glucose levels [56]. Rezania et al. provided evidence that hESC-derived pancreatic progenitor cells successfully matured into functional islets *in vivo* and controlled glycemia of STZ-treated immunodeficient mice. The study demonstrated that the differentiation protocol has generated a highly enriched Pdx1+ pancreatic progenitor cell population *in vitro* without cell sorting. These progenitor cells were remarkably developed similar to human fetal pancreas development and resulted in the formation of insulin-producing cells that closely resembled matured human β-cells. The implanted cells contributed to protect mice against STZ-induced diabetes with robust glucose-stimulated human C-peptide secretion *in vivo* [57]. Currently, cellular encapsulation has been provided immunoprotection in host with the potential to reduce or eliminate the need for chronic immunosuppression. In a recent study, Kirk and colleagues used a bilaminar device (Theracyte) to investigate the kinetics of cellular engraftment and the maturation dynamics of hESC-derived pancreatic epithelium (PE). They found that *in vitro* derived hESC-PE cells generated through glucose-responsive insulin-producing cells inside an encapsulation device with no increase in cell mass and without cell escape. The encapsulated hESC-derived PE exhibited full physiological function *in vivo* and resulted in amelioration of alloxan-induced diabetes following implantation [58]. Despite promising findings in a test of therapeutic potential, both safety and efficacy of the hESC-based therapy for insulin-dependent diabetes must be further investigated.

3. Induced pluripotent stem cells

3.1. The generation of iPSCs

The induced pluripotent stem cells (iPSCs) also provided an alternative approach to produce autologous cell-based therapy (**Figure 2**). The iPSCs have been shown the properties similar to hESCs including morphology, self-renewal capacity, gene expression profiles, and retained a normal karyotype. The differentiation potential of these cells would allow researchers to study disease mechanisms, drug screening and provide another autologous cell sources for transplantation [59–61].

Based on the knowledge of transcriptional regulators that maintain the stem cell state, researchers have been developed a technique that can be reprogrammed adult cells into pluripotent stage. In 2006, Takahashi and Yamanaka discovered that the introduction of four transcription factors (Oct4, Sox2, c-Myc, and Klf4) could induce pluripotency in mouse embryonic or adult fibroblasts. These iPSCs exhibited the morphology and growth properties of ESCs and expressed ESC marker genes. The resultant cells resulted in tumors containing a variety of tissues from all three germ layers when transplantation into nude mice. Consequently, Yu and colleagues designed a reprogramming strategy in which a set of four factors (Oct4, Sox2, Nanog, and Lin28) were sufficient to induce pluripotency in human somatic cells. These cells displayed normal karyotypes, expressed telomerase activity, expressed the hESC surface markers/genes, and maintained the developmental potential to differentiate into advanced derivatives of all three primary germ layers [62]. Additionally, it has been found that the miR-302 family is specifically expressed in undifferentiated ESCs. The miR-302 has a role in regulating ESC pluripotency and differentiation [63, 64]. Also, several studies have demonstrated that miR-302 can directly reprogram somatic cells. Lin et al. reported that miR-302 inhibits stem cell tumorigenicity by enhancing G1 phase arrests pathway [65]. This evidence suggests that the miR-302-reprogrammed iPSCs may provide the potential applications of iPSC technology.

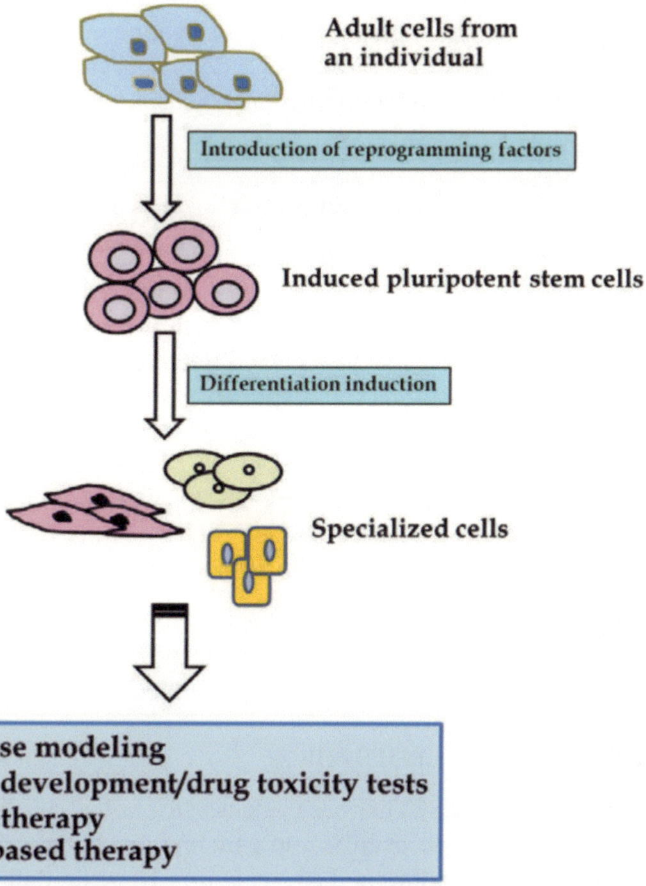

Figure 2. Induced pluripotent stem cells (iPSCs) and the potential use in regenerative medicine. Adult cells or somatic cells can be reprogrammed into pluripotent stem cells by introducing a specific set of reprogramming factors. These iPSCs can then be differentiated to specialized cell types, which can be used as a clinical tool for disease modeling, drug development/drug toxicity tests, gene therapy, and cell-based therapy.

The approach to reprogram somatic cells involved the use of genome-integrating retroviruses to transfer transcription factors. However, the usage of integrating retroviral vector has major limitation due to the potential risk relating to tumor formation. Thus, researchers have also used the different strategies with non-integrating system that provided the solution to make it clinically applicable [66]. Although the reprogramming efficiencies with non-integrating method are lower (~0.001%) than those achieved with integrating method (0.1–1%), this method increases the safety of generating and using iPSCs [67]. Alternative methods have been developed to avoid genetic modification as the reprogramming proteins, or mRNA was delivered directly into the cells. This has been successfully demonstrated that high reprogramming efficiencies, however, can be more complicated to perform [66].

Interestingly, iPSCs technology has been possible to create disease-specific cells from individual patients that could be analyzed the disease pathology, provided treatment methods and drug development [65]. In addition, it has been documented that iPSCs can be generated into the desired cell types, which hold great promise for treating many diseases. There have been reported in iPSC-derived cells in clinical applications in several types of disease including neurological, immune system, endocrinology/metabolism, muscle skeletal, genetic, hematological, and inherited liver disease [68].

3.2. Differentiation of iPSCs into insulin-producing cells and transplantation

The achievement of hESCs differentiation into insulin-producing cells has raised up the possibility of generation of iPSCs-derived insulin-secreted cells by adopted the same protocols that have been used in hESCs works into iPSCs studies with or without modification the protocols. Recently, researchers have also been differentiated iPSCs into functional insulin-producing cells *in vitro*. The resulting cells expressed pancreatic lineage-related genes, which further normalized blood glucose levels and restored insulin secretion when transplantation into diabetic models. As such, the potential of iPSC-derived insulin-producing cells is also considered as useful tool for diabetes therapy [69–72]. Currently, it has been demonstrated that iPSCs differentiated into posterior foregut and endocrine cells can be transplant and secrete insulin in immunodeficient mouse model [73].

These findings suggested that iPSC-derived insulin-producing cells will provide a potential therapeutic source for treating diabetes. It should be noted that iPSC technology is a possible alternative to induce patient-derived iPSCs for autologous cell transplantation therapy.

3.3. The limitations of hESCs and iPS cells

The potential application of hESCs has been extensively studied due to their high differentiation capacity. However, the use of hESCs to generate functional cells raises the problems associated with technical limitations and ethical issues. These limitations including

(1) the ethical controversies associated with the use of fresh human embryos, (2) developing xenogeneic products-free culture systems both for culturing hESCs and insulin-producing cell differentiation, (3) teratoma formation has become a critical obstacle for the therapeutic applications of hESCs, and (4) destruction of transplanted cells *in vivo*.

The ethical controversies represent another issue associated the use of fresh human embryos. There is concern about guidelines on the use of fresh embryos as a source of new hESC lines.

http://dx.doi.org/10.5772/62850

In this regard, induced pluripotent stem cells (iPSCs) technology provides a solution to the ethical debate surrounding hESCs since it does not require both the destruction of an embryo and the use of human oocytes [74].

It is generally believed that highly purified progenitors or terminally differentiated cell types derived from hESCs results in prevention of teratoma formation [75]. While subpopulations of hESCs have been characterized by the expression of distinct surface markers, their fates have provided a valuable tool for generating tissue-specific reagents for cell-based therapy [76]. Moreover, the combined gene transfer/hESCs therapies can generate a pure population of genetically modified differentiated cells with the selection using lineage-specific markers [77]. Furthermore, the encapsulation procedure has the potential to prevent the formation of tumors [78].

The possible destruction of transplanted hESCs derivatives by the patient's immune system should also be considered before the transplantation of these cells. Some solutions to prevent the rejection of hESC-derived cells are the use of hematopoietic chimerism for tolerance induction and the creation of universal donor cell line [79, 80] and encapsulation technology [81].

The great promise of iPSCs is based on their properties of self-renewal capacity and differentiation into specialized cell types. In addition, the possibility to obtain patient-specific or disease-specific pluripotent stem cells is a promising approach for medical applications. Moreover, iPSCs are not associated with ethical issues as hESCs regarding the use of human oocytes or embryos in research. However, there are still some limitations including inadequate cell number, immune rejection, and teratoma formation upon transplantation.

3.4. Encapsulation technology: a power tool for hESCs and iPS cells applications

In the context of therapeutic applications, the encapsulation technology represents a powerful tool toward the implementation of hESCs and iPSCs in clinical and industrial applications. Generally, there are two types of encapsulation technologies available at the present time including microencapsulation and macroencapsulation. Microencapsulation technology aims to generate a small size semipermeable bag which popularly made of hydrogel polymers to cover small group of cells. On the other hand, macroencapsulation technology has been used to generate semipermeable membrane and hydrogel sheets to hold large scale cell quantity. In the aspect of oxygen diffusion into the encapsulated environment, the microencapsulated sphere allows higher oxygen diffusion capacity than macroencapsulated sphere.

In stem cell research, the encapsulated hESC results in high expansion ratio and high cell recovery yields after cryopreservation. This method also improves the culture of hESC aggregates by protecting cells from hydrodynamic shear stress, controlling aggregate size, and maintaining cell pluripotency [82]. It has been demonstrated that hESCs encapsulated in alginate hydrogels maintain the undifferentiated state and retain their pluripotent capabilities without any enzymatic treatment, mechanical expansion, or manipulation in a feeder-free environment. This approach is well-suited for providing automated culture scale-up process and the opportunity of long-term culture, feeder-free, and non-labor-intensive culture of hESCs [83].

Alginate encapsulation systems have been shown to support the ability of ES cells to differentiate into specific cell types. The researchers use an alginate encapsulation process for the proliferation and growth of mESC aggregates, which further supports the differentiation of insulin-positive cells from mESCs [84]. In addition, the other group demonstrates 3D model to culture and differentiate hESCs that are encapsulated in calcium alginate microcapsules. This system promotes cellular interactions that are essential for both maintaining pluripotency and differentiation. In addition, encapsulated hESCs are separated from feeder cells during the process of differentiation, which mimics *in vivo* microenvironment and bypass the EBs formation step in a controlled manner. Thus, this 3D culturing of hESCs using alginate microcapsules may be useful for direct differentiation of hESCs toward particular cell types and also has potential for immunoisolation and prevention of teratoma formation of hESCs during transplantation [85].

Cell encapsulation has been proposed to be a solution for treatment of diabetes since it potentially allows the cell protection from host immune system by a concept of immunoisolation. In particular, the microcapsules of islets provide a delicate balance of characteristics including physical strength, immunocompatibility, and selective permeability that will block large immune components. Additionally, its membrane allows the passage of smaller molecules such as oxygen, glucose, water, and insulin [86]. In addition, encapsulated islets in a biocompatible alginates have protected the islets against immune rejection, which is confirmed by prolonged survival of encapsulated islet allografts up to 200 days [87]. Furthermore, Schneider et al. [88] have developed a microcapsule system that protects adult rat and human islets against xenogeneic rejection in immunocompetent diabetic mice without immunosuppression.

3.5. Mechanism of insulin secretion by glucose stimulation

Glucose in the blood is the strong stimulator for the insulin secretion from the islets pancreatic beta cells. Besides glucose, amino acids, ketones, some nutrients, gastrointestinal peptides, and neurotransmitters can also influences the insulin secretion from the beta cells. The glucose-regulated insulin secretion is the complex process and involved several key proteins. Normally, the blood glucose level of more than 70 mg/dl (3.9 mmol/L) enhances insulin production by promoting protein translation and processing. This stimulatory process starts with the binding of glucose to the GLUT2 glucose transporter on the cell membrane of the beta cells. The cytoplasmic glucose is subjected to metabolic process by the action of glucokinase to produce glucose-6-phosphate, which is the rate-limiting step of insulin secretory pathway. Further glycolysis of glucose-6-phosphate finally produces ATP, which is the key molecule for inhibition of the ATP-sensitive K^+ channel protein on the beta cell membrane. The inhibition of ATP-sensitive K^+ channel protein results in beta cell membrane depolarization. This depolarization of the cell membrane induces the opening of voltage-dependent calcium channels with subsequently influx of calcium. High level of cytoplasmic calcium will drive the secretion of insulin from the secretory granules. Insulin secretory profile shows pulsatile

http://dx.doi.org/10.5772/62850

pattern of insulin release, with small bursts occurring about every 10 min. Superimposed upon greater amplitude oscillations of about 80–150 min. This glucose-induced insulin secretion can be enhanced by incretins, released from the neuroendocrine cells of the gastrointestinal tract following food ingestion (**Figure 3**) [89, 90]. However, the mechanism of secretion of insulin with regard to the concentration of glucose from insulin-producing cells derived from hESCs and iPSCs remains unclear.

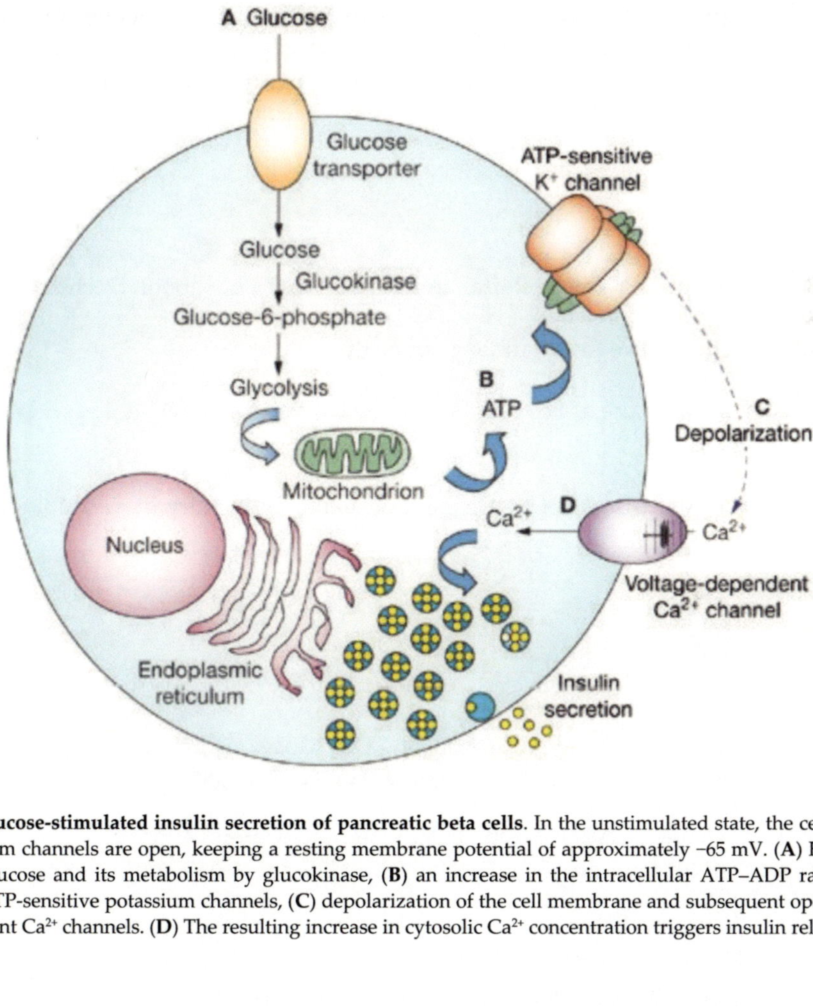

Figure 3. Glucose-stimulated insulin secretion of pancreatic beta cells. In the unstimulated state, the cell ATP-sensitive potassium channels are open, keeping a resting membrane potential of approximately −65 mV. (**A**) Following the uptake of glucose and its metabolism by glucokinase, (**B**) an increase in the intracellular ATP–ADP ratio results in closure of ATP-sensitive potassium channels, (**C**) depolarization of the cell membrane and subsequent opening of voltage-dependent Ca^{2+} channels. (**D**) The resulting increase in cytosolic Ca^{2+} concentration triggers insulin release [90].

4. Conclusion

To date, the researchers have discovered a process that can generate insulin-producing cells from both hESCs and iPSCs. The step toward generating the insulin-producing cells required a differentiation protocol in a manner that mimics differentiation *in vivo*. In order to promote the mature cells, the expression of a key transcription factors should be achieved *in vitro*. Based on these studies, the differentiation processes were successful at generating functional insulin-producing cells *in vitro*. Although there may be some concerns about hESC and iPSCs research, there are emerging evidences that both hESCs and iPSCs were successfully engrafted, secreted insulin, and regulated blood glucose level in animal models. The encapsulation technology also improves the transplantation efficiency by prevention of encapsulated cells from immune destruction, reduce risks of cancer generation by stem cells, and decrease chronic immunosuppression health risks.

However, current prospects for hESC- and iPSCs-based therapy for diabetes treatment still be requires to investigated further more in early phase and subsequent trials in animal models to generate more safety and effective treatment prior to apply to human therapy.

Author details

Wilairat Leeanansaksiri[1*], Piyaporn Rattananinsruang[1] and Chavaboon Dechsukhum[2]

*Address all correspondence to: wilairat@g.sut.ac.th

1 School of Preclinic, Institute of Science, Suranaree University of Technology, Nakhon Ratchasima, Thailand

2 School of Pathology, Institute of Medicine, Suranaree University of Technology, Nakhon Ratchasima, Thailand

References

[1] Kahn CR, Weir GC, King GL, Jacobson AM, Moses AC, Smith RJ. Joslin's diabetes mellitus. 14th ed. Philadelphia: Lippincott Williams & Wilkins, 2005.

[2] American Diabetes Association. Diagnosis and classification of diabetes mellitus. *Diabetes Care* 2013; 36: S67–S74. [PMID: 23264425 doi:10.2337/dc13-S067]

[3] Mutlu F, Bener A, Eliyan A, Delghan H, Nofal E, Shalabi L, Wadi N. Projection of diabetes burden through 2025 and contributing risk factors of changing disease prevalence: an emerging public health problem. *J Diabetes Metab* 2014; 5: 341–347. [doi: 10.4172/2155-6156.1000341]

[4] Norris SL, Nichols PJ, Caspersen CJ, Glasgow RE, Engelgau MM, Jack L, Jr., Isham G, Snyder SR, Carande-Kulis VG, Garfield S, Briss P, McCulloch D, the Task Force on Community Preventive Services. The effectiveness of disease and case management for people with diabetes. *Am J Prev Med* 2002; 22: 15–38. [PMID: 11985933]

[5] Fowler MJ. Diabetes treatment, part 1: diet and exercise. *Clin Diabetes* 2007; 25: 105–109.

[6] White JR, Davis SN, Davidson MB, Mulcahy K, Manko GA, the Diabetes Consortium Medical Advisory Board. Clarifying the role of insulin in type 2 diabetes management. *Clin Diabetes* 2003; 21: 14–21.

[7] Hirsch IB, Bergenstal RM, Parkin CG, Wright E, Jr., Buse JB. A real-world approach to insulin therapy in primary care practice. *Clin Diabetes* 2005; 23: 78–86.

[8] Qaseem A, Humphrey LL, Chou R, Snow V, Shekelle P, Clinical guidelines committee of the American College of Physicians. Use of intensive insulin therapy for the management of glycemic control in hospitalized patients: a clinical practice guideline from the American College of Physicians. *Ann Intern Med* 2011; 154: 260–267. [PMID: 21320941 doi:10.7326/0003-4819-154-4-201102150-00007]

[9] Bernroider E, Brehm A, Krssak M, Anderwald C, Trajanoski Z, Cline G, Shulman GI, Roden M. The role of intramyocellular lipids during hypoglycemia in patients with intensively treated type 1 diabetes. *J Clin Endocrinol Metab* 2005; 90: 5559–5565. [PMID: 15998784]

[10] Ganiats T. Variability in insulin action: mechanisms, implications, and recent advances. *Internet J Fam Pract* 2006; 5(2):1–9.

[11] Nyenwe EA, Jerkins TW, Umpierrez GE, Kitabchi AE. Management of type 2 diabetes: evolving strategies for the treatment of patients with type 2 diabetes. *Metabolism* 2011; 60: 1–23. [PMID: 21134520 doi:10.1016/j.metabol.2010.09.010]

[12] Weir GC, Cavelti-Weder C, Bonner-Weir S. Stem cell approaches for diabetes: towards beta cell replacement. *Genome Med.* 2011; 3: 61.

[13] Rother KI, Harlan DM. Challenges facing islet transplantation for the treatment of type 1 diabetes mellitus. *J Clin Investig* 2004; 114: 877–883. [PMID: 15467822 doi:10.1172/JCI23235]

[14] Ren J, Jin P, Wang E, Liu E, Harlan DM, Li X, Stroncek DF. Pancreatic islet cell therapy for type I diabetes: understanding the effects of glucose stimulation on islets in order to produce better islets for transplantation. *J Transl Med* 2007; 5: 1–15. [PMID: 17201925 doi:10.1186/1479-5876-5-1]

[15] Gunasekaran S. Human pancreatic islet transplantation. *Int J Diabetes Dev Ctries* 2003; 23: 55–57.

[16] Fung M, Thompson D, Warnock G. Pancreatic islet transplantation: a review. *BC Med J* 2004; 46: 457–460.

[17] Pepper AR, Gala-Lopez B, Ziff O, Shapiro AMJ. Current status of clinical islet transplantation. *World J Transpl* 2013; 24: 48–53. [PMID: 24392308 doi:10.5500/wjt.v3.i4.48]

[18] Bruni A, Gala-Lopez B, Pepper AR, Abualhassan NS, Shapiro AMJ. Islet cell transplantation for the treatment of type 1 diabetes: recent advances and future challenges.

Diabetes Metab Syndr Obes 2014; 7: 211–223. [PMID: 25018643 doi:10.2147/DMSO.S50789]

[19] Meier JJ, Bhushan A, Butler PC. The potential for stem cell therapy in diabetes. *Pediatr Res* 2006; 59: 65R–73R. [PMID: 16549551 doi:10.1203/01.pdr.0000206857.38581.49]

[20] Shi Y. Generation of functional insulin-producing cells from human embryonic stem cells in vitro. *Methods Mol Biol* 2010; 636: 79–85. [PMID: 20336517 doi: 10.1007/978-1-60761-691-7_5]

[21] Bose B, Shenoy SP, Konda S, Wangikar P. Human embryonic stem cell differentiation into insulin secreting β-cells for diabetes. *Cell Biol Int* 2012; 36: 1013–1020. [PMID: 22897387 doi:10.1042/CBI20120210]

[22] Skottman H, Hovatta O. Culture conditions for human embryonic stem cells. *Reproduction* 2006; 132: 691–698. [PMID: 17071770]

[23] Kim HS, Oh SK, Park YB, Ahn HJ, Sung KC, Kang MJ, Lee LA, Suh CS, Kim SH, Kim DW, Moon SY. Methods for derivation of human embryonic stem cells. *Stem Cells* 2005; 23: 1228–1233. [PMID: 16051988]

[24] Trounson A. The production and directed differentiation of human embryonic stem cells. *Endocr Rev* 2006; 27: 208–219. [PMID: 16434509]

[25] Stojkovic M, Lako M, Strachan T, Murdoch A. Derivation, growth and applications of human embryonic stem cells. *Reproduction* 2004; 128: 259–267.

[26] Thomson JA, Itskovitz-Eldor J, Shapiro SS, Waknitz MA, Swiergiel JJ, Marshall VS, Jones JM. Embryonic stem cell lines derived from human blastocysts. *Science* 1998; 282: 1145–1147.

[27] Mallon BS, Park KY, Chen KG, Hamilton RS, McKay RD. Toward xeno-free culture of human embryonic stem cells. *Int J Biochem Cell Biol* 2006; 38: 1063–1075. [PMID: 16469522]

[28] Cheng L, Hammond H, Ye Z, Zhan X, Dravid, G. Human adult marrow cells support prolonged expansion of human embryonic stem cells in culture. *Stem Cells* 2003; 21: 131–142. [PMID: 12634409]

[29] Richards M, Tan S, Fong CY, Biswas A, Chan WK, Bongso, A. Comparative evaluation of various human feeders for prolonged undifferentiated growth of human embryonic stem cells. *Stem Cells* 2003; 21: 546–556. [PMID: 12968109]

[30] Lee JB, Song JM, Lee JE, Park JH, Kim SJ, Kang SM, Kwon JN, Kim MK, Roh SI, Yoon HS. Available human feeder cells for the maintenance of human embryonic stem cells. *Reproduction* 2004; 128: 727–735. [PMID: 15579590]

[31] Kibschull M, Mileikovsky M, Michael IP, Lye SJ, Nagy A. Human embryonic fibroblasts support single cell enzymatic expansion of human embryonic stem cells in xeno-free cultures. *Stem Cell Res* 2011; 6: 70–82. [PMID: 20934930 doi:10.1016/j.scr.2010.08.002]

[32] Stojkovic P, Lako M, Stewart R, Przyborski S, Armstrong L, Evans J, Murdoch A, Strachan T, Stojkovic M. An autogeneic feeder cell system that efficiently supports growth of undifferentiated human embryonic stem cells. *Stem Cells* 2005; 23: 306–314. [PMID: 15749925]

[33] Li Z, Leung M, Hopper R, Ellenbogen R, Zhang M. Feeder-free self-renewal of human embryonic stem cells in 3D porous natural polymer scaffolds. *Biomaterials* 2010; 31: 404–412. [PMID: 19819007 doi:10.1016/j.biomaterials.2009.09.070]

[34] Lu HF, Narayanan K, Lim SX, Gao S, Leong MF, Wan AC. A 3D microfibrous scaffold for long-term human pluripotent stem cell self-renewal under chemically defined conditions. *Biomaterials* 2012; 33: 2419–2430. [PMID: 22196900 doi:10.1016/j.biomaterials.2011.11.077]

[35] Chen VC, Couture SM, Ye J, Lin Z, Hua G, Huang HI, Wu J, Hsu D, Carpenter MK, Couture LA. Scalable GMP compliant suspension culture system for human ES cells. *Stem Cell Res* 2012; 8: 388–402. [PMID: 22459095 doi:10.1016/j.scr.2012.02.001]

[36] Raikwar SP, Mueller T, Zavazava N. Strategies for developing therapeutic application of human embryonic stem cells. *Physiology (Bethesda)* 2006; 21: 19–28. [PMID: 16443819]

[37] Hwang NS, Varghese S, Elisseeff J. Controlled differentiation of stem cells. *Adv Drug Deliv Rev* 2008; 60: 199–214. [PMID: 18006108]

[38] Kurosawa H. Methods for inducing embryoid body formation: *in vitro* differentiation system of embryonic stem cells. *J Biosci Bioeng* 2007; 103: 389–398. [PMID: 17609152]

[39] Karp JM, Yeh J, Eng G, Fukuda J, Blumling J, Suh KY, Cheng J, Mahdavi A, Borenstein J, Langer R, Khademhosseini A. Controlling size, shape and homogeneity of embryoid bodies using poly(ethylene glycol) microwells. *Lab Chip* 2007; 7: 786–794. [PMID: 17538722]

[40] Maria R, Nicholas B, Austin S. Towards consistent generation of pancreatic lineage progenitors from human pluripotent stem cells. *Philos Trans R R Soc B* 2015; 370: 1–11.

[41] Camilla HR, Dorthe RP, Jonas BM, Mattias H, Martin D. Collagen type I improve the differentiation of human embryonic stem cells towards definitive endroderm. *Plos One* 2015; 10 (12): 1–21.

[42] Segev H, Fishman B, Ziskind A, Shulman M, Itskovitz-Eldor J. Differentiation of human embryonic stem cells into insulin-producing clusters. *Stem Cells* 2004; 22: 265–274. [PMID: 15153604]

[43] Jiang J, Au M, Lu K, Eshpeter A, Korbutt G, Fisk G, Majumdar AS. Generation of insulin-producing islet-like clusters from human embryonic stem cells. *Stem Cells* 2007; 25: 1940–1953. [PMID: 17510217]

[44] Bruin JE, Erener S, Vela J, Hu X, Johnson JD, Kurata HT, Lynn FC, Piret JM, Asadi A, Rezania A, Kieffer TJ. Characterization of polyhormonal insulin-producing cells

derived in vitro from human embryonic stem cells. *Stem Cell Res* 2014; 12: 194–208. [PMID: 24257076 doi:10.1016/j.scr.2013.10.003]

[45] Assady S, Maor G, Amit M, Itskovitz-Eldor J, Skorecki KL, Tzukerman M. Insulin production by human embryonic stem cells. *Diabetes* 2001; 50: 1691–1697. [PMID: 11473026]

[46] Baharvand H, Jafary H, Massumi M, Ashtiani SK. Generation of insulin-secreting cells from human embryonic stem cells. *Dev Growth Differ* 2006; 48: 323–332. [PMID: 16759282]

[47] Jones MP, Burns CJ, Persaud S. Beta-cell replacement technologies: the potential of stem cells. *Drug Discov Today Ther Strateg* 2004; 1: 213–217.

[48] Jacqueline VS, Suzanne JM, Susan Hawes, Andrew GE, Edouard GS. Derivation of insulin-producing beta-cells from human pluripotent stem cells. *Rev Diabetic Stud* 2014; 11: 6–18. [doi:10.1900/RDS.2014.11.6]

[49] Wei R, Yang J, Hou W, Liu G, Gao M, Zhang L, Wang H, Mao G, Gao H, Chen G, Hong T. Insulin-producing cells derived from human embryonic stem cells: comparison of definitive endoderm- and nestin-positive progenitor-based differentiation strategies. *Plos One* 2013; 8: e72513. [PMID: 23951327 doi:10.1371/journal.pone.0072513]

[50] Talitha VDM, Mark OH. Maturation of stem cell-derived beta-cells guided by the expression of Urocortin 3. *Rev Diabet Stud* 2014; 11: 115–132.

[51] Felicia WP, Jeffrey RM, Mads G, Michael S, Alana VD, Jennifer HR, Quinn PP, Dale G, Douglas AM. Generation of functional human pancreatic β cells in vitro. *Cell* 2014; 159: 428–439.

[52] Ahmed E, Nagwa E. The cell cycle as a brake for β-cell regeneration from embryonic stem cells. *Stem cell Res Ther* 2016; 7: 1–9.

[53] Fujikawa T, Oh SH, Pi L, Hatch HM, Shupe T, Petersen BE. Teratoma formation leads to failure of treatment for type I diabetes using embryonic stem cell-derived insulin-producing cells. *Am J Pathol* 2005; 166: 1781–1791. [PMID: 15920163]

[54] Eshpeter A, Jiang J, Au M, Rajotte RV, Lu K, Lebkowski JS, Majumdar AS, Korbutt GS. In vivo characterization of transplanted human embryonic stem cell-derived pancreatic endocrine islet cells. *Cell Prolif* 2008; 41: 843–858. [PMID: 19040565 doi:10.1111/j.1365-2184.2008.00564.x]

[55] Hua XF, Wang YW, Tang YX, Yu SQ, Jin SH, Meng XM, Li HF, Liu FJ, Sun Q, Wang HY, Li JY. Pancreatic insulin-producing cells differentiated from human embryonic stem cells correct hyperglycemia in SCID/NOD mice, an animal model of diabetes. *Plos One* 2014; 10: e102198. [PMID: 25009980 doi:10.1371/journal.pone.0102198]

[56] Kroon E, Martinson LA, Kadoya K, Bang AG, Kelly OG, Eliazer S, Young H, Richardson M, Smart NG, Cunningham J, Agulnick AD, D'Amour KA, Carpenter MK, Baetge EE. Pancreatic endoderm derived from human embryonic stem cells generates glucose-

responsive insulin-secreting cells *in vivo*. *Nat Biotechnol* 2008; 26: 443–452. [PMID: 18288110 doi:10.1038/nbt1393]

[57] Rezania A, Bruin JE, Riedel MJ, Mojibian M, Asadi A, Xu J, Gauvin R, Narayan K, Karanu F, O'Neil JJ, Ao Z, Warnock GL, Kieffer TJ. Maturation of human embryonic stem cell-derived pancreatic progenitors into functional islets capable of treating pre-existing diabetes in mice. *Diabetes* 2012; 61: 2016–2029. [PMID: 22740171 doi:10.2337/db11-1711]

[58] Kirk K, Hao E, Lahmy R, Itkin-Ansari P. Human embryonic stem cell derived islet progenitors mature inside an encapsulation device without evidence of increased biomass or cell escape. *Stem Cell Res* 2014; 12: 807–814. [PMID: 24788136 doi:10.1016/j.scr.2014.03.003]

[59] Teoh HK, Cheong SK. Induced pluripotent stem cells in research and therapy. *Malays J Pathol* 2012; 34: 1–13. [PMID: 22870592]

[60] Takahashi K, Yamanaka S. Induced pluripotent stem cells in medicine and biology. *Development* 2013; 140: 2457–2461. [PMID: 23715538 doi:10.1242/dev.092551]

[61] Wu SM, Hochedlinger K. Harnessing the potential of induced pluripotent stem cells for regenerative medicine. *Nat Cell Biol* 2013; 13: 497–505. [PMID: 21540845 doi:10.1038/ncb0511-497]

[62] Lin SL, Chang DC, Ying SY, Leu D, Wu DT. MicroRNA miR-302 inhibits the tumorigenecity of human pluripotent stem cells by coordinate suppression of the CDK2 and CDK4/6 cell cycle pathways. *Cancer Res* 2010; 70: 9473–9482. [PMID: 21062975 doi:10.1158/0008-5472.CAN-10-2746]

[63] Kuo CH, Ying SY. Advances in microRNA-mediated reprogramming technology. *Stem Cells Int* 2012; 2012: 823709. [PMID: 22550519 doi:10.1155/2012/823709]

[64] Rosa A, Brivanlou AH. Regulatory non-coding RNAs in pluripotent stem cells. *Int J Mol Sci* 2013; 14: 14346–14373. [PMID: 23852015 doi:10.3390/ijms140714346]

[65] Park IH, Arora N, Huo H, Maherali N, Ahfeldt T, Shimamura A, Lensch MW, Cowan C, Hochedlinger K, Daley GQ. Disease-specific induced pluripotent stem cells. *Cell* 2008; 134: 877–886. [PMID: 18691744 doi:10.1016/j.cell.2008.07.041]

[66] Zhou Y, Zeng F. Integration-free methods for generating induced pluripotent stem cells. *Genom Proteom Bioinform* 2013; 11: 284–287. [doi:10.1016/j.gpb.2013.09.008]

[67] Stadtfeld M, Hochedlinger K. Induced pluripotency: history, mechanisms, and applications. *Genes Dev* 2010; 24: 2239–2263. [PMID: 20952534 doi:10.1101/gad.1963910]

[68] Chun YS, Chaudhari P, Jang YY. Applications of patient-specific induced pluripotent stem cells; focused on disease modeling, drug screening and therapeutic potentials for liver disease. *Int J Biol Sci* 2010; 6: 796–805. [PMID: 21179587]

[69] Zhu FF, Zhang PB, Zhang DH, Sui X, Yin M, Xiang TT, Shi Y, Ding MX, Deng H. Generation of pancreatic insulin-producing cells from rhesus monkey induced

pluripotent stem cells. *Diabetologia* 2011; 54: 2325–2336. [PMID: 21755313 doi:10.1007/s00125-011-2246-x]

[70] Jeon K, Lim H, Kim JH, Thuan NV, Park SH, Lim YM, Choi HY, Lee ER, Kim JH, Lee MS, Cho SG. Differentiation and transplantation of functional pancreatic beta cells generated from induced pluripotent stem cells derived from a type 1 diabetes mouse model. *Stem Cells Dev* 2012; 21: 2642–2655. [PMID: 22512788 doi:10.1089/scd.2011.0665]

[71] Wang L, Huang Y, Guo Q, Fan X, Lu Y, Zhu S, Wang Y, Bo X, Chang X, Zhu M, Wang Z. Differentiation of iPSCs into insulin-producing cells via adenoviral transfection of PDX-1, NeuroD1 and MafA. *Diabetes Res Clin Pract* 2014; 104: 383–392. [PMID: 24794627 doi:10.1016/j.diabres.2014.03.017]

[72] Rezania A, Bruin JE, Arora P, Rubin A, Batushansky I, Asadi A, O'Dwyer S, Quiskamp N, Mojibian M, Albrecht T, Yang YH, Johnson JD, Kieffer TJ. Reversal of diabetes with insulin-producing cells derived in vitro from human pluripotent stem cells. *Nat Biotechnol* 2014; 32(11): 1121–1133. [doi:10.1038/nbt.3033]

[73] Pellegrini S, Ungaro F, Mercalli A, Melzi R, Sebastiani G, Dotta F, Broccoli V, Piemonti L, Sordi V. Human induced pluripotent stem cells differentiation into insulin-producing cells able to engraft in vivo. *Acta Diabetol.* 2015; 52(6): 1025–1035. [doi:10.1007/s00592-015-0726-z]

[74] Kastenberg ZJ, Odorico JS. Alternative sources of pluripotency: science, ethics, and stem cells. *Transplant Rev (Orlando)* 2008; 22: 215–222. [PMID: 18631882 doi:10.1016/j.trre.2008.04.002]

[75] Hentze H, Soong PL, Wang ST, Phillips BW, Putti TC, Dunn NR. Teratoma formation by human embryonic stem cells: evaluation of essential parameters for future safety studies. *Stem Cell Res* 2009; 2: 198–210. [PMID: 19393593 doi:10.1016/j.scr.2009.02.002]

[76] King FW, Ritner C, Liszewski W, Kwan HC, Pedersen A, Leavitt AD, Bernstein HS. Subpopulations of human embryonic stem cells with distinct tissue-specific fates can be selected from pluripotent cultures. *Stem Cells Dev* 2009; 18: 1441–1450. [PMID: 19254177 doi:10.1089/scd.2009.0012]

[77] Strulovici Y, Leopold PL, O'Connor TP, Pergolizzi RG, Crystal RG. Human embryonic stem cells and gene therapy. *Mol Ther* 2007; 15: 850–866. [PMID: 17356540]

[78] Fong CY, Gauthaman K, Bongso A. Teratomas from pluripotent stem cells: a clinical hurdle. *J Cell Biochem* 2010; 111: 769–781. [PMID: 20665544 doi:10.1002/jcb.22775]

[79] Drukker M, Benvenisty N. The immunogenicity of human embryonic stem-derived cells. *Trends Biotechnol* 2004; 22: 136–141. [PMID: 15036864]

[80] Boyd AS, Higashi Y, Wood KJ. Transplanting stem cells: potential targets for immune attack. Modulating the immune response against embryonic stem cell transplantation. *Adv Drug Deliv Rev* 2005; 57: 1944–1969. [PMID: 16289432]

[81] Arturo JV, Omid V, Mads G, Jeffrey RM, Felicia WP, Andrew RB, Joshua CD, Jie L, Michael C, Karsten O, Hok HT, Siddharth J, Erin L, Stephanie A, Srujan G, James JM, Matthew AB, Jennifer H, Jose O, Dale LG, Gordon CW, Douglas AM, Robert L, Daniel GA. Long-term glycemic control using polymer-encapsulated human stem cell-derived beta cells in immune-competent mice. *Nat Med* 2016; 22: 306–311[doi:10.1038/nm.4030]

[82] Serra M, Correia C, Malpique R, Brito C, Jensen J, Bjorquist P, Carrondo MJ, Alves PM. Microencapsulation technology: a powerful tool for integrating expansion and cryopreservation of human embryonic stem cells. *Plos One* 2011; 6: e23212. [PMID: 21850261 doi:10.1371/journal.pone.0023212]

[83] Siti-Ismail N, Bishop AE, Polak JM, Mantalaris A. The benefit of human embryonic stem cell encapsulation for prolonged feeder-free maintenance. *Biomaterials* 2008; 29: 3946–3952. [PMID: 18639332 doi:10.1016/j.biomaterials.2008.04.027]

[84] Wang N, Adams G, Buttery L, Falcone FH, Stolnik S. Alginate encapsulation technology supports embryonic stem cells differentiation into insulin-producing cells. *J Biotechnol* 2009; 144: 304–312. [PMID: 19686786 doi:10.1016/j.jbiotec.2009.08.008]

[85] Chayosumrit M, Tuch B, Sidhu K. Alginate microcapsule for propagation and directed differentiation of hESCs to definitive endoderm. *Biomaterials* 2010; 31: 505–514. [PMID: 19833385 doi:10.1016/j.biomaterials.2009.09.071]

[86] Lee MK, Bae YH. Cell transplantation for endocrine disorders. *Adv Drug Deliv Rev* 2000; 42: 103–120. [PMID: 10942817]

[87] de Vos P, Faas MM, Strand B, Calafiore R. Alginate-based microcapsules for immunoisolation of pancreatic islets. *Biomaterials* 2006; 27: 5603–5617. [PMID: 16879864]

[88] Schneider S, Feilen PJ, Brunnenmeier F, Minnemann T, Zimmermann H, Zimmermann U, Weber MM. Long-term graft function of adult rat and human islets encapsulated in novel alginate-based microcapsules after transplantation in immunocompetent diabetic mice. *Diabetes* 2005; 54: 687–693. [PMID: 15734844]

[89] Kasper D, Fauci A, Hauser S, Longo D, Jameson J, Loscalzo J. Harrison's Principles of Internal Medicine 19/E 19th Edition. McGraw-Hill Medical, New York, USA, 2015.

[90] De León DD, Stanley CA. Mechanisms of disease: advances in diagnosis and treatment of hyperinsulinism in neonates. *Nat Clin Pract Endocrinol Metab* 2007; 3: 57–68. [doi:10.1038/ncpendmet0368]

Chapter 22

Stem Cell-Derived Regulatory T Cells for Therapeutic Use

Mohammad Haque, Praneet Sandhu and
Jianxun Song

Additional information is available at the end of the chapter

http://dx.doi.org/10.5772/62695

Abstract

CD4$^+$ regulatory T cells (Tregs) are essential for normal immune surveillance, and their dysfunction can lead the development of autoimmune diseases. Pluripotent stem cells (PSCs) can be utilized to obtain a renewable source of healthy Tregs to treat autoimmune disorders as they have the ability to produce almost all cell types in the body, including Tregs. However, the right conditions for the development of antigen (Ag)-specific Tregs from PSCs (i.e., PSC-Tregs) have not been fully defined, especially the signaling mechanisms that the direct differentiation of such Tregs. Ag-specific PSC-Tregs can be tissue-associated and infiltrate to local inflamed tissue to suppress autoimmune responses after adoptive transfer, thereby avoiding potential overall immunosuppression from non-specific Tregs. Development of cell-based therapies using Ag-specific PSC-Tregs will provide an important step toward personalized therapies for autoimmune disorders.

Keywords: regulatory T cells, pluripotent state cells, cell differentiation, immunotherapy, autoimmunity

1. Introduction

Regulatory T cells (Tregs) are a component of the normal immune system and contribute to the maintenance of peripheral tolerance. Tregs are defined phenotypically by the expression of the interleukin (IL)-2 receptor α-chain (CD25) and the transcriptional factor, forkhead box P3 (FoxP3), which is required for the Treg development and controls a genetic program specifying cell fate. Tregs can down-regulate immune responses and are essential for immune homeostasis [1]. Tregs are key effectors in preventing and treating autoimmune disorders, for example, rheumatoid arthritis (RA), type 1 diabetes (T1D), and inflammatory bowel diseases (IBD), and controlling both the transplant rejection graft-versus-host disease (GVHD) [2–5].

Tregs or suppressors of T cells, as they had been originally named, have been studied since early 1970s. In 1970, Gershon and Kondo reported that T cells not only augmented but also have the ability to dampen immune responses. The immune dampening was mediated by a special class of T cells that were different from helper T cells; this special type of T-cell population is known as suppressor T cells and is found in $CD4^+$ population [6]. It is well understood that $CD4^+$ Tregs are essential for normal immunological self-tolerance and immune homeostasis. Failure of immunologic self-tolerance often leads to the development of autoimmune disorders, which are estimated to afflict up to 5% of the population. Although there are a number of debates regarding the etiology of autoimmune diseases, it is well documented that T cells are the key mediators of many autoimmune disorders, such as autoimmune arthritis, autoimmune thyroiditis, and insulin-dependent diabetes mellitus (IDDM). There are various mechanisms for establishing and sustaining immunological self-tolerance and immune homeostasis. In addition to these, T cell-mediated suppression of immune responses toward self and non-self antigens (Ags) has recently attracted enormous interest [7]. T-cell–mediated suppression is mainly achieved by $FoxP3^+$ Tregs, which play an important role in the prevention and suppression of autoimmunity. The deficiencies in Treg function have been identified in a wide variety of human autoimmune disorders, such as IPEX (immunodysregulation polyendocrinopathy enteropathy X-linked syndrome) [8–10]. As the Treg numbers are important for effective suppressive function, adoptive transfer of exogenous Tregs would be ideal for the treatment of autoimmunity or Treg-deficient diseases.

Tregs comprise of 5–10% of the mature $CD4^+$ T helper cell subpopulation in mice and about 1–2% of $CD4^+$ T cells in human. As a result, it is crucial to develop large numbers of Tregs *in vitro* and adoptively transfer them for cell-based therapies. There has been no approach described for isolating a large number of Tregs with % specificity. Therefore, the attention has now been focused on utilizing pluripotent stem cells (PSCs) as a source for obtaining Ag-specific Tregs. PSCs can be utilized to obtain a renewable source of healthy Tregs to treat autoimmune disorders as they have the ability to produce almost all cell types in the body, including Tregs. However, optimal conditions and signaling network for the development and differentiation of Ag-specific Tregs from PSCs (*i.e.*, PSC-Tregs) have not been fully defined [11]. Ag-specific PSC-Tregs serve as a better choice of Tregs for cell-based therapies as they can accumulate in local inflamed tissues to suppress autoimmune responses after adoptive transfer, thereby avoiding potential overall immunosuppression from non-specific Tregs.

2. $CD4^+$ and $CD8^+$ Tregs

Tregs are an integral component of the normal immune system and contribute to the maintenance of peripheral tolerance. Tregs can down-regulate immune responses and are essential for immune homeostasis. There are two major classes of Tregs: $CD4^+$ and $CD8^+$ Tregs. $CD4^+$ Tregs consists of two types: Naturally occurring Tregs (nTregs) that is characterized by

http://dx.doi.org/10.5772/62695

constitutively expression of CD25 and FoxP3, and adaptive or inducible Tregs (iTregs) that are induced upon persistent Ag exposure in the periphery.

nTregs develop in the thymus. nTregs originate as $CD4^+$ cells which highly express CD25, the alpha chain of the IL-2 receptor, and the transcription factor FoxP3. Of the total $CD4^+$ T-cell population, approximately 5–10% are nTregs, which can be visualized at the single-positive (SP) stage of development in the thymus [12]. nTregs are positively selected in the thymus and represent the thymocytes with a relatively high avidity for self-Ags. The signals for the development of Tregs are the interaction between the T-cell receptor (TCR) and the complex of *MHC* (major histocompatibility complex) class *II* molecules with self-peptide expressed on the thymic stroma [13]. It is also known that nTregs are essentially cytokine independent during the development.

iTregs originate from the thymus as a single-positive $CD4^+$ cells. They differentiate into CD25- and FoxP3-expressing Tregs following adequate antigenic stimulation in the presence of cognate Ag and specialized immune-regulatory cytokines, such as TGF-β, IL-10, and IL-4 [14].

3. Treg development

Till today, several models have been proposed for the development of Tregs in the thymus. The interaction of costimulatory molecule CD28 and its B7 family member ligands is important for the development of nTreg in the thymus. CD28 ligand B7.1 and B7.2 are primarily expressed on thymic *antigen-presenting cells* (APCs), including dendritic cells (DCs), and epithelial cells. Although the role of this costimulation in the development of nTregs is not clear, it has been reported that mice deficient in the costimulatory molecule CD28 or CD28 ligands B7.1/B7.2 has decreased numbers and percentages of thymic nTregs [15–17]. One possible function is that they provide a quantitative signal along with TCR stimulation that drives T cells to develop into nTregs. This costimulation has the ability to function in preventing negative selection and supporting nTreg development [18]. NF-κB is a transcription factor that functions the downstream of the CD28/B7 costimulation. NF-κB family member c-Rel has been shown to be a critical factor for the development of nTregs in the thymus [19]. TCR engagement leads to the expression of high-affinity IL-2 receptor (CD25), which causes IL-2-induced FoxP3 expression and nTreg commitment [20]. Published evidence showed that IL-2-deficient mice exhibited 50% reduction of nTregs in the thymus [21]. Therefore, IL-2 is an important cytokine required for the development of nTregs. FoxP3 is another important transcription factor for the development of nTregs. Mice deficient in FoxP3 or harboring a mutated FoxP3 gene developed lethal multi-organ inflammation. Adoptive transfer of $FoxP3^+$ T cells into neonates protected FoxP3-deficient mice from their autoimmune pathology [22, 23]. Clearly, FoxP3 is required for the nTreg development and affects its functional activity.

Hematopoietic stem cell (HSC)-derived hematopoietic progenitors migrate into the thymus and develop into different types of T cells. The development of *αβ* T cells in the thymus is a highly ordered process. The most immature thymocyte population ($CD4^-CD8^-$) is referred to as *double negative* (DN) cells. DN precursors are subdivided into sequential developmental subsets as follows: DN1 ($CD44^+CD25^-$), DN2 ($CD44^+CD25^+$), DN3 ($CD44^-CD25^+$), and DN4 ($CD44^-CD25^-$). Recombination activating genes (*Rag*), including *Rag1 and Rag2* catalyze the TCRβ locus for rearrangement, which is initiated during cell transit from the DN2 to the DN3 stage. A functional TCR β chain generated only in DN3 cells has the ability to pair with the invariant pre-Tα/CD3 and create a pre-TCR. The DNs cells with pre-TCR are selected and can continue to differentiate to DN4. This episode is called β-selection, which is the initial checkpoint of T lymphocyte differentiation in the thymus. Pre-TCR formation in DN4 cells drives cell differentiation and ends the rearrangement of TCRβ locus, resulting in the development of the $CD4^+CD8^+$ *double positive* (DP) cells from DN4 cells [24].

In summary, the Treg development within the thymus includes a series of processes—positive selection (*e.g.*, TCR rearrangement) and negative selection (*e.g.*, clonal deletion) [25]. The autoimmune regulator Aire (largely expressed in thymic medullary epithelial cells—TECs) [26] and FoxP3 have key functions in clonal deletion and Treg selection [27, 28]. There are links among Aire expression, FoxP3 up-regulation, and Treg selection [29–31]. Evidence suggests that Aire deficiency affects the negative selection of self-reactive T cells and FoxP3 controls the development and function of the nTregs [29].

4. Ag-specific induction of Tregs

The mechanisms of acquisition of self-tolerance by the immune system are still being investigated. However, a widely accepted mechanism is the deletion of immature thymocytes before acquiring functional maturity in the thymic cortex and medulla [32, 33]. However, some self-reactive cells can escape to the periphery by breaking central tolerance [34, 35]. It is possible that such cells contain TCRs that recognize weak self-epitopes and as such, these autoreactive cells require peripheral tolerance to counteract them. Maintenance of peripheral tolerance, such as the deletion or reversible anergy of T cells, is performed by Tregs [36, 37]. The identification and characterization of Tregs definitively confirmed the existence of dominant tolerance [23, 38, 39]. Generating the peripheral Tregs in diverse microenvironments, for example, the gut, as well as the specific locations with tumors or microbes maintains local homeostasis. Tregs engendered in the peripheral tissue, external of the thymus, stay by way of resting cells on inter-mitotic phase, self-regulating further supply of agonist ligand which drives the formation of Tregs. This critical feature of the immune system lets the Tregs an approaching production to conquer undesirable immunity. As soon as encountering agonist TCR ligand, Tregs have the ability to migrate to Ag-draining lymph nodes in which Tregs undertake substantial expansion [40, 41]. The specificity of Treg-mediated suppression results from the corecruitment of Tregs and other T cells in Ag-draining lymph nodes. As a result, Tregs with certain specificity have the ability to suppress various effector cells through distinctive specificity when restricted in the area of identical APCs [42, 43]. At locations of inflammation, suppressive function, including the suppression of Th1 and Th17 responses, needs Tregs' trafficking and migration in tissues and secondary or peripheral lymphoid organs, such as lymph nodes and the spleen [44]. In addition, compared with naive or activated T cells, FoxP3-expressing Tregs have a distinct transcriptional profiling, showing a different

http://dx.doi.org/10.5772/62695

number of differentially expressed genes, including certain genes generally up-regulated in activated T cells, for example, *IL2ra*, *Ctla4*, and *Tnfrsf18*, which individually contribute CD25, CTLA4, and glucocorticoid-induced TNF receptor (GITR). Therefore, these target genes represent the transcriptional induction of the Treg signature by FoxP3 [45, 46].

5. Treg-mediated suppression

Tregs play an important role in the maintenance of immune tolerance. Sakaguchi and colleagues first described the importance of Tregs in the prevention of autoimmune disease development. They demonstrated that adoptive transfer of $CD4^+$ T cells, depleted of $CD25^+$ T cells by a specific monoclonal antibody against CD25 into BALB/c athymic nude mice, caused spontaneous development of T-cell–mediated autoimmune disease [47]. When these mice were reconstituted with $CD4^+CD25^+$ T cells, the autoimmune disease development was successfully ameliorated within a brief period post transfer. The discovery of the X chromosome-encoded transcription factor FoxP3 as a specification and maintenance factor subsequently confirmed $CD4^+CD25^+$ T cells as a unique thymus-derived lineage. Examining their role in the murine immune system revealed that Tregs have a central role in immune homeostasis: Genetic defects resulting in the dysfunctional Tregs result in multi-organ autoimmune disease, and the depletion of nTregs induces autoimmunity. The next question to be addressed is to define the mechanisms of Treg-mediated suppression. It has already been established that $Foxp3^+CD25^+CD4^+$ nTregs suppress the proliferation of naive T cells and their differentiation to effector T cells *in vivo*. They also have the ability to suppress effector activities of differentiated $CD4^+$ and $CD8^+$ T cells and the function of natural killer (NK) cells, natural killer T (NKT) cells, B cells, macrophages, osteoclasts, and DCs [48–50]. Tregs suppress the proliferation and cytokine production (in particular of IL-2) of responder T cells when the two populations are cocultured *in vitro* and stimulated by Ag in the presence of APCs [51]. Once activated by a particular Ag, Tregs can suppress responder T cells irrespective of whether they share Ag specificity with the Treg [52].

Several mechanisms of Treg-mediated suppression have been proposed, including the secretion of immunosuppressive cytokines, cell contact-dependent suppression, and functional modification or killing of APCs. For example, IL-10 and TGF-β contribute to the suppression of arthritis in Ag-induced arthritis mice [11]. Another study showed that IL-10 and TGF-β contribute to the suppression of IBD induced in mice by Treg depletion [53]. Another mechanism for the killing of responder T cells or APCs is cell-to-cell contact. Tregs can secrete granzyme or perforin to destroy target T cells or APCs by cell-to-cell interaction method, or deliver a negative signal by CTLA-4 or FasL to responder T cells. Potential critical signals involve up-regulation of intracellular cyclic adenosine monophosphate (cAMP) that results in the suppression of T-cell proliferation and cytokine production, such as IL-2, or the production of pericellular adenosine catalyzed by CD39, that is, ectonucleoside triphosphate diphosphohydrolase 1, and CD73, that is, ecto-5′-nucleotidase, which are presented through Tregs [50]. Activated Tregs can cause down-regulation of CD80/86 expression on APCs or stimulate DCs to form the enzyme indoleamine 2,3-dioxygenase, which catabolizes the essential amino acid tryptophan to kynurenines, causing toxicity to T cells. All these functions are dependent on the expression of CTLA-4 on Tregs. Treg-mediated suppression can occur (1) *via* Ag-specific Tregs upon antigenic stimulation, which is highly mobile and swiftly recruited at the site of inflammation; (2) Ag-activated Tregs contacting DCs that restrict DC function, thereby hindering the activation of other T cells; and (3) through the secretion of granzyme/perforin, IL-10, or other immune suppressive cytokines, such as IL-35, depending on the strength and duration of antigenic stimulation and the local milieu of cytokines and chemokine.

CTLA-4 is particularly critical for Treg function in spite of a number of distinct molecules that are associated with Treg suppression. CTLA-4 is constitutively expressed on FoxP3$^+$ Tregs, and FoxP3 regulates its expression. The blockage of CTLA-4 abolishes Treg-mediated inhibition. Moreover, fatal autoimmunity and inflammation appear in the germline removal as well as Treg-specific conditional deficit of CTLA-4. The fate of responder T cells that are suppressed by Tregs is also unclear, that is, whether they remain non-activated, die by apoptosis, or become anergic. Additional studies are required to elucidate the molecular basis of suppression mediated by Tregs.

Recent advances in the use of large-scale *in vitro* expansion of Tregs followed by *in vivo* re-infusion of these cells raises the possibility that this strategy may be successfully utilized for the treatments of autoimmune disorders. While cell-based therapies using Tregs are currently largely recognized in animal experimental tests, up to the present time, cell-based therapies using Tregs have not been clinically utilized in the suppression of autoimmune disorders.

There are numerous issues to be solved for using Tregs in humans, such as the requirement for vigorous methods to separate and grow these cells. *First*, only low numbers of Tregs can be harvested from the peripheral blood mononuclear cells (PBMCs). CD4 and CD25 have been used to isolate Tregs for *ex vivo* expansion. CD4$^+$CD25$^+$ T cells are not homogenous and contain both Tregs and conventional effector T cells (Teffs). Current expansion protocols activate both Tregs and Teffs, and because it takes a longer time for Tregs to enter the S phase of cell cycle, Teffs outgrow Tregs [54]. In addition, Tregs can lose suppressive activity after repetitive stimulation with α-CD3 plus α-CD28 Abs with or without rIL-2 *in vitro* [55–57]. *Second*, despite a growing number of published purification protocols for isolating subsets of Tregs, no approach to date has demonstrated the capacity to isolate the entire Treg population with 100% specificity from patients (the current clinical approach). Even FoxP3 or more recently Eos, a transcriptional factor, considered the gold standard for identification of Tregs, is expressed transiently in some activated non-regulatory human T cells [58], highlighting the difficulty in both identifying and isolating a pure Treg population. Adoptive transfer of non-regulatory Teffs with Tregs has a potential to worsen diseases. *Third*, gene transduction of CD4$^+$ T cells from PBMCs with Ag-specific TCR [59–61] or chimeric Ag receptor (CAR) [59–61, 63] elicits generation of suppressive T-cell populations [64, 65] and overcomes the hurdle of the limited numbers of Ag-specific T cells. Moreover, gene transduction of human PBMC with Ag-specific TCR generated functional Ag-specific T cells, which targeted tissue-associated inflammation [63]. However, the engineered Tregs express endogenous and exogenous polyclonal TCRs, which may reduce their therapeutic potential (the current experimental approach) [66, 67]. Also, TCR mispairing is a concern with regard to the safety of TCR gene-transferred Tregs for clinical use, because the formation of new heterodimers of TCR can induce immunopathology [68]. Therefore, there is a need to improve this strategy and generate monoclonal Tregs. *Fourth*, the differentiation state of Tregs is inversely related to their capacity to proliferate and persist [69, 70]. The "right" Tregs resist terminal differentiation, maintain high replicative potential (*e.g.*, expression of common-γ chain – γc, CD132), are less prone to apoptosis (*e.g.*, low expression of PD-1), and have a greater ability to respond to homeostatic cytokines [71], which facilitates their survival. In addition, the "right" Tregs express high levels of molecules that facilitate their homing to lymph nodes (LNs), such as CD62L and CC-chemokine receptors (*e.g.*, CCR4, CCR7), and maintain stability or plasticity under certain inflammatory conditions. Furthermore, after an effective immune response, the "right" Tregs persist in a variety of differentiation states, providing protective immunity. Thus, the "right" Tregs are the superior subsets for use in cell-based therapies. *Finally*, because there are too few cells, harvesting sufficient numbers of tissue-associated Tregs from patients' PBMC for TCR gene transduction can be problematic.

Taken together, strong arguments support the development of Treg-based therapies in autoimmune disorders using engineered Tregs. While clinical trials show the safety, feasibility, and potential therapeutic activity of Treg-based therapies using this approach, concerns about autoimmunity due to cross-reactivity with healthy tissues remain a major safety issue [72, 73]. In addition, genetically modified Tregs using current approaches are usually intermediate or later effector Tregs [74], which only have short-term persistence *in vivo*.

6. Stem cell-derived Tregs

To date, stem cells are the only source available to generate a high number of the "right" Tregs [75, 76]. It has been already demonstrated that induced pluripotent stem cell (iPSCs) are like embryonic stem cells (ESCs) in different aspects, including the expression of definite stem cell genes and proteins, doubling time, embryoid body formation, viable chimera formation, potency and differentiability, chromatin methylation patterns, and teratoma formation [77]. However, the similarity between iPSCs and ESCs is still being assessed [78]. The generation of iPSCs from the mouse and human somatic cells has garnered considerable attention. Research has shown that iPSCs could be generated from the mouse and human somatic cells by introducing Oct3/4 and Sox2 with either Klf4 and c-Myc or Nanog and Lin28 using retroviruses or lentiviruses-mediated transduction [79–82]. Thus, iPSC technology continues to progress rapidly, and clinically applicable iPSCs can be generated from patients with noninvasive medical procedures. Many genetic methods as well as protein-based approaches have been developed to produce iPSCs with potentially reduced risks, including that of immunogenicity and tumorigenicity [83]. Because of the plasticity and the potential for an unlimited capacity for self-renewal, iPSCs have great potential to be used in cell-based therapies for autoimmune disorders comparable to ESCs and HSCs.

Previously, T cells have been demonstrated to be differentiated from ESCs and HSCs; recently T cells [84] and Ag-specific $CD8^+$ cytotoxic T lymphocytes (CTLs) [85] have been confirmed to be differentiated from iPSCs. In these investigations, the researchers genetically modified mouse iPSCs with Ag (ovalbumin, OVA)-specific MHC II (I-A^b)-restricted TCR and FoxP3, and then *in vivo* differentiated the iPSCs into functional Ag-specific $CD4^+$ Tregs, which dramatically prevented the mice from Ag-induced arthritis. Thus, a new approach to generate a high number of functional Tregs from iPSCs may be used for the treatments of autoimmune disorders. In fact, Ag-specific Tregs can be *in vitro* generated from iPSCs through a Notch-mediated signaling. It has been shown that Ag-specific Tregs were generated from iPSCs or ESCs genetically modified with the FoxP3 and Ag-specific TCR followed by stimulation with an *in vitro* Notch signaling. Furthermore, adoptive transfer of these stem cell-derived Ag-specific Tregs had the ability to secrete a large amount of suppressive cytokines, including TGF-β and IL-10, and suppressed autoimmunity [11, 75]. Additionally, forced expression of Ag-specific TCR can suppress the expression of recombination-activating (*Rag*) genes, resulting in a uniform expression of Ag-specific TCR on iPSC or ESC-derived Tregs. As a result, this method has a potential to develop a great number of single-type Ag-specific Tregs.

Collectively, a large number of single-type Ag-specific Tregs can be generated by gene transduction of iPSCs with Ag-specific TCR with FoxP3 followed by T lineage differentiation through an *in vitro* Notch signaling or an *in vivo* approach. The Ag-specific Tregs may be applied for cell-based therapies of autoimmune disorders, such as T1D, and RA. Of note, these stem cell-derived Ag-specific Tregs are less differentiated and have the ability to persist *in vivo* after adoptive transfer. It can be predicted that the use of iPSCs as a mean to develop disease-specific immune cells for immunotherapy has a great potential in the prevention of many diseases. Therefore, iPSC-derived Ag-specific Tregs can be used in cell-based therapies for autoimmune disorders.

7. Stem cell-derived Ag-specific Tregs for therapeutic use

Treatments of autoimmune disorders with Tregs have been shown to work in a number of mouse models, such as T1D and RA. Tregs were activated with its cognate Ag and can suppress the conventional Tregs within the immediate vicinity regardless of specificity [86]. Additionally, this suppressive function extends to a wide range of other immune cells such as B cells, DCs, and monocytes, including naive, effector, and memory T cells. Ag specificity of Tregs is important to counteract the ongoing autoimmunity because high doses of polyclonal antibody may fail to keep in check autoimmune responses. Tregs require Ag specificity to home/be retained at the appropriate site or tissue, and exert active suppression where Ag specificity is chiefly determined by the individual TCR expressed. Under usual circumstances, a small population of Ag-specific Tregs exists within a polyclonal population. For therapeutic interventions, the generation of a large number of Ag-specific Tregs becomes essential. Furthermore, existing Tregs in patients are insufficient to prevent the initiation of autoimmunity, it is questionable whether simply putting a large number of these cells back into the patient, without modifying specificity or function, would have the desired effect. In addition, Tregs specific to tissue/organ (*e.g.*, joints, pancreas, intestine) facilitate stable FoxP3 expression and avoid the induction of a potentially harmful systemic immunosuppression [87, 88].

http://dx.doi.org/10.5772/62695

Therefore, in order for Tregs to be a viable treatment for autoimmune disorders, approaches for generating the populations of Ag-specific Tregs are essential.

Previous studies have demonstrated the broad application of genetic manipulation of PSCs for immunotherapy and have provided proof-of-principle data for using TCR gene-transduced PSCs for cell-based therapies [85, 89–91]. We also showed that functional iPSC-Treg differentiates *in vitro* mediated through the Notch signaling [11, 75]. Murine iPSCs were genetically modified with OVA-specific MHCII-restricted TCR (OTII) and FoxP3 by retrovirus-mediated transduction. Genetically modified iPSCs were stimulated with an *in vitro* Notch ligand to direct iPSC differentiation into functional OVA-specific Tregs, which were able to produce suppressive cytokines (TGF-β and IL-10), and inhibit other immune cell activities *in*

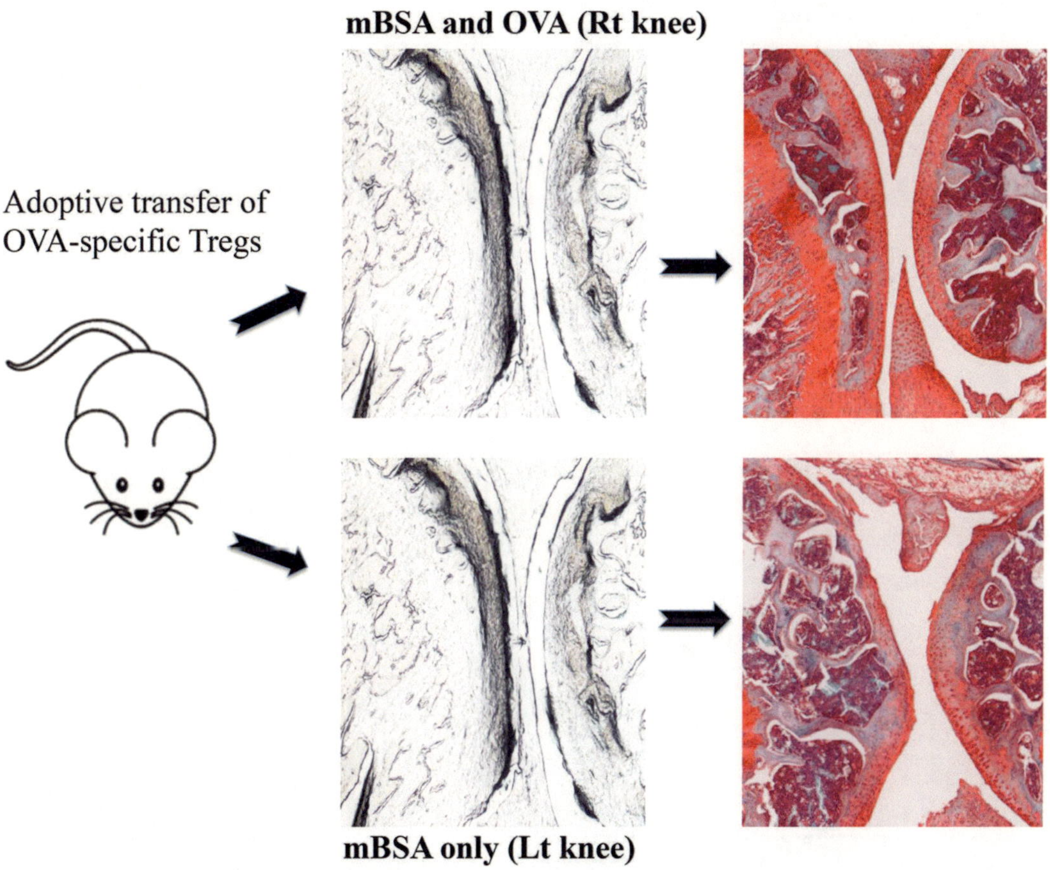

Figure 1. AIA in mice. Mice were immunized with methylated BSA (mBSA) followed by intra-articular knee re-challenge with mBSA to induce T cell-mediated tissue damage. In each animal, one knee (right, Rt) was injected with mBSA and OVA, and the contralateral control knee (left, Lt) was injected with mBSA only. As a result, the Ag-specific Tregs just recognized the OVA Ag in the Rt knee. Conversely, the arthritis-inducing T cells recognized the mBSA Ag in both knees. The data presented that while OVA was present (Rt knee), the transferred iPSC-derived Tregs essentially reduced the inflammatory knee swelling, however, did not protect the control Lt knee in which only mBSA was injected.

vitro. In addition, adoptive transfer of Ag-specific iPSC-Tregs significantly suppressed the development of autoimmunity in murine models.

Adoptive transfer of OVA-specific iPSC-Tregs in a well-established mouse model of Ag-induced arthritis (AIA) inhibited the development of arthritis [11]. In this murine model, arthritis was induced by intra-articular injection of methylated bovine serum albumin (mBSA) into the knee (Rt). To direct the transferred cells to the knees, OVA was injected into one knee and phosphate-buffered saline (PBS) was injected into the contralateral knee (Lt). Arthritis was characterized by swelling of the synovium and damage of the cartilage around the joints leading to the joint destruction (**Figure 1**). OVA-injected knees were protected from developing arthritis, where PBS injected the knee developed severe arthritis (**Figure 2**). Particularly, OVA-specific iPSC-Tregs infiltrate into the knee joints and maintain the Treg phenotype *in vivo* (**Figure 3**). These results indicate that genetically modified iPSC-derived Tregs are tissue-associated and are able to suppress autoimmune arthritis.

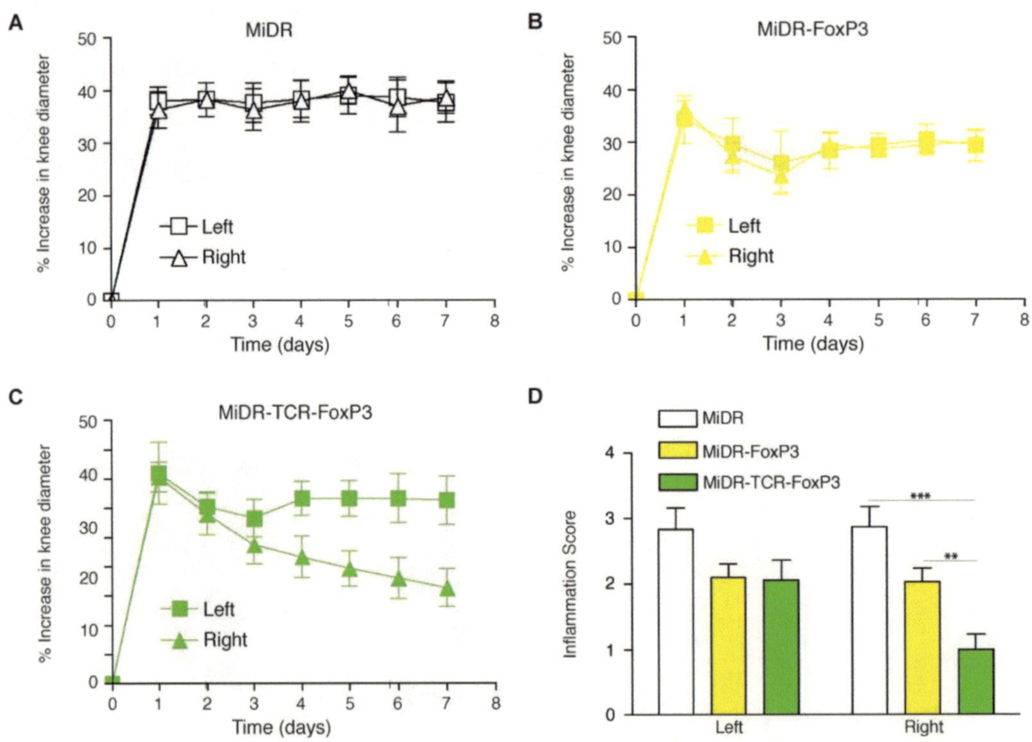

Figure 2. Ag-specific iPSC-Tregs ameliorate AIA in mice. Murine iPSCs transduced with the retroviral construct MiDR, MiDR-FoxP3, or MiDR-TCR-FoxP3 and were cocultured on the OP9-DL1/DL4/I-Ab cells. On day 7, the gene-transduced cells (3 × 106/mouse) were *i.v.* adoptively transferred into C57BL/6 mice that were induced AIA 2 weeks later after the cell transfer. On the following day of arthritis induction, arthritis severity was monitored by the measurement of knee diameter. (**A–C**) % increase in knee diameter. (**D**) The mean scoring on day 7 for both knees from five mice. Data are represented as the mean ± s.d. from three independent experiments (** $p < 0.01$, *** $p < 0.001$, two-way ANOVA).

http://dx.doi.org/10.5772/62695

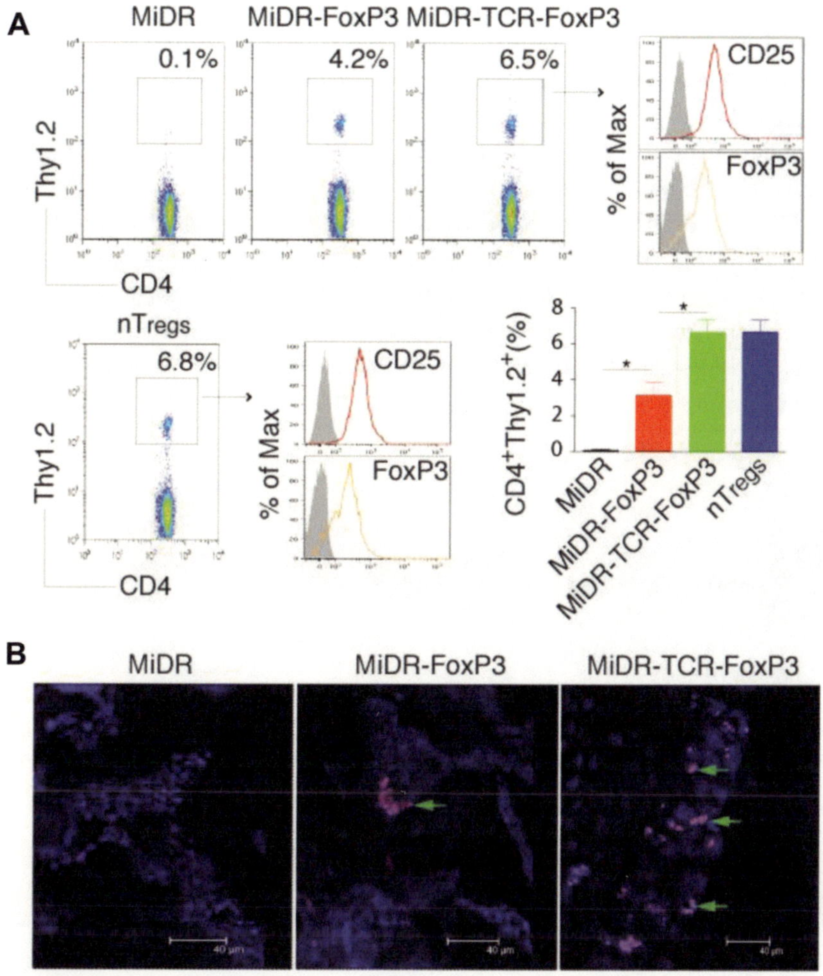

Figure 3. Ag-specific iPSC-Tregs infiltrate into the knee joints and maintain the Treg phenotype *in vivo*. OVA-specific iPSC-Tregs or nTregs ($CD4^+CD25^+$) from OT-II TCR transgenic mice ($Thy1.2^+$) were *i.v.* adoptively transferred into C57BL/6 congenic mice ($Thy1.1^+$) with AIA. (**A**) Six weeks later, the popliteal lymph nodes (LNs) from the inflammatory right side were analyzed for $CD4^+$ $Thy1.2^+$ cells. The mean ± s.d. from three independent experiments is shown (* $p < 0.05$, one-way ANOVA). (**B**) On days 7–14 after arthritis induction, knees were removed and stained for immunohistology with FoxP3. The $FoxP3^+$ cells (←) are indicated.

Adoptive transfer of OVA-specific iPSC-Tregs in a well-established mouse model of autoimmune diabetes suppressed the development of diabetes. T1D is driven by self-reactive T cells that infiltrate the pancreatic islets of Langerhans and induce the destruction of beta cells and the loss of insulin production. This gradually causes pancreas to be unable to control blood glucose levels. During the development of diabetes, the pancreatic islets release the beta cell component that is occupied by immature DCs (IDC) in pancreatic islets. IDC carried the beta cell component to the draining pancreatic lymph node, process the Ag, and present to $CD4^+$ T cells. T cell priming in lymph node leads to the expansion of circulating autoreactive T cells. Following clonal expansion, autoreactive T cells express a number of adhesion molecules, including ICAM 1, the intercellular adhesion molecule 1, and LFA1, the lymphocyte function-associated Ag 1 [92]. This allows the effector cells to home to the pancreatic islets. Once they are in the pancreas, they activate inflammatory cells and causing insulitis (**Figure 4**). PSC-derived Ag-specific Tregs have been used to cell-based therapies of autoimmune diabetes in a murine model, RIP-mOVA × OT-I TCR F1 double transgenic mice. Mice will develop autoimmune diabetes (blood glucose levels >250 mg/dl) when challenged with vaccina viruses expressing OVA (VV-OVA). Adoptive transfer of OVA-specific iPSC-Trergs significantly affected clinical outcome (**Figure 5**) by secreting IL-10 and TGF-β in the pancreas and reducing the expression of ICAM 1. Particularly, adoptive transfer of OVA-specific iPSC-Tregs reduces the number of inflammatory cells and protects beta cell destruction in the pancreas (**Figure 6**). These results also suggest that genetically modified iPSC-derived Tregs are tissue-associated and are able to suppress autoimmune diabetes.

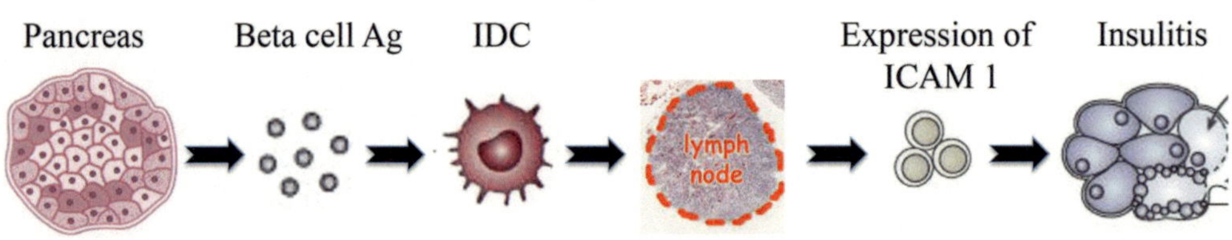

Figure 4. Insulitis in autoimmune arthritis. During the development of diabetes, the pancreatic islets release the beta cell component that is taken-up by IDC in pancreatic islets. IDC carried the beta cell component to the draining pancreatic lymph node, process the Ag, and present to CD4+ T cells. T-cell priming in lymph nodes leads to the expansion of circulating autoreactive T cells. Following clonal expansion, autoreactive T cells express a number of adhesion molecules, including ICAM 1, which allows the effector cells to home to the pancreatic islets. Once they are in the pancreas, they activate inflammatory cells and causing insulitis.

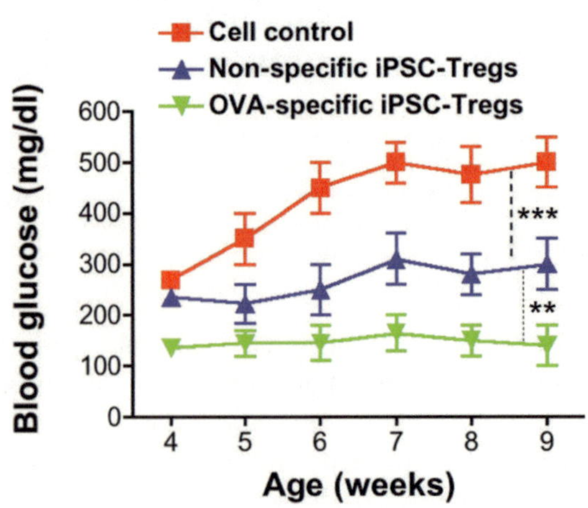

Figure 5. Ag-specific iPSC-Tregs ameliorate autoimmune diabetes in mice. 3-week-old RIP-mOVA × OT-I F1 transgenic mice (n = 5/group) were *i.p.* injected with VV-OVA viruses and adoptively transferred with OVA-specific iPSC-Tregs, non-specific iPSC-Tregs, or iPSC control. In the following weeks, the blood glucose levels were monitored by measurement of blood glucose (*** $p < 0.001$, ** $p < 0.01$, one-way ANOVA).

http://dx.doi.org/10.5772/62695

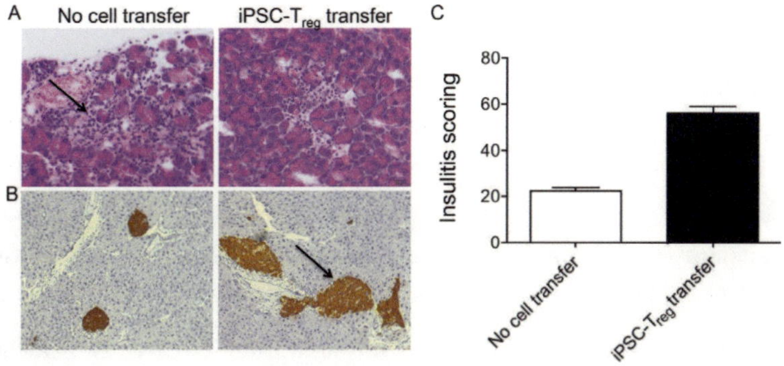

Figure 6. Adoptive transfer of Ag-specific iPSC-Tregs reduces the number of inflammatory cells and protects beta cell destruction in the pancreas. At week 2 following T1D induction, the pancreases were removed, sectioned, and stained with HE or insulin immunofluorescence. (**A**) Representative photomicrographs of HE staining. The cellular infiltrations (–) of inflammatory cells are indicated. (**B**) Representative photomicrographs by insulin immunoflurescent staining. Insulin-secreting cells (–) are indicated. (**C**) Quantitation of beta cell colonies (insulitis scoring) in each group.

8. Future perspectives

T-cell–mediated suppression in immunologic tolerance still remains an exciting area of active research in immunology. It has already been established that a unique Treg population is engaged in the maintenance of immunologic self-tolerance. The natural development of such Tregs will be the crucial for mediating the self-tolerance. However, Tregs are hard to define phenotypically due to the lack of characteristic surface markers. Investigating the function and development of Tregs will contribute to the understanding of immunologic self-tolerance and shed light on the acquisition of autoimmune disorders. Published evidence showed that human Tregs constitutively express high levels of FoxP3 and that mutations in *FOXP3* results in severe autoimmunity. This demonstrates that the expression of this transcription factor has a key role in Treg function. Moreover, genetic modification of stem cells with FoxP3 for the differentiation of Ag-specific Tregs can pave a way for new strategies for the treatment or prevention of autoimmune diseases. However, preclinical data supporting the safety and efficacy of gene therapy approaches is required to allow the transition from the bench to the clinic.

Author details

Mohammad Haque, Praneet Sandhu and Jianxun Song*

*Address all correspondence to: jsong2@hmc.psu.edu

The Pennsylvania State University College of Medicine, Hershey, PA, USA

References

[1] Huynh A, DuPage M, Priyadharshini B, Sage PT, Quiros J, Borges CM, et al. Control of PI(3) kinase in Treg cells maintains homeostasis and lineage stability. Nature Immunology. 2015;16(2):188–96.

[2] Liu Y, Wang L, Han R, Beier UH, Akimova T, Bhatti T, et al. Two histone/protein acetyltransferases, CBP and p300, are indispensable for Foxp3+ T-regulatory cell development and function. Molecular and Cellular Biology. 2014;34(21):3993–4007.

[3] Baca Jones C, Pagni PP, Fousteri G, Sachithanantham S, Dave A, Rodriguez-Calvo T, et al. Regulatory T cells control diabetes without compromising acute anti-viral defense. Clinical Immunology. 2014;153(2):298-307.

[4] Wing JB, Ise W, Kurosaki T, Sakaguchi S. Regulatory T cells control antigen-specific expansion of Tfh cell number and humoral immune responses via the coreceptor CTLA-4. Immunity. 2014;41(6):1013–25.

[5] Maeda Y, Nishikawa H, Sugiyama D, Ha D, Hamaguchi M, Saito T, et al. Detection of self-reactive CD8(+) T cells with an anergic phenotype in healthy individuals. Science. 2014;346(6216):1536–40.

[6] Gershon RK, Kondo K. Cell interactions in the induction of tolerance: the role of thymic lymphocytes. Immunology. 1970;18(5):723–37.

[7] Sakaguchi S. Regulatory T cells: key controllers of immunologic self-tolerance. Cell. 2000;101(5):455–8.

[8] Ehrenstein MR, Evans JG, Singh A, Moore S, Warnes G, Isenberg DA, et al. Compromised function of regulatory T cells in rheumatoid arthritis and reversal by anti-TNFalpha therapy. The Journal of Experimental Medicine. 2004;200(3):277–85.

[9] Valencia X, Stephens G, Goldbach-Mansky R, Wilson M, Shevach EM, Lipsky PE. TNF downmodulates the function of human CD4+CD25hi T-regulatory cells. Blood. 2006;108(1):253–61.

[10] Viglietta V, Baecher-Allan C, Weiner HL, Hafler DA. Loss of functional suppression by CD4+CD25+ regulatory T cells in patients with multiple sclerosis. The Journal of Experimental Medicine. 2004;199(7):971–9.

[11] Haque M, Song J, Fino K, Sandhu P, Song X, Lei F, et al. Stem cell-derived tissue-associated regulatory T cells ameliorate the development of autoimmunity. Scientific Reports. 2016;6:20588.

[12] Piccirillo CA, Thornton AM. Cornerstone of peripheral tolerance: naturally occurring CD4+CD25+ regulatory T cells. Trends in Immunology. 2004;25(7):374–80.

[13] Fehervari Z, Sakaguchi S. Development and function of CD25+CD4+ regulatory T cells. Current Opinion in Immunology. 2004;16(2):203–8.

[14] Chatenoud L, Bach JF. Adaptive human regulatory T cells: myth or reality? The Journal of Clinical Investigation. 2006;116(9):2325–7.

[15] Salomon B, Lenschow DJ, Rhee L, Ashourian N, Singh B, Sharpe A, et al. B7/CD28 costimulation is essential for the homeostasis of the $CD4^{+}CD25^{+}$ immunoregulatory T cells that control autoimmune diabetes. Immunity. 2000;12(4):431–40.

[16] Tang Q, Henriksen KJ, Boden EK, Tooley AJ, Ye J, Subudhi SK, et al. Cutting edge: CD28 controls peripheral homeostasis of $CD4^{+}CD25^{+}$ regulatory T cells. The Journal of Immunology. 2003;171(7):3348–52.

[17] Lohr J, Knoechel B, Kahn EC, Abbas AK. Role of B7 in T cell tolerance. The Journal of Immunology. 2004;173(8):5028–35.

[18] Wirnsberger G, Hinterberger M, Klein L. Regulatory T-cell differentiation versus clonal deletion of autoreactive thymocytes. Immunology and Cell Biology. 2011;89(1):45–53.

[19] Hori S. c-Rel: a pioneer in directing regulatory T-cell lineage commitment? European Journal of Immunology. 2010;40(3):664–7.

[20] Lio CW, Hsieh CS. A two-step process for thymic regulatory T cell development. Immunity. 2008;28(1):100–11.

[21] Fontenot JD, Rasmussen JP, Gavin MA, Rudensky AY. A function for interleukin 2 in Foxp3-expressing regulatory T cells. Nature Immunology. 2005;6(11):1142–51.

[22] Fontenot JD, Rasmussen JP, Williams LM, Dooley JL, Farr AG, Rudensky AY. Regulatory T cell lineage specification by the forkhead transcription factor foxp3. Immunity. 2005;22(3):329–41.

[23] Fontenot JD, Gavin MA, Rudensky AY. Foxp3 programs the development and function of $CD4^{+}CD25^{+}$ regulatory T cells. Nature Immunology. 2003;4(4):330–6.

[24] Tanigaki K, Honjo T. Regulation of lymphocyte development by Notch signaling. Nature Immunology. 2007;8(5):451–6. Epub 2007/04/19.

[25] Yi Z, Lin WW, Stunz LL, Bishop GA. The adaptor TRAF3 restrains the lineage determination of thymic regulatory T cells by modulating signaling via the receptor for IL-2. Nature Immunology. 2014;15(9):866–74.

[26] Pomie C, Vicente R, Vuddamalay Y, Lundgren BA, van der Hoek M, Enault G, et al. Autoimmune regulator (AIRE)-deficient $CD8^{+}CD28$low regulatory T lymphocytes fail to control experimental colitis. Proceedings of the National Academy of Sciences of the United States of America. 2011;108(30):12437–42.

[27] Hossain DM, Panda AK, Manna A, Mohanty S, Bhattacharjee P, Bhattacharyya S, et al. FoxP3 acts as a cotranscription factor with STAT3 in tumor-induced regulatory T cells. Immunity. 2013;39(6):1057–69.

[28] Passerini L, Rossi Mel E, Sartirana C, Fousteri G, Bondanza A, Naldini L, et al. CD4(+) T cells from IPEX patients convert into functional and stable regulatory T cells by FOXP3 gene transfer. Science Translational Medicine. 2013;5(215):215ra174.

[29] Aschenbrenner K, D'Cruz LM, Vollmann EH, Hinterberger M, Emmerich J, Swee LK, et al. Selection of Foxp3+ regulatory T cells specific for self antigen expressed and presented by Aire+ medullary thymic epithelial cells. Nature Immunology. 2007;8(4): 351–8. Epub 2007/02/27.

[30] Lin W, Haribhai D, Relland LM, Truong N, Carlson MR, Williams CB, et al. Regulatory T cell development in the absence of functional Foxp3. Nature Immunology. 2007;8(4):359–68.

[31] Gavin MA, Rasmussen JP, Fontenot JD, Vasta V, Manganiello VC, Beavo JA, et al. Foxp3-dependent programme of regulatory T-cell differentiation. Nature. 2007;445(7129):771–5.

[32] Kappler JW, Roehm N, Marrack P. T cell tolerance by clonal elimination in the thymus. Cell. 1987;49(2):273–80.

[33] Kisielow P, Bluthmann H, Staerz UD, Steinmetz M, von Boehmer H. Tolerance in T-cell-receptor transgenic mice involves deletion of nonmature CD4+8+ thymocytes. Nature. 1988;333(6175):742–6.

[34] Kishimoto H, Sprent J. A defect in central tolerance in NOD mice. Nature Immunology. 2001;2(11):1025–31.

[35] Liston A, Hardy K, Pittelkow Y, Wilson SR, Makaroff LE, Fahrer AM, et al. Impairment of organ-specific T cell negative selection by diabetes susceptibility genes: genomic analysis by mRNA profiling. Genome Biology. 2007;8(1):R12.

[36] Rocha B, von Boehmer H. Peripheral selection of the T cell repertoire. Science. 1991;251(4998):1225–8.

[37] Rocha B, Tanchot C, Von Boehmer H. Clonal anergy blocks in vivo growth of mature T cells and can be reversed in the absence of antigen. The Journal of Experimental Medicine. 1993;177(5):1517–21.

[38] Sakaguchi S. Naturally arising Foxp3-expressing CD25+CD4+ regulatory T cells in immunological tolerance to self and non-self. Nature Immunology. 2005;6(4):345–52.

[39] Sakaguchi S, Yamaguchi T, Nomura T, Ono M. Regulatory T cells and immune tolerance. Cell. 2008;133(5):775–87.

[40] Apostolou I, Sarukhan A, Klein L, von Boehmer H. Origin of regulatory T cells with known specificity for antigen. Nature Immunology. 2002;3(8):756–63.

[41] Kretschmer K, Apostolou I, Hawiger D, Khazaie K, Nussenzweig MC, von Boehmer H. Inducing and expanding regulatory T cell populations by foreign antigen. Nature Immunology. 2005;6(12):1219–27.

[42] Jaeckel E, von Boehmer H, Manns MP. Antigen-specific FoxP3-transduced T-cells can control established type 1 diabetes. Diabetes. 2005;54(2):306–10.

[43] Tarbell KV, Yamazaki S, Olson K, Toy P, Steinman RM. $CD25^+$ $CD4^+$ T cells, expanded with dendritic cells presenting a single autoantigenic peptide, suppress autoimmune diabetes. The Journal of Experimental Medicine. 2004;199(11):1467–77.

[44] Rudensky AY, Campbell DJ. In vivo sites and cellular mechanisms of T reg cell-mediated suppression. The Journal of Experimental Medicine. 2006;203(3):489–92.

[45] Marson A, Kretschmer K, Frampton GM, Jacobsen ES, Polansky JK, MacIsaac KD, et al. Foxp3 occupancy and regulation of key target genes during T-cell stimulation. Nature. 2007;445(7130):931–5.

[46] Chen C, Rowell EA, Thomas RM, Hancock WW, Wells AD. Transcriptional regulation by Foxp3 is associated with direct promoter occupancy and modulation of histone acetylation. The Journal of Biological Chemistry. 2006;281(48):36828–34.

[47] Itoh M, Takahashi T, Sakaguchi N, Kuniyasu Y, Shimizu J, Otsuka F, et al. Thymus and autoimmunity: production of $CD25^+CD4^+$ naturally anergic and suppressive T cells as a key function of the thymus in maintaining immunologic self-tolerance. The Journal of Immunology. 1999;162(9):5317–26.

[48] Shevach EM. From vanilla to 28 flavors: multiple varieties of T regulatory cells. Immunity. 2006;25(2):195–201.

[49] von Boehmer H. Mechanisms of suppression by suppressor T cells. Nature Immunology. 2005;6(4):338–44.

[50] Tang Q, Bluestone JA. The $Foxp3^+$ regulatory T cell: a jack of all trades, master of regulation. Nature Immunology. 2008;9(3):239–44.

[51] Takahashi T, Kuniyasu Y, Toda M, Sakaguchi N, Itoh M, Iwata M, et al. Immunologic self-tolerance maintained by $CD25^+CD4^+$ naturally anergic and suppressive T cells: induction of autoimmune disease by breaking their anergic/suppressive state. International Immunology. 1998;10(12):1969–80.

[52] Takahashi T, Tagami T, Yamazaki S, Uede T, Shimizu J, Sakaguchi N, et al. Immunologic self-tolerance maintained by CD25(+)CD4(+) regulatory T cells constitutively expressing cytotoxic T lymphocyte-associated antigen 4. The Journal of Experimental Medicine. 2000;192(2):303–10.

[53] Read S, Malmstrom V, Powrie F. Cytotoxic T lymphocyte-associated antigen 4 plays an essential role in the function of CD25(+)CD4(+) regulatory cells that control intestinal inflammation. The Journal of Experimental Medicine. 2000;192(2):295–302.

[54] Vogtenhuber C, O'Shaughnessy MJ, Vignali DA, Blazar BR. Outgrowth of CD4low/neg$CD25^+$ T cells with suppressor function in $CD4^+CD25^+$ T cell cultures upon polyclonal stimulation ex vivo. The Journal of Immunology. 2008;181(12):8767–75.

[55] Hoffmann P, Boeld TJ, Eder R, Huehn J, Floess S, Wieczorek G, et al. Loss of FOXP3 expression in natural human $CD4^{+}CD25^{+}$ regulatory T cells upon repetitive in vitro stimulation. European Journal of Immunology. 2009;39(4):1088–97.

[56] Choi J, Ritchey J, Prior JL, Holt M, Shannon WD, Deych E, et al. In vivo administration of hypomethylating agents mitigate graft-versus-host disease without sacrificing graft-versus-leukemia. Blood. 2010;116(1):129–39.

[57] Xu W, Lan Q, Chen M, Chen H, Zhu N, Zhou X, et al. Adoptive transfer of induced-Treg cells effectively attenuates murine airway allergic inflammation. Plos One. 2012;7(7):e40314.

[58] Sharma MD, Huang L, Choi JH, Lee EJ, Wilson JM, Lemos H, et al. An inherently bifunctional subset of $Foxp3^{+}$ T helper cells is controlled by the transcription factor eos. Immunity. 2013;38(5):998–1012.

[59] Perro M, Tsang J, Xue SA, Escors D, Cesco-Gaspere M, Pospori C, et al. Generation of multi-functional antigen-specific human T-cells by lentiviral TCR gene transfer. Gene Therapy. 2010;17(6):721–32.

[60] Johnson LA, Morgan RA, Dudley ME, Cassard L, Yang JC, Hughes MS, et al. Gene therapy with human and mouse T-cell receptors mediates cancer regression and targets normal tissues expressing cognate antigen. Blood. 2009;114(3):535–46.

[61] Bobisse S, Rondina M, Merlo A, Tisato V, Mandruzzato S, Amendola M, et al. Reprogramming T lymphocytes for melanoma adoptive immunotherapy by T-cell receptor gene transfer with lentiviral vectors. Cancer Research. 2009;69(24):9385–94.

[62] Porter DL, Levine BL, Kalos M, Bagg A, June CH. Chimeric antigen receptor-modified T cells in chronic lymphoid leukemia. The New England Journal of Medicine. 2011;365(8):725–33.

[63] Gehring AJ, Xue SA, Ho ZZ, Teoh D, Ruedl C, Chia A, et al. Engineering virus-specific T cells that target HBV infected hepatocytes and hepatocellular carcinoma cell lines. Journal of Hepatology. 2011;55(1):103–10.

[64] Brusko TM, Koya RC, Zhu S, Lee MR, Putnam AL, McClymont SA, et al. Human antigen-specific regulatory T cells generated by T cell receptor gene transfer. Plos One. 2010;5(7):e11726.

[65] Wright GP, Notley CA, Xue SA, Bendle GM, Holler A, Schumacher TN, et al. Adoptive therapy with redirected primary regulatory T cells results in antigen-specific suppression of arthritis. Proceedings of the National Academy of Sciences of the United States of America. 2009;106(45):19078–83.

[66] Hill JA, Feuerer M, Tash K, Haxhinasto S, Perez J, Melamed R, et al. Foxp3 transcription-factor-dependent and-independent regulation of the regulatory T cell transcriptional signature. Immunity. 2007;27(5):786–800.

[67] Stauss HJ, Cesco-Gaspere M, Thomas S, Hart DP, Xue SA, Holler A, et al. Monoclonal T-cell receptors: new reagents for cancer therapy. Molecular Therapy: The Journal of the American Society of Gene Therapy. 2007;15(10):1744–50.

[68] Bendle GM, Linnemann C, Hooijkaas AI, Bies L, de Witte MA, Jorritsma A, et al. Lethal graft-versus-host disease in mouse models of T cell receptor gene therapy. Nature Medicine. 2010;16(5):565–70, 1p following 70.

[69] Sanchez AM, Zhu J, Huang X, Yang Y. The development and function of memory regulatory T cells after acute viral infections. The Journal of Immunology. 2012;189(6):2805–14.

[70] Svenson U, Gronlund E, Soderstrom I, Sitaram RT, Ljungberg B, Roos G. Telomere length in relation to immunological parameters in patients with renal cell carcinoma. Plos One. 2013;8(2):e55543.

[71] Gratz IK, Truong HA, Yang SH, Maurano MM, Lee K, Abbas AK, et al. Cutting Edge: Memory regulatory T cells require IL-7 and Not IL-2 for their maintenance in peripheral tissues. The Journal of Immunology. 2013;190(9):4483–7.

[72] Kuball J, Dossett ML, Wolfl M, Ho WY, Voss RH, Fowler C, et al. Facilitating matched pairing and expression of TCR chains introduced into human T cells. Blood. 2007;109(6):2331–8.

[73] van Loenen MM, de Boer R, Amir AL, Hagedoorn RS, Volbeda GL, Willemze R, et al. Mixed T cell receptor dimers harbor potentially harmful neoreactivity. Proceedings of the National Academy of Sciences of the United States of America. 2010;107(24):10972–7.

[74] Kim YC, Zhang AH, Su Y, Rieder SA, Rossi RJ, Ettinger RA, et al. Engineered antigen-specific human regulatory T cells: immunosuppression of FVIII-specific T- and B-cell responses. Blood. 2014.

[75] Haque R, Lei F, Xiong X, Bian Y, Zhao B, Wu Y, et al. Programming of regulatory T cells from pluripotent stem cells and prevention of autoimmunity. The Journal of Immunology. 2012;189(3):1228–36.

[76] Lei F, Haque R, Xiong X, Song J. Directed differentiation of induced pluripotent stem cells towards T lymphocytes. Journal of Visualized Experiments: JoVE. 2012(63):e3986.

[77] Chin MH, Mason MJ, Xie W, Volinia S, Singer M, Peterson C, et al. Induced pluripotent stem cells and embryonic stem cells are distinguished by gene expression signatures. Cell Stem Cell. 2009;5(1):111–23.

[78] Kiskinis E, Eggan K. Progress toward the clinical application of patient-specific pluripotent stem cells. The Journal of Clinical Investigation. 2010;120(1):51–9.

[79] Park IH, Zhao R, West JA, Yabuuchi A, Huo H, Ince TA, et al. Reprogramming of human somatic cells to pluripotency with defined factors. Nature. 2008;451(7175):141–6.

[80] Takahashi K, Tanabe K, Ohnuki M, Narita M, Ichisaka T, Tomoda K, et al. Induction of pluripotent stem cells from adult human fibroblasts by defined factors. Cell. 2007;131(5):861–72.

[81] Takahashi K, Yamanaka S. Induction of pluripotent stem cells from mouse embryonic and adult fibroblast cultures by defined factors. Cell. 2006;126(4):663–76.

[82] Yu J, Vodyanik MA, Smuga-Otto K, Antosiewicz-Bourget J, Frane JL, Tian S, et al. Induced pluripotent stem cell lines derived from human somatic cells. Science. 2007;318(5858):1917–20.

[83] Zhao T, Zhang ZN, Rong Z, Xu Y. Immunogenicity of induced pluripotent stem cells. Nature. 2011;474(7350):212–5.

[84] Lei F, Haque R, Weiler L, Vrana KE, Song J. T lineage differentiation from induced pluripotent stem cells. Cellular Immunology. 2009;260(1):1–5.

[85] Lei F, Zhao B, Haque R, Xiong X, Budgeon L, Christensen ND, et al. In vivo programming of tumor antigen-specific T lymphocytes from pluripotent stem cells to promote cancer immunosurveillance. Cancer Research. 2011;71(14):4742–7.

[86] Thornton AM, Shevach EM. Suppressor effector function of $CD4^+CD25^+$ immunoregulatory T cells is antigen nonspecific. The Journal of Immunology. 2000;164(1):183–90.

[87] Komatsu N, Okamoto K, Sawa S, Nakashima T, Oh-hora M, Kodama T, et al. Pathogenic conversion of $Foxp3^+$ T cells into TH17 cells in autoimmune arthritis. Nature Medicine. 2014;20(1):62–8.

[88] Bailey-Bucktrout SL, Martinez-Llordella M, Zhou X, Anthony B, Rosenthal W, Luche H, et al. Self-antigen-driven activation induces instability of regulatory T cells during an inflammatory autoimmune response. Immunity. 2013;39(5):949–62.

[89] Yang L, Baltimore D. Long-term in vivo provision of antigen-specific T cell immunity by programming hematopoietic stem cells. Proceedings of the National Academy of Sciences of the United States of America. 2005;102(12):4518–23.

[90] Zhao Y, Parkhurst MR, Zheng Z, Cohen CJ, Riley JP, Gattinoni L, et al. Extrathymic generation of tumor-specific T cells from genetically engineered human hematopoietic stem cells via Notch signaling. Cancer Research. 2007;67(6):2425–9.

[91] Alajez NM, Schmielau J, Alter MD, Cascio M, Finn OJ. Therapeutic potential of a tumor-specific, MHC-unrestricted T-cell receptor expressed on effector cells of the innate and the adaptive immune system through bone marrow transduction and immune reconstitution. Blood. 2005;105(12):4583–9.

[92] Roncarolo MG, Battaglia M. Regulatory T-cell immunotherapy for tolerance to self antigens and alloantigens in humans. Nature Reviews Immunology. 2007;7(8):585–98.

Chapter 23

Mesenchymal Stem Cells: Biological Characteristics and Potential Clinical Applications for Haematopoietic Stem Cell Transplantation

Teresa de Souza Fernandez and
Cecília de Souza Fernandez

Additional information is available at the end of the chapter

http://dx.doi.org/10.5772/63772

Abstract

Mesenchymal stem cells (MSCs) are multipotent cells that can be expanded and manipulated *ex vivo*. These cells demonstrated three biological characteristics that qualify them for the use in cellular therapy: (1) potential of differentiation, (2) secretion of trophic factors and (3) immunoregulatory properties. The bone marrow (BM) has been considered as the traditional source of MSCs and much knowledge for potential clinical applications has been obtained from studies using MSCs derived from adult bone marrow. MSCs need to be expanded *in vitro* for the purpose of cell therapy. However, sometimes, the culture expansion could generate cytogenetic and molecular alterations. Accumulation of these alterations during many passages of culture could lead to malignant cell transformation. So, it is important to perform a rigorous control using different methods to test the safety and efficacy of MSCs for cell therapies. BM-MSCs have potential clinical applications in haematopoietic stem cell transplantation (HSCT) as an adjuvant cellular therapy. This chapter reviews the advances in the study of MSCs and the potential clinical applications of MSCs in haematopoietic stem cell transplantation (HSCT). We also describe the importance of statistical methods to aid the analysis of the efficacy and safety for the clinical use of MSCs for HSCT.

Keywords: mesenchymal stem cells, biological characteristics, cell therapy, haematopóietic stem cell transplantation, statistical methods.

1. Introduction

Mesenchymal stem cells (MSCs) are multipotent cells that can differentiate *in vitro* into mesenchymal cell lineages such as adipocytes, osteocytes and chondrocytes. MSCs can be expanded and manipulated *ex vivo*. According to the minimal criteria of the International Society for Cellular Therapy, MSCs are defined by their growth pattern *in vitro* (the cultured plastic-adherent cells), the specific surface antigen expression (CD73, CD90 and CD105, in the absence of lineage commitment markers such as CD14, CD19, CD34, CD45 and HLA-DR) and multilineage potential (these cells must be able to differentiate into osteoblasts, adipocytes and chondroblasts *in vitro*) [1]. MSCs can be derived from adult bone marrow (BM), adipose tissue and several fetal tissues as umbilical cord. The bone marrow has been considered as the conventional source of MSCs and most knowledge for potential clinical applications has been obtained from studies using MSCs derived from adult bone marrow [2].

MSCs can be expanded and manipulated *ex vivo* and can demonstrate immunomodulatory functions *in vitro* and *in vivo*. Thus, they represent promising tools to be used in immunoregulatory and regenerative cell therapies. Recently, many studies have revealed the clinical use of MSCs as an emerging field for treating cardiovascular disorders, neurodegenerative diseases, bone defects and fractures, inflammatory arthritis and in the field of haematopoietic stem cell transplantation (HSCT) [3]. Bone marrow MSCs constitute approximately 0.01% of mononuclear cells in the bone marrow [4]. Hence, MSCs have to be expanded *in vitro* on tissue culture plastic for the purpose of cell therapy. An extensive amplification *in vitro* is necessary without affecting the cells' genomic characteristics and differentiation properties. But, sometimes, the culture expansion could generate cytogenetic and molecular alterations. Accumulation of these alterations during many passages of culture could lead to malignant cell transformation. Hence, it is important to perform a quality control using different methods to test the safety and efficacy of MSCs for cell therapies.

MSCs generate most of the stromal cells present in the bone marrow (BM). They form part of the haematopoietic stem cell (HSC) niche and produce various factors regulating haematopoiesis. It has been proposed that BM-MSCs are useful as an adjuvant cellular therapy for promoting rapid haematopoietic recovery in the HSCT patients. This chapter reviews the advances in the study of MSCs and the potential clinical applications of MSCs in haematopoietic stem cell transplantation (HSCT). We will also describe the importance of statistical methods to aid the analysis of the efficacy and safety for the clinical use of MSC for HSCT.

2. Definition and biological characteristics of mesenchymal stem cells (MSCs)

MSCs were first described in 1966 by Friedenstein and colleagues. They reported the presence of fibroblastoid cells that could be obtained from bone marrow of adult mice and when transplanted subcutaneously, they could differentiate toward osteogenesis [5]. After this discovery, several studies have been done using human mesenchymal stem cells. These studies confirmed that it is possible to culture and do sub-passages of the whole bone marrow into plastic culture dishes and after discarding the non-adherent cells a few hours later, the cells adhered to the plastic were capable of forming colonies (colony-forming unit: fibroblastic, CFU-F). It was observed that the MSCs have two important properties. First, they can differentiate into distinctive end-stage cell types, including bone, cartilage, muscle, bone marrow stroma, tendon/ligament, fat, dermis and other connective tissues. Second, MSCs, themselves, secrete a broad spectrum of bioactive macromolecules that are immunoregulatory and serve to structure regenerative microenvironments in an injured tissue [6].

MSCs are not only found in bone marrow. MSCs have been isolated from multiple tissues such as skeletal muscle, adipose tissue, synovial membranes, dental pulp, periodontal ligaments, cervical tissue, umbilical cord, amniotic fluid and placenta. However, much knowledge regarding the biological characteristics and clinical experiences has been obtained from studies of MSCs derived from adult bone marrow [2, 7–9]. MSCs, also known as multipotent cells, are found in adult tissues of different sources. They are self-renewable, multipotent, easily accessible, and culturally expandable *in vitro* [10].

When cultivated *in vitro*, MSCs have three biological characteristics that qualify them for use in cellular therapy: (1) potential of differentiation, (2) secretion of trophic factors that help tissue remodelling and (3) immunoregulatory properties [2]. Therapeutic benefits of MSCs are dependent on their capacity to act as a trophic factor pool. After MSCs home to damaged tissue sites for repair, they interact with local stimuli, such as inflammatory cytokines, ligands of Toll-like receptors (TLRs) and hypoxia, which can stimulate MSCs to produce a large amount of growth factors that act with multiple functions for tissue regeneration. Many of these factors are critical mediators in angiogenesis and prevention of cell apoptosis such as vascular endothelial growth factor (VEGF), insulin-like growth factor 1 (IGF-1) and IL-6 [11].

Many studies have demonstrated the immunoregulatory properties of MSCs. These cells affect the immune response through their interactions with the cellular components of the immune system: T lymphocytes, B lymphocytes, natural killer (NK) cells and dendritic cells (DCs). MSC immunoregulation can occur through cellular contact and/or by secretion of diverse factors. Because of these properties, MSCs can prevent the inappropriate activation of T lymphocytes and generate a tolerogenic environment during repair or stop an immune response during healing, thus contributing to the maintenance of immune homeostasis [12,13]. Immunomodulatory properties of MSCs can be grouped into three categories: being hypoimmunogenic, modulating T cell phenotype and immunosuppressing the local environment [14, 15].

MSCs have decreased the expression of surface molecules including low levels of MHC class I and costimulatory CD40, CD80, and CD86 and no MHC class II molecules. This distribution of surface markers allows MSCs to evade detection from certain immune cells and contributes to their hypoimmunogenicity. MSCs also have the ability to immunosuppress the local environment and this can be attributed to their effect on cytokine secretion profiles. Specifically, in co-cultures with immune cells, MSCs had an indirect effect on T cell maturation and proliferation by up regulating the secretion of suppressive cytokines (IL-4 and IL-10) to decrease the secretion of proinflammatory cytokines (TNF-α and IFN-γ) from dendritic cells, T helper cells and macrophages. MSCs have the ability to induce regulatory T cells, which ultimately inhibit the proliferation and function of T cells, B cells and natural killer cells. Several soluble mediators, such as transforming growth factor β1, prostaglandin E2 (PGE2), human leukocyte antigen G5, haemoxygenase I, nitric oxide, IL-6 and indoleamine 2,3-dioxygenase

(IDO), are important for this process [13]. Indoleamine 2,3-dioxygenase, which is induced by IFN-γ, catalyzes the conversion from tryptophan to kynurenine and inhibits T-cell responses [16,17].

As we can observe, the immunomodulatory characteristics of MSCs are important for cell therapy. But, approximately 2×10^6 cells/kg are required for clinical application of MSCs [3]. Therefore, for cell therapy, it is necessary to expand the MSCs using culture methods.

2.1. Isolation and culture expansion of MSCs for cell therapy

Clinical protocols employ cell culture technologies that use a small fraction of primary MSCs isolated from a selected tissue source and expanded by multiple passages in order to generate a clinically relevant number of cells. Consequently, once the tissue source of MSCs is determined for a specific clinical application, the safety and efficacy may be significantly influenced by cell bioprocessing protocols [18, 19].

There are no standard culture protocols for isolation and expansion of MSCs. Hence, the way in which these cells are cultured *in vitro* varies considerably between research groups. Consequently, it is difficult to compare results from different studies [19–21]. But, the Mesenchymal and Tissue Stem Cell Committee of the International Society for Cellular Therapy proposed minimal criteria to define MSCs *in vitro*: (1) MSCs must be adherent to plastic under standard tissue culture conditions; (2) MSCs must express certain cell surface markers such as CD73, CD90 and CD105 and lack expression of other markers including CD45, CD34, CD14, or CD11b, CD79alpha or CD19 and HLA-DR surface molecules; (3) MSCs must have the capacity to differentiate into osteoblasts, adipocytes and chondroblasts under *in vitro* conditions [1].

The procedures used to isolate MSCs, for example, from bone marrow, usually use density centrifugation (with Ficoll™, Lymphoprep™ or Percoll™ density mediums) to separate the mononuclear cell fraction from the other marrow constituents such as red blood cells, plasma and lipids. This mononuclear cell fraction contains an enriched population of T cells, B cells, monocytes, hematopoietic stem cells (HSCs), endothelial progenitor cells and MSCs. Following plating onto tissue culture flasks, MSCs, which represent the adherent cell population, form colonies. The adherent cells remain in culture and the other non-adherent cells are discarded while changing the medium [19, 21]. The MSCs are anchorage-dependent cells that expand when maintained in culture conditions such as medium DMEN supplemented with 10% of FBS. Initial growth of MSCs in primary BM cell culture on a plastic surface is characterized by the formation of single cell-derived colonies. The efficiency with which they form colonies still remains an important assay for the quality of cell preparations. In general, MSCs have a great propensity for expansion in culture, although their proliferation potential is highly variable, mainly between young and older donors who retain reduced proliferative potential [19,22].

MSCs seeding densities range between 2.000 and 5.000/cm^2; however, there is evidence that lower seeding densities enhance proliferation, which is thought to be attributed to a reduction in contact inhibition. Some studies demonstrated that MSCs proliferate more rapidly when passaged by plating the cells at low densities as 10–100 cells/cm^2 [19, 23, 24]. MSCs are most commonly expanded in a basal media such as Dulbecco's modified Eagle's medium (DMEM)/ DMEM F-12 or alpha-MEM with 10% fetal bovine serum (FBS) [21]. All current protocols for *in vitro* culture of MSCs include FBS as a nutritional supplement [25]. However, some problems are associated with the use of FBS, for example, risk of contamination associated with harmful pathogens such as viruses, mycoplasma, prions or unidentified zoonotic agents. The chance of contamination or immunological reaction towards xenogeneic compounds must be also taken into consideration [19, 26]. Hence, for using FBS, tests are necessary for providing optimal growth conditions [21].

Successful expansion techniques aim to facilitate significant increases in cell number without affecting the MSC therapeutic potential. MSCs can be cultured *in vitro* for 8–15 passages, corresponding to approximately 25–40 population doublings and 80–120 days. MSCs demonstrate a marked decrease in proliferation as a function of duration in culture and passage number, thereby becoming senescent and ceasing to proliferate [4, 7, 27]. MSCs may lose differentiation capacity during the time in culture to assess multilineage potential. The potency of MSCs rapidly decline as a function of 2D expansion and this shows a demand for alternative expansion techniques [4].

Bone is a 3D substrate composed of water, organic collagen and inorganic hydroxyapatite. MSCs reside within crevices of the blood-submerged bone and among several cell types with which they engage in a complex orchestra of crosstalk. Numerous aspects of the bone marrow niche, which regulate MSC behavior, are absent in 2D culture. Hence, it is necessary to develop new techniques to recreate characteristics of the elaborate niche to preserve MSC progenitor potency through 3D expansion [4, 28–30]. Some studies have been demonstrated that MSCs can be expanded using scaffolds or scaffoldless approaches, usually in combination with a bioreactor. 3D MSC expansion has been performed on hydroxyapatite (HA), chitosan gelatin and HA/chitosan gelatin and gelatin microcarriers [4, 28, 31].

Bioreactors are devices that facilitate the development of biological and/or biochemical processes through operating parameters such as pH, temperature, nutrient supply and waste removal. Bioreactor systems are essential tools to achieve the goals in the clinical-scale expansion and tissue engineering applications [4]. They maintain the minimal criteria to define MSCs, which include plastic adhesion, expression of a set of specific surface markers and the ability of differentiation along osteogenic, adipogenic and chondrogenic lineages [1]. MSCs also bear broad regenerative and trophic activities, including the secretion of extracellular matrix (ECM), pro-mitotic and pro-angiogenic factors, anti-inflammatory and immune-regulatory factors, and other bioactive molecules that stimulate tissue regeneration by reconstructing a pro-regeneration microenvironment and by modulating immune and inflammatory responses. Thus, these unique properties play a central role to the success of MSC-based therapeutic applications [32–34].

2.1.1. *Multilineage potential*

MSCs have the potential for multilineage differentiation. This property has been studied for the development of MSC transplantation as a regenerative therapy. Multilineage potential is a criterion to define the MSCs *in vitro*. The multilineage potential may be observed under culture conditions that induce cell differentiation into three lineages: osteogenic, adipogenic and chondrogenic [35]. A number of *in vitro* assays can be used to assess the multipotentiality of these cell preparations. Osteogenic differentiation of MSCs can be induced using dexamethasone, ascorbate-2-phosphate and beta glycerolphosphate. Osteoblasts may be identified using alizarine red S staining. Adipogenic differentiation can be induced with the medium containing dexamethasone, indomethacin, isobutylmethylxanthine and insulin. Oil red O staining may be used to detect lipid accumulation. Chondrogenic differentiation can be induced in a defined medium containing dexamethasone, ascorbate-2-phosphate, insulin, selenious acid, transferrin, sodium pyruvate and transforming growth factor-beta [3, 36, 37]. The ability of MSCs to differentiate along these lineages is strongly associated with their multipotency and stem cell nature. However, MSCs do not maintain these characteristics indefinitely and MSCs senescence with extensive subcultivation *in vitro* whereby they lose their proliferation and differentiation potential [37]. Such culture expansion could also generate genetic and epigenetic instability, including chromosome alterations. The accumulation of genetic changes during cell culturing and subsequent risk of cell transformation are other important points of stem cell therapy [3, 38].

2.1.2. Cytogenetic and molecular characteristics of MSCs

The utilization of MSCs for cell therapy requires large-scale *in vitro* expansion, thus increasing the probability of cytogenetic and molecular instabilities [38]. The expansion of MSCs in culture could generate chromosomal abnormalities such as aneuploidy (the presence of an abnormal number of chromosomes in a cell) or structural chromosomal alterations, reflecting the chromosomal instability. But, it is not clear how many passages can be performed before these cells acquire chromosome instability or lose their multipotency [3, 39, 40]. Some studies have been shown that *in vitro* culture of MSCs from bone marrow and adipose tissue retained normal karyotypes between passages 1 and 5 [3, 39, 41]. At later passage cultures, MSCs began to show chromosomal abnormalities such as aneuploidy. However, other studies observed that MSC cultures derived from bone marrow and adipose tissue had normal karyotypes up to passage 20 [42, 43]. Although these results are debatable, they show the necessity of cytogenetic analyses for safety before therapeutic application of mesenchymal stromal cells.

Molecular studies also have an important role to determine appropriate MSCs to be used for cell therapy. All primary human cells, including MSCs, undergo only a limited number of cell divisions under standard culture conditions, in a process called cellular senescence. Senescence is considered to be a stress response triggered by activation of some mechanisms as telomere erosion and accumulation of DNA damage [44, 45]. *In vitro* cultures cause significant telomere shortening. Telomeres are the termini of eukaryotic chromosomes and their principal function is to protect chromosomes from illegitimate fusion and recombination, thereby preserving genome integrity [45, 46]. Since MSCs have only a finite ability for self-renewal like most somatic cells, assaying for telomere length in hMSCs provides critical information on the replicative capacity of the cells, an important criterion in the selection of MSCs for therapy. Telomere length is generally quantified by Southern blotting and fluorescence *in situ* hybridization and more recently by polymerase chain reaction (PCR)-based methods [47].

The multipotency of MSCs has allowed a significant progress in our understanding about differentiation pathways of various lineages for tissue engineering and therapeutic purposes [48]. Runt-related transcription factor 2 (Runx2) has been considered as a master regulatory gene responsible for early osteogenic differentiation [49]. While Runx2 acts to promote osteoblastic differentiation, another important osteogenic inducer, osterix, suppresses chondrogenesis and promotes osteoblastic differentiation at a later stage. Low levels of osterix are sufficient to inhibit chondrogenesis, while a high expression level is necessary for osteogenic differentiation [48, 50]. Furthermore, *ex vivo* MSCs have successfully differentiated into osteoblasts in osteogenic media supplemented by dexamethasone and ascorbate. The selective capability to promote osteogenic differentiation has potential clinical implications in bone repair and regeneration [48, 51].

In-vitro differentiation of MSCs into a chondrogenic lineage has been studied through exposure to growth factors, co-culture with cartilage and overexpression of specific genes such as SRY-box 9 (Sox9) to promote chondrocytic differentiation. Sox9 cooperates with its downstream proteins Sox5 and Sox6 to promote chondroycte proliferation and maturation and matrix formation [48, 52]. MSCs also have the capacity to differentiate into an adipogenic lineage. PPAR-γ plays a critical role in this process by regulating the function of many adipocyte-specific genes. In addition, PPARγ interacts with members of the CCAAT/enhancer-binding protein (C/EBP) family to regulate adipogenesis. Cells can also be induced to undergo adipogenesis through exposure to exogenous factors or by culturing them in adipogenic media containing insulin and dexamethasone [48, 53]. As the multilineage potential is one of the three criteria to define the MSCs in vitro according to the International Society of Cellular Therapy [1], the use of molecular tests to analyze the expression of the genes involved in the differentiation of the osteogeneic, chondrogenic and adipogenic are important to associate the biological function of the MSCs for their clinical use.

The expansion of MSCs *in vitro* is associated with genetic instability. Hence, molecular studies comparing the molecular profile during the culture passages are important to acquire knowledge about molecular modifications and potential risks for cell therapy. In this sense, proteomic and transcriptomic approaches have been used to verify molecular modifications of MSCs from different culture passages [3, 54].

We need to be careful before using BM-MSCs for clinical applications. Some changes may be analyzed such as enlarged morphology, decreased number of cell divisions, random loss of genomic regions and telomere shortening. These modifications process could lead to a reduction in the multipotent state of MSCs and might lead to tumour formation under specific conditions. It is very important to characterize the cytogenetic and molecular profiles during expansion *in vitro* of BM-MSCs; thus, appropriate tests should be applied to ensure the integrity of the genome and epigenome [54].

2.1.3. Quality control for cell therapy

There are many challenges associated with characterizing and quantifying cells for use in cell- and tissue-based therapies. From a regulatory perspective, these advanced treatments must not only be safe and effective, but also must be made by high-quality manufacturing processes [55]. Prolonged exposure to stressful conditions during the cell enrichment and differentiation processes has raised concerns about the safety of stem cell therapy. The International Society for Stem Cell Research has created "Guidelines for the Clinical Translation of Stem Cells" [56]. Some cytogenetic tests that may be performed to ensure the safety of the stem cells include: G-banding, fluorescence in situ hybridization (FISH) and array comparative genomic hybridization (array CGH) [38, 57]. Molecular genetic tests may be performed as the analysis of the telomere length, the expression of genes involved in the osteogenic, adipogenic and chondrogenic differentiation. Some characteristics and tests, which may be considered as a quality control for the use of MSCs for cell therapy, are shown in **Figure 1**. The practical application of these recommended tests can be standardized for the sensitivity and specificity between laboratories.

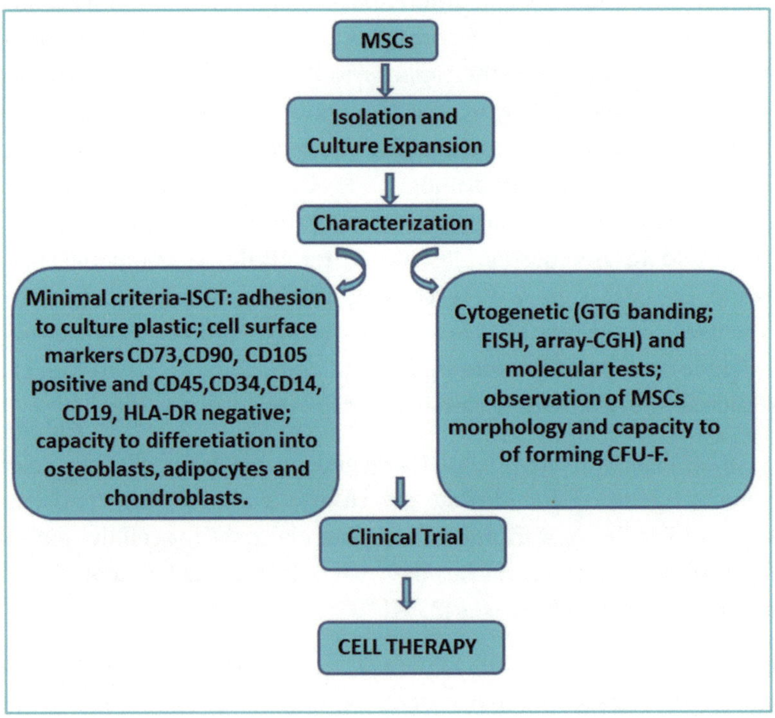

Figure 1. Some characteristics and tests that may be considered as a quality control for the use of MSCs for cell therapy.

The first clinical trial using culture-expanded MSCs was performed in 1995 and bone marrow samples were obtained from 23 patients with haematologic malignancies in complete remission. In this study, as no adverse reactions were observed with the infusion of the MSCs, Lazarus and colleagues concluded that MSCs obtained from cancer patients can be collected, expanded *in vitro* and infused intravenously without toxicity [58]. Many completed clinical trials have demonstrated the efficacy of MSC infusion for diseases including acute myocardial ischaemia, liver cirrhosis, amyotrophic lateral sclerosis and graft versus host disease (GVHD) [12, 59]. The statistical methods are important tools to evaluate the quality, safety and efficiency of MSCs for cellular therapy as we will observe in the last section.

3. Potential clinical applications of MSCs in haematopoietic stem cell transplantation

Haematopoietic stem cell transplantation (HSCT) is the first field wherein human stem cell therapy was successful. Allogeneic HSCT has been an important modality to cure various diseases, including haematologic malignancies, various non-malignant haematologic diseases and primary immunodeficiency diseases. However, autologous HSCT is generally performed to rescue bone marrow aplasia following high-dose chemotherapy for solid tumour or multiple myeloma [60].

The first successful HSCT using bone marrow from a relative donor was performed in 1968 in a boy with X-linked severe combined immunodeficiency disease (SCID). Since the first successful achievement in human, numerous trials and errors were repeated until 1980, when allogeneic HSCT began to be actively performed under a better recognition of transplant immunology. Dr. Donnall Thomas received Nobel Prize for his pioneering work in bone marrow transplantation to cure leukaemia and other haematologic malignancies [60].

Results from basic and clinical research have allowed the improvement of HSCT. Some of these improvements were the use of haematopoietic stem cells from peripheral blood or cord blood, which promoted a change in the terminology from bone marrow transplantation to haematopoietic stem cell transplantation [61]; a better understanding of the complexities of the human leukocyte (HLA) system, which has allowed selecting compatible sibling donors and the establishment of larger registries of HLA-typed volunteers; advances in the immunogenetics of HLA, especially typing of molecular techniques; the development of the preparative regimen of cyclophosphamide and busulfan, which avoids the use of irradiation for some diseases [62]; the development of non-myeloblative conditioning regimens for allogeneic HSCT, avoiding regimen-related toxicity and death and that opened the way to include elderly patients [61, 63].

These advances make HSCT one curative treatment modality for many patients mainly with malignant haematological diseases. But, remaining challenges include further advances in the prevention and treatment of both infections and graft-versus-host disease (GVHD). Despite significant progress in HSCT, GVHD remains a significant cause of morbidity and mortality after allogeneic HSCT [63–67].

Homeostasis of the haematopoietic system, which is a balance between self-renewal and differentiation, is thought to be tightly regulated by interactions among haematopoietic stem cells (HSCs) and the specialized microenvironment where they reside, the haematopoietic niche. The haematopoietic niche consists of a heterogeneous cellular population of non-haematopoietic and haematopoietic origin as well as of extracellular matrix, which collectively provide the structural scaffold, the spatial framework and the appropriate physiological and trophic cues to control HSC maintenance and function [68]. The key component of the haematopoietic microenvironment is bone marrow mesenchymal stromal cells (BM-MSCs).

Many studies have reported that MSCs can promote HSC expansion *in vitro*. Koc and colleagues (2000) first reported rapid haematopoietic recovery after co-infusion of autologous BM-MSCs at the time of HSCT without significant side effects [69]. Lazarus and colleagues (2005) showed, in a multicenter trial with 46 patients receiving allogeneic HSCT and MSCs from HLA-identical siblings, a rapid haematopoietic recovery in most patients [58]. These studies showed the beneficial effects of MSC on engraftment after HSCT [2].

BM-MSCs have potential clinical applications in HSCT as an adjuvant cellular therapy for promoting the rapid reconstitution of haematopoiesis after HSCT, to prevent and treat of graft failure, in graft versus tumour effect and in GVHD [2, 70].

3.1. Pathophysiology of graft-versus-host disease (GVHD) and clinical use of MSC for GVHD

Allogeneic HSCT is an effective treatment for many haematologic and genetic diseases. However, donor-derived cells may also recognize recipient organs as foreign and mount an immune attack against the patient's own tissues, knows as graft-versus-host disease (GVHD) [71].

Graft-versus-host disease (GVHD) is the major cause of morbidity and mortality after an allogeneic HSCT. GVHD has traditionally been classified by the time of its clinical manifestations. Acute GVHD occurs within the first 100 days after haematopoietic stem cell transplantation, whereas chronic GVHD occurs after day 100. This simple classification is increasingly unsatisfactory particularly as reduced-intensity condition regimens gain wider acceptance. The clinical manifestations of acute GVHD after such conditioning often occur much later, sometimes coinciding with the day 100, the demarcation of chronic GVHD [72]. New recommendations that emphasize the importance of qualitative differences, as opposed to the time of onset after HSCT, are being used to standardize the diagnosis and clinical assessment of chronic GVHD [73].

The pathophysiology of acute GVHD after HSCT can be considered as a three-step process where the innate and adaptive immune system interacts. These three-step processes are as follows: (1) initiation of tissue damage; (2) activation and proliferation of donor T cells; and (3) the effector phase involving cellular and inflammatory factors. The pathophysiology of GVHD is a complex process. Chemotherapy and radiation cause tissue damage, producing pro-inflammatory cytokines, resulting in donor T-cell activation through the host antigen-presenting cell (APC) interaction via MHC-T cell receptor binding and co-stimulatory signals. This leads to T cell expansion and differetiation into various subtypes which traffic through blood vessels to target organs, where they cause tissue destruction and recruitment of other inflammatory cells through pathways such as perforin/grazyme and cytokine release. Moreover, these inflammatory cells and cytokines can propagate the cycle of GVHD [71, 72, 74]. The main step of GVHD reaction is step 2, where donor T cells are activated by the host antigen-presenting cells (APCs). The GVHD reaction is amplified by the intensity of the recipient-conditioning regimen, which can result in an intensive tissue damage with the release of various cytokines and augment of inflammatory response [75].

Acute GVHD affects mainly skin, liver and gastrointestinal tract. Approximately 50% of the patients treated with HSCT subsequently developed acute GVHD and required systematic treatment. Chronic GVHD occurs in 40% of patients treated with allogeneic HSCT from HLA-identical sibling and more than 50% of patients treated with HSCT from an HLA non-identical-related donor [67, 75].

Chronic GVHD is one of the most significant complications of long-term survivors after allogeneic haematologic stem cell transplantation. Experimental studies and clinical observations have elucidated the mechanisms of acute GVHD, but the biology of chronic GVHD is not well understood. Experimental studies generated at least four theories to explain the pathophysiology of chronic GVHD: (1) thymic damage and the defective negative selection of Tcells; (2) regulatory T cells deficiences; (3) auto-antibody production by aberrant B cells; and (4) the formation of profibrotic lesions [73]. The immunopathology of chronic GVHD is mediated in part by helper T lymphocyte 2 (Th2) cells, with a syndrome of immunodeficiency and an autoimmune disorder [76].

GVHD remains associated with significant morbidity and mortality in allogeneic HSCT. Improving outcomes in HSCT will require additional therapeutic modalities such as the use of MSCS. MSCs may be used to modulate the immune system, as prophylaxis to prevent GVHD and as treatment for established GVHD. Some studies have been performed demonstrating that the MSCs can act in GVHD [76–81]. In a multicenter clinical trial, HSCs and MSCs derived from HLA-identical sibling donors were infused and promoted haematopoietic engraftment and limited GVHD. In this study, 31 patients received myeloblative conditionig and HLA-identical sibling bone marrow or peripheral blood stem cells. Escalating doses of MSCs from 1 to 5 $\times 10^6$/kg were given. Toxicity related to MSC infusion was not observed. The incidence of acute GVHD was 15% in the co-transplanted group compared with 40% in a matched control group [17, 77]. Le Blanc and colleagues (2004) reported a child with acute lymphoblastic leukaemia receiving haploidentical MSC infusion for severe acute GVHD with a satisfactory clinical outcome [78]. The immunomodulatory abitily of BM-MSCs shows promise in treating GVHD, especially acute GVHD. Zhou and colleagues (2010) also showed the potential clinical application of MSCs for chronic GVHD. In this study, four patients with sclerodermatous chronic GVHD were reported, who received MSCs expanded *ex vivo* from unrelated donors by intra BM injection. After MSC infusion, the symptoms gradually improved in all four patients. During the course of MSC treatment, the patient's vital signs and laboratory studies remained normal. None of the four patients had recurrence of leukaemia. This study, despite its limited number of patients, suggests a benefit of MSC infusion therapy in treating sclerodermatous chronic GVHD [76]. MSCs are capable of escaping recognition by the alloreactive immune system and can exert immunomodulatory and anti-inflammatory effects. Hence, these cells represent a promising tool in the prevention and treatment of GVHD [2].

3.2. Prevention and treatment of graft failure using MSC

Graft failure or graft rejection after HSCT may occur as either a lack of initial engrafment of donor cells or loss of donor cells after initial engraftment. In the later case, autologous recovery may appear or, alternatively, marrow aplasia may be developed. Rejection is a major cause of graft failure and it is due to recipient immune response against donor Immuno-haematopoietic cells. Graft failure may also occur to other causes, such as viral infections, specifically, cytomegalovirus (CMV) [82]. In patients with leukaemia receiving myeloblative conditioning, the rejection rate was 0.1% in patients given HLA-identical sibling transplants compared to 5% in those given HLA-mismatched grafts. Another risk factor for graft failure is the reduced intensity conditioning that is used in the elderly patients, with lower doses of chemoradiation therapy, the host immune system may persist, resulting in an increased risk of allograft rejection. Hence, the main risk factors associated with graft failure include HLA disparity in the donor/recipient pair, viral infections and the type of conditioning regimen [82, 83].

Some studies have demonstrated that the use of MSCs is efficient in prevention and treatment of graft failure [2, 9, 70]. Ball and colleagues carried out a pilot study of co-transplanted of BM-derived, *ex vivo*-expanded MSCs of donor origin in 14 children undergoing transplantation of granulocytic colony stimulating factor (G-CSF) mobilized, CD34-selected progenitor cells from HLA disparate relative. In this study, a graft failure rate of 15% was observed in 47 controls and all patients, who received MSCs, showed sustained haematopoietic engraftment without any adverse reaction. These results suggested that MSC co-transplantation may modulate host alloreactivity and/or promote better engraftment of donor haematopoiesis, thus reducing the risk of graft failure [9].

3.3. Graft-versus-tumour effect and MSCs

Two mechanisms are involved in the cure of a malignant disease by stem cell transplantation. The first is the conditioning regimen, which confers a powerful anti-tumour effect from myeloblative doses of chemotherapy or radiotherapy. The second is the graft-versus-leukaemia (GVL) effect or the graft-versus-tumor (GVT) effect exerted by transplanted donor T cells and NK cells against malignant tissue. The T lymphocytes recognize antigens presented by HLA molecules on malignant cells. They destroy tumour cells by direct cytotoxicity inducing death by lysis though the perforin-granzyme pathway and by apoptosis though activation of Fas on the cell surface. The graft-versus leukaemia effect requires donor immune competence that often accompanies graft-versus-host disease (GVHD) [84].

Despite HLA identity between a patient and donor, approximately 40% of patients receiving HLA-identical grafts develop acute GVHD due to genetic differences that lie outside the HLA loci, or minor histocompatibility antigens (HA). Some of such antigens, such as HY and HA-3, are expressed on all tissues and are targets for both GVHD and GVL. Others, such as HA-1 and HA-2, are expressed most abundantly on haematopoietic cells (including leukaemic cells) and may therefore induce a greater GVL effect with less GVHD [75].

Bearing in mind two fundamental aspects to the success of allogeneic HSCT, the GVL effect and the GVHD, the co-infusion of MSCs in transplantation should aim at reducing the severity of GVHD while preserving the GVL. In this sense, the study of Baron and colleagues (2010) demonstrated that the MSC co-infusion appeared to be safe; furthermore, MSC co-infusion might have prevented death from GVHD without abrogating GVT effects. In this study, 20 patients with haematologic malignancies received MSCs from HLA-mismatched donors after conditioning with TBI (total body irradiation) and fludarabine. The HLA-mismatched non-myeloblative HSCT with MSC co-infusion had a therapeutic effect on the haematologic malignancies. MSCs may have important beneficial characteristics in terms of promoting GVT effects due to their immunomodulatory properties after HSCT and their tropism towards the microenvironment [85, 86]. **Figure 2** shows the potential clinical applications of MSCs for haematopoietic stem cell transplantation.

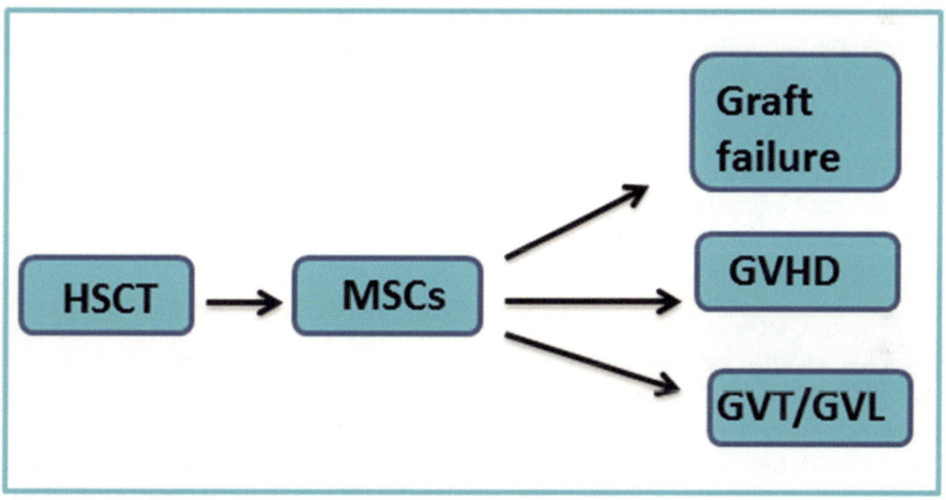

Figure 2. Potential clinical applications of MSCs for HSCT.

4. Using probability techniques to study clinical applications of mesenchymal stem cells

Nowadays, people recognize the importance of Mathematics in Medicine. Many statements in this area can be better understood using mathematical concepts and results. In this section, we will present mathematical techniques based on the concept and results of probability, as we are interested in making decisions in the face of uncertainty. In fact, in Medicine, clinical outcomes, such as the occurrence of disease, death, symptoms or functional impairment, can be counted and expressed as numbers, but in most clinical situations, the diagnosis, prognosis and treatment outcomes are uncertain for an individual patient. A person will experience a clinical outcome or not: the prediction is rarely exact. Therefore, the prediction must be expressed as a *probability*.

Although earlier work on probability was done by the Italian mathematician Giralamo Cardano (1501–1576), the investigation of probability as a branch of Mathematics sprang about 1654 with two great French mathematicians: Blaise Pascal (1623–1662) and Pierre Fermat (1601–1665) [87]. Of course, we shall not do a discourse on probability. But, we need to say that the theory of probability underlies the procedures in testing hypotheses, which are very useful to Medicine and other disciplines in the health field.

In this section, we will avoid mathematical formulas and theorems.

4.1. Descriptive statistics and inferential statistics

Statistics is a branch of Mathematics. The word "statistics" derives from the Latin word *status*, meaning "manner of standing" or "position." Statistics were first used by tax assessors to collect information for determining assets and assessing taxes. Statistics applied to Medicine and other health disciplines is called biostatistics or biometrics. For those who would like to study this subject, we recommend the book of J. H. Zar [88].

Statistics is divided into two branches: descriptive and inferential. *Descriptive statistics* is used to organize and summarize data. *Inferential statistics* is used to draw inferences and reach conclusions about a population, when only a sample from that population has been studied. A population is a complete set of observations, patients, measurements and so forth. A sample is a subset of a certain population.

The mathematical techniques used on descriptive statistics are graphic (*tables and graphs*) and numerical (*quantitative indices*). Tables are often used to present qualitative and quantitative data. Graphs are used widely to provide a visual display data. The bar diagram, histogram and frequency polygon are three graphic formats that are commonly used to present medical data. Quantitative indices are numbers that describe the centre and the variation of a distribution, which are called *parameters* if they are referred to a population and called *statistics* if they are referred to a sample. Quantitative indices that describe the centre of a distribution are referred to as *measures of central tendency*. The mean, known also as the arithmetic mean, median and mode are three common measures of central tendency. Quantitative indices that describe the variation or dispersion of a distribution are referred to as *measures of dispersion*. The range, variance and standard deviation are three common measures of dispersion. Other quantitative indices such as risk difference, relative risk and odds ratio are also used in Medicine.

One mathematical technique used on inferential statistics is hypothesis tests. They are much more sophisticated than the techniques used in descriptive analysis, because hypothesis tests are based on probability models.

A probability model for a particular experiment is a probability distribution that *predicts* the relative frequency of each outcome if the experiment is performed a large number of times. A

http://dx.doi.org/10.5772/63772

probability distribution is a table, a graph or a formula that describes what probably will happen instead of describing what really happened.

A probability distribution can be discrete or continuous. An important example of a discrete probability distribution is the *binomial distribution* and an important example of a continuous distribution is the *normal distribution*.

4.2. Hypothesis tests

A hypothesis test is a method used to determine if there is enough evidence in a sample to infer that a certain property holds for the entire population. It works with two hypothesis: the null hypothesis (designated H_0) and the alternative hypothesis (designated H_A).

In the null hypothesis we use the words *no difference* or *equal to* and in the alternative hypothesis we use the words *different from, less than* or *greater than*. But let us mention that, in fact, we should say *no statistical difference, statistically equal to, statistically different from, statistically less than* or *statistically greater than,* because we are dealing with the probabilities of an event happens or not. When we retain H_A (equivalently reject H_0), we say the results are significant and when we retain H_0 (equivalently reject H_A), we say the results are not significant. Because we are dealing with probabilities, this implies in making two possible errors from four possible relations between the conclusions obtained using a hypothesis test and real situations, as shown in **Table 1**.

		Real difference	
		Presence	Absence
Conclusion of the statistical test	Results are significant	True	Type I error
	Results are not significant	Type II error	True

Table 1. Relations between statistical conclusions and real situations.

The two errors mentioned in the previous paragraph are known as Type I error and Type II error. A Type I error leads to a *false positive* conclusion. The probability of such an error occurs is noted by α. Mathematically, α is a conditional probability: α is the probability of reject H_0 when there is no real difference. A Type II error leads to a *false negative* conclusion. The probability of such an error occurs is noted by β. Mathematically, β is a conditional probability: β is the probability of retain H_0 when there is a real difference.

Hypothesis tests are used to estimate the *probability* of a Type I error. So, they are based on probability models. In the literature, we usually use $\alpha < 0.05$. This means we are assuming a probability less than 0.05 of rejecting H_0 when there is no real difference between treatments, drugs or procedures. In other words, if the study were repeated one-hundred times, we *probably* would find five outcomes showing H_0 should be accepted.

The hypothesis tests commonly used in the medical literature are presented in **Table 2**. They can be parametric or nonparametric. The choice in using one of them depends on the purpose of the study, the size of the sample and the type of the variables involved at the study, for instance. If we can guarantee that the sampling distribution is normal or approximately normal, we can use a parametric test. Because a normal distribution has good mathematical properties (bell-shaped, symmetric, and so on), using a parametric test leads to better results compared with a nonparametric test. In other words, we say that nonparametric tests are less powerful, in the sense that, they lead to a small probability to reject H_0, when H_0 is false.

To test the statistical significance of the difference between ...		
Two or more proportions	Chi-square	Nonparametric
Two proportions	Fisher's exact	Parametric
Two medians	Mann–Whitney	Nonparametric
Two means	Student's *t*	Parametric
More than two means	Kruskal–Wallis (one-factor)	Nonparametric
More than two means	ANOVA (one-factor)	Parametric
More than two means	ANOVA (more factors)	Parametric

Table 2. Statistical tests usually used in the medical literature.

When we use a hypothesis test, we compute a *p*-value. The *p*-value is the probability of obtaining a result as extreme or more extreme than the sample value, assuming that the null hypothesis is true. The sample value is calculated. Depending on the test we use, there is a specific formula to calculate the sample value. An appropriate computer software can do such a calculation.

We finish this subsection noting that many methods described above are univariate methods, because they are only concerned with analyzing only one variable. The point in using a multivariate analysis is to consider several variables simultaneously in the study. With multivariate methods, we can reduce data information, we can classify objects or variables, we can investigate the dependence among variables and, finally, we can make predictions. Principal component analysis, factor analysis and clustering methods can be used in order to reduce data and investigate the dependence among variables. To make predictions, we can use Hotelling's T^2 test, which allows inference about one mean vector, and we can use MANOVA, which allows inference about a finite number of mean vectors. It is important to say that both Hotelling's T^2 test and MANOVA are generalizations of Student *t*-test and ANOVA, respectively. So, they are based on the multivariate normal distribution. For those who would like to study this subject, we recommend the book of R.A. Johnson and D.W. Wichern [89].

4.3. Probability techniques on the study of mesenchymal stem cells

In cellular therapy, safety remains one of the main characteristic and refers to validation tests. The culture process should be reproducible, robust and efficient. The increasing use of MSCs

http://dx.doi.org/10.5772/63772

has led to a production of processes which needs to be in accordance with current Good Manufacturing Practice (cGMP) [90, 91]. For the validation of the tests, it is important to use probability techniques comparing the results of the tests and the results between different laboratories, with the main to standardize the procedures to characterize the MSCs at the stage of production for cell therapy. The evidence of clinical efficacy is also required. In this case, the application of probability methods is also an important tool in clinical trials and clinical outcomes revealing the impact on the use of MSCS in cellular therapy.

5. Conclusion

Mesenchymal stem cells (MSCs) represent promising tools to be used in immunoregulatory and regenerative cell therapies. For this purpose, an extensive amplification *in vitro* of MSCs is necessary without affecting the cells' genomic characteristics and differentiation properties. However, sometimes, the culture expansion could generate cytogenetic and molecular alterations. Accumulation of these alterations during many passages of culture could lead to malignant cell transformation. Hence, it is important to do a quality control using different methods to test the safe and efficacy of MSCs for cell therapies. Many studies have revealed the clinical use of MSCs as an emerging field for treating cardiovascular disorders, neurodegenerative diseases, bone defects and fractures, inflammatory arthritis and in the field of haematopoietic stem cell transplantation. The BM-MSCs have potential clinical applications in HSCT as an adjuvant cellular therapy for promoting the rapid reconstitution of haematopoiesis after HSCT, to prevent and treat of graft failure, in graft-versus tumor effect and in GVHD. Although these studies showed positive results, it is necessary to continue the scientific and clinical research to clarify some points as: the characterization of the appropriate cell passage during the culture of MSCSs to ensure the genomic stability; the definition of the tests for quality control to ensure the safety of the MSCS for clinical practice; the practical application of these recommended tests can be standardized for the sensitivity and specificity between the laboratories; it is necessary to define the optimum cell dose and the number of infusions of MSCs during the treatment; a long follow-up to characterize the positive clinical effects and also the adverse clinical effects that may occur with the use of MSCs. With the advancement of basic and clinical research, we hope that the use of MSCs in cell therapy brings excellent results especially for patients treated with HSCT.

Author details

Teresa de Souza Fernandez[1*] and Cecília de Souza Fernandez[2]

*Address all correspondence to: teresafernandez@inca.gov.br

1 National Cancer Institute (INCA), Bone Marrow Transplantation Center (CEMO), Laboratory Division, Rio de Janeiro, RJ, Brazil

2 Mathematics and Statistics Institute, Federal Fluminense University (UFF), Niterói, RJ, Brazil

References

[1] Dominici M, Le Blanc K, Mueller I, Slaper-Cortenbach I, Marini F, Krause D: Minimal criteria for defining multipotent mesenchymal stromal cells. The International Society for Cellular Therapy position statement. Cytotherapy 2006; 8:315–317.

[2] Wu KH, Wu HP, Chan CK, Hwang SM, Peng CT, Chao YH: The role of mesenchymal stem cells in hematopoietic stem cell transplantation: from bench to besides. Cell Transpl 2013; 22:723–729. DOI: 10.3727/096368912X655217.

[3] Binato R, de Souza Fernandez T, Lazzarotto-Silva C, Du Rocher B, Mencalha A, Pizzatti L: Stability of human mesenchymal stem cells during in vitro culture: considerations for cell therapy. Cell Prolif. 2013; 46:10–22. DOI: 10.1111/cpr.12002.

[4] Hoch AI, Leach JK: Optimizing expansion of bone marrow mesenchymal stem/stromal cells for clinical applications. Stem Cells Transl Med 2014; 3:643–652. DOI: 10.5966/sctm.2013-0196.

[5] Friedenstein AJ, Piatetzky-Shapiro II, Petrakova KV: Osteogenesis in transplants of bone marrow cells. J Embryol Exp Morphol 1966; 16:381–390.

[6] Caplan AI: Adult mesnechymal stem cells for tissue engineering versus regenerative medicine. J Cell Physiol 2007; 213:341–347.

[7] Banfi A, Muraglia A, Dozin B, Mastrogiacomo M, Cancedda R, Quarto R: Proliferation kinetics and differentiation potential of ex vivo expanded human bone marrow stromal cells: implications for their use in cell therapy. Exp Hematol 2000; 28:707–715.

[8] Deans RJ, Moseley AB: Mesenchymal stem cells: biology and potential clinical uses. Exp Hematol 2000; 28:875–884. DOI: 10.1016/S0301-472X(00)00160-0.

[9] Ball LM, Bernardo M E, Roelofs H, Lankester A, Cometa A, Egeler RM: Cotransplantation of ex vivo expanded mesenchymal stem cells accelerates lymphocyte recovery and may reduce the risk of graft failure in haploidentical hematopoietic stem cell transplantation. Blood 2007; 110:2764–2767. DOI:10.1182/blood-2007-04-087056.

[10] Ullah I, Subbarao RB, Rho GJ: Human mesenchymal stem cells – current trends and future prospective. Biosci Rep 2015; 35: e00191. DOI: 10.1042/BSR20150025.

[11] Wei X, Yang X, Han ZP, Qu FF, Shao L, Shi YF: Mesenchymal stem cells: a new trend for cell therapy. Acta Pharmacol Sin 2013; 34:747–754. DOI: 10.1038/aps.2013.50.

[12] Wang S, Qu X, Zhao RC: Clinical applications of mesenchymal stem cells. J Hematol Oncol 2012; 5:19.

[13] Castro-Manrreza ME, Montesinos JJ: Immunoregulation by mesenchymal stem cells: biological aspects and clinical applications. J Immunol Res2015; 2015:394917. DOI: 10.1155/2015/394917.

[14] Atoui R, Chiu RCJ: Concise review: immunomodulatory properties of mesenchymal stem cells in cellular transplantation: update, controversies, and unknowns. Stem Cells Transl Med 2012; 1:200–205. DOI: 10.5966/sctm.2011-0012.

[15] Faiella W, Atoui R: Immunotolerant properties of mesenchymal stem cells: updated review. Stem Cells Int 2016; 2016:1859567. DOI: 10.1155/2016/1859567.

[16] Munn DH, Zhou M, Attwood JT, Bondarev I, Conway SJ, Marshall B: Prevention of allogeneic fetal rejection by tryptophan catabolism. Science. 1998; 281:1191–1193. DOI: 10.1126/science.281.5380.1191.

[17] Le Blanc K, Ringdén O: Immunobiology of human mesenchymal stem cells and future use in hematopoietic stem cell transplantation. Biol Blood Marrow Transplant 2005; 11:321–334. DOI: 10.1016/j.bbmt.2005.01.005.

[18] Horwitz EM, Maziarz RT, Kebriae P: MSCs in hematopoietic cell transplantation. Biol Blood Marrow Transplant 2011; 17: S21-S29. DOI: 10.1016/j.bbmt.2010.11.026.

[19] Jung S, Panchalingam KM, Rosenberg L, Behie LA: Ex vivo expansion of human mesenchymal stem cells in defined serum-free media. Stem Cells Int 2012; 2012:123030. DOI:10.1155/2012/123030.

[20] Penfornis P, Pochampally R: Isolation and expansion of mesenchymal stem cells/ multipotential stromal cells from human bone marrow. Methods Mol Biol 2011; 698:11–21. DOI: 10.1007/978-1-60761-999-4_2.

[21] Bara JJ, Richards RG, Alini M, Stoddart MJ: Concise review: bone marrow derived mesenchymal stem cells change phenotype following in vitro culture: implications for basic research and the clinic. Stem Cells 2014; 32:1713–1723. DOI: 10.1002/stem.1649.

[22] Kassem M: Stem cells: potential therapy for age-related disease. Ann N Y Acad Sci 2006; 1067:436–442.

[23] Sekiya I, Larson BL, Smith JR, Pochampally R, Cui J-G, Prockop D: Expansion of human adult stem cells from bone marrow stroma: conditions that maximize the yields of early progenitors and evaluate their quality. Stem Cells 2002; 20:530–54. DOI: 10.1634/ stemcells.20-6-530.

[24] Both SK, van der Muijsenberg AJ, van Blitterswijk CA, de Boer J, de Bruijn JD: A rapid and efficient method for expansion of human mesenchymal stem cells. Tissue Eng 2007; 13:3–9. DOI:10.1089/ten.2005.0513.

[25] Shahdadfar A, Katrine Frønsdal K, Haug T, Reinholt FP, Brinchmanna JE: In vitro expansion of human mesenchymal stem cells: choice of serum is a determinant of cell

proliferation, differentiation, gene expression, and transcriptome stability. Stem Cells 2005; 23:1357–1366. DOI: 10.1634/stemcells.2005-0094.

[26] Tonti GA, Mannello F: From bone marrow to therapeutic applications: different behaviour and genetic/epigenetic stability during mesenchymal stem cell expansion in autologous and foetal bovine sera? Int J Dev Biol 2008; 52:1023–1032. DOI: 10.1387/ijdb.082725gt.

[27] Bonab MM, AlimoghaddamK, Talebian F, Ghaffari SH, Ghavamzadeh A, Nikbin B: Aging of mesenchymal stem cell in vitro. BMC Cell Biol 2006; 7:14. DOI: 10.1186/1471-2121-7-14.

[28] Zhao F, Chella R, Ma T: Effects of shear stress on 3-D human mesenchymal stem cell construct development in a perfusion bioreactor system: experiments and hydrodynamic modeling. Biotechnol Bioeng 2007; 96:584–595.

[29] Keung AJ, Kumar S, Schaffer DV: Presentation counts: microenvironmental regulation of stem cells by biophysical and material cues. Annu Rev Cell Dev Biol 2010; 26:533–556. DOI: 10.1146/annurev-cellbio-100109-104042.

[30] Frith JE, Thomson B, Genever PG: Dynamic three-dimensional culture methods enhance mesenchymal stem cell properties and increase therapeutic potential. Tissue Eng Part C Methods 2010; 16:735–749. DOI: 10.1089/ten.TEC.2009.0432.

[31] Braccini A, Wendt D, Farhadi J, Schaeren S, Heberer M, Martin I: The osteogenicity of implanted engineered bone constructs is related to the density of clonogenic bone marrow stromal cells. J Tissue Eng Regen Med 2007; 1:60–65. DOI: 10.1002/term.11.

[32] Bartosh TJ, Ylostalo JH, Mohammadipoor A, Bazhanov N, Coble K, Claypool K: Aggregation of human mesenchymal stromal cells (MSCs) into 3D spheroids enhances their anti-inflammatory properties. Proc Natl Acad Sci U S A 2010; 107:13724–13729. DOI: 10.1073/pnas.1008117107.

[33] Bianco P, Cao X, Frenette PS, Mao JJ, Robey PG, Simmons PJ: The meaning, the sense, the significance: Translating the science of mesenchymal stem cells into medicine. Nat Med 2013; 19:35–42. DOI:10.1038/nm.3028.

[34] Sart S, Agathos SN, Li Y, Ma T: Regulation of mesenchymal stem cell 3D microenvironment: From macro to microfluidic bioreactors. Biotechnol J 2016; 11:43–57. DOI: 10.1002/biot.201500191.

[35] Shi C. Recent progress toward understanding the physiological function of bone marrow mesenchymal stem cells. Immunology 2012; 136:133–138. DOI: 10.1111/j.1365-2567.2012.03567.x.

[36] Bobis S, Jarocha D, Majka M. Mesenchymal stem cells: characteristics and clinical applications. Folia Histochem Cytobiol 2006; 44:215–230.

[37] Solchaga LA, Penick KJ, Welter JF: Chondrogenic differentiation of bone marrow-derived mesenchymal stem cells: tips and tricks. Methods Mol Biol 2011; 698:253–278. DOI: 10.1007/978-1-60761-999-4_20.

[38] Kim JA, Im KO, Park SN, Kwon JS, Kim SY, Oh K: Cytogenetic heterogeneity and their serial dynamic changes during acquisition of cytogenetic aberrations in cultured mesenchymal stem cells. Mutat Res 2015; 777:60–68. DOI: 10.1016/j.mrfmmm.2015.04.003.

[39] Zhang ZX, Guan LX, Zhang K, Wang S, Cao PC, Wang YH: Cytogenetic analysis of human bone marrow-derived mesenchymal stem cells passaged in vitro. Cell Biol Int 2007; 31:645–648. DOI: 10.1016/j.cellbi.2006.11.025.

[40] Nikitina VA, Osipova EY, Katosova LD, Rumyantsev SA, Skorobogatova EV, Shamanskaya TV: Study of genetic stability of human bone marrow multipotent mesenchymal stromal cells. Bull Exp Biol Med 2011; 150:627–631.

[41] Bochkov NP, Voronina ES, Kosyakova NV, Liehr T, Rzhaninova AA, Katosova LD: Chromosome variability of human multipotent mesenchymal stromal cells. Bull Exp Biol Med 2007; 143:122–126.

[42] Bernardo ME, Zaffaroni N, Novara F, Cometa AM, Avanzini MA, Moretta A: Human bone marrow derived mesenchymal stem cells do not undergo transformation after long-term in vitro culture and do not exhibit telomere maintenance mechanisms. Cancer Res 2007; 67:9142–9149. DOI: 10.1158/0008-5472.CAN-06-4690.

[43] Izadpanah R, Kaushal D, Kriedt C, Tsien F, Patel B, Dufour J: Long-term in vitro expansion alters the biology of adult mesenchymal stem cells. Cancer Res 2008; 68:4229–4238. DOI: 10.1158/0008-5472.CAN-07-5272.

[44] Collado M, Blasco MA, Serrano M: Cellular senescence in cancer and aging. Cell 2007; 130:223–233.

[45] Estrada JC, Torres Y, Benguría A, Dopazo A, Roche E, Carrera-Quintanar L: Human mesenchymal stem cell-replicative senescence and oxidative stress are closely linked to aneuploidy. Cell Death Dis 2013; 4: e691. DOI: 10.1038/cddis.2013.211.

[46] Zimmermann S, Voss M, Kaiser S, Kapp U, Waller CF, Martens UM: Lack of telomerase activity in human mesenchymal stem cells. Leukemia 2003; 17:1146–1149. DOI: 10.1038/sj.leu.2402962.

[47] Samsonraj RM, Raghunath M, Hui JH, Ling L, Nurcombe V, Cool SM: Telomere length analysis of human mesenchymal stem cells by quantitative PCR. Gene. 2013; 519:348–55. DOI:10.1016/j.gene.2013.01.039.

[48] Rastegar F, Shenaq D, Huang J, Zhang W, Zhang BQ, He BC: Mesenchymal stem cells: molecular characteristics and clinical applications. World J Stem Cells 2010; 2:67–80. DOI:10.4252/wjsc.v2.i4.67.

[49] Fujita T, Azuma Y, Fukuyama R, Hattori Y, Yoshida C, Koida M: Runx2 induces osteoblast and chondrocyte differentiation and enhances their migration by coupling with PI3K-Akt signaling. J Cell Biol 2004; 166:85–95. DOI: 10.1083/jcb.200401138.

[50] Tominaga H, Maeda S, Miyoshi H, Miyazono K, Komiya S, Imamura T: Expression of osterix inhibits bone morphogenetic protein-induced chondrogenic differentiation of mesenchymal progenitor cells. J Bone Miner Metab 2009; 27:36–45.

[51] Griffin M, Iqbal SA, Bayat A: Exploring the application of mesenchymal stem cells in bone repair and regeneration. Bone Joint Surg Br 2011; 93:427–34. DOI: 10.1302/0301-620X.93B4.25249.

[52] Akiyama H: Control of chondrogenesis by the transcription factor Sox9. Mod Rheumatol 2008; 18:213–219. DOI: 10.1007/s10165-008-0048-x.

[53] Farmer SR: Regulation of PPARgamma activity during adipogenesis. Int J Obes (Lond) 2005; 29 Suppl 1:S13–S16.

[54] Redaelli S; Bentivegna A; Foudah D; Miloso M; Redondo J; Riva G: From cytogenomic to epigenomic profiles: monitoring the biologic behavior of in vitro cultured human bone marrow mesenchymal stem cells. Stem Cell Res Ther 2012; 3:47. DOI: 10.1186/scrt138.

[55] Rayment EA, Williams DJ: Concise review: mind the gap: challenges in characterizing and quantifying cell- and tissue-based therapies for clinical translation. Stem Cells 2010; 28:996–1004. DOI: 10.1002/stem.416.

[56] ISSCR. Guidelines for the Clinical Translation of Stem Cells, International Society for Stem Cell Research, 2008; 1–19. Available from: www.isscr.org.

[57] Kim SY, Im K, Park SN, Kwon J, Kim JA, Choi Q: Asymmetric aneuploidy in mesenchymal stromal cells detected by in situ karyotyping and fluorescence in situ hybridization: suggestions for reference values for stem cells. Stem Cells Dev 2015; 24:77–92. DOI: 10.1089/scd.2014.0137.

[58] Lazarus HM, Haynesworth SE, Gerson SL, Rosenthal NS, Caplan AI: Ex vivo expansion and subsequent infusion of human bone marrow-derived stromal progenitors cells (mesenchymal progenitors cells): implications for therapeutic use. Bone Marrow Transplant 1995; 16:557–564.

[59] Otto WR, Wright NA: Mesenchymal stem cells: from experiment to clinic. Fibrogenesis Tissue Repair 2011; 4:20. DOI: 10.1186/1755-1536-4-20.

[60] Park B, Yoo KH, Kim C: Hematopoietic stem cell expansion and generation: the ways to make a breakthrough. Blood Res 2015; 50:194–203. DOI: 10.5045/br.2015.50.4.194.

[61] Thomas ED, Blume KG: Historical markers in the development of allogeneic hematopoietic cell transplantation . Biol Blood Marrow Transplant 1999; 5:341–346. DOI: 10.1016/S1083-8791(99)70010-8.

[62] Santos GW: Busulfan (Bu) and cyclophosphamide (Cy) for marrow transplantation. Bone Marrow Transplant 1989; 4:236.

[63] Baron F, Storb R: Allogeneic hematopoietic cell transplantation as treatment for hematological malignancies: a review. Springer Semin Immunopathol 2004; 26:71–94.

[64] HoVT, Soiffer RJ: The history and future of T-cell depletion as graft-versus-host-disease prophylaxis for allogeneic hematopoietic stem cell transplantation. Blood 2001; 98:3192–3204. DOI:10.1182/blood.V98.12.3192.

[65] Komanduri KV, Couriel D, Champlin RF: Graft-versus-host disease after allogeneic stem cell transplantation: evolving concepts and novel therapies including photopheresis. Biol Blood Marrow Transplant 2006; 12:1–6. DOI:10.1016/j.bbmt.2005.11.003.

[66] Ball LM, Egeler RM: On behalf of the EBMT Paediatric Working Party. Acute GvHD: pathogenesis and classification. Bone Marrow Transplant 2008; 41: S58–S64.

[67] Skoczen S, Bik-Multanowski M, Pietrzyk JJ, Grabowska A, Fijorek K, Strojny W: Genetic background of immune complications after allogeneic hematopoietic stem cell transplantation in children. Stem Cells Int 2016, 2016:2626081. DOI: 10.1155/2016/2626081.

[68] Pontikoglou C, Deschaseaux F, Sensebé L, Papadaki HA: Bone marrow mesenchymal stem cells: biological properties and their role in hematopoiesis and hematopoietic stem cell transplantation. Stem Cell Rev. 2011; 7:569–589. DOI: 10.1007/s12015-011-9228-8.

[69] Koç ON, Gerson SL, Cooper BW, Dyhouse SM, Haynesworth SE, Caplan AI: Rapid hematopoietic recovery after coinfusion of autologous blood stem cells and culture-expanded marrow mesenchymal stem cells in advanced breast cancer patients receiving high-dose chemotherapy. J Clin Oncol 2000; 18:307–316.

[70] Le Blanc K, Samuelsson H, Gustafsson B, Remberger M, Sundberg B, Arvidson J: Transplantation of mesenchymal stem cells to enhance engraftment of hematopoietic stem cells. Leukemia 2007; 21:1733–1738.

[71] Sung AD,Chao NJ: Concise Review: acute graft-versus-host disease: immunobiology, prevention and treatment. Stem Cells Transl Med 2013; 2:25–32. DOI: 10.5966/sctm.2012-0115.

[72] Ferrera JMLM, Yanik G, Valdez R. Graft-versus-host-disease. In: Young NS, Gerson SL, High KA, editors. Clinical Hematology. Elsevier; Philadelphia, 2006, 1235–1251.

[73] Min CK. The pathophysiology of chronic graft-versus host disease: the unrevealing of an enigma. Korean J Hematol 2011; 46:80–87. DOI: 10.5045/kjh.2011.46.2.80.

[74] Ichiki Y, Bowlus CL, Shimoda S, Ishibashi H, Vierling JM, Gershwin ME: T cell immunity and graft-versus-host-disease (GVHD). Autoimmun Rev 2006; 5:1–9. DOI: 10.1016/j.autrev.2005.02.006.

[75] Ferrera JLM, Levine JE, Reddy P, Holler E: Graft-versus-host-disease. Lancet 2009; 373:1550–1561.

[76] Zhou H, Guo M, Bian C, Sun Z, Yang Z, Zeng Y: Efficacy of bone marrow derived mesenchymal stem cells in the treatment of Sclerodermatous chronic graft-versus-host disease: clinical report. Biol Blood Marrow Transplant 2010; 16:403–412. DOI:10.1016/j.bbmt.2009.11.006.

[77] Frassoni F, Labopin M, Bacigalupo A: Expanded mesenchymal stem cells (MSC), co-infused with HLA-identical hematopoietic stem cell transplants, reduce acute and chronic graft-vs-host disease: a matched pair analysis. Bone Marrow Transplant 2002; 29 (suppl 2):S2.

[78] Le Blanc K, Rasmusson I, Sundberg B, Gotherstrom C, Hassan M, Uzunel M: Treatment of severe acute graft-versus-host disease with third party haploidentical mesenchymal stem cells. Lancet 2004; 363:1439–1441.

[79] Kebriaei P, Isola L, Bahceci E, Holland K, Rowley S, McGuirk J: Adult human mesenchymal stem cells added to corticosteroid therapy for the treatment of acute graft-versus-host disease. Biol Blood Marrow Transplant 2009; 15:804–811 DOI:10.1016/j.bbmt.2008.03.012.

[80] Lucchini G, Introna M, Dander E, Rovelli A, Balduzzi A, Bonanomi S: Platelet-lysate-expanded mesenchymal stromal cells as a salvage therapy for severe resistant graft-versus-host disease in a pediatric population. Biol Blood Marrow Transplant 2010; 16:1293–1301. DOI: 10.1016/j.bbmt.2010.03.017.

[81] von Dalowski F, Kramer M, Wermke M, Wehner R, Röllig C, Alakel N: Mesenchymal stromal cells for treatment of acute steroid-refractory graft versus host disease: clinical responses and long-term outcome. Stem Cells 2016; 34:357–366. DOI: 10.1002/stem.2224.

[82] Mattsson J, Ringdén O, Storb R: Graft failure after allogeneic hematopoietic cell transplantation. Biol Blood Marrow Transplant 2008; 14:165–170. DOI: 10.1016/j.bbmt.2007.10.025.

[83] Locatelli F, Lucarelli B, Merli P: Current and future approaches to treat graft failure after allogeneic hematopoietic stem cell transplantation. Exp Opin Pharmacother 2014; 15:23–36. DOI: 10.1517/14656566.2014.852537.

[84] Barrett J. Allogeneic Bone Marrow Transplantation. In: Young NS, Gerson SL, High KA. editors. Clinical Hematology. Elsevier; Philadelphia, 2006; 1190–1210.

[85] Baron F, Lechanteur C, Willems E, Bruck F, Baudoux E, Seidel L: Cotransplantation of mesnchymal stem cells might prevent death from graft-versus-host disease (GVHD) without abrogating graft-versus-tumor effects after HLA-mismatched allogeneic transplantation following non-myeloablative conditioning. Biol Blood Marrow Trnasplant 2010; 16:838–847.

[86] Kim EJ, Kim N, Cho SG. The potential use of mesenchymal stem cells in hematopoietic stem cell transplantation. Exp Mol Med 2013; 45: e2. DOI: 10.1038/emm.2013.2

[87] Katz VJ. A History of Mathematics – An Introduction. HarperCollins College Publishers, New York, 1993.

[88] Zar JH. Biostatistical Analysis. 5th Ed. Pearson Prentice Hall, New Jersey, 2010.

[89] Johnson RA, Wichern DW. Applied Multivariate Statistical Analysis. 6th Ed. Pearson Prentice Hall, New Jersey, 2007.

[90] Sensebé L, Bourin P, Tarte K: Good manufacturing practice production of mesenchymal stem /stromal cells. Hum Gen Ther 2011; 22:19–26.

[91] Kolkundkar U, Gottipamula S, Majumdar A S: Cell therapy manufacturing and quality control: current process and regulatory challenges. Stem Cell Res Ther 2014; 4:1–10.